ENVIRONMENTAL MONITORING

Edited by **Ema O. Ekundayo**

INTECHOPEN.COM

Environmental Monitoring
Edited by Ema O. Ekundayo

CBS, Edition 2016

Published by InTech

Janeza Trdine 9, 51000 Rijeka, Croatia

Publishing Process Manager Ivana Zec

Technical Editor Teodora Smiljanic

Cover Designer InTech Design Team

Image Copyright jaimaa, 2011. Used under license from Shutterstock.com

Additional hard copies can be obtained from orders@intechopen.com

Environmental Monitoring, Edited by Ema O. Ekundayo
p. cm.
ISBN 978-953-307-724-6

INTECH

open science | open minds

Contents

Preface

Environmental Monitoring is a book designed by InTech - Open Access Publisher in collaboration with scientists and researchers all over the world with a proven record of scientific accomplishment and knowledge in the field of environmental monitoring in particular, and environmental sciences in general. The book is designed to present recent research developments and advances in environmental monitoring to a global audience of scientists, researchers, environmental educators, administrators, technicians, managers, students and the general public.

A series of chapters addressing varied topics like the monitoring of heavy metal contaminants in atmospheric, terrestrial and aquatic environments; biological monitoring using wildlife/ecotoxicological monitoring; and the use of wireless sensor networks in environmental monitoring are included in this book. The book's concepts, ideas, sampling/analytical techniques described, results and research findings reflect what leading environmental scientistes and researchers around the world have done, and are currently doing in the field of environmental monitoring.

Special words of appreciation are due to Ms Ivana Zec, the Publishing Process Manager who oversaw and coordinated the publishing of all materials and assisted me and the authors in completing our work easily and in a timely manner. My profound thanks also to the technical editor who prepared these manuscripts for publication in InTech - Open Access Publisher.

Dr. E.O. Ekundayo
Alberta Institute of Agrologists,
Canada

Part 1

Biological Monitoring/Ecotoxicology

1

Analysis of Environmental Samples with Yeast-Based Bioluminescent Bioreporters

Melanie Eldridge[1], John Sanseverino[1], Gisela de Arãgao Umbuzeiro[2] and Gary S. Sayler[1]
[1]*University of Tennessee*
[2]*University of Campinas*
[1]*United States of America*
[2]*Brazil*

1. Introduction

Extensive research over the past decade has found the widespread presence of organic wastewater contaminants (OWC) in surface waters around the globe including the United States, (Alvarez et al., 2009; Focazio et al., 2008; Kolpin et al., 2002; Owens et al., 2007; Zheng et al., 2008), Asia (Ma et al., 2007), Europe (Cargouet et al., 2007; Cespedes et al., 2005; Gros et al., 2009; Reemtsma et al., 2006) and South America (Bergamasco et al., submitted; Jardim et al., 2011; Kuster et al., 2009). These OWC include pesticides, plasticizers, pharmaceuticals, and natural and synthetic hormones as well as pollutants from chemical spills into the environment. These compounds may be introduced into surface waters by runoff from land application of biosolids, through leaking sewer lines and septic systems, or by incomplete removal from wastewater treatment systems. Further, a wide variety of these chemicals have been implicated in endocrine disruption in invertebrates and vertebrates (Cooper & Kavlock, 1997; Fang et al., 2000; Folmar et al., 2002; Fossi & Marsili, 2003; Guillette et al., 1999; Hayes et al. 2010; Kavlock et al., 1996; Kidd et al. 2007; Ropstad et al., 2006; Sonne et al., 2006; Tyler et al., 1998).

An endocrine disruptor is an exogenous substance that causes adverse health effects in an organism or its offspring by way of alteration in the function of the endocrine system. As such endocrine disruption is a mechanism leading to a variety of adverse health effects, most of which are considered as reproductive or developmental toxicities (OECD, 2002). The complex nature of reproductive and developmental effects suggests that *in vivo* tests are necessary to detect endocrine disruption. Several *in vivo* mammalian assays (e.g. O'Connor et al., 2002) and *in vitro* assays (e.g. Fang et al., 2000; Zacharewski, 1997) exist for measuring estrogenic effects in various biological systems. However, these are not suitable for rapid, high-throughput screening of chemicals or necessarily screening of environmental samples. Yeast-based *in vitro* estrogen and androgen screens have been firmly established as a means for rapidly identifying chemicals with potential endocrine disrupting activity. This chapter will review the development and use of yeast-based bacterial bioluminescent bioreporters for the detection of endocrine disruption compounds.

1.1 Bioreporters

Reporter gene fusions have been widely used for the detection and quantification of chemical, biological, and physical agents (Daunert et al., 2000). The principle is to fuse a specific genetic promoter or response element with a reporter gene. Induction by a specific target chemical initiates transcription/translation of the bioreporter molecule, which generates a measurable signal. There are three widely-used classes of bioreporters: colorimetric (e.g. *lacZ, cat*), fluorescent (e.g. *gfp*), and bioluminescent (e.g. *luc, lux*). One example of a colorimetric-based bioreporter is the *lacZ* gene which encodes the β-galactosidase enzyme. β-Galactosidase mediates the breakdown of lactose to glucose + galactose. As a bioreporter, β-galactosidase is widely used in molecular biology in the blue-white screening assay. The chromophore X-gal (bromo-chloro-indolyl-galactopyranoside) is cleaved into galactose and an indole moiety that turns the medium blue. For chemical detection, *lacZ* is fused to a chemical-responsive promoter and when the cells are exposed to chromophores, such as chlorophenol red-β-D-galactopyranoside (CPRG), the assay medium changes from yellow to red. This type of colorimetric bioreporter is inexpensive and can be used in a qualitative or quantitative type of assay. Color density can be measured on a standard spectrophotometer.

Fluorescent assays take advantage of the green fluorescent protein (GFP). GFP was originally isolated from the jellyfish *Aequorea victoria* (Johnson et al., 1962; Shimomura et al., 1962). GFP is widely used as a bioreporter in eukaryotic systems for its simplicity to clone and no requirement for an organic substrate other than excitation with either UV or blue light. Quantification of the signal is by a fluorescent spectrophotometer or plate reader. There are different versions of *gfp* including blue-, red-, and yellow-shifted variants each requiring different excitation wavelengths and each of which fluoresce at different wavelengths (Hein & Tsien, 1996; Kendall & Badminton, 1998). In some cases this may be advantageous, especially when multiple bioreporters will be used simultaneously. These genes have been used extensively since they were first employed as gene expression biomarkers (Chalfie et al., 1994).

Firefly luciferase is another well-used bioreporter in eukaryotic systems. The luciferase, encoded by the *luc* gene (*lucFF*), was originally isolated from *Photinus pyralis* (firefly) and generates luciferase by a two-step conversion of D-luciferin to oxyluciferin (de Wet et al., 1985). This reaction generates light at 560 nm. However, the gene does not encode for the D-luciferin substrate and therefore substrate addition in any assay is required, which adds processing time and expense to the assay. Luc-based assays may also be constrained by the requirement for a cell lysis step followed by addition of the D-luciferin, adding both time and expense to the assay.

Bacterial bioluminescence has been widely used as a bioreporter in prokaryotic systems. The *lux* operon (*lux*CDABE) was originally isolated from *Vibrio fischeri* (Engebrecht et al., 1983), *Vibrio harveyi* (Cohn et al., 1983), and *Photorhabdus luminescens* (Szittner & Meighen, 1990). The *lux* operon encodes for the luciferase enzyme (*luxAB*) and the long-chain aldehyde substrate (*luxCDE*) for that reaction. An assay employing bacterial bioluminescence does not require an external organic substrate; the only requirement is for oxygen (O_2). A long chain aldehyde and a reduced flavin mononucleotide ($FMNH_2$) are converted by luciferase (LuxAB) to a long chain carboxylic acid and FMN, producing light at 490 nm wavelength (Meighen & Dunlap, 1993). The *luxAB* (without *luxCDE*) can also be used as a bioreporter and while these strains also produce light at 490 nm, they are less suited for high

throughput analysis due to additional handling steps (costly substrate addition) and additional cost.
The *luc* genes have been reported to be more sensitive than *lux*-based systems, however in a recent comparison of *luc*- and *lux*-based hormone-sensing bioreporters, Svobodova and Cajthaml (2010) determined that some *lux*-based bioreporters (BLYES/BLYAS bioassays, discussed below) are of comparable sensitivity and in some cases much more sensitive than *luc*-based bioreporters.
Several reviews are available on the properties and use of *luc, luxAB, luxCDABE, gfp,* and *gfp*-derived reporter genes in environmental systems (Hakkila et al., 2002; Keane et al., 2002; Ripp et al., 2010). Each of these reporter technologies has advantages and disadvantages depending on the application. For high throughput analysis of samples, bioreporters with the *luxCDABE* genes expressed are particularly well-suited for screening large numbers of samples. For both *luxAB*- and *lucFF*-based bioreporters, costly substrates must be continually added to the cells for visualization of the reaction. This increases not only handling difficulty but also costs to perform the assay. For GFP-based bioreporters, no exogenous substrates are necessary but fluorescent molecules must be excited by a light source to fluoresce. Each of these types of bioreporters produces signals for different lengths of time and has different light emission maxima and optimum temperatures. For example, while the *Photorhabdus luminescens* luciferase (Lux) is stable up to 42°C, firefly luciferase (Luc) has a temperature optimum at 25°C and is thermally inactivated above 30°C (Keane et al., 2002). Bioreporter fusions incorporating the full *lux* cassette are advantageous in that they do not require exogenous substrates, cell lysis is not required, the signal is quantitative and reproducible (King et al., 1990). Further, continuous on-line monitoring is possible (e.g. DiGrazia et al., 1991; Heitzer et al., 1994; Heitzer et al., 1992; King et al., 1990).

1.2 Bacterial *lux* expression in *Saccharomyces cerevisiae*

Prior to 2003, the *lux* genetic system was previously limited only to expression in prokaryotic systems. However, Gupta et al. (2003) were successful in expressing the *P. luminescens lux* cassette in the yeast *S. cerevisiae*. Specifically, the *luxA, -B, -C, -D,* and *-E* genes from *P. luminescens* and the *frp* gene from *Vibrio harveyi* were re-engineered for expression in *Saccharomyces cerevisiae*. The *lux* operon was engineered using two pBEVY yeast expression vectors (Miller et al., 1998), which allowed bidirectional, constitutive expression of the individual *luxA, -B, -C, -D,* and *-E* genes. The *luxA* and *luxB* genes were independently expressed from divergent yeast constitutive promoters GPD and ADH1 on pBEVY-U (Figure 1). The *luxCD* and *luxE-frp* genes were independently expressed from a second plasmid (pBEVY-L), also using the GPD and ADH1 promoters. An internal ribosome entry site (IRES) was inserted between the *luxC* and *luxD* genes and the *luxE* and *frp* genes. The IRES allows translation of multiple genes from a single promoter in eukaryotes (Hellen & Sarnow, 2001).
Constitutive expression of the *luxCDABEfrp* genes in *S. cerevisiae* W303a generated approximately 9,000,000 photons per second per unit optical density (Gupta et al., 2003). This is comparable to similar expression in prokaryotic systems. This was a significant milestone in expression of bacterial operons in lower eukaryotic systems and created possibilities for screening organic wastewater contaminants with mammalian health significance.

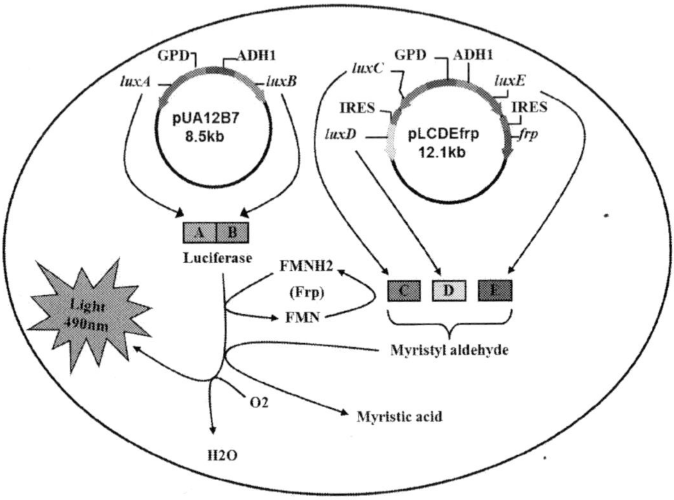

Fig. 1. Schematic representation of *S. cerevisiae* BLYEV (currently known as BLYR). This strain produces light continuously by constitutive expression of the *luxCDABE* genes from *Photorhabdus luminescens* and the *frp* gene from *Vibrio harveyi.*

2. Chemical detection using *S. cerevisiae*-based bioluminescent bioreporters

Yeast-based bioassays containing human receptors for estrogens and androgens fall into the recombinant receptor/reporter gene assay category. Estrogen or androgen response elements linked to a bioreporter molecule offer a low-cost method for screening samples rapidly for determining the presence of possible endocrine disruptors. Two widely used receptor/reporter assays for detecting estrogenic and androgenic compounds are the Yeast Estrogen Screen (YES) (Routledge & Sumpter, 1996) and the Yeast Androgen Screen (YAS) (Purvis et al., 1991). The *S. cerevisiae* YES and YAS bioreporters are colorimetric *lacZ*-based estrogen and androgen-sensing strains, respectively. The *S. cerevisiae* host strain for YES and YAS, contains the human estrogen receptor (hER-α) and human androgen receptor, respectively (Purvis et al., 1991; Routledge & Sumpter, 1996). Further, each host strain contains a series of either human estrogen response elements (EREs) or human androgen response elements (AREs) fused to the *lacZ* gene. The *lacZ* gene product, β-galactosidase, transforms the chromogenic substrate CPRG to a red product, measured by absorbance at 540 nm. These were the first widely used assays for yeast-based detection of estrogenic compounds.

The YES and YAS assays have been used extensively to measure endocrine responses to specific chemicals including polychlorinated biphenyls (PCBs) and hydroxylated derivatives (Layton et al., 2000; Schultz, 2002; Schultz et al., 1998), polynuclear aromatic hydrocarbons (PAH) (Schultz & Sinks, 2002), pesticides (Sohoni et al., 2001) and other compounds (Schultz et al., 2002). These assays have been adapted to environmental matrices including environmental waterways (Thomas et al., 2002), aquifers (Conroy et al., 2005), wastewater treatment systems (Layton et al., 2000) and dairy manure (Raman et al., 2004). Additional yeast-based bioreporters have been developed using either a colorimetric detection (Bovee et al., 2004; Gaido et al., 1997; Le Guevel & Pakdel, 2001; Rehmann et al., 1999), green fluorescent protein (Bovee et al., 2007; Bovee et al., 2004) or the firefly luciferase bioreporter (Bovee et al., 2004; Leskinen et al., 2005; Michelini et al., 2005).

While the YES and YAS assays were highly specific for their target compounds, the colorimetric assays have disadvantages including addition of the chromophore for color development and a 3-5 day reaction time. This latter requirement hindered their ability for high-throughput analysis. Further, after 3 -5 days of incubation, it was unknown if any oxidation reactions were occurring that may activate the target compound. Some newer colorimetric assays have dramatically shortened the time required for color development (4-6 h) through the use of alternative substrates but have the disadvantage of requiring cell lysis steps (Jaio et al., 2008).

To overcome these limitations, bioluminescent version of the YES and YAS reporters were developed by modifying the plasmid constructs of Gupta et al. (2003). Triple repeats of the human ERE were inserted in between the GPD and ADH1 constitutive promoters regulating the *luxA* and *luxB* genes, respectively (Figure 2) generating strain BLYES (Sanseverino et al., 2005). A similar strategy was used for strain BLYAS (Eldridge et al., 2007), which functions in the same way except that it contains the human androgen receptor gene on its genome and *luxAB* are under control of four androgen response elements (AREs), while the constitutive strain (BLYR) has both the *luxAB* and *luxCDEfrp* genes constitutively produced therefore it makes light constantly. The BLYR strain is used to determine whether samples or chemicals are toxic to the yeast, preventing false negatives. If a chemical is highly toxic, killing or inhibiting the cells, no light will be produced and it would be easy to mistake toxicity for no estrogenic response. However, if bioluminescence of the BLYR strain is reduced, since it produces light constitutively, it is obvious that toxicity exists in the sample.

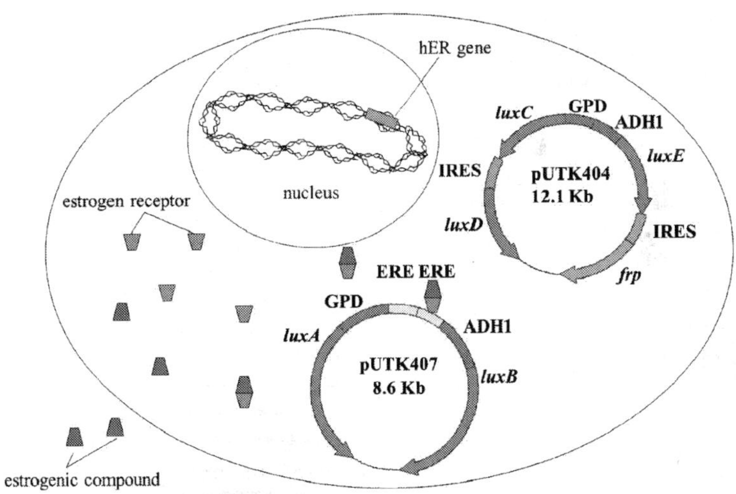

Fig. 2. Schematic representation of *S. cerevisiae* BLYES. Estrogenic compounds cross the cell membrane and bind to the human estrogen receptor (hER). This complex interacts with estrogen response elements (RE) initiating transcription of *luxA* and *luxB*. *S. cerevisiae* BLYES contains the human estrogen receptor in its genome, while *S. cerevisiae* BLYAS has the human androgen receptor in the genome.

Comparison of the BLYES and BLYAS strains to their colorimetric counterparts and proof-of-concept as to their utility has been established (Eldridge et al., 2007; Sanseverino et al., 2005). The BLYES and BLYAS assays are consistent with previously published yeast-based reporter assays (Sanseverino et al., 2009). The 40 - 50% variability of the EC_{50} values shown in Figure 3 reaffirms the suggestion that no single assay should be used to determine an absolute EC_{50} value but rather as a first step in estimating the hormonal activity of a chemical (Beresford et al., 2000).

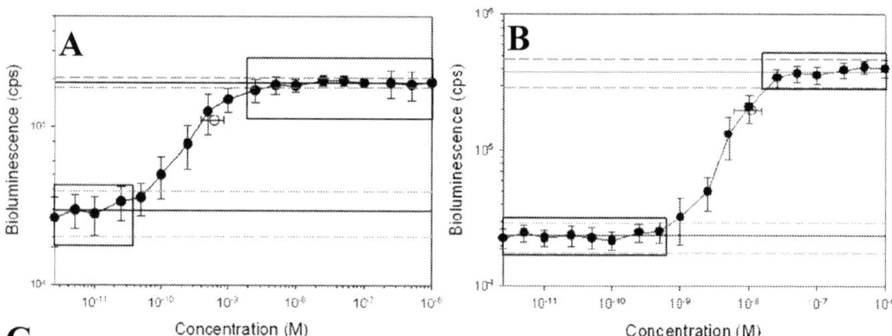

C

Assay	Chemical Standard	EC_{50} (M)	Upper Limit of Detection (M)	Lower Limit of Detection (M)
BLYES	17β-Estradiol	$6.3 \pm 2.4 \times 10^{-10}$	5.0×10^{-9}	2.5×10^{-11}
BLYAS	5α-Dihydrotestosterone	$1.1 \pm 4.6 \times 10^{-8}$	5.0×10^{-8}	1.0×10^{-10}

Fig. 3. **A.** *S. cerevisiae* BLYES standard curve (n = 13) using 17β-estradiol. **B.** *S. cerevisiae* BLYAS standard curve (n = 13) using dihydrotestosterone as a standard. Open circle: calculated EC_{50} values with error bars. A 50% effective concentration (EC_{50}) value was determined from the midpoint of the linear portion of the sigmoidal dose response curve. The mean and standard deviation values were calculated from replicate EC_{50} values for each standard to determine the variability between assays. **C.** Summary of EC_{50} values for BLYES and BLYAS strain with upper and lower limits of detection.

S. cerevisiae BLYES, *S. cerevisiae* BLYAS, *S. cerevisiae* BLYR, were used to assess their reproducibility and utility in screening 69, 68, and 71 chemicals for estrogenic, androgenic, and toxic effects, respectively (Sanseverino et al., 2009). This screening was part of an assessment of the United States Environmental Protection Agency's Tiered screening of chemicals for endocrine-disrupting ability. The 3-tier system includes (i) priority setting, (ii) Tier 1 screening, and (iii) Tier 2 screening. Priority setting focuses on identifying chemicals that require further testing; i.e., excluding chemicals with little or no known hormonal activity and that are generally regarded as safe. The intent of Tier I screening is to rapidly identify chemicals that interact with the estrogen, androgen, and thyroid systems while Tier 2 screenings provide a more in-depth study of how each chemical interacts with each endocrine system. In this study, EC_{50} values were $6.3 \pm 2.4 \times 10^{-10}$ M (n = 18) and $1.1 \pm 0.5 \times 10^{-8}$ M (n = 13) for BLYES and BLYAS, using 17β-estradiol and 5α-dihydrotestosterone

(DHT) over concentration ranges of 2.5×10^{-12} thru 1.0×10^{-6} M, respectively. Based on analysis of replicate standard curves, comparison to background controls, and screening a variety of chemicals, a set of quantitative rules was formulated to interpret data and determine if a chemical is potentially hormonally active, toxic, both, or neither (Sanseverino et al., 2009). The results demonstrated that these assays were applicable for Tier I chemical screening in EPA's Endocrine Disruptor Screening and Testing Program as well as for monitoring endocrine disrupting activity of unknown chemicals in water.

Additional *S. cerevisiae* bioluminescent bioreporters for estrogens and androgens have been developed using the firefly luciferase as the reporter molecule. The bioreporters of Leskinen et al., (2005) each contain the firefly luciferase gene (*lucFF*) under control of hormone-responsive promoters. The four strains, designated BMAEREluc/ERα, BMAEREluc/ERβ, BMAEREluc/AR, and BMA64/luc were used to detect estrogens (two versions), androgens, and toxicity, respectively. This bioassay is unique in that it uses two estrogen-sensing bioreporters; one contains the alpha form of the estrogen receptor and one contains the beta form (ERα, ERβ). These bioreporters were used by Svobodova et al. (2009) to test commercially available PCB mixtures and triclosan for estrogenic and androgenic activity but did not detect any activity with these samples (estrogenic or androgenic). This lack of estrogenic response in the bioluminescent assays may be due to the different mode of action of chemicals like triclosan (Stoker et al., 2010). In a study that examined the effects of triclosan exposure on female Wistar rats, triclosan advanced the onset of puberty symptoms. Also, a combination of ethinyl estradiol (EE2) and triclosan increased uterine weight significantly more than EE2 alone while triclosan alone had no effect. Therefore the mode of action of triclosan appears to have a synergistic effect on EE2 activity in Wistar rats. This effect appears to be independent of estrogen receptor binding given that bioluminescent yeast bioassays (Svobodova et al. 2009, Eldridge et al. unpublished data), which measure binding to the hER and then EREs, did not respond to triclosan.

In addition to hormone-mimicking chemicals, several other types of contaminants are also detectable with *S. cerevisiae*-based bioluminescent bioreporters. For example, the aryl hydrocarbon-sensing strain of Leskinen et al. (2008) contains genomically integrated human aryl hydrocarbon receptor and human aryl hydrocarbon nuclear translocator genes. In addition, it carries a plasmid-encoded copy of the firefly luciferase gene (*lucFF*) that is regulated by a series of aryl hydrocarbon receptor complex (AHRC) response elements (also called dioxin response elements or xenobiotic response elements, AhREs/DREs/XREs). Aryl hydrocarbon receptor proteins interact with both their AH ligand and the nuclear translocator protein then bind to the AhRE region of the *luc*-containing plasmid, activating transcription of luciferase, similarly to the receptor-response element system present in the BLYES bioassay. Since this bioassay is *luc*-based, D-luciferin must be added.

Another *S. cerevisiae*-based bioreporter has been created to measure arsenate and also UV damage (Bakhrat et al., 2011). This strain is based on the BLYES strain of Sanseverino et al. (2005), containing a constitutive *luxCDEfrp* plasmid and a *luxAB* plasmid that has been re-engineered to be under control of the UFO1 promoter, which specifically responds to DNA damage by UV light and also arsenate. The strain is able to detect very low concentrations of arsenate (1×10^{-12} to 1×10^{-6} M), which makes them useful for environmental monitoring. It was also used to evaluate the level of UV protection in commercial sunscreens. When films of Saran wrap were placed between the cells with SPF100 or SPF15 sunscreen on them, the sunscreen provided 100% and 90% protection, respectively, in comparison to a control in which samples were shielded with only Saran wrap. Studies of this type demonstrate this bioassay's usefulness on complex samples.

3. Analysis of aqueous environmental samples

For use on environmental samples, the BLYES/BLYAS/BLYR bioreporter suite is particularly well-suited. They require no substrate addition or illumination source, are inexpensive to use, and are optimized for 96-well plate formats. For water samples where OWC are typically found in the ppb range, a concentration step is necessary. Figure 4 outlines the procedures for analysis of aqueous samples. For wastewater effluents and source drinking water samples, solid phase extraction is performed to isolate and concentrate any chemical contaminants.

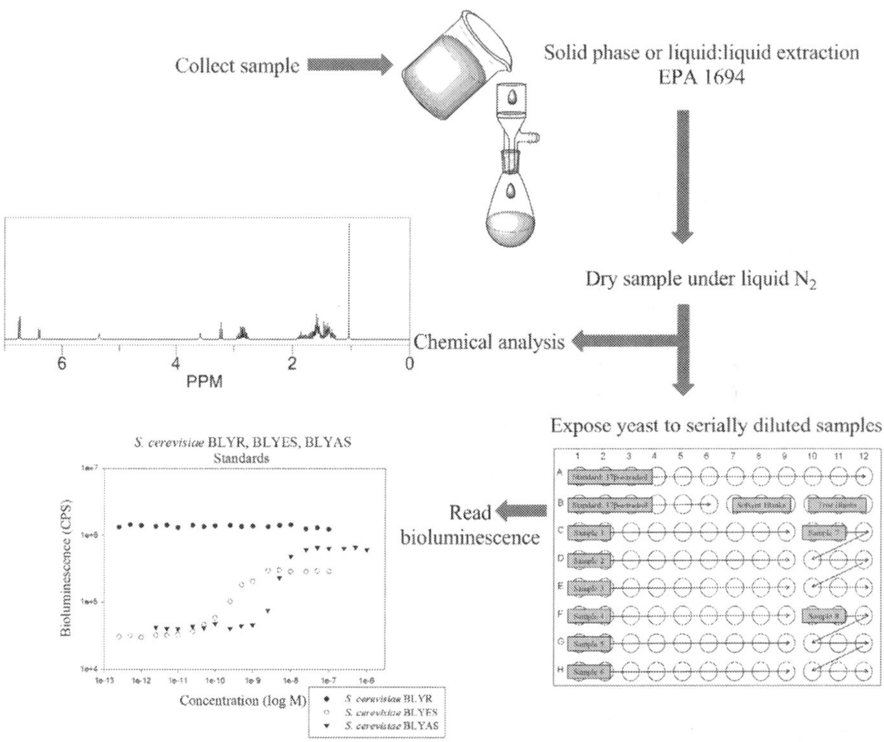

Fig. 4. Schematic of sample preparation and analysis. Typically, 1 L samples are collected aseptically and passed through a solid phase extraction unit. After elution by an appropriate solvent, concentrating the sample 1,000-fold, the sample is analyzed by the bioassays and/or with chemical analysis such as GC/MS or LC/MS. Typically, eight samples are analyzed on a 96-well plate, including standards and control wells (both solvent control and no treatment controls). By combining multiple plates in one assay run, numerous samples are processed at one time. Bioluminescence is monitored and recorded over time using a photon-counting system.

Numerous methods for solid phase extraction exist but commonly a modification of United States EPA 1694 (2007) is used. Briefly, Oasis filters (Waters, Inc.) or cartridges are conditioned with methanol and water, then the sample (typically ~1 L) is passed though the membrane slowly under a small amount of pressure. Chemicals are eluted in a solvent, either singly or in combinations, such as methanol or a methanol:acetone mixture. The solvents are evaporated to dryness and may be used immediately or stored at -20°C for future use. For the BLYES/BLYAS/BLYR assays, samples are resuspended in methanol (or DMSO) such that they are 1000x concentrated compared to the original sample, e.g. 1 L of sample is concentrated, dried, and resuspended in 1 mL of solvent, yielding an effective concentration factor of 1000x. This may then be split for chemical analysis and bioassays.

In the bioassays, samples are serially diluted in methanol to achieve a range of concentrations (1000x-2.5x). In addition, standard chemicals (17β-estradiol (E2) for BLYES/BLYR and 5α-dihydrotestosterone (DHT) for BLYAS) are suspended in methanol at 0.01 M and then serially diluted 18 times to generate a concentration range of 4×10^{-7} M to 1×10^{-12} M for E2 and 4×10^{-6} M to 1×10^{-11} M for DHT. Samples and standards (50 μL) are then spotted into the wells of 96-well plates (Figure 4). Triplicate plates are made (one for each of three strains) and then methanol is evaporated at room temperature.

For preparation of the bioassay, each yeast strain is grown overnight at 28°C with shaking (150 rpm) in Yeast Minimal Media (YMM) without leucine or uracil (Routledge & Sumpter, 1996) to an OD_{600} of 1.0. Yeast strains (200 μL) are spotted into the wells of 96-well plates containing dry samples and standards, beginning the exposure. This generates a concentration range of 250x-0.625x for environmental samples, 1×10^{-7} M to 2.5×10^{-13} M for E2, and 1×10^{-6} M to 2.5×10^{-12} M for DHT. Negative controls included wells with (i) medium + cells and (ii) medium + cells + evaporated methanol, to monitor whether estrogenic or androgenic substances are present in the solvent. Plates are then placed into a plate reader (such as Perkin-Elmer Victor2 Multilabel Counter) with an integration time of 1 s/well. Bioluminescence is measured every 30-60 min for four hours. Relative light unit data (as counts per second) is plotted versus the log of concentration in SigmaPlot (or similar statistical software) (Figure 5).

For each chemical, the log of bioluminescence (counts per second) versus the log of chemical concentration (M) is plotted, generating a sigmoidal curve for hormonally active compounds. A 50% effective concentration (EC_{50}) value is determined from the midpoint of the linear portion of the sigmoidal curve. The mean and standard deviation values are calculated from replicate EC_{50} values for standards to determine the variability between assays. Detection limits are determined by calculating the concentration of chemical at background bioluminescence plus three standards deviations. Toxicity is calculated as the concentration of sample that reduces the signal from the constitutively bioluminescent strain (BLYR) by 20% (IC_{20}). For environmental samples, the concentration factor that yields 50% maximal response is considered the EC_{50} and when this value is divided by the EC_{50} for that assay's standard, estrogenic or androgenic equivalents are calculated (in terms of E2 or DHT, respectively); this determines the amount of potentially estrogenic substances that are present in a sample relative to the standard.

For samples in which DMSO is the preferred solvent, a 4% solution of DMSO is used for the serial dilutions of environmental samples and standards (by incubating the sample in a small volume of 100% DMSO for 15 minutes then adding ultra-pure water to achieve a final DMSO concentration of 4%). Next, 100 μL of sample or standard are spotted into 96-well plates along with 100 μL of yeast cells (without drying the samples/standards), yielding a final DMSO concentration of 2% in all wells. Negative controls should consist of wells with (i) medium + cells and (ii) medium + cells + DMSO to monitor whether DMSO is toxic to yeast cells and whether the solvent contains potentially estrogenic substances.

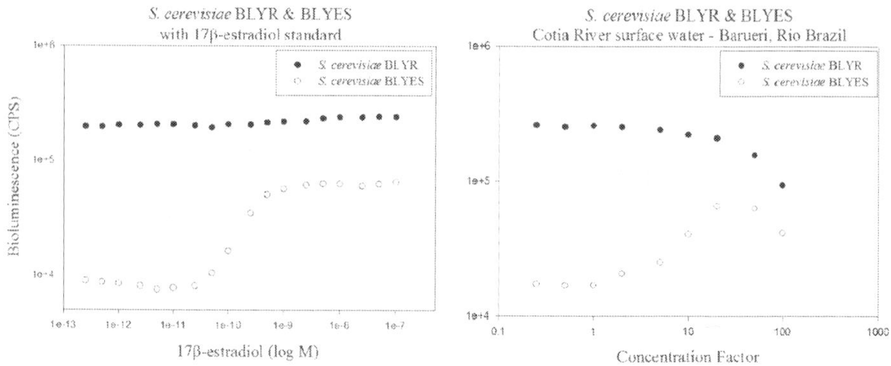

Fig. 5. Yeast assay data using environmental samples. The graphs show the responses of the yeast strains S. *cerevisiae* BLYR and BLYES in response to the 17β-estradiol standard and serial dilutions of a solid phase extracted sample. The sample consisted of surface water from the Cotia River in Brazil, which has a high concentration of estrogenic substances present (1.2 ng E2 equivalents/L) and exhibits marked toxicity. Analysis of surface and groundwater are of particular interest to regulatory agencies (and the public) because they are source waters for drinking water treatment plants.

Using this bioassay, surface water samples were surveyed from the U.S. and Brazil (Eldridge-Umbuzeiro, unpublished data, Figures 5 and 6), with both studies determining that the estrogen-sensing strain detects more estrogen-like activity than predicted through chemical analysis alone. This is expected however, given that chemical analysis targets certain contaminants and cannot be expected to screen samples for all known estrogens. In addition, it is relatively unknown if/how chemicals act synergistically to promote estrogen- or androgen-like activity. The assay can provide a clear evaluation on the levels of potential estrogenicity in monitoring studies of surface water samples as can be seen from Figure 6 (unpublished data). The levels varied from 0.01 to 19.3 ng/L of E2 equivalents per liter of water. In this particular case, the river water that was monitored was from Brazil, with the highest levels of pollution expected to occur in the dry season (corresponding to June-October).

In Jardim et al. (2011), surface water samples from Brazil were also examined. The samples were collected from sites classified by the São Paulo State Environmental Agency (CETESB) as excellent, good, medium, fair, and poor. Both bioassays and chemical analysis were performed on samples following solid phase extraction. The authors targeted estrone (E1), 17β-estradiol (E2), ethinyl estradiol (EE2), estriol (E3), bisphenol A (BPA), 4-n-octophenol (OP), and 4-n-nonylphenol (NP) in their chemical analysis and used the estrogen-sensing BLYES as the bioassay. From this data, the authors determined that the bioassay data is not fully explained by the amount and strength of the detected estrogens. For example, the highest estrogenic response determined by the yeast bioassay (BLYES) was also determined to have the highest concentrations of estrogen by chemical analysis. Also, in drinking water samples in which targeted estrogens were not detected by chemical analysis, the yeast bioassay also did not detect any estrogenic activity. However, in some samples from different surface water intake points, often yeast bioassays detected estrogenic activity at levels that chemical analysis data did not predict. For example, BPA detected at 3.53 ng/L (according to chemical analysis) is not a sufficient concentration to elicit an estrogenic response. However, the same sample elicited a response equivalent to 0.7 ng/L of E2 equivalents (according to the yeast bioassay). This suggests that *S. cerevisiae* BLYES was responding to a) something that was not recognized by chemical analysis, b) by-products of the targeted chemicals, or c) that a mixture effect is causing a synergistic estrogen response. The reader is cautioned that these assays should be a first determination of estrogenic activity. *S. cerevisiae* does not have an endocrine system and cannot explicitly identify endocrine disruptors. Advanced testing with alternate assays (i.e. mammalian-based assays) should be used for confirmation of endocrine-disrupting activity.

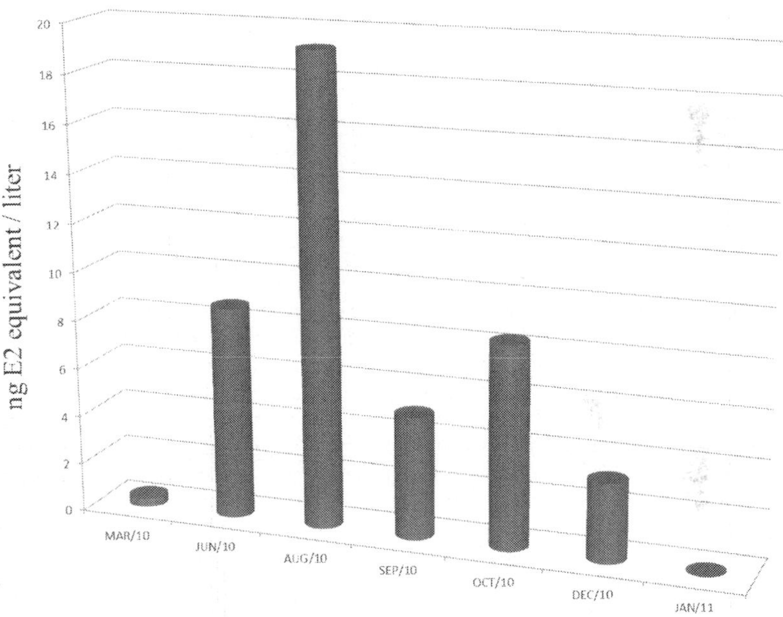

Fig. 6. Surface water samples from Brazil assessed with the BLYES bioassay. Surface water samples were solid phase extracted and then processed with the *S. cerevisiae* BLYR and BLYES.

Alvarez et al. (2009) used BLYES for the analysis of Potomac River water samples in a study on the reproductive health of bass. The authors criticized the collection of single grab samples, in favor of using passive samplers to concentrate contaminants. They examined extracts from passive samplers that had been deployed for 31 days in the Potomac River and its tributaries, which receive significant amounts of flow from WWTP effluents. Samples were collected once yearly, for two years (2005-2006), both upstream and downstream of known WWTP discharge sites. They also performed both chemical analysis (targeting E1, E2, EE2, and E3) and bioassays (using BLYES and BLYR). They were able to detect potentially estrogenic compounds at levels statistically different than the field blanks. Levels of E2 equivalents were detected in the nanomolar range. The authors were able to measure estrogenic responses with BLYES but they were not able to detect a seasonal difference in estrogenicity (some chemicals were detected seasonally via chemical analysis but others were not) though it is unclear whether there was no seasonal effect or whether the estrogens were detected at such a low concentration that a conclusion cannot be drawn.

In both studies (Alvarez et al., 2009; Jardim et al., 2011), the expected response with bioassays was lower than the actual response determined with the bioassay. Expected responses are calculated by multiplying a chemical's concentration (determined through chemical analysis) by its potency relative to a reference estrogen, such as E2. It is expected that if all contaminating estrogenic molecules are detected by chemical analysis then the expected responses should match actual bioassay responses. However, it is difficult to anticipate (and therefore target) all possible endocrine-active contaminants that are present in environmental samples. In addition, prediction of the effects of mixtures of chemicals, especially at low concentrations, has proven to be problematic. Moreover, bioassays are likely to detect metabolites of estrogenic chemicals, as long these molecules continue to interact with the human hormone receptor/response element-sensing systems. Given these reasons, it is natural to expect that chemical analysis is unlikely to ever fully predict actual bioassay responses.

The androgen-sensing strain of Leskinen et al. (2005) has been used to monitor wastewater before and after treatment in wastewater treatment plants in several cities in Italy (Michelini et al. 2005). It was determined that both samples (pre- and post-treatment) contained chemicals with androgenic activity, however treatment decreased this activity. They determined that approximately 30% of androgenic activity was typically removed but occasionally activity was reduced by 90%. They attributed the decreased activity to the presence of carbon-based filters, which should bind chemicals, thereby removing them from wastewater. This study illustrates the effectiveness of yeast-based bioreporters for the rapid analysis of samples before and after water treatment. It also demonstrates that wastewater treatment does not necessarily remove chemicals associated with potential endocrine disrupting activity.

In addition, the strains of Leskinen et al. (2005) (BMAEREluc/ERα, BMAEREluc/ERβ, BMAEREluc/AR, and BMA64/luc) were used to test several lotion samples, as a simulation of using the strains on complex sample matrices. Five of the seven lotion samples demonstrated estrogenic activity, even at dilutions as low as 1:175. The authors attributed this activity to parabens present in the lotions, given that samples with no parabens were not estrogenic but samples with mixtures of parabens were. The authors state that parabens are present in many cosmetic products and are generally considered safe (Soni et al., 2002), despite having been demonstrated to produce an estrogenic response (Routledge et al., 1998) and being present in breast cancer tumors (Darbre et al., 2004). No androgenic activity was found for any of the samples.

More recently, Svobodova et al. (2009) examined the endocrine disrupting potential of a commercial PCB mixture (Delor 103) and a series of potential PCB degradation metabolites (chlorobenzoic acids and cholorophenols). The authors did not detect any estrogenic activity with any of the chemicals or mixtures tested using bioluminescent yeast, except that 5 mg/L chlorophenol caused a response. This is in contrast to the results obtained with the Yeast Estrogen Screen (YES) colorimetric bioreporter, which detected estrogenic activity with all the tested chemicals except chlorophenols. One reason for this difference may be the different length of exposure time between the two bioassays. The YES was incubated with chemicals for three days whereas the Leskinen strains were only incubated with the chemicals for 2.5 h prior to sample processing. It is possible that over three days' time, the PCBs may have oxidized (yeast are incubated in aerobic conditions) to forms that are more likely estrogenic. Indeed, hydroxylated PCBs have been demonstrated to harbor estrogenic activity (Korach et al., 1988; Schultz, 2002; Schultz et al., 1998). Interestingly, using the bioluminescent androgen-sensing bioreporter (BMAEREluc/AR), androgenic activity was detected with the commercial PCB mixture, but not with chlorobenzoic acids, chlorophenols, or triclosan. Triclosan has been demonstrated to have no activity with the BLYES and BLYAS bioassays as well (data not shown).

4. Future applications

Saccharomyces cerevisiae-based bioluminescent bioreporters offer excellent opportunities beyond bacterial bioreporters for rapid analysis of chemicals with human and environmental significance. Expression of the bacterial *lux* cassette in a lower eukaryote offers many opportunities not only for high-throughput screening systems but also bioprocess monitoring, diagnostic applications, fungal gene expression analysis, and *in vivo* sensing of fungal infections (Gupta et al., 2003). Expression in *S. cerevisiae* has led to advances in transferring this system to mammalian cell lines (Close et al., 2010; Patterson et al., 2005).

The advantages for detection of endocrine-disrupting chemicals in water by *S. cerevisiae lux*-based bioreporters are numerous including accuracy, ease of use, not expensive, and amenable to automation in performing and collection of data. In addition to screening aqueous samples, BLYES, BLYAS, and BLYR, and other variants described in the literature are useful for Tier I screening as proposed by the EPA, analysis of wastewater influent and effluent, chemical leaching from manufactured products, for example. In fact, the State Environmental Agency of São Paulo (CETESB) in Brazil is considering using the *S. cerevisiae* BLYES bioassay for routine monitoring of surface and ground water samples for the presence of potentially estrogenic substances. Two of the authors (M.E. and G.S.) have begun routine monitoring of wastewater treatment plant effluents from a treatment facility in TN as well as screening 250 water samples across the state of Tennessee in a broad survey.

Ideally, detection of potential endocrine disruptors (or any other chemical of interest) by bioluminescent bioreporter strains would be coupled to remote detection systems for continuous real-time monitoring. Bioluminescent bioreporter integrated circuits fuse reporter cells to an integrated circuit containing a photodetector (e.g. Sayler et al., 2001; Nivens et al., 2004; Sayler et al., 2004). These devices could be distributed in networks and coupled with wireless communications would send signals indicating the presence/absence of chemical contaminants. Roda et al. (2011) have developed a device that couples estrogen- or androgen-sensing *S. cerevisiae* expressing firefly bioluminescence to fiber optics with detection by a CCD sensor, yielding a fully functional biosensor. While this device resulted in strains whose detection limit was approximately 10-fold higher than bioassays performed in the lab and was larger than previously reported remote detection systems, it does demonstrate this device's usefulness in future environmental monitoring. Remotely deployed devices may allow long integration times to account for chronic exposure to low-levels (ppb) of a potential endocrine disrupter that may not be captured in a single grab sample.

5. Acknowledgments

The authors would like to thank the Center for Environmental Biotechnology and the São Paulo Research Foundation (FAPESP) for support of our work (FAPESP 2007/58449-2).

6. References

Alvarez, D.A., Cranor, W.L., Perkins, S.D., Schroeder, V.L., Iwanowicz, L.R., Clark, R.C., Guy, C.P., Pinkney, A.E., Blazer, V.S. & Mullican, J.E. (2009). Reproductive health of bass in the Potomac, USA, drainage: part 2. Seasonal occurrence of persistent and emerging organic contaminants. *Environmental Toxicology and Chemistry* 28: 1084-1095.

Bakhrat, A., Eltzov, E., Finkelstein, Y., Marks, R.S. & Raveh, D. (2011). UV and arsenate toxicity: a specific and sensitive yeast bioluminescence assay. *Cell Biology and Toxicology* 27: 227-236.

Beresford, N., Routledge, E.J., Harris, C.A. & Sumpter, J.P. (2000). Issues arising when interpreting results from an in vitro assay for estrogenic activity. *Toxicology and Applied Pharmacology* 162: 22-33.

Bergamasco, A.M., Eldridge, M.L., Sanseverino, J., Sodré, F.F., Montagner, C.C., Pescara, I.C., Jardim, W.D. & Umbuzeiro, G.D. Bioluminescent yeast estrogen assay (BLYES) as a promising tool to monitor surface and drinking water for estrogenicity. *Journal of Environmental Monitoring* (in press).

Bovee, T.F.H., Helsdingen, R.J.R., Hamers, A.R.M., van Duursen, M.B.M., Nielen, M.W.F. & Hoogenboom, R. (2007). A new highly specific and robust yeast androgen bioassay for the detection of agonists and antagonists. *Analytical and Bioanalytical Chemistry* 389: 1549-1558.

Bovee, T.F.H., Helsdingen, R.J.R., Koks, P.D., Kuiper, H.A., Hoogenboom, R. & Keijer, J. (2004). Development of a rapid yeast estrogen bioassay, based on the expression of green fluorescent protein. *Gene* 325: 187-200.

Cargouet, M., Perdiz, D. & Levi, Y. (2007). Evaluation of the estrogenic potential of river and treated waters in the Paris area (France) using in vivo and in vitro assays. *Ecotoxicology and Environmental Safety* 67(1): 149-156.

Cespedes, R., Lacorte, S., Raldua, D., Ginebreda, A., Barcelo, D. & Pina, B. (2005). Distribution of endocrine disruptors in the Llobregat River basin (Catalonia, NE Spain). *Chemosphere* 61: 1710-1719.

Chalfie, M., Tu, Y., Euskirchen, G., Ward, W.W. & Prasher, D.C. (1994). Green fluorescent protein as a marker for gene-expression. *Science* 263: 802-805.

Close, D.M., Patterson, S.S., Ripp, S., Baek, S.J., Sanseverino, J. & Sayler, G.S. (2010). Autonomous bioluminescent expression of the bacterial luciferase gene cassette (*lux*) in a mammalian cell line. *Plos One* 5(8).

Cohn, D.H., Ogden, R.C., Abelson, J.N., Baldwin, T.O., Nealson, K.H., Simon, M.I. & Mileham, A.J. (1983). Cloning of the *Vibrio harveyi* luciferase genes - use of a synthetic oligonucleotide probe. *Proceedings of the National Academy of Sciences of the United States of America-Biological Sciences* 80: 120-123.

Conroy, O., Quanrud, D.M., Ela, W.P., Wicke, D., Lansey, K.E. & Arnold, R.G. (2005). Fate of wastewater effluent hER-agonists and hER-antagonists during soil aquifer treatment. *Environmental Science & Technology* 39: 2287-2293.

Cooper, R.L. & Kavlock, R.J. (1997). Endocrine disruptors and reproductive development: a weight-of-evidence overview. *Journal of Endocrinology* 152: 159-166.

Darbre, P.D., Aljarrah, A., Miller, W.R., Coldham, N.G., Sauer, M.J. & Pope, G.S. (2004). Concentrations of parabens in human breast tumours. *Journal of Applied Toxicology* 24: 5-13.

Daunert, S., Barrett, G., Feliciano, J.S., Shetty, R.S., Shrestha, S. & Smith-Spencer, W. (2000). Genetically engineered whole-cell sensing systems: coupling biological recognition with reporter genes. *Chemical Reviews* 100: 2705-2738.

de Wet, J.R., Wood, K.V., Helinski, D.R. & Deluca, M. (1985). Cloning of firefly luciferase cDNA and the expression of active luciferase in *Escherichia coli*. *Proceedings of the National Academy of Sciences of the United States of America* 82: 7870-7873.

DiGrazia, P., King, J., Blackburn, J., Applegate, B., Bienkowski, P., Hilton, B. & Sayler, G. (1991). Dynamic response of naphthalene biodegradation in a continuous flow soil slurry reactor. *Biodegradation* 2: 81-91.

Eldridge, M.L., Sanseverino, J., Layton, A.C., Easter, J.P., Schultz, T.W. & Sayler, G.S. (2007). *Saccharomyces cerevisiae* BLYAS, a new bioluminescent bioreporter for detection of androgenic compounds. *Applied and Environmental Microbiology* 73: 6012-6018.

Engebrecht, J., Nealson, K. & Silverman, M. (1983). Bacterial bioluminescence - isolation and genetic-analysis of functions from *Vibrio fischeri*. *Cell* 32: 773-781.

Fang, H., Tong, W.D., Perkins, R., Soto, A.M., Prechtl, N.V. & Sheehan, D.M. (2000). Quantitative comparisons of *in vitro* assays for estrogenic activities. *Environmental Health Perspectives* 108: 723-729.

Focazio, M.J., Kolpin, D.W., Barnes, K.K., Furlong, E.T., Meyer, M.T., Zaugg, S.D., Barber, L.B. & Thurman, M.E. (2008). A national reconnaissance for pharmaceuticals and other organic wastewater contaminants in the United States - II) Untreated drinking water sources. *Science of the Total Environment* 402: 201-216.

Folmar, L.C., Hemmer, M.J., Denslow, N.D., Kroll, K., Chen, J., Cheek, A., Richman, H., Meredith, H. & Grau, E.G. (2002). A comparison of the estrogenic potencies of estradiol, ethynylestradiol, diethylstilbestrol, nonylphenol and methoxychlor in vivo and in vitro. *Aquatic Toxicology* 60: 101-110.

Fossi, M.C. & Marsili, L. (2003). Effects of endocrine disruptors in aquatic mammals. *Pure and Applied Chemistry* 75: 2235-2247.

Gaido, K.W., Leonard, L.S., Lovell, S., Gould, J.C., Babai, D., Portier, C.J. & McDonnell, D.P. (1997). Evaluation of chemicals with endocrine modulating activity in a yeast-based steroid hormone receptor gene transcription assay. *Toxicology and Applied Pharmacology* 143: 205-212.

Gros, M., Petrovic, M. & Barcelo, D. (2009). Tracing pharmaceutical residues of different therapeutic classes in environmental waters by using liquid chromatography/quadrupole-linear ion trap mass spectrometry and automated library searching. *Analytical Chemistry* 81: 898-912.

Guillette, L.J., Brock, J.W., Rooney, A.A. & Woodward, A.R. (1999). Serum concentrations of various environmental contaminants and their relationship to sex steroid concentrations and phallus size in juvenile American alligators. *Archives of Environmental Contamination and Toxicology* 36: 447-455.

Gupta, R.K., Patterson, S.S., Ripp, S., Simpson, M.L. & Sayler, G.S. (2003). Expression of the *Photorhabdus luminescens lux* genes (*luxA, B, C, D*, and *E*) in *Saccharomyces cerevisiae. Fems Yeast Research* 4: 305-313.

Hakkila, K., Maksimow, M., Karp, M. & Virta, M. (2002). Reporter genes *lucFF, luxCDABE, gfp,* and *dsred* have different characteristics in whole-cell bacterial sensors. *Analytical Biochemistry* 301: 235-242.

Hayes, T.B., V. Khoury, A. Narayan, M. Nazir, A. Park, t. Brown, L. Adame, E. Chan, D. Bucholz, T. Stueve, & S. Gallipeau. (2010). Atrazine induces complete feminization and chemical castration in make African clawed frogs (*Xenopus laevis*). *Proceeding of the National Academy of Science*, USA, 107:4612-4617.

Hein, R. & Tsien, R.Y. (1996). Engineering green fluorescent protein for improved brightness, longer wavelengths and fluorescence resonance energy transfer. *Current Biology* 6: 178-182.

Heitzer, A., Malachowsky, K., Thonnard, J.E., Bienkowski, P.R., White, D.C. & Sayler, G.S. (1994). Optical biosensor for environmental online monitoring of naphthalene and salicylate bioavailability with an immobilized bioluminescent catabolic reporter bacterium. *Applied and Environmental Microbiology* 60: 1487-1494.

Heitzer, A., Webb, O.F., Thonnard, J.E. & Sayler, G.S. (1992). Specific and quantitative assessment of naphthalene and salicylate bioavailability by using a bioluminescent catabolic reporter bacterium. *Applied and Environmental Microbiology* 58: 1839-1846.

Hellen, C.U.T. & Sarnow, P. (2001). Internal ribosome entry sites in eukaryotic mRNA molecules. *Genes & Development* 15: 1593-1612.

Jaio, B.W., Yeung, E.K.C., Chan, C.B. & Cheng, C.H.K. (2008). Establishment of a transgenic yeast screening system for estrogenicity and identification of the anti-estrogenic activity of malachite green. *Journal of Cellular Biochemistry* 105: 1399-1409.

Jardim, W., Montagner, C., Pescara, I., Umbuzeiro, G., Bergamasco, A., Eldridge, M. & Sodre, F. (2011). An intergrated approach to evaluate emerging contaminants in drinking water. *Separation and Purification Technology. In press.*

Johnson, F.H., Gershman, L.C., Waters, J.R., Reynolds, G.T., Saiga, Y. & Shimomura, O. (1962). Quantum efficiency of *Cypridina* luminescence, with a note on that of *Aequorea. Journal of Cellular and Comparative Physiology* 60: 85-103.

Kavlock, R.J., Daston, G.P., DeRosa, C., FennerCrisp, P., Gray, L.E., Kaattari, S., Lucier, G., Luster, M., Mac, M.J., Maczka, C., Miller, R., Moore, J., Rolland, R., Scott, G., Sheehan, D.M., Sinks, T. & Tilson, H.A. (1996). Research needs for the risk assessment of health and environmental effects of endocrine disruptors: a report of the US EPA-sponsored workshop. *Environmental Health Perspectives* 104: 715-740.

Keane, A., Phoenix, P., Ghoshal, S. & Lau, P.C.K. (2002). Exposing culprit organic pollutants: a review. *Journal of Microbiological Methods* 49: 103-119.

Kendall, J.M. & Badminton, M.N. (1998). *Aequorea victoria* bioluminescence moves into an exciting new era. *Trends in Biotechnology* 16: 216-224.

Kidd, K. A., Blanchfield, P. J., Mills, K. H., Palace, V. P., Evans, R. E., Lazorchak, J. M., & Flick, R. W. (2007). Collapse of a fish population after exposure to a synthetic estrogen. *Proceeding of the National Academy of Science*, USA, 104:8897–8901.

King, J.M.H., Digrazia, P.M., Applegate, B., Burlage, R., Sanseverino, J., Dunbar, P., Larimer, F. & Sayler, G.S. (1990). Rapid, sensitive bioluminescent reporter technology for naphthalene exposure and biodegradation. *Science* 249: 778-781.

Kolpin, D.W., Furlong, E.T., Meyer, M.T., Thurman, E.M., Zaugg, S.D., Barber, L.B. & Buxton, H.T. (2002). Pharmaceuticals, hormones, and other organic wastewater contaminants in US streams, 1999-2000: A national reconnaissance. *Environmental Science & Technology* 36: 1202-1211.

Korach, K.S., Sarver, P., Chae, K., McLachlan, J.A. & McKinney, J.D. (1988). Estrogen receptor-binding activity of polychlorinated hydroxybiphenyls - conformationally restricted structural probes. *Molecular Pharmacology* 33: 120-126.

Kuster, M., Azevedo, D.A., de Alda, M.J.L., Neto, F.R.A. & Barcelo, D. (2009). Analysis of phytoestrogens, progestogens and estrogens in environmental waters from Rio de Janeiro (Brazil). *Environment International* 35: 997-1003.

Layton, A.C., Gregory, B.W., Seward, J.R., Schultz, T.W. & Sayler, G.S. (2000). Mineralization of steroidal hormones by biosolids in wastewater treatment systems in Tennessee USA. *Environmental Science & Technology* 34: 3925-3931.

Le Guevel, R. & Pakdel, F. (2001). Streamlined beta-galactosidase assay for analysis of recombinant yeast response to estrogens. *Biotechniques* 30: 1000-1004.

Leskinen, P., Hilscherova, K., Sidlova, T., Kiviranta, H., Pessala, P., Salo, S., Verta, M. & Virta, M. (2008). Detecting AhR ligands in sediments using bioluminescent reporter yeast. *Biosensors & Bioelectronics* 23: 1850-1855.

Leskinen, P., Michelini, E., Picard, D., Karp, M. & Virta, M. (2005). Bioluminescent yeast assays for detecting estrogenic and androgenic activity in different matrices. *Chemosphere* 61: 259-266.

Ma, M., Rao, K.F. & Wang, Z.J. (2007). Occurrence of estrogenic effects in sewage and industrial wastewaters in Beijing, China. *Environmental Pollution* 147: 331-336.

Meighen, E.A. & Dunlap, P.V. (1993). Physiological, biochemical and genetic-control of bacterial bioluminescence. In *Advances in Microbial Physiology, Vol 34*. London, Academic Press Ltd. 34: 1-67.

Michelini, E., Leskinen, P., Virta, M., Karp, M. & Roda, A. (2005). A new recombinant cell-based bioluminescent assay for sensitive androgen-like compound detection. *Biosensors & Bioelectronics* 20: 2261-2267.

Miller, C.A., Martinat, M.A. & Hyman, L.E. (1998). Assessment of aryl hydrocarbon receptor complex interactions using pBEVY plasmids: expression vectors with bi-directional promoters for use in *Saccharomyces cerevisiae*. *Nucleic Acids Research* 26: 3577-3583.

Nivens, D.E., McKnight, T.E., Moser, S.A., Osbourn, S.J., Simpson, M.L., & Sayler, G.S. (2004). Bioluminescent bioreporter integrated circuits: potentially small, rugged and inexpensive whole-cell biosensors for remote environmental monitoring. *Journal of Applied Microbiology* 96:33-46.

O'Connor, J.C., Cook, J.C., Marty, M.S., Davis, L.G., Kaplan, A.M. & Carney, E.W. (2002). Evaluation of Tier I screening approaches for detecting endocrine-active compounds (EACs). *Critical Reviews in Toxicology* 32: 521-549.

OECD (2002). Appraisal of test methods for sex-hormone disrupting chemicals. Detailed Review Paper. Series on testing andassessments. No. 21. Paris, France, Organization for Economic Co-operation and Development (OECD).

Owens, C.V., Lambright, C., Bobseine, K., Ryan, B., Gray, L.E., Gullett, B.K. & Wilson, V.S. (2007). Identification of estrogenic compounds emitted from the combustion of computer printed circuit boards in electronic waste. *Environmental Science & Technology* 41: 8506-8511.

Patterson, S.S., Dionisi, H.M., Gupta, R.K. & Sayler, G.S. (2005). Codon optimization of bacterial luciferase (*lux*) for expression in mammalian cells. *Journal of Industrial Microbiology & Biotechnology* 32: 115-123.

Purvis, I.J., Chotai, D., Dykes, C.W., Lubahn, D.B., French, F.S., Wilson, E.M. & Hobden, A.N. (1991). An androgen-inducible expression system for *Saccharomyces cerevisiae*. *Gene* 106(1): 35-42.

Raman, D.R., Williams, E.L., Layton, A.C., Burns, R.T., Easter, J.P., Daugherty, A.S., Mullen, M.D. & Sayler, G.S. (2004). Estrogen content of dairy and swine wastes. *Environmental Science & Technology* 38: 3567-3573.

Reemtsma, T., Weiss, S., Mueller, J., Petrovic, M., Gonzalez, S., Barcelo, D., Ventura, F. & Knepper, T.P. (2006). Polar pollutants entry into the water cycle by municipal wastewater: a European perspective. *Environmental Science & Technology* 40: 5451-5458.

Rehmann, K., Schramm, K.W. & Kettrup, A.A. (1999). Applicability of a yeast oestrogen screen for the detection of oestrogen-like activities in environmental samples. *Chemosphere* 38: 3303-3312.

Ripp, S., DiClaudio-Eldridge, M.L. & Sayler, G.S. (2010). Biosensors as environmental monitors. In *Environmental Microbiology, 2nd Edition*. R. Mitchell and J. D. Gu. Hoboken, NJ, Wiley-Blackwell: 213-233.

Roda, A., Cevenini, L., Michelini, E. & Branchini, B.R. (2011). A portable bioluminescence engineered cell-based biosensor for on-site applications. *Biosensors & Bioelectronics* 26: 3647-3653.

Ropstad, E., Oskam, I.C., Lyche, J.L., Larsen, H.J., Lie, E., Haave, M., Dahl, E., Wiger, R. & Skaare, J.U. (2006). Endocrine disruption induced by organochlorines (OCs): field studies and experimental models. *Journal of Toxicology and Environmental Health-Part a-Current Issues* 69: 53-76.

Routledge, E.J., Parker, J., Odum, J., Ashby, J. & Sumpter, J.P. (1998). Some alkyl hydroxy benzoate preservatives (parabens) are estrogenic. *Toxicology and Applied Pharmacology* 153: 12-19.

Routledge, E.J. & Sumpter, J.P. (1996). Estrogenic activity of surfactants and some of their degradation products assessed using a recombinant yeast screen. *Environmental Toxicology and Chemistry* 15: 241-248.

Sanseverino, J., Eldridge, M.L., Layton, A.C., Easter, J.P., Yarbrough, J., Schultz, T.W. & Sayler, G.S. (2009). Screening of potentially hormonally active chemicals using bioluminescent yeast bioreporters. *Toxicological Sciences* 107: 122-134.

Sanseverino, J., Gupta, R.K., Layton, A.C., Patterson, S.S., Ripp, S.A., Saidak, L., Simpson, M.L., Schultz, T.W. & Sayler, G.S. (2005). Use of *Saccharomyces cerevisiae* BLYES expressing bacterial bioluminescence for rapid, sensitive detection of estrogenic compounds. *Applied and Environmental Microbiology* 71: 4455-4460.

Sayler, G.S., Ripp, S., Nivens, D., & Simpson, M. (2001). Bioluminescent Bioreporter Integrated Circuits: Sensing Analytes and Organisms with Living Microorganisms. *Journal of Environmental Biotechnology* 1: 33-39.

Sayler, G.S., Simpson, M.L., & Cox, C.D. (2004). Emerging foundations: nano-engineering and bio-microelectronics for environmental biotechnology. *Current Opinion in Microbiology* 7:267-273.

Schultz, T.W. (2002). Estrogenicity of biphenylols: activity in the yeast gene activation assay. *Bulletin of Environmental Contamination and Toxicology* 68: 332-338.

Schultz, T.W., Kraut, D.H., Sayler, G.S. & Layton, A.C. (1998). Estrogenicity of selected biphenyls evaluated using a recombinant yeast assay. *Environmental Toxicology and Chemistry* 17: 1727-1729.

Schultz, T.W. & Sinks, G.D. (2002). Xenoestrogenic gene exression: structural features of active polycyclic aromatic hydrocarbons. *Environmental Toxicology and Chemistry* 21: 783-786.

Schultz, T.W., Sinks, G.D. & Cronin, M.T.D. (2002). Structure-activity relationships for gene activation oestrogenicity: Evaluation of a diverse set of aromatic chemicals. *Environmental Toxicology* 17: 14-23.

Shimomura, O., Johnson, F.H. & Saiga, Y. (1962). Extraction, purification and properties of aequorin, a bioluminescent protein from luminous hydromedusan, *Aequorea*. *Journal of Cellular and Comparative Physiology* 59: 223-239.

Sohoni, P., Lefevre, P.A., Ashby, J. & Sumpter, J.P. (2001). Possible androgenic/anti-androgenic activity of the insecticide fenitrothion. *Journal of Applied Toxicology* 21: 173-178.

Soni, M.G., Taylor, S.L., Greenberg, N.A. & Burdock, G.A. (2002). Evaluation of the health aspects of methyl paraben: a review of the published literature. *Food and Chemical Toxicology* 40: 1335-1373.

Sonne, C., Leifsson, P.S., Dietz, R., Born, E.W., Letcher, R.J., Hyldstrup, L., Riget, F.F., Kirkegaard, M. & Muir, D.C.G. (2006). Xenoendocrine pollutants may reduce size of sexual organs in East Greenland polar bears (*Ursus maritimus*). *Environmental Science & Technology* 40: 5668-5674.

Stoker, T.E., Gibson, E.K. & Zorrilla, L.M. (2010). Triclosan exposure modulates estrogen-dependent responses in the female Wistar rat. *Toxicological Sciences* 117: 45-53.

Svobodova, K. & Cajthaml, T. (2010). New in vitro reporter gene bioassays for screening of hormonal active compounds in the environment. *Applied Microbiology and Biotechnology* 88: 839-847.

Svobodova, K., Plackova, M., Novotna, V. & Cajthaml, T. (2009). Estrogenic and androgenic activity of PCBs, their chlorinated metabolites and other endocrine disruptors estimated with two in vitro yeast assays. *Science of the Total Environment* 407: 5921-5925.

Szittner, R. & Meighen, E. (1990). Nucleotide-sequence, expression, and properties of luciferase coded by *lux* genes from a terrestrial bacterium. *Journal of Biological Chemistry* 265: 16581-16587.

Thomas, K.V., Hurst, M.R., Matthiessen, P., McHugh, M., Smith, A. & Waldock, M.J. (2002). An assessment of in vitro androgenic activity and the identification of environmental androgens in United Kingdom estuaries. *Environmental Toxicology and Chemistry* 21: 1456-1461.

Tyler, C.R., Jobling, S. & Sumpter, J.P. (1998). Endocrine disruption in wildlife: a critical review of the evidence. *Critical Reviews in Toxicology* 28: 319-361.

United States-EPA (2007). Pharmaceuticals and personal care products in water, soil, sediment, and biosolids by HPLC/MS/MS. *Method 1694.*

Zacharewski, T. (1997). *In vitro* bioassays for assessing estrogenic substances. *Environmental Science & Technology* 31: 613-623.

Zheng, W., Yates, S. & Bradford, S. (2008). Analysis of steroid hormones in a typical dairy waste disposal system. *Environmental Science & Technology* 42: 530-535.

2

Physical Mechanisms of "Poisoning" the Living Organism by Heavy Metals

G.P. Petrova
Lomonosov Moscow State University, Russia

1. Introduction

The toxic affect of heavy metals over the living organisms is known to arouse from the alternation of the course for biological reactions in cells. One of such violations appears to be the process of supra-molecular structures formation, for example, the dipole protein nano-clusters in blood. This phenomenon can be well studied in the biological solutions, such as blood serum, or a widely adopted normal saline solution of albumin [1-3].

In the works devoted to problems of pollution of the surrounding environment and ecological monitoring, for today to heavy metals carry more than 40 metals of periodic system D.I. Mendeleyev with nuclear weight over 50 nuclear units: V, Cr, Mn, Fe, Co, Ni, Cu, Zn, Mo, Cd, Sn, Hg, Pb, Bi, etc. Thus the important role in categorize heavy metals is played by following conditions: their high toxicity for live organisms in rather low concentration, and also ability to bioaccumulation. Almost all metals getting under this definition (except for lead, mercury, cadmium and the bismuth which biological role currently is not clear), actively participate in biological processes, are a part some many enzymes. On N.Rejmersa's classification, heavy it is necessary to consider metals with density more than 8 g/cm^3 .Thus, heavy metals concern Pb, Cu, Zn, Ni, Cd, Co, Sb, Sn, Bi, Hg.

The results of our investigations can to conclude that the process of cluster formation depends on the value of ionic radius metal.

In our works the interaction of some proteins - albumins, globulins, collagen, lisozym, collagenase, creatin cenase, pepsin with heavy metal ions like Cs, Rb, Cu, Cd, Pb in water solutions was studied.

Especial influence on some protein like albumins, globulins, collagen, lisozym and so on has the potassium. K^+ ions presence in the protein solutions also induced appearance of dipole protein nano-clusters.

The appearance of cluster cans disturbance metabolic processes in the cells, membranes, tissue.

The interaction of proteins with ions of alkaline heavy metals like Cs^+, Rb^+, and Cu^{2+}, Cd^{2+}, Pb^{2+} in aqueous solutions was studied in our earlier works [4-9] with the Rayleigh-Debye light scattering (RDLS), the photon correlation spectroscopy (PCS) and the fluorescence polarization (FP) methods. It should be noted, that the nano-sized cluster formation process was also registered in some proteins and enzymes solutions like albumin, globulin, collagen, lysozyme, collagenase and so on in the presence of K^+ ions which does not belong to the heavy metals group [8, 11].

METALS and its IONIC RADII

Metals	Na^+ 11	K^+ 19	Cs^+ 55	Rb^+ 37	Cd^+ 48	Pb^{2+} 82	Eu^{3+} 63	Ce^+ 58
Ionic radii, Å	0,87	1,3	1,67	1,47	1,14	1,2	0,95	1,27

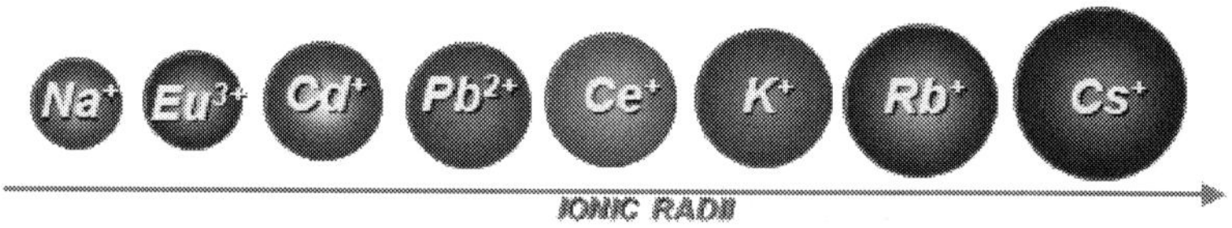

Fig. 1. Metals and its ionic radii

2. Physical model

It was found earlier that cluster formation depends on metal ion radius [2–4]. Interaction of these ions with a protein surface involves, as a rule, their hydrated shells. In cases where protein solutions contain small ions like Na+ (the ion radius equals 0.87 Å), dipole clusters are not formed, because sodium ions are located near the protein surface surrounded by water molecules and cannot bind directly with the negative charges on the protein.

$$E_{pq} = \frac{q^2 p_w^2}{12\pi\varepsilon r_0^4}\frac{1}{kT}.$$

The energy of the ion and the water dipole molecule binding, determined by equation (q- is the charge of the ion, p_w - water molecule dipole moment, ε -is dielectrical permeability of water, r_0- the ionic radius), is inversely proportional to the fourth power of the ionic radius. In the heavy ions case, it may be of the same order or less than the heat energy kT, and the water shell cannot stay on ion surfaces. This is observed for ions with large radii, such as Cs^+, Rb^+, Cd^+, Ce^+, Pb^{2+}, and Eu^{3+}, as well as K^+. In interacting with the protein surface directly, a metal ion with a large radius is bound more strongly to negatively charged groups on the protein and can form a Coulomb complex on a protein macromolecule with a common hydrated shell. In this case, the metal ions compensate completely for the local surface charge of the protein molecule [5].

The effective decrease of the protein surface charge that takes place as a result of strong binding of metal ions with a large radius and the macromolecule can lead to a situation where the main type of interaction between the protein molecules is a dipole–dipole attraction instead of Coulomb repulsion, because the proteins have abnormally high dipole moments (several hundred Debye uniques). So the protein molecules can go closely to each other and forming aggregates - dipole clusters (see fig 2)

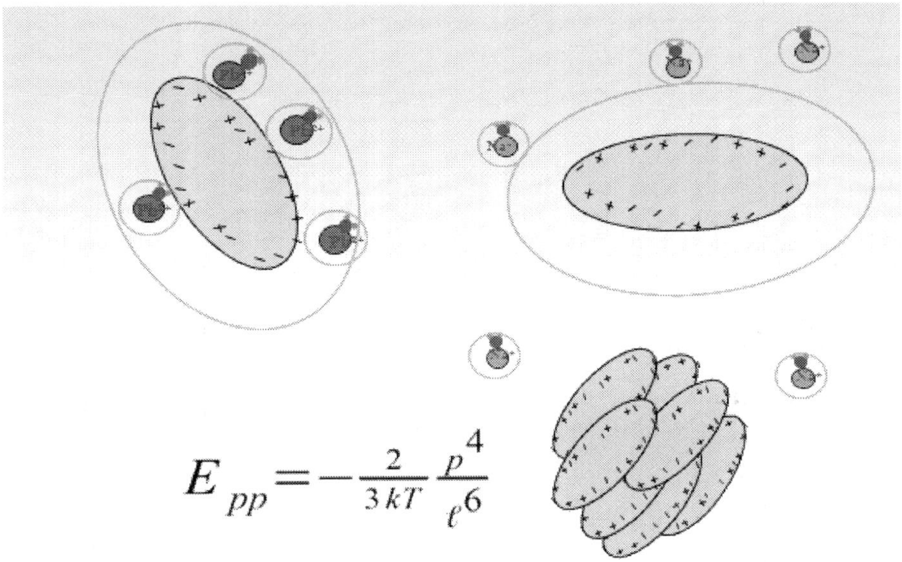

Fig. 2. The scheme of interaction processes small and big ions with protein surface. (In the formulae , E_{pp} – the energy of dipole-dipole interaction, p is the protein molecule dipole moment, l- the distance between macromolecules)

3. Different optical methods of heavy metals interaction with biological macromolecules investigations

3.1 Rayleigh-Debye light scattering (RDLS) method

The RDLS method is reliable for the determination of static parameters of the solid compounds in the suspension. The way how the light is scattered in the solution bears the information on the effective mass and molecular interaction coefficient of the particles. In case of the diluted solutions the measured experimental value of the Rayleigh scattering coefficient R_{90} is related within Debye's theory to the mass of a macromolecule M, according to the virial expansion of the osmotic pressure by concentration c:

$$\frac{cH}{R_{90}} = \frac{1}{M} + 2Bc + \dots, \qquad H = \frac{2\pi^2 n_0^2 \left(\frac{dn}{dc}\right)^2}{\lambda^4 N_A},$$

where λ_0 is the incident beam wavelength, n_0 and n are the refractive indexes of the dissolvent and the solution.

Fig. 3 presents data for pH-dependences of particle mass values obtained by RDLS method for the cases of the *Egg* albumin solution with Cesium (a), the bovine serum albumin (BSA) and the Gamma-globulin solutions with Potassium (b,c). All three graphs reveal the formation of large particles, one order heavier than the initial protein molecule. It should be noted that the maximum mass of nano-clusters in case of the K^+ ions in the solutions relates to the physiological pH values.

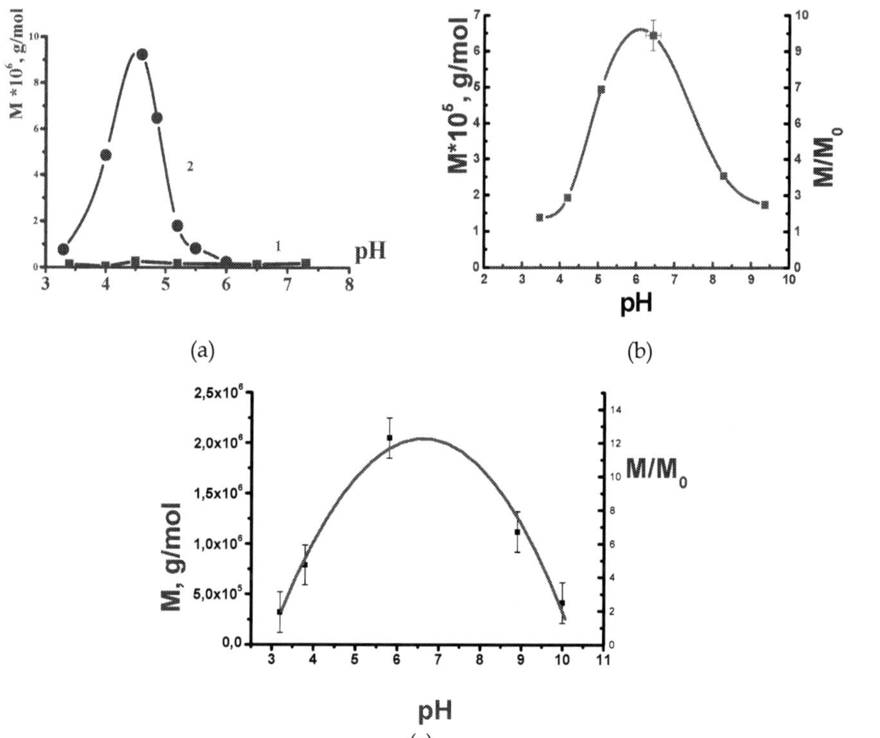

Fig. 3. (a) pH-dependencies of scattered particle mass for Egg albumin in water solution in presence of Cs ions (2) (μ = 0,00105 mol/l), (1) - Egg albumin in pure water solution. (b) pH-dependences of scattering particle mass for albumin, , containing ions K+. (c) pH-dependences of scattering particle mass for γ-globulin water solutions, containing ions K+.

3.2 Photon-correlation spectroscopy (PCS)

The PCS method was suggested to investigate the dynamic parameters of proteins in the aqueous solutions containing heavy metals [4, 5]. The translational diffusion coefficient D_t is described by the Stocks-Einstein-Debye formula as:

$$D_t = \frac{kT}{6\pi\eta r_h}$$

In this formulae η_h is viscosity, r_h - hydrodynamic radius of the particle. The normalized experimental autocorrelation function of the scattered light intensity relates to the translational diffusion coefficient D_t as:

$$g^{(1)}(\tau) = \exp(-D_t q^2 \tau),$$

where, q is wave-vector, τ - correlation time.

Fig. 4 shows the dependences of translation diffusion coefficient on pH for the pure gamma-globulin solution (a) and the one containing K^+ ions (b).

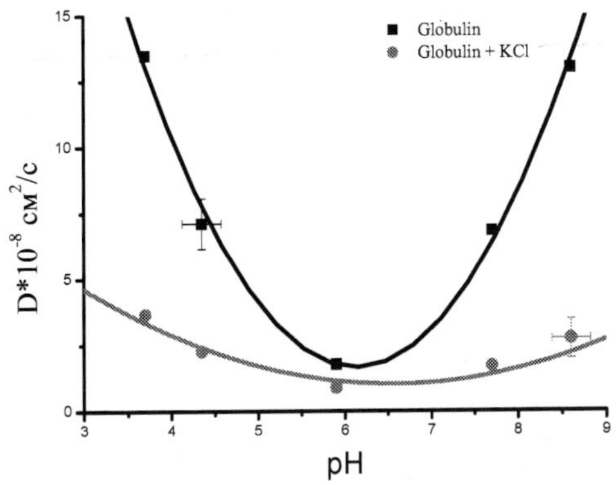

Fig. 4.Translation diffusion coefficient as function of pH for γ-Globulin water solutions with and without K^+ ions

The D_t value is twice less in the latter case when studied in the isoelectric point area of pH~6. It means that the mass of the particles in the solution with K^+ ions is one order greater than that of the gamma-globulin molecule:

$$\left(\frac{D_0}{D_K}\right)^3 = 11 \sim \frac{M_{cluster}}{M_{protein}},$$

where, $M_{protein}$ is the molecular mass of protein and $M_{cluster}$ - the mass of scattering particle.

3.3 Polarized fluorescence method

The fluorescence polarization (FP) method was used to determine the orientation correlation time t_{rot} of albumin in the solutions containing Pb^{2+} and Na^+ ions. This parameter is based on the fluorescence polarization experimental data [6] and is calculated according to the Levshin-Perrin relation [7]:

$$\frac{1}{P} = \frac{1}{P_0} + \left(\frac{1}{P_0} - \frac{1}{3}\right)\frac{t_{fl}}{t_{rot}}, \qquad t_{rot} = \frac{V\eta}{kT} = \frac{M\eta}{\rho kT},$$

where t_{fl} is the lifetime of the excited state. The latter proportion determines linear dependence of the t_{rot} on the mass M of the particle.

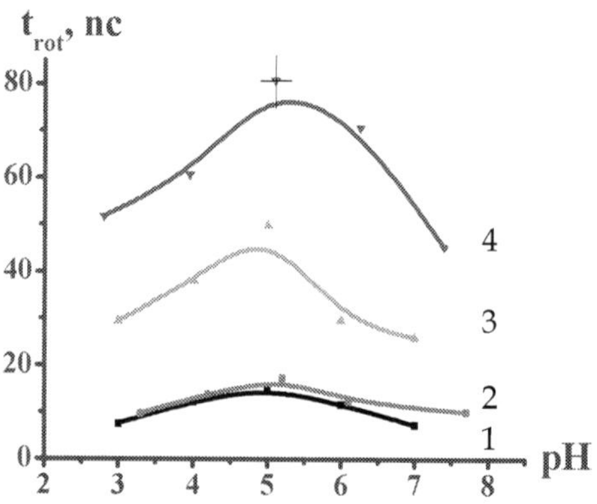

Fig. 5. pH-dependence of time rotation of albumin in the water solutions with Pb^{2+} and Na^{+} ions.

1. BSA 6,4·10^{-6}M + Na+ 5,6·10^{-3} M
2. BSA 6,4·10^{-6}M + Pb^{2+}+8,3·10^{-10}M
3. BSA 6,4·10^{-6}M + Pb^{2+}+1,7·10^{-7} M
4. BSA 6,4·10^{-6}M + Pb^{2+}+6,3·10^{-5} M

As Fig. 5 shows the orientation correlation time increases along with the concentration of the heavy metal Pb^{2+} ions.

For comparison, fig 6 shows the plot of relative clusters mass depends on relative concentration - metal/protein for BSA solutions with potassium and lead ions.

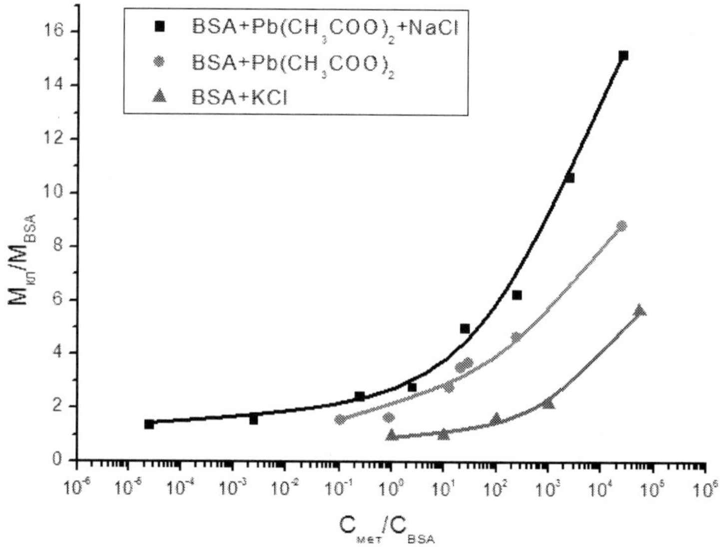

Fig. 6. Relative clusters mass dependences on relative concentration - metal/protein for BSA solutions with potassium and lead ions.

Thus, the FP method confirms the formation of the nano-sized clusters in the protein solutions with presence of heavy metal ions.

4. Sorption of the ions with various ionic radii on protein surface in the process of nano-clusters formation

In this part the sorption process of ions with various radii on the serum blood protein surface during the nano-clusters formation stage was study. A number of static parameters were achieved by Rayleigh-Debye light scattering, including effective masses and molecular interaction coefficient of the particles in the proteins aqueous solution containing ions of Na^+, K^+ and Pb^{2+} at different ionic strength. It was found that the nano-cluster formation process depends on the ionic radius of the metal.

4.1 Results and discussion

The following table represents the metal ions as studied in this investigation:

Metal	Mass, a. u.	Nuclear charge	Ionic radius, Å	Relative mass of cluster
Na_{23}^{11}	23	11	0,87	<2
K_{39}^{19}	39	19	1,33	20-35
Pb_{207}^{82}	207	82	1,2	>20

Table 1.

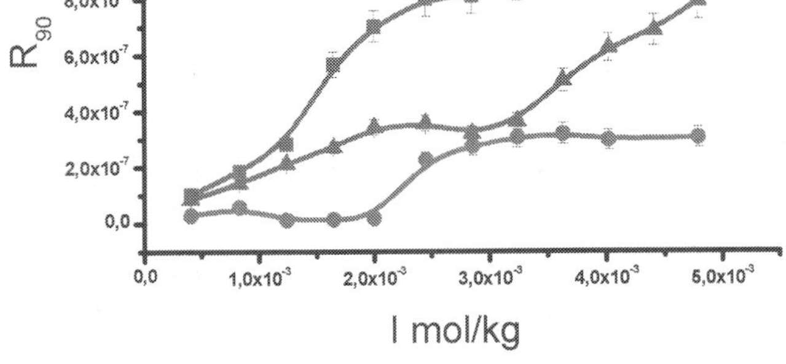

Fig. 7. Rayleigh scattering coefficient (R_{90}) as function of ionic strength of albumin water solution containing Na^+ ions.

The mentioned above metals were used to study the dependence of the Rayleigh scattering coefficient R_{90} on the value of the ionic strength I in the aqueous solutions of albumin produced by "Sigma Inc." (USA).
Fig. 7 shows the dependence of R_{90} on I for the solution with Na^+ ions, whereas Fig.8 shows the relative masses of scattering particles dependence for this solution at pH=7.0 on I, which is the concentration of Na^+ ions in this case.

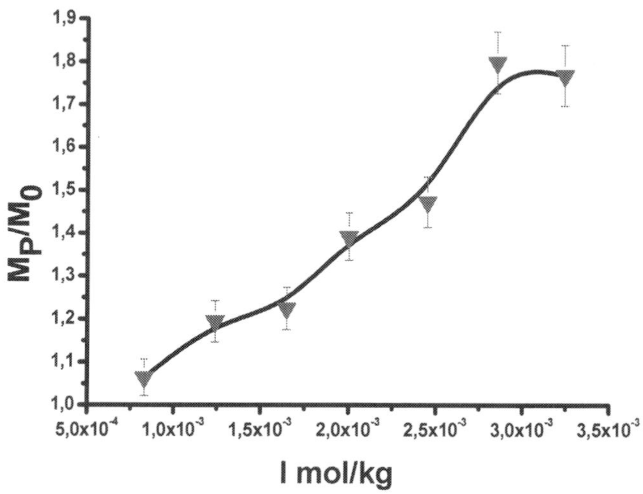

Fig. 8. Scattering particles (M_P) relative mass in albumin (M_0) solution as a function of ionic strength Na^+.

As follows from these graphs the presence of Na^+ ions in this solution at higher ionic strength slightly increases the masses of the scattering particles. Compared to the mass of the albumin molecule the masses of these particles are less than twice heavier, approx. ~ 1,8. Probably, a number of protein molecules in the albumin solution with Na^+ ions can form dimers.
Contrary to that the effect is absolutely different with K^+ and Pb^{2+} ions in the albumin solution.
Fig. 9 shows the dependences of R_{90} on ionic strength in the BSA solution, containing K^+ ions for a number of pH values. The dependence of relative masses of scattering particles for this solution at pH=7 is shown on Fig.10.
In this case the value of the relative mass $M_{cluster}/M_{protein}$,which represents the mass ratio of the nano-sized cluster to the albumin molecule, lies in the area of 20-35 for the ionic strength around 2-3 mmol/l.
The concentration variations of the Pb^{2+} ions in the albumin solution leads to a dramatic decrease of the molecular interaction coefficient, which is the second virial coefficient B upon the increase of the ionic strength.

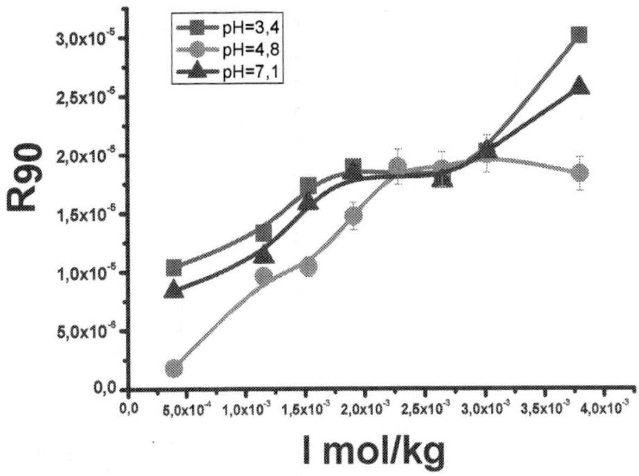

Fig. 9. R_{90} as the function of ionic strength in albumin water solution containing K^+ ions.

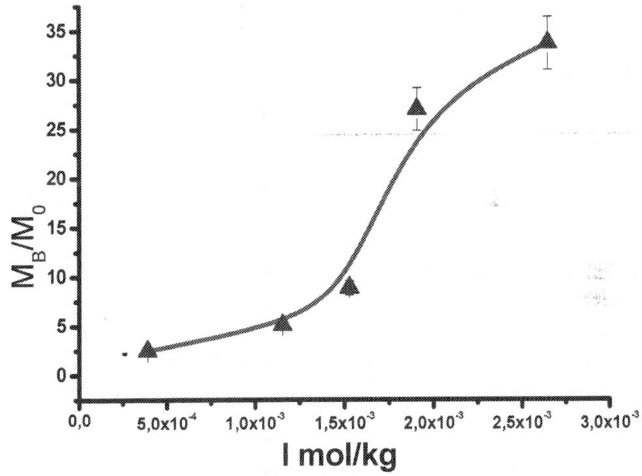

Fig. 10. Dependence of relative masses of scattering particles for BSA solution, containing K^+ at pH=7.

As Fig. 11 shows the former changes its sign and becomes negative when the latter reaches the values in the area of 10-15 mmol/l. This effect is due to the change in the type of molecular interaction which is caused by the increment of the Pb^{2+} ions concentration. In this case the Coulomb repulsion between protein macromolecules, when *B* is positive, diminishes, the pure dipole attraction takes over, and *B* descends below zero.

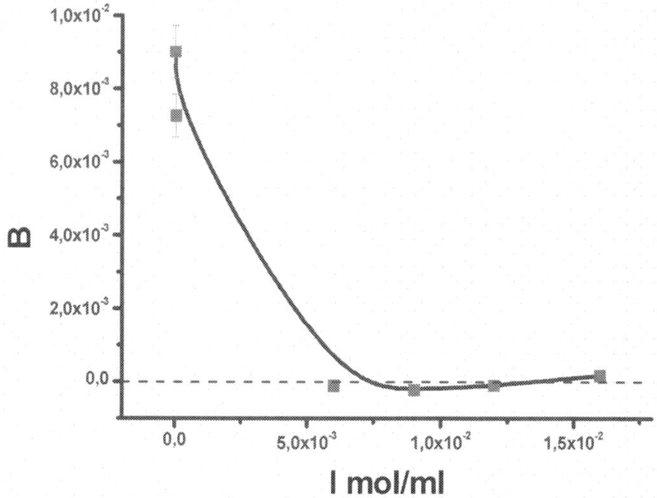

Fig. 11. Dependence of B (the second virial coefficient) from ionic strength in albumin solution with Pb^{++} ions.

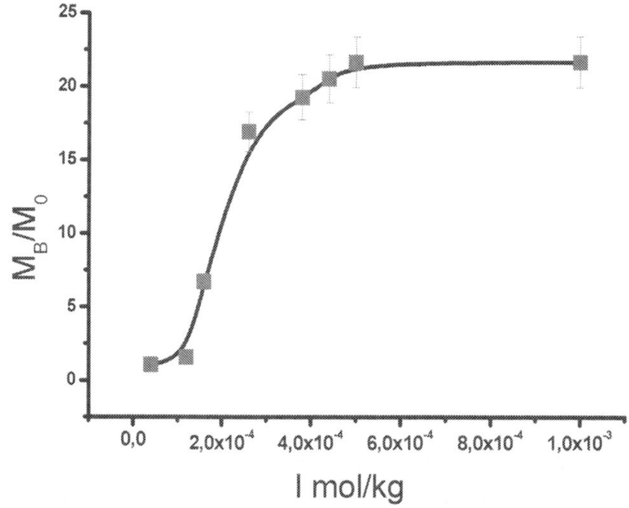

Fig. 12. Dependence of relative mass value from ionic strength of albumin solution with Pb^{++} ions (pH=7, 5)

Fig. 12 shows the dependence of the relative scattering particles mass on the ionic strength of the solution. The curve possesses a small slope rise of the relative mass. The ionic strength

values in the range from 0,05 mmol/kg to 0,17 mol/kg relate to the process of monolayer formation which takes place until the Langmuir saturation is achieved.
As graph data shows that the scattering particles masses are more than 20 times greater than the mass of the albumin molecule. It depicts the process of the formation of the larger particles which appear to be the nano-sized clusters generated by a number of the original macromolecules. With the presence of Pb^{2+} ions in the solution the cluster formation process occurs at the significantly smaller ionic strength values of 0,15 mmol/kg, as compared to the case of K^{+} ions of 1,5 mmol/kg. Nonetheless, the cluster formation process runs faster in case of Pb^{2+} ions although the generated particles appear to be lighter than in the case with K^{+} ions.

5. Conclusions

- The interaction of the metal ions with the charged surface of the protein in the solution is studied by the measurement of the light scattering coefficient along with the concentration variation of the former.
- The dependence of masses of the scattering particles on the ionic strength and pH of the solution shows the Langmuir sorption process which leads upon the monolayer saturation to the dipole cluster formation.
- The nano-sized clusters form as a result of the phase transition when the Coulomb repulsion forces diminish and the pure dipole attraction forces take over.
- The nano-cluster formation process in the protein solution depends on the ionic radii of metal. The clusters are formed in case of the solutions containing K^{+} and Pb^{2+} ions, whereas the presence of Na^{+} ions in the solution reveals no effect.
- Cluster formation process can explain toxic influence of heavy metal ions at the very small concentration on the living organisms.

The work was supported by the Russian Foundation for Fundamental Research, grant No. 09-02-00438-a.

6. Acknowledgements

In memoriam of professor Yuriy M. Petrusevich (1935-2010).
I would like to thank my colleagues Yu.M. Petrusevich, K.V. Fedorova, M.A.Gurova, M.S. Ivanova, V.P. Khlapov, A.M. Makurenkov, I.A. Sergeeva, T.N. Tikhonova, E.A.Papish, N.V.Sokol for taking part in these investigations.

7. References

[1] Edsall J.T. et al. "Light Scattering in Solutions of Serum Albumin: effects of charge and ionic strength" // J. of American Chem. Soc., 1950, V.72, P.4641.
[2] P.Debye. Light scattering in solutions. Journal Appl.Phys. 15, 338-349, 1944
[3] Scathard G., Batchelder A.C., Brown A. J. Am.Chem.Soc.68 2610 (1946)
[4] Petrova G.P., Petrusevich Yu. M., Evseevicheva A.N. //General Physiology and Biophysics, V.17(2),P.97,(1998).
[5] Petrova G.P., at al.// Proceedings of SPIE, V.4263, p.150, (2001),
[6] Petrova G.P., Petrusevich Yu.M., Ten D.I.// Quantum Electronics, 32(10), p.897 (2002).

[7] G.P. Petrova G.P., Yu.M. Petrusevich, A.V. Boiko, D.I. Ten, I.V. Dombrovskaya, G.N. Dombrovskii" // Proceedings of Int. Conf. Advanced Laser Technologies, ALT-05, SPIE, V. 6344, 63441R (2006).

[8] Sergeeva I.A. et al.//Moscow University Phys.Bull. V.64,(4), P.446 (2009)

[9] Petrova G.P., Sokol N.V. The fluorescence of serum albumin solutions containing Pb and Na ions. Moscow University Physics Bulletin, Vol. 62, Number 1, 62-64.

[10] Joseph R. Lakowicz . Principles of fluorescence spectroscopy, Plenum Press. New York, London,1983

[11] T. N. Tikhonova, G. P. Petrova, Yu. M. Petrusevich, K. V. Fedorova, and V. V. Kashin //Moscow University Physics Bulletin, 2011, Vol. 66, No. 2, pp. 190–195. © Allerton Press, Inc., 2011.

3

Histological Biomarker as Diagnostic Tool for Evaluating the Environmental Quality of Guajará Bay – PA - Brazil

Caroline da Silva Montes,
José Souto Rosa Filho and Rossineide Martins Rocha
Universidade Federal do Pará,
Brazil

1. Introduction

It has been reported that in recent decades the level of foreign compounds known as xenobiotics in aquatic ecosystems has increased alarmingly as a result of domestic, industrial and agricultural effluents. In the 20th century, many thousands of organic trace pollutants, such as polychlorinated biphenyls (PCBs),organochlorine pesticides (OCPs), polycyclic aromatic hydrocarbons (PAHs), and dybenzon - p - dioxins (PCDDs) have been produced and in part, released into the environment (van der Oost et al., 2003). This has led to substantial reduction in environmental quality, adding to the deterioration of human health and living organisms that depend on these ecosystems (Cajaravlle et al., 2000). However, the presence of a foreign compound in a segment of an aquatic ecosystem does not, by it self, indicate injurious effects. Connections must be established between external levels of exposure, internal levels of tissue contamination and early adverse effects and determining the extent and severity of such contamination only by the results of water chemical analysis is insufficient and often overestimates the proportion and duration of exposure to the toxic agent (van der Oost et al., 2003 & Giari et al., 2008). Thus, studies using biomarkers are essential to complement such environmental monitoring, given that in order to control pollution effects of effluents on the animals that inhabit the water bodies must be understood (Martinez & Colus, 2002; Camargo & Martinez, 2006). Biomarkers are defined as responses to any exposure evidenced in histological, physiological, biochemistry, genetic and behavioral modification (Leonzio & Fossi, 1993). More recent, van der Oost et al. 2003 defined biomark as a biological indicator from an expousure to a stressor responding in various ways such a response can be seen and adaptation as a defense. Some authors note that biomarkers are used as a warning sign to emerging environmental problems (Au, 2004). In this type of environmental assessment, the health of an ecosystem can be measured by the health of its individual components (Hugget et al., 1992). It is essential to this study, as there is a variety of responses that can be used as tools to assess the health of animals exposed to certain chemicals, to provide information on spatial and temporal changes in pollutant concentrations and indicate the occurrence of environmental quality or adverse ecological consequences (Kammenga et al., 2000). In Brazil there are few studies about impact of contaminants on tropical ecosystems, therefore tropical ecotoxicology needs further studies on the effect of pollution on native aquatic organisms (Monserrat et al., 2007). The biological communities of Amazonian aquatic environments are poorly known, despite its economic and ecological importance. Belém and its surrounding areas are part of the Amazon estuary in northern Brazil. The Combú Island, near Belém, is included on Combú Environmental Protection Area (Law 6.083 of 11.13.1997) and corresponds to a lowland environment region, according to the daily tidal flooding, especially during the lunar cycles and rainy season (Ribeiro, 2004). The island's population depends on aquatic resources (fish and shrimp) as a source of food and income, and poses an imminent threat to the conservation of natural resources. The species *Plagioscion squamosissimus, Hypophthalmus marginatus* and *Lithodoras dorsalis* are economically important to the Amazon region, since in some areas this represents the main protein source for families. These animals occur in different types of environments, suggesting they are tolerant of a wide range of physico-chemical variables (de La Torre et al., 2005). Thus, they are suitable for environmental monitoring. The objective of this study was to evaluate the histological alterations in gills and liver of the species *P. squamosissimus, H. marginatu* and *L. dorsalis,* as well as assess the environmental influence on fish health from amazon estuary, Guajará bay.

2. Material and methods

2.1 Study area

The study area is situated around the island of Combú, near Belém-PA-Brazil, located between the coordinates 01 ° 25 'S and 48 ° 25' W. This island is inserted in the Area of Environmental Protection Combú (Law 6.083 of 11.13.1997). This area undergoes severe impacts that modify water quality due to increased population and its proximity to the metropolitan area of Belém-PA-Brazil. A total of ninety-one (91) specimens were captured in Guajará Bay and Guamá river during the dry period (July 2009). Samples were collected in three areas (Figure 1): Area A - away from pollution sources; Area B and C - considered impacted by the presence of domestic sewage and urban influence.

2.2 Biotic and abiotics data

During the study the physicochemical variables such as: pH, temperature, Dissolved oxygen (DO), nitrite, nitrate and phosphate were obtained. The pH and temperature were measured *in situ* using an Orion pH-meter, model 210 and a mercury thermometer. To determine the other variables, water samples were collected at the surface layer using a Van Dorn-type bottle. They were later processed (filtered and cooled) and taken to laboratory for analysis. We used three fish species of interest to the local population, *P. squamosissimus, L. dorsalis* and *H. marginatus*. These were caught by artisanal fishing, using gill nets with different mesh sizes (25 mm, 40 mm and 50 mm). After captured, the fish were placed in plastic bags, appropriately refrigerated in isothermal boxes and transported to the laboratory. The fish were then examined internally and externally for gross lesions, removing a fragment of the gills and liver. The tissue samples were fixed in Bouin's solution. After fixation, the tissues were dehydrated in increasing concentrations of alcohol, cleared in xylene and embedded in paraffin, obtained from 5mm thick sections and stained with HE (hematoxylin and eosin solution). The sections were examined and photographed using Carl Zeiss optical microscope (Axiostar Plus1169-151).

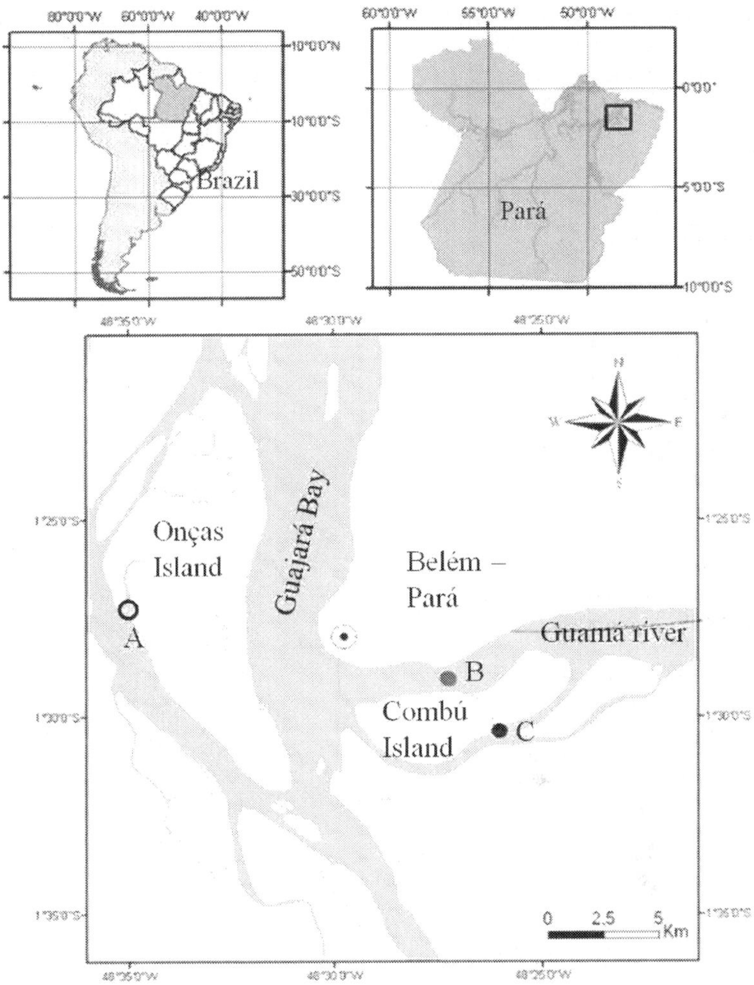

Fig. 1. Map of study area and collection points. A (away from sources polution); B and C (impacted).

2.3 Diagnostic histopathology

The histopathological changes were evaluated semi-quantitatively in two ways: The first one was modified according to Schwaiger et al. (1997), which assigned a numerical value to each animal according a degree of change: 1 (initial stage of change in some points with a chance of recovery), 2 (occasional occurrence of localized lesions with little chance of recovery) and 3 (widely distributed lesions in the body without chance of recovery). The second one was adapted from Poleksic & Mitrovic - Tutundzic (1994) that examines the calculation of the histopathological alteration index (HAI). For this, the changes were classified as progressive stages for the deterioration of organ functions: I (do not compromise the functioning of the organ) II (severe, affecting normal body functions) and III (very severe and irreversible) table 1. A value of HAI was calculated for each animal using the formula.

$$HAI = 10^0 \sum I + 10^1 \sum II + 10^2 \sum III \quad (1)$$

Since I, II, III correspond to the number of stages of change, the mean HAI was divided into five categories: 0-10 = normal tissue; 10-30 = mild to moderate damage to the tissue, 31-60 = moderate to severe damage to the tissue, 61-100 = severe damage to the tissue , greater than 100 = irreparable damage to the tissue.

GILL/LIVER HISTOPATHOLOGY	STAGE
1. **Hypertrophy and hyperplasia of gill epithelium**	
Hypertrophy of respiratory epithelium	I
Lifting of respiratory epithelium	I
lamellar epithelial hyperplasia	I
lamellar disarray	I
Incomplete fusion of some lamellae	I
Complete fusion of all lamellae	II
Lamellar epithelium disruption	II
Uncontrolled proliferation of tissue	III
2. **Changes in blood vessels**	
Dilation of sinus blood	I
Constriction of sinus blood	I
Vascular congestion	II
Disruption of pillars cells	II
Lamellar aneurism	III
1. **Changes in hepatocytes**	
cell hypertrophy	I
cell atrophy	I
Melanomacrophage centers	II - III
Inflammation	II
Fatty degeneration	II
Necrosis	II - III
2. **Changes in blood vessels**	
Hepatitis	II
Vascular congestion	II

Table 1. Classification of histopathological changes of gill and liver in relation to the type, location and stage of lesions in which they operate. Modified Poleksić and Mitrovic - Tutundzic (1994).

2.4 Statistical analysis

The frequency of altered animals and the mean HAI for each fish caught at each site were calculated. The occurrence of histopathological lesions and HAI were compared between areas using the nonparametric Kruskal-Wallis tests. The differences were considered significant $p < 0.05$.

3. Results

Table 2. corresponds the total number of animals captured in the different study areas (A, B and C). The results of physico-chemical variables during the study are analyzed in Table 3. The temperature values observed are within the normal range for the tropics. Regarding pH, it was observed that this was slightly acid in areas B and C, while the DO was lower than what is recommended in all areas. The results of gill and liver changes are displayed in Tables 4 and 5 and Figures 2 - 8. The gills of the specimens were normal as described for teleosts, consisting of four arches, supported by partially calcified cartilaginous tissue, each gill arch has two rows of primary lamellae, which in turn support the secondary lamellae. The branchial lamellar epithelium is a mosaic of primary paviment cells, mucus-secreting cells and chloride cells. The chloride cells were less evident in light microscopy because of the color used. The secondary lamella formed by the epithelium has a single layer of paviment cells, supported by the basement membrane lining the pillar cells, which surround the space through which blood circulates (Figure 5). The liver tissue of teleost fish is composed of two lobes, the right lobe which is adjacent to the gallbladder and the left lobe near the spleen. The liver is composed of hepatocytes, epithelial cells of the bile ducts, macrophages, blood cells and endothelial cells. The hepatocytes are polyhedral cells with one or two large, spherical and centrally nuclei located with evident nucleolus, and granular cytoplasm and vacuolated appearance (Figure 7). Changes in these organizations were considered to be alterations. Several changes were observed in gill and liver that differed significantly from the animals caught in the impacted areas (B and C). The area A was the only one which had healthy animals, and fish with soft lesions of type I and II and no animals with severe lesions of type III (Table 4). It was also found that they had the lowest histopathological changes index (HAI) in the 0 to 10 range (Table 5). Unlike the fish collected in areas B and C, where they all had some kind of change, many were classified as degree 3 lesions, showing the most severe type III and the highest values of HAI ranging from 41 to 91, considered moderate to severe damage, such as lamellar aneurysm characterized by blood leakage inside the lamellae, causing disruption of pillar cells and consequent dilation of blood vessels; lifting epithelium which is the detachment of the lamellar epithelium; lamellar fusion, characterized by an increase in the number chloride cells between the secondary lamellae in the respiratory tract causing reduction in the gills (Figure 6). In liver were evident such diseases: cellular hypertrophy, necrosis, presence of centers of melanomacrophages, hepatitis and inflammation (Figure 8). Regarding the responses of different species, it was observed that the species *H. marginatus* showed the lowest values while the HAI *P. squamossissimus* presented the highest values. *L.dorsalis* and *P. squamossissimus* showed more type III lesions and were therefore classified as degree 3 (Figure 4-6).

Species	Number of fish caught		
	A	B	C
H. marginatus	10	6	10
L. dorsalis	14	15	9
P. squamossissimus	14	8	5

Table 2. Number of fish caught in different areas (A, B e C).

Variables	A	B	C	Recommended
T (°)	30	30	31	-
pH	6.1	5.8	5.9	6.0 – 9.0
DO (mg/L)	4	4.2	4.5	> 5 (mg/L)
Phosphate (mg/L)	0.01	0.01	0	0.01
Nitrite (mg/L)	0.001	0.001	0.001	0.001
Nitrate (mg/L)	1.3	1.2	1	>1

Table 3. Physico-chemical variables observed in different study areas and the value recommended.

species	Types	Gill			Liver		
		A	B	C	A	B	C
H. marginatus	I	12	17	29	12	18	24
	II	3 [a, b]	11	13	3 [a]	12	9
	III	-	2	4	-	2	3
L. dorsalis	I	13 [a]	49	33	13 [a]	47	27
	II	5 [a, b]	23	20	5 [a, b]	27	14
	III	-	7	4	-	7	4
P. squamossissimus	I	4 [a, b]	26	14	13	28	16
	II	1 [a, b]	15	10	5 [a]	20	11
	III	-	5	3	-	5	3

Table 4. Total number of different types of histopathological lesions in gill and liver from three fish species in study areas.
Note: Significant difference (p<0,05): [a] between A and B ; [b] between A and C.

	Gill			Liver		
Species	A	B	C	A	B	C
H. marginatus	4.2 ± [a, b] 0.3	54.5 ± 9.6	55.9 ± 8.3	4.2 ± [a, b] 1.3	56.33± 8.7	41.4 ± 5.5
L. dorsalis	4.5 ± [a, b] 2.1	65.27 ± 7.8	70.33 ± 10.6	3.94 ± [a, b] 2.1	67.8 ± 14.4	63 ± 6.5
P. squamossissimus	1 ± [a, b] 1.1	84.5 ± 16.5	82.8 ± 15.7	4.2 ± [a, b] 0.5	91 ± 19.9	85.2 ± 24.5

Table 5. Mean and standard deviation of HAI calculated from histological alterations in gill and liver tissue from three fish species in study areas.
Note: Significant difference (p<0,05): [a] between A and B ; [b] between A and C.

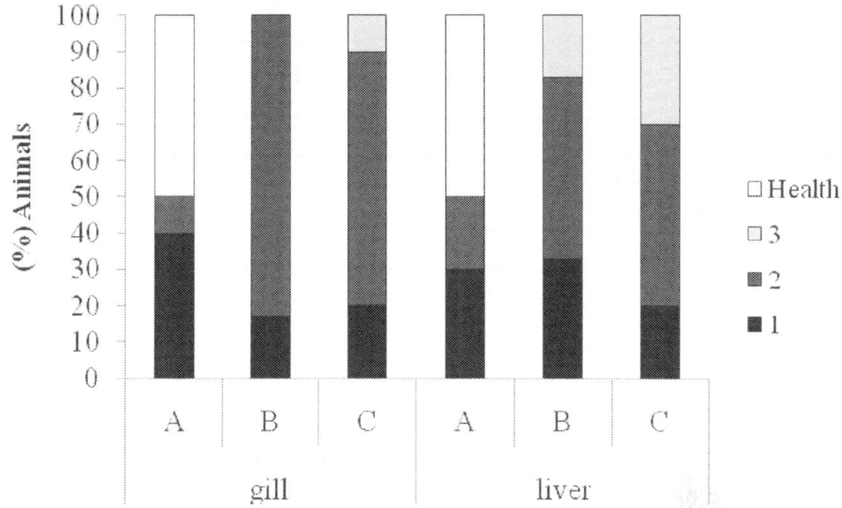

Fig. 2. Percentage of the species *H. marginatus* with gill and liver changes captured in the study areas (A, B and C). 1, 2 and 3 correspond to the different degrees of alteration of animals and health corresponds to those with no alteration.

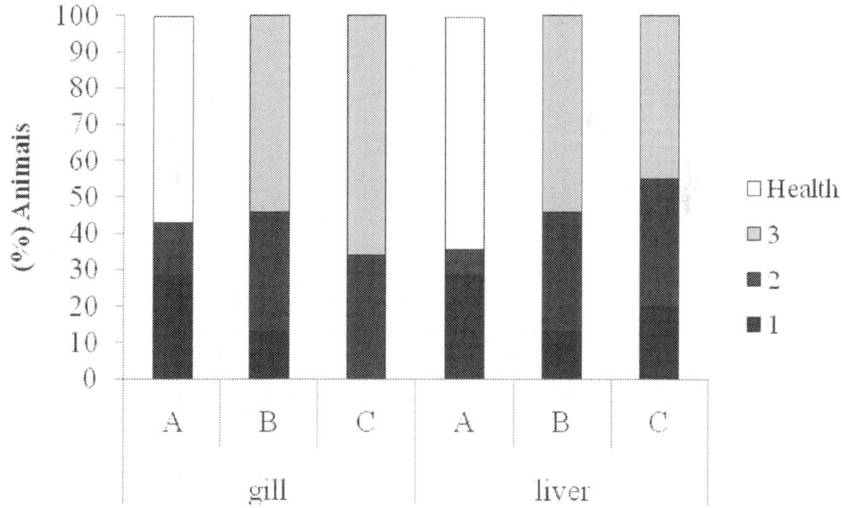

Fig. 3. Percentage of the species *L. dorsalis* with gill and liver changes captured in the study areas (A, B and C). 1, 2 and 3 correspond to different degrees of alteration of animals and health corresponds those with no alteration.

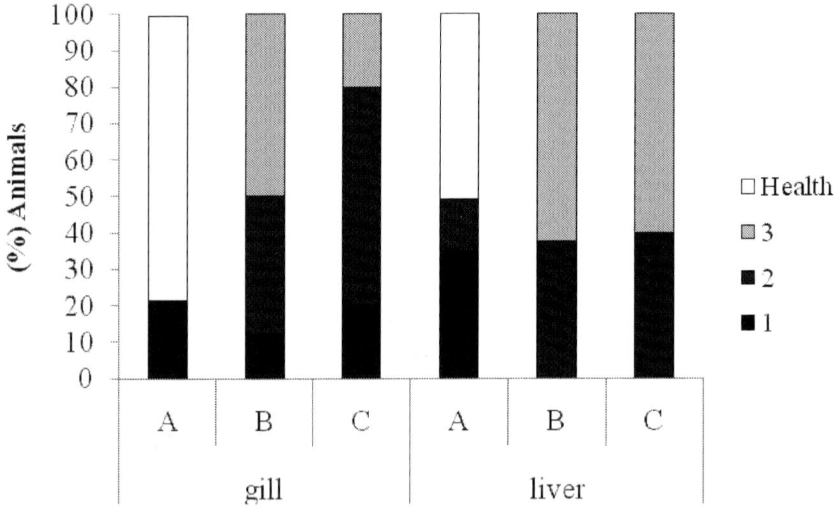

Fig. 4. Percentage of the species *P. Squamossissimus* with gill and liver changes captured in the study areas (A, B and C). 1, 2 and 3 correspond to different degrees of alteration of animals and health corresponds to those with no alteration.

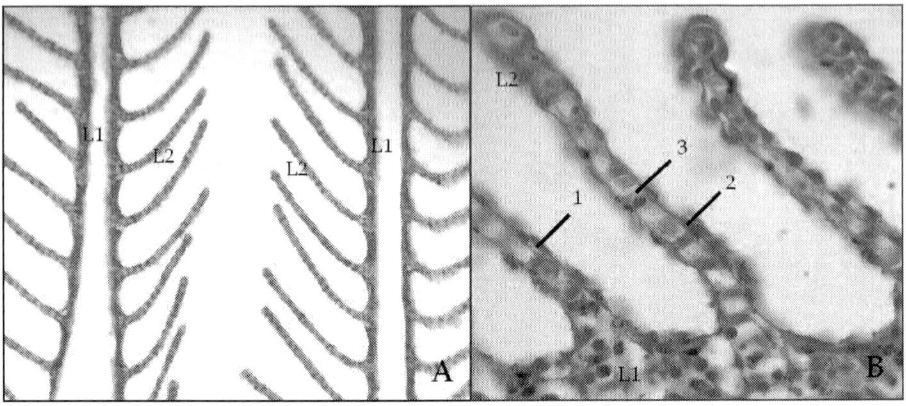

Fig. 5. Photomicrography of the gills tissue of animals captured in area. A - Normal gill structure with primary lamella (L1) and secondary (L2) with a single layer of pavement cells of slender appearance. 400X. B - Detail of a normal secondary lamella showingall cell types, 1 - squamous cell, 2 - interlamellar cells and 3 - pillars cells 1000X. HE.

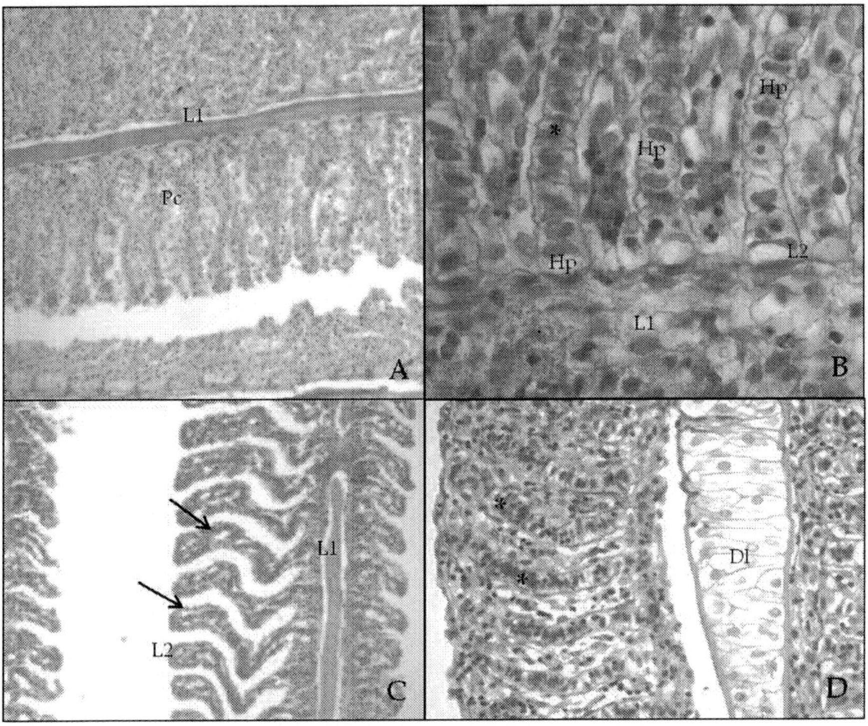

Fig. 6. Photomicrograph of branchial histopathology of animals captured in areas B and C. A - Changed gill tissue with intense celular proliferation (Pc) causing severe lamellar fusion 200X. B – hypertrophy (Hp) 1000X. C – Epithelium Lifting (Arrow) 400X. D – Dilation of sinus blood (Dl) and early aneurysm (*) 400X. HE.

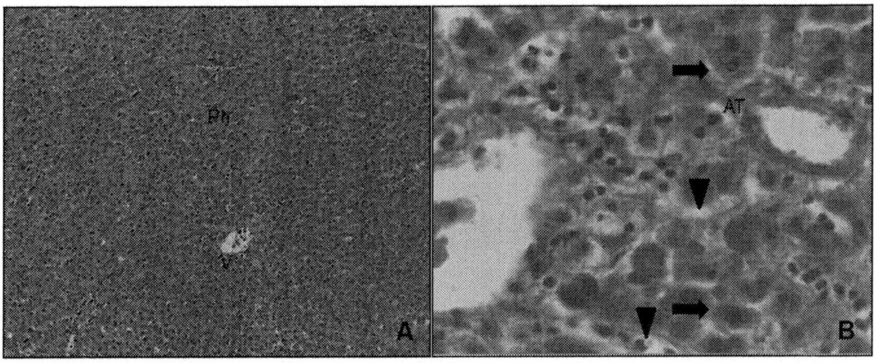

Fig. 7. Photomicrography of liver tissues of animals captured in control area - Normal liver structure with parenchyma (Ph) and veins (V) well defined 50X. B - Detail of a normal parenchyma, Hepatocytes (Thick Arrow) and sinusoids (head Arrow) 1000X. HE

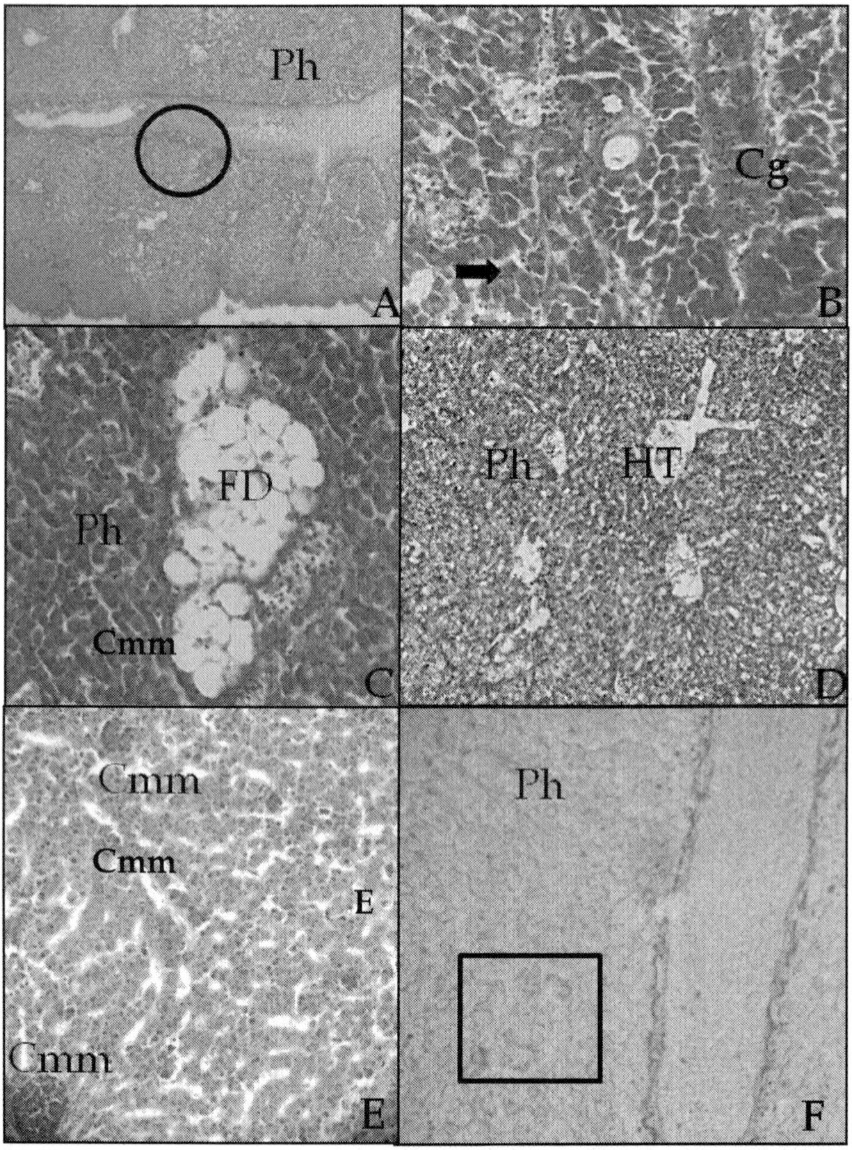

Fig. 8. Photomicrograph of hepatic histopathology of animals caught in areas B and C. A - Changed liver tissue with intense inflamation (circle) 200X. B - Congested vein (Cg) 400X. C - fatty degeneration (FD) 400X. D - Hepatites (HT). 200X. E - Intense melanomacrophages centers (Cmm). 400X. F - Detail of a hepatic parenchima with necrosis (□). 400X. HE

4. Discussion

In this study the temperature values were considered normal and slightly acidic pH. According to Campagna (2005), there is a close relationship between pH and carbon dioxide levels in the water, because with the discharge of sewage into rivers the water quality is altered by pathogenic bacteria and degradable organic substances and the decomposition of microorganisms involves the release of carbon dioxide and a consequent increase in the acidity of water. A similar situation may have occurred in areas B and C, since these areas have anthropogenic interference. These areas around the island showed a rapid population occupancy process, evidenced by domestic and industrial waste (Pinheiro, 1987; COHAB, 1997), particularly worrying since this island is inserted in an environmental protection area. The low values of DO in area A it is a typical feature of the region due to high turbidity and low incidence of light, hence causing a decrease in DO. Histology is a sensitive tool for the diagnosis of direct and indirect toxic effects that affect animal tissues (Braunbeck & Volk, 1993; Heath, 1995;Ferreira et al., 2005). Therefore it is considered an excellent method for assessing environmental quality (Freire et al., 2008). Thus this study used the histological responses of gill and liver of fish native to the Amazon as biomarker tools. The tissue damage could clearly differentiate the areas compromised as well as the control of the area, because healthy animals or mild changes of type I and II were found only in area A, result totaly different from those observed in areas B and C, where several animals showed gill lesions and hepatic diseases classified as type III. Both gill and liver are extremely important organs because they serve for respiration and osmoregulation and for regulating ion concentrations (Hinton et al., 1992), hence they play a central role in metabolism (Arellano et al., 1999). Because the gills are in direct contact with water, toxic substances can easily interfere the morphophysiology of these organs, as for instance the use of organic pesticides (Laurent & Perry, 1991), detergents (Bolis & Rankin, 1980), acids (McDonald, 1983), salt (Fanta et al., 1995), industrial waste (Lindesjöö & Thulin, 1994), ammonia (Miron et al., 2008) and heavy metals (Oliveira Ribeiro et al., 1996). During the breathing process, to prevent secondary lamellae, solid agents cross the filaments during the inflow of water, however, high concentrations of irritants dissolved in water inevitably come into contact with the outer surface of the gill filaments and secondary lamellae of the current circulation, which can alter the normal gill morphology, causing cell proliferation, epithelial lifting, hypertrophy, infiltration, and aneurysm (Simonato et al., 2008). When fish are subjected to stress, the proliferation of epithelial cells is one of the earliest changes that occurs rapidly in order to eliminate toxic agents (Laurent & Perry., 1991). Similar results were observed in this study since cell proliferation was the most found in all animals from areas B and C and some from area A. The epithelial lifting, which is a more severe injury, is caused by the change in distance of the respiratory epithelium basement membrane, causing the inefficient absorption of oxygen (Hibiya, 1992;Nowak, 1992). Result observed by Montes et al., 2010. Berrêdo et al. (2000) in a study conducted in Guajará Bay, showed high lead and chromium concentrations, metals that can undermine the tissue structure in the exposed animals. Thus we can infer that the animals evaluated in this study were responding to the effects of toxic substances. The histological changes found in the liver were: congestion, inflamation, hepatites, hemorrhage and necrosis. Vacuolated hepatocytes were also observed. The first effects of the contaminants usually occur at the cellular or intracellular level (Stephan & Mount, 1973). The melanomacrophage, also know as pigmented cells are related with the first segmento of organism defense, therefore are responsible for storing foreign material by capturing and processing of exogenous antigens and products of cell degradation (Truslé et al.,

1996; Bombonato et al., 2007), as a result the increase in the amount of pigment indicate the increasing expousure (Fernandes et al., 2009), result observed in this study since in impacted areas the number of pigments were greater. Baldisserotto (2002) developed a classification that relates the degree of changes in liver tissues according to the degree of pollution in aquatic systems. These authors consider that a compromised parenchyma with several melano-macrophages centers of already exposed tissue can be considered as highly polluted environments, the situation seen in the specimens collected in areas B and C. Necrosis is induced by high concentrations of toxic substances (Rocha et al., 2010), it is then considered as a type III alteration. Some animals in the study areas had such an injury, thus we can infer a certain degree of pollution to the area. There was no significant difference among the animals, and despite their similar feeding habits, they responded similar to the same degree of toxicity. The morphological changes observed in the gill and liver of juveniles evidences an early sign of contamination in the Guajará Bay, thus indicating that these species can be used in environmental monitoring programs. However more studies should be performed as this is a protected area.

5. Conclusion

This study presented significant results and was effective in showing that human action can be mischievous if not properly controlled and the proximity of the Combú island with the urban area may be affecting water quality. In addition both the results of the pathology and the species were excellent tools for diagnosing and determining and such data may be used by managers as a form of environmental monitoring and possible remediation of the impacted area as this island is a protected area.

6. Acknowledgments

We would like to thank the CNPq (conselho nacional de desenvolvimento cientifico e tecnologico) for financial support to the project n°(552952/2007-2009)

7. References

Arellano, J.M..; Storch, V. & Sarasquete, C. (1999). Histological changes and copper accumulation in liver and gills of the senegales *Sole solea senegalensis*. *Ecotoxicology and Environmental Safety*, Vol. 44, No. 1, pp 62-72.

Au, D.W.T. (2004). The application of histo-cytopathological biomarkers in marine pollution monitoring: a review. *Marine pollution bulletin*, Vol. 48, No. 9-10, pp 817-834.

Baldisseroto, B. (2002). *Fisiologia de peixe aplicada a piscicultura*, 211p. Santa Maria, SC, Brazil.

Berrêdo, J.F ; Mendes, E.P.C.B. ; Mendes, A.C. ; Corrêa, G.C.S. & Neves, F.C.O. (2000). Transporte e comportamento geoquímico de metais pesados no estuário guajarino/pa - brasil. *v workshop ecolab (ecossistemas costeiros amazônicos)*, Vol. 1, Macapá, AP, Brazil.

Bolis, L.; Rankin, J.C. (1980). Interactions between vascular actions of detergent and catecholamines in perfused gills of european eel, *Anguilla anguilla L.* and brown trout, *Salmo trutta L. Journal of fish biology*, Vol. 16, pp. 61-73.

Bombonato, M.T.S.; Rochel, S.S.; Vicentini, C.A. & Vicentini, I.B.F. (2007). Estudo morfológico do tecido hepático de leporinus macrocephalus. *acta scientiarum – biological sciences*, Vol. 29, No. 1, pp. 81-85, 2007.

Braunbeck, T. & Volkl, A. (1993). Toxicant-induced cytological alterations in fish liver as biomarkers of environmental pollution? a case study on hepatocellular effects of dinitro-o-cresol in golden ide (*Leuciscus idus melanotus*). in: *fish ecotoxicology and ecophysiology* (Ed.) 55–80, Braunbeck, T.; Hanke, W. & Segner, H. Birkhäuser verlag, Basel.

Bruslè, J. & Anadon, G.G. (1996). The structure and fuction of fsh liver. in: *Fish morphology horizon of new research.* (Ed.), p 16, Munshi, J.S.D. & Dutta, H.M, Beirute.

Cajaravlle, M.P.; Benianno, J.M.; Blasco, J.; Porte, C.; Sarasquete, C. & Viarengo, A. (2000) The use of biomarkers to assess the impact of pollution in coastal environments of the iberian península: a pratical approach. *The science of the total environment.* No.247, pp. 295-311.

Camargo, M.M.P. &, Martinez, C.B.R. (2007) Histopathology of gills, kidney and liver of a neotropical fish caged in na urbam stream. *Neotropical ichthyology,* Vol. 5, No.3, pp. 327-336.

Campagna, A.F. (2005) Toxicidade dos sedimentos da bacia hidrográfica do rio monjolinho (são carlos-sp): ênfase nas substâncias cobre, aldrin e heptacloro. *Dissertação (mestrado em zootecnia),* universidade de são paulo, Pirassununga, Brazil.

COHAB. (1997) *Relatório ambiental da região metropolitana de belém.* Belém.

Conama. (2005). *Ministério do meio ambiente. conselho nacional de meio ambiente.* portaria n° 357 de 17 de março de 2005. Brasília.

Fanta, E. (1995). Gill structure of antartic fishes notothenia (gobinotothen) gibberifrons and trematomusnewnesi (notothenidae) stressed by salinity changes and some behavioral consequences. *anta. rec.,* Vol. 39, pp. 25-39.

Fernandes, C., Fontaínhas-fernandes, A.; Ferreira, M. & Salgado, M.A. (2009) Oxidative stress response in gill and liver of *liza saliens*, from the esmoriz-paramos coastal lagoon, portugal. *Archives of environmental contamination toxicology,* No.52, pp. 262-269.

Ferreira, M.; Moradas-ferreira, P. & Reis-henriques, M.A. (2005) Oxidative stress biomarkers in two resident species, mullet (*Mugil cephalus*) and flouder (*Platichthys flesus*), from a polluted site in river douro estuary, portugal. *aquatic toxicology,* No. 71, pp. 39-48.

Freire, M.M..; Santos, V.G.; Ginuino, I.S.F.; Arias, A.R.L. (2008) Biomarcadores na avaliação da saúde ambiental dos ecossistemas aquáticos. *oecologia brasilienses,* Vol. 12, No. 3, pp. 347-354.

Giari, L.; Simoni, E.; Manera, M. & Dezfuli, B.S. (2008) histo-cytological responses of *Dicentrarchus labrax* (l.) following mercury exposure. Ecotoxicology and environmental safety, Vol. 70, pp. 400-410.

Heath, A.G. (1995) *Water pollution and fish physiology.*(ed. 2). Boca Raton.

Hibiya, T. (1982). *An atlas of fish histology, normal and pathological features.* New york, kodansha tokio.

Hinton, D.E.; Baumen, P.C.; Gardener, G.C.; Hawkins, W.E.; Hendricks, J.D.; Murchelano, R.A. & Okhiro, M.S. (1992). Histopathological biomarker . in: *biomarkers: biochemical, physiological and histological markers na anthropogenic stress society of environmental toxicology and chemistry special publication series* (eds), p 155-210. Huggett, R.J.; Kimerle, R.A.; Merhle, P.M. & Bergman, H.L. Chelsea, MI, USA.

Huggett, R.J.; Kimerle, R.A. & Mehrle, P.M. (1992). Biomarkers biochemical, physiological, and histological markers of anthropogenic stress. Bergman, h.l. (ed). Boca Raton, FL, USA.

Kammenga, J.E.; Dalliner, R.; Donker, M.H.; Kohler, H.R.; Simonsen V.; Triebskorn, R. & Weeks, J.M. (2000) Biomarkers in terrestrial invertebrates for ecotoxicological Soil risk assessment. *Revist of Environmental Contamination Toxicology.* Vol. 164, pp. 93-147.

Laurent, P. & Perry, S.F. (1991) Environmental effects on fish gill morphology. *physiology zoology.* Vol.64, pp. 4-25.

Leonzio, C. & Fossi, M.C. (1993) Nondestructive biomarkers strategy: perspectives and applications: in: *Nondestructive biomarkres in vertebrates* (eds). 297-312. Fossi, M.C & Leonzio, C. London.

Lindesjöö, E. & Thulin, J. (1994). Histopathology of skin and gills of fish in pulp mill effluents. *Aquatic organisms*, Vol. 18, pp. 81-93.

Martinez, C.B.R. & Souza, M.M. (2002) Acute effects of nitrite on ion regulation in two neotropical fish species. *Comparative biochemistry and physiology*,Vol. 133ª, pp. 151-160.

McDonald, D.G. (1983). The effects of h^+ upon the gills pf freshwater fish. *Canadian journal of zoology*, Vol. 61, pp. 691-703.

Miron, D.S.; Moraes, B.; Becker, A.G.; Crestani, M.; Spanevello, R.; Loro, V.L. & Baldisserotto, B. (2008) Ammonia and ph effects on some metabolic parameters and gill histology of silver catfish, *Rhamdia quelen* (Heptapteriadae). Aquaculture, Vol. 277, pp. 192-196.

Monserrat, J.M.; Martinez, P.E.; Geracitano, L.A.; Amado, L.L.; Martins, C.M.G.M.; Pinho, G.L.L.; Chaves, I.S.; Ferreira-cravo, M.; Ventura-lima, J. & Bianchini, A. (2007) Pollution biomarkers in estuarine animals: critical review and new perspectives. *Comparative biochemistry and physiology part* c, No.146, pp. 221-234.

Montes, C.S.; Ferreira, M.A.P; Santos, S.S.D.; von Ledebur, E.I.C.F. & Rocha, R.M. (2010) Branchial histopathological study of brachyplatystoma rousseauxii (castelnau, 1855) in the guajará bay, belém, pará state, brazil. *acta scientiarium biolical science,* Vol. 32, No. 1, pp. 87-92.

Nowak, B. (1992) Histological changes in gill induced by residues of endossulfan. *Aquatic toxicology*, Vol. 23, pp. 65-84.

Oliveira-ribeiro, C.A. (1996) Lethal effects of inorganic mercury on cells and tissues of trichomycterus brasilienseis (pisces; siluroidei). *Biocellular*. Vol.20, No.3,pp.171-178.

van der Oost, R..; Beyer, J.; Vermeulen, N.P.E. (2003) Fish bioaccumulation and biomarkers in environmental risk assessment: a review. *Enviromental toxicology and pharmacology*, Vol. 13, pp. 57-149.

Pinheiro, R.V.L. (1987) Estudo hidrodinâmico e sedimentológico do estuário guajará - belém (pa). *Dissertação (mestrado em geociências)* universidade federal do pará, belém, PA, Brazil.

Poleksic, V. & Mitrovic-Tutundzic, V. (1994) Fish gills a monitor of sublethal and choronic effects of pollution. in: muller, r.; lloyd, r. sublethal and chronic effects of pollutants of freshwater fish. oxford: fishing news books. No. 30, pp. 339-352.

Ribeiro, K.T.S. (2004) Água e a saúde humana em belém. belém. cejup.

Rocha, R.M.; Coelho, R.P.; Montes, C.S.; Santos, S.S.D. & Ferreira, M.A.P. (2010) Avaliação histopatológica do fígado de *Brachyplatystoma rousseauxii* (castelnau, 1855) da Baía do guajará, belém, pará. *Ciência animal brasileira (ufg)*, Vol. 11, pp. 101-109.

Schwaiger, J.; Wanke, R.; Adam, S.; Pawert, M.; Honnen, W. & Triebskorn, R. (1997) The use of histopathological indicators to evaluate contaminant related stress in fish. Dordretch. *Journal Aquatic Ecosystem Stress Recovery*, Vol.6, No 1, pp. 75-86.

Simonato J.D.; Guedes, C.L.B. & Martinez, C.B.M. (2008) biochemical, physiological and histological changes in the neotropical fish *prochilodus lineatus* exposed to diesel oil. *Ecotoxicol environmental safety* No. 69, pp. 112-120.

Stephan, C.E. & Mount, D.J. (1973) Use of toxicity tests with fish in water polution control, In: *Biological methods for the assessment of water quality*. Vol 528, pp. 164-177, Philadephia.

de la Torre, F.R.; Ferrari, L. & Salibián, A. (2005) Biomarkers of a native fsh species (*Cnesterodon decemmaculatus*) applicationto the water toxicity assessment of a peri-urban polluted river of Argentina. *Chemosphere*, Vol.59, No. 4, pp. 577-583.

Part 2

Advances in Environmental Monitoring Research and Technologies

4

Air Pollution Analysis with a Possibilistic and Fuzzy Clustering Algorithm Applied in a Real Database of Salamanca (México)

B. Ojeda-Magaña,[1] R. Ruelas[1], L. Gómez-Barba[1], M. A. Corona-Nakamura[1], J. M. Barrón-Adame[2], M. G. Cortina-Januchs[2], J. Quintanilla-Domínguez[2] and A. Vega-Corona[2]

[1]*University of Guadalajara*
[2]*University of Guanajuato*
México

1. Introduction

Air pollution is one of the most important environmental problems in developed and undeveloped countries and it is associated with significant adverse health effects. Air pollution is characterized by the presence of a heterogeneous, complex mixture of gases, liquids and particulate matter in air. Pollution is caused by both natural and man-made sources, and it may greatly vary from one region to another according to the geography, demography, climate, and topography of these ones. For example, pollutant concentrations decrease significantly when the urban area meets certain characteristics as topography or large rain season (Celik & Kadi, 2007). Forest fires, volcanic eruptions, wind erosion, pollen dispersal, evaporation of organic compounds, and natural radioactivity are among natural causes of air pollution. Major man-made sources of air pollution include: industries, transportation, agriculture, power generation, and unplanned urban areas (Fenger, 2009).

Air pollutants exert a wide range of impacts on biological, physical, and ecosystems. Their effects on human health are of particular concern. The World Health Organization (WHO) consider air pollution as the mayor environmental risk to health and is estimated to cause approximately 2 million premature deaths worldwide per year (WHO, 2008).

This type of pollution is classified in criterio and non-criterio pollutants, the firsts are considered dangerous to human and animal health, its name was given after the result of various evaluations regarding air pollution published by the United States of America (EPA, 2008). Six criteria of pollutants are defined: Nitrogen Dioxide (NO_2), Sulfur Dioxide (SO_2), Carbon Monoxide (CO), Particulate Matter (PM), Lead (Pb), and Ozone (O_3). The objective of this classification is to establish permissible levels to protect human and animal health and for the preservation of the environment. Human health is one of the most important concerns due to the short-term consequences of air pollution, especially in metropolitan areas, health effects are dependent on the type of pollutant, its concentration in air, length of exposure to the pollutant and individual susceptibility. Several groups of individuals react differently to air pollution, Children and elderly people are the most affected by this kind of pollution. Global warming and the greenhouse effect are among long term consequences of the global climate.

Examine and study air pollutant information is very important for a better understanding of the human exposure and its potential impacts in health and welfare.
In recent years, the city of Salamanca has been catalogued as one of the most polluted cities in Mexico (Zuk et al., 2007). Sulphur Dioxide (SO_2), and Particular Matter (PM_{10}) are the criteria for searching air pollutants with the highest concentration in Salamanca, where three monitoring stations have been installed in order to know the level of air pollution; measure records of each monitoring station are handled separately. Actually an environmental contingency alarm is activated when the daily average pollutant concentration exceeds an established threshold (in a single monitoring station).
In this work, we propose to apply the PFCM (*Possibilistic Fuzzy c Means*) clustering algorithm to the measured data obtained from three monitoring stations so that a local environmental contingency alarm can be taken, according to the pollutant concentration reported by each monitoring station, general (or city) environmental contingency alarms will depend on the levels provided by the combined measure. So, the PFCM algorithm is used to find the prototypes of patterns that represent the relation between SO_2 and PM_{10} air pollutants. For this relation analysis we use records from January 2007.
Once the prototypes have been estimated, a comparison is made between the average pollution of each monitoring station and the prototypes. In the analysis is used a data set from January to December 2007. The analysis include pollutant concentration as SO_2 , PM_{10}, meteorological variables, wind speed, wind direction, temperature, and relative humidity.
It is also analyzed the impact of meteorological variables on the dispersion of pollutants, this is done through the calculus of correlation coefficients. This important correlation analysis is very simple and it is intended for improving decision making in environmental programs. Only the data gathered by the *Nativitas monitoring* station is used for the correlation analysis.
This paper is organized as follow: In Section 2 is presented the features, and explain the air pollution problem in Salamanca. In Section 3 is introduced the PFCM (*Possibilistic Fuzzy c Means*) clustering algorithm and the correlation coefficients. Section 4 presents the obtained results. And finally, in Section 5 we present our conclusions.

2. Study case

Salamanca is located in the state of Guanajuato, Mexico, and it has an approximate population of 234,000 inhabitants INEGI (2005). The city is 340 km northwest from Mexico City, with coordinates 20°34'22" North latitude, and 101°11'39" West longitude. It is located on a valley surrounded by the *Sierra Codornices*, where there are elevations with an average height of 2,000 meters Above Mean Sea Level (AMSL).
Salamanca has been one of the Mexican cities with more important industrial development in the last fifty years. Refinery and Power Generation Industries settled down in the fifty and seventy decades, respectively. These industries constitute the main and most important energy source for local, regional and national economy. However, the increase of population, quantity of vehicles, and the industry, refinery and thermoelectric activities, as well as orography and climatic characteristics have propitiated the increment in SO_2 and PM_{10} concentrations INE (2004). The existent orography difficults the dispersion of pollutants by the wind, which produces the worst pollutant concentrations. SO_2 emissions are bigger than those in the Metropolitan area of Mexico City or Guadalajara city, the two biggest cities of Mexico, even when these ones have a bigger population than the city of Salamanca Cortina-Januchs et al. (2009). Orography hinders the dispersion of the worst pollutants by winds.

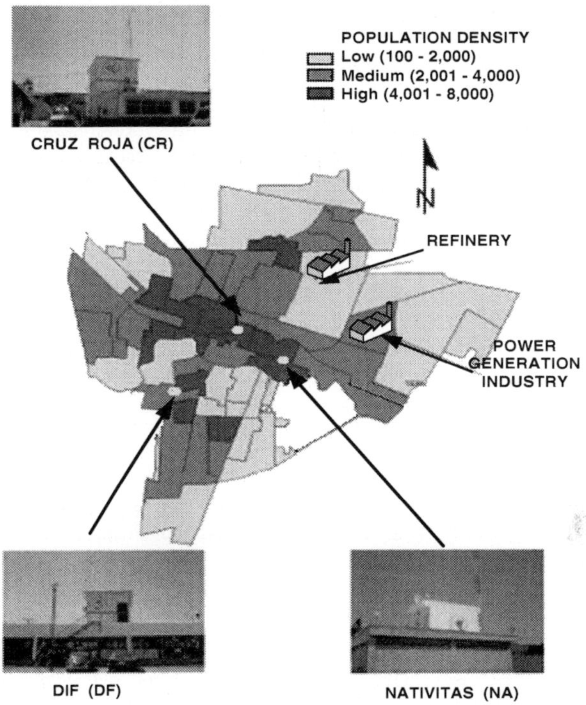

Fig. 1. Location of monitoring stations in the city of Salamanca.

Sulfur dioxide is produced fundamentally by the combustion of fossil fuels, and it has the energy generation sector as the main source of pollution. That is, the industrial sector generates 99.3 % of this pollutant, and only an approximate percentage of 0.06 % is generated by the transport sector. Particles produced by electric power generation represent 29 % of the total emissions, it follows the vehicular traffic in the roads without paving with 27 %, next the agriculture burns with 17 %, transport sector with 10 %,and the remaining 17 % is emitted by other sub-sectors.

Authorities of the city have made important efforts to measure and record on concentrations of pollutants Zamarripa & Sainez (2007). In 1999 the *Air Quality Monitoring Patronage* (AQMP) was formed. Since then the AQMP has been in charge of running the *Automatic Environmental Monitoring Network* (AEMN), and disseminate information. This information is validated by the *Institute of Ecology* (IE), which constantly analyzes the levels of pollutants INE (2004). The AEMN consists of three fixed and one mobile stations. The fixed stations are: *Cruz Roja* (CR), *Nativitas* (NA), and *DIF*.

The fixed stations cover approximately 80 % of the urban area while the mobile station covers the remaining 20 %. Fig. 1 illustrates the location of the three fixed stations. Each station has the necessary instrumentation to automatically track concentration of pollutants and meteorological variables every minute. Table 1 contains a sample of the concentration of pollutants and meteorological variables in each of the three fixed stations.

Pollutants

	Cruz Roja	Nativitas	DIF
Ozone (O_3)	√	√	√
Sulfur Dioxide(SO_2)	√	√	√
Carbon Monoxide (CO)	√	√	√
Nitrogen Dioxide (NO_x)	√	√	√
Particulate Matter less than 10 micrometer in diameter (PM_{10})		√	√

Meteorological variables

	Cruz Roja	Nativitas	DIF
Wind Direction (WD)	√	√	√
Wind speed (WS)	√	√	√
Temperature (T)		√	√
Relative Humidity (RH)		√	√
Barometric Pressure (BP)		√	√
Solar Radiation (SR)		√	√

√ Measured

Table 1. Pollutants concentrations and meteorological variables recorder in the monitoring stations

3. Clustering algorithms

In this work we take advantage of the qualities of fuzzy and possibilistic clustering algorithms in order to find c groups in a set of unlabeled data set $Z = \{z_1, z_2, \ldots, z_k, \ldots, z_N\}$ in an M-dimensional space, where the nearest z_k to a prototype, or group center v_i, belong to the group i among c possible groups. The membership of each z_k to the different groups depends on the kind of partition of the M-dimensional space where data set is defined. This way, a *c-partition* can be either: hard (or crisp), fuzzy, and possibilistic Bezdek et al. (1999). The hard c-partition of the space for a data set $Z(k) = \{z_k | k = 1, 2, ..., N\}$, of finite dimension and c groups, where $2 \leq c < N$, is defined by (1), (2) defines the fuzzy c-partition, whereas (3) defines the possibilistic c-partition.

$$M_{hcm} = \Big\{\mathbf{U} \in \Re^{c \times N} | \mu_{ik} \in \{0,1\}, \forall i \text{ and } k; \sum_{i=1}^{c} \mu_{ik} = 1, \forall k; \quad 0 < \sum_{k=1}^{N} \mu_{ik} < N, \forall i\Big\}; \tag{1}$$

$$M_{fcm} = \Big\{\mathbf{U} \in \Re^{c \times N} | \mu_{ik} \in [0,1], \forall i \text{ and } k; \sum_{i=1}^{c} \mu_{ik} = 1, \forall k; \quad 0 < \sum_{k=1}^{N} \mu_{ik} < N, \forall i\Big\}; \tag{2}$$

$$M_{pcm} = \Big\{ \mathbf{U} \in \Re^{c \times N} | \mu_{ik} \in [0,1], \forall i \ and \ k;$$
$$\forall k, \exists i, \mu_{ik} > 0; \quad 0 < \sum_{k=1}^{N} \mu_{ik} < N, \forall i \Big\}. \tag{3}$$

3.1 Fuzzy c-Means algorithm

The Fuzzy c-Means clustering algorithm (FCM) was initially developed by Dunn Dunn (1973), and generalized later by Bezdek Bezdek (1981). This algorithm is based on the optimization of the objective function given by (4),

$$J_{fcm}(\mathbf{Z}; \mathbf{U}, \mathbf{V}) = \sum_{i=1}^{c} \sum_{k=1}^{N} (\mu_{ik})^m \|z_k - v_i\|^2, \tag{4}$$

where the membership matrix $U = [\mu_{ik}] \in M_{fmc}$, is a fuzzy c-partition of the space where Z is defined, $V = [v_1, v_2, ..., v_c]$ is the vector of prototypes of the c groups, which are calculated according to $D_{ikA_i} = \|z_k - v_i\|^2$, a squared inner-product distance norm, and $m \in [1, \infty]$ is a weighting exponent which determines the fuzziness of the partition. The optimal c-partition for a Fuzzy c-Means algorithm, is reached through the couple (U^*, V^*) which minimizes locally the objective function J_{fcm}, according to the *alternating optimization* (AO).

Theorem FCM Bezdek (1981): If $D_{ikA_i} = \|z_k - v_i\| > 0$, for every $i, k, m > 1$, and Z contains at least c distinct data points, then $(U, V) \in M_{fcm} \times \Re^{c \times N}$ may minimize J_{fcm} only if

$$\mu_{ik} = \left(\sum_{j=1}^{c} \left(\frac{D_{ikA_i}}{D_{jkA_i}} \right)^{2/(m-1)} \right)^{-1} \tag{5}$$
$$1 \leq i \leq c; \quad 1 \leq k \leq N$$

$$v_i = \sum_{k=1}^{N} \mu_{ik}^m z_k \Big/ \sum_{k=1}^{N} \mu_{ik}^m \tag{6}$$
$$1 \leq i \leq c.$$

Following the previous equations of the FCM algorithm, the solution can be reached with the next steps:

FCM-AO-V

Given the data set Z choose the number of clusters $1 < c < N$, the weighting exponent $m > 1$, as well as the ending tolerance $\delta > 0$.

I Provide an initial value to each one of the prototypes $v_i, i = 1, .., c$. These values are generally given in a random way.

II Calculate the distance of z_k to each one of the prototypes v_i, using $D^2_{ikA_i} = (z_k - v_i)^T A_i (z_k - v_i), \ 1 \leq i \leq c, \ 1 \leq k \leq N$.

III Calculate the membership values of the matrix $U = [\mu_{ik}]$, if $D_{ik\mathbf{A}} > 0$, using equation (5).

IV Update the new values of the prototypes v_i using equation (6).

V Verify if the error is equal or lower than δ,

$$\|V_{k+1} - V_k\|_{err} \leq \delta,$$

If this is truth, stop. Else, go to step **II**.

The FCM is an algorithm that calculates a membership value μ_{ik} for each point z_k in function of all prototypes v_i. The sum of the membership values of z_k to the c groups must be equal to one. However, a problem arises when there are several equidistant points from the prototypes of the groups, because the FCM is not able to detect noise points or nearest and furthest points from the prototypes. Pal *et al* Pal et al. (2004) show an example with two points located in the boundary of two groups, one point near to the prototypes and the other one far away from them. This must be handled with care, as both points are not *equally representative* of the groups, even if they have the same membership values. One way to overcome this inconvenience is to use a possibilistic algorithm.

3.2 Possibilistic c-Means algorithm

The Possibilistic c-Means clustering algorithm (PCM) Krishnapuram & Keller (1993) is based on *typicality values* and relaxes the constraint of the FCM concerning the sum of membership values of a point to all the c groups, which must be equal to one. Thus, the PCM identifies the similarity of data points with an alone prototype v_i using a typicality values that takes values in [0,1]. The nearest data points to the prototypes are considered *typical*, further data points are *atypical* and data points with zero, or almost zero, typicality values are considered *noise* Ojeda-Magaña et al. (2009a). The objective function J_{pcm} proposed by Krishnapuram Krishnapuram & Keller (1993) for this algorithms is given by

$$J_{pcm}(\mathbf{Z};\mathbf{T},\mathbf{V},\gamma) = \left\{ \sum_{k=1}^{N}\sum_{i=1}^{c}(t_{ik})^m \|z_k - v_i\|_A^2 + \sum_{i=1}^{c}\gamma_i \sum_{k=1}^{N}(1-t_{ik})^m \right\}, \tag{7}$$

where

$$T \in M_{pcm}, \quad \gamma_i > 0, \quad 1 \leq i \leq c. \tag{8}$$

The first term of J_{pcm} is identical to that of the FCM objective function, which is based on the distance of the points to the prototypes. The second term, that includes a penalty γ_i, tries to bring t_{ik} toward 1.

Theorem PCM Krishnapuram & Keller (1993): if $\gamma_i > 0, 1 \leq i \leq c$, $m > 1$ and Z has at least c distinct data points, then $(T, V) \in M_{pcm} \times \Re^{c\times N}$ may minimize J_{pcm} only if

$$t_{ik} = \frac{1}{1 + \left(\frac{\|z_k - v_i\|^2}{\gamma_i}\right)^{1/(m-1)}}, \tag{9}$$

$$1 \leq i \leq c; \qquad 1 \leq k \leq N$$

$$v_i = \sum_{k=1}^{N} t_{ik}^m z_k \Big/ \sum_{k=1}^{N} t_{ik}^m, \tag{10}$$

$$1 \leq i \leq c; \qquad 1 \leq k \leq N.$$

Krishnapuram and Keller Krishnapuram & Keller (1993) Krishnapuram & Keller (1996) recommend to apply the FCM at a first time, such that the initial values of the PCM algorithm can be estimated. They also suggest the calculus of the penalty γ_i with equation (11)

$$\gamma_i = K \frac{\sum_{k=1}^{N} \mu_{ik}^m \|z_k - v_i\|_A^2}{\sum_{k=1}^{N} \mu_{ik}^m} \tag{11}$$

where $K > 0$, although the most common value is $K = 1$, and the membership values $\{\mu_{ik}\}$ are those calculated with the FCM algorithm in order to reduce the influence of noise.

The PCM algorithm is very sensitive to the $\{\gamma_i\}$ values, and the typicality values depend directly on it. For example, if the value of γ_i is small, the typicality values t_{ik} of T are also small, whereas if the value of γ_i is high, the t_{ik} are also high. For this work, the $\{\gamma_i\}$ values are obtained from equation (11).

In order to avoid a problem with the initial PCM algorithm, as sometimes the prototypes of different groups coincided Hoppener et al. (2000), even if the natural structure of data has well delimited different groups, Tim *et al* Timm et al. (2004); Timm & Kruse. (2002) have modified the objective function to include a constraint based on the repulsion among groups, thus avoiding identical groups when they must be different.

The objective of the fuzzy clustering algorithms is to find an internal structure in a numerical data set into n different subgroups, where the members of each subgroup have a high similarity with its prototype (centroid, cluster center, signature, template, code vector) and a high dissimilarity with the prototypes of the other subgroups. This justifies the existence of each one of the subgroups Andina & Pham (2007).

A simplified representation of a numerical data set into n subgroups, help us to get a better comprehension and knowledge of the data set Barron-Adame et al. (2007). Besides, the particional clustering algorithms (hard, fuzzy, probabilistic or possibilistic) provide, after a learning process, a set of prototypes as the most representative elements of each subgroups.

Ruspini was the first one to use fuzzy sets for clustering Ruspini (1970). After that, Dunn Dunn (1973) developed in 1973 the first fuzzy clustering algorithm, named Fuzzy c-Means (FCM), with a parameter of fuzziness m equal to 2. Later on Bezdek Bezdek (1981) generalized this algorithm. The FCM is an algorithm where the membership degree of each point to each fuzzy set A_i is calculated according to its prototype. The sum of all the membership degrees of each individual point to all the fuzzy sets must be equal to one.

Krishnapuram and Keller Krishnapuram & Keller (1993) developed the Possibilistic c-Means (PCM) clustering algorithm, where the principal characteristic is the relaxation of the restriction that gives the relative typicality property of the FCM. The PCM provides a similarity degree between data points and each one of the prototypes, value known as absolute typicality or simply typicality Pal et al. (1997). So, the nearest points to a prototype are identified as typical, whereas the furthest points as atypical, and noise Ojeda-Magaña et al. (2009a)Ojeda-Magaña et al. (2009b).

3.3 PFCM clustering algorithm

Pal *et al.* Pal et al. (1997) have proposed to use the membership degrees as well as the typicality values, looking for a better clustering algorithm. They called it *Fuzzy Possibilistic c-Means* (FPCM). However, the sum equal to one of the typicality values for each point was the origin of a problem, particularly when the algorithm uses a lot of data. In order to avoid this problem, Pal *et al* Pal et al. (2005) proposed to relax this constraint and they developed the PFCM clustering algorithm, where the function to be optimized is given by (12)

$$J_{pfcm}(\mathbf{Z};\mathbf{U},\mathbf{T},\mathbf{V}) = \sum_{i=1}^{c}\sum_{k=1}^{N}(a\mu_{ik}^{m} + bt_{ik}^{\eta}) \times \|z_k - v_i\|^2 + \sum_{i=1}^{c}\gamma_i\sum_{k=1}^{N}(1-t_{ik})^{\eta}, \tag{12}$$

and subject to the constraints $\sum_{i=1}^{c}\mu_{ik} = 1\forall k;\ 0 \leq \mu_{ik}, t_{ik} \leq 1$ and the constants $a > 0$, $b > 0$, $m > 1$ and $\eta > 1$. The parameters a and b define a relative importance between the membership degrees and the typicality values. The parameter μ_{ik} in (12) has the same meaning as in the FCM. The same happens for the t_{ik} values with respect to the PCM algorithm.
emphTheorem PFCM Pal et al. (2005): If $D_{ikA} = \|z_k - v_i\| > 0$, for every $i, k, m, \eta > 1$, and Z contains at least c different patterns, then $(U,T,V) \in M_{fcm} \times M_{pcm} \times \Re^p$ and J_{pfcm} can be minimized if and only if

$$\mu_{ik} = \left(\sum_{j=1}^{c}\left(\frac{D_{ikA_i}}{D_{jkA_i}}\right)^{2/(m-1)}\right)^{-1} \tag{13}$$

$$1 \leq i \leq c;\quad 1 \leq k \leq n$$

$$t_{ik} = \frac{1}{1 + \left(\frac{b}{\gamma_i}D_{ikA_i}^2\right)^{1/(\eta-1)}} \tag{14}$$

$$1 \leq i \leq c;\quad 1 \leq k \leq n$$

$$v_i = \sum_{k=1}^{N}(a\mu_{ik}^{m} + bt_{ik}^{m})z_k \Big/ \sum_{k=1}^{N}(a\mu_{ik}^{m} + bt_{ik}^{m}), \tag{15}$$

$$1 \leq i \leq c.$$

The membership degrees are calculated with equation (13), the typicality values with (14) and for the prototypes the equation (15) is used.
The iterative process of this algorithm follows the next steps:

PFCM-AO-V

Given the data set Z choose the number of clusters $1 < c < N$, the weighting exponents $m > 1, \eta > 1$, and the values of the constants $a > 0$, and $b > 0$.

I Provide an initial value to each one of the prototypes $v_i, i = 1,..,c$. These values are generally given in a random way.

II Run the FCM-AO-V algorithm.

III With these results, calculate the penalty parameter γ_i for each cluster i. Take $K = 1$.

IV Calculate the distance of z_k to each one of the prototypes v_i using $D^2_{ikA_i} = (z_k - v_i)^T A_i (z_k - v_i)$, $1 \le i \le c$, $1 \le k \le N$.

V Calculate the membership values of the matrix $U = [\mu_{ik}]$ if $D_{ik\mathbf{A}} > 0$, use equation (13).

VI Calculate the typicality values of the matrix $T = [t_{ik}]$, if $D_{ik\mathbf{A}} > 0$, use equation (14).

VII Update the value of the prototypes v_i using equation (15).

VIII Verify if the error is equal or lower than δ,

$$\|V_{k+1} - V_k\|_{err} \le \delta,$$

if this is truth, stop. Else, go to step **IV**.

3.4 PFCM clustering algorithm in the AEMN

As it is known, in the partition clustering algorithms is necessary a minimum of two groups. However, in our problem we only have one group, this group is formed by patterns $[SO_2;PM_{10}]$ pollutant concentrations. Therefore, is proposed a synthetic cloud of patterns with the following covariance matrix and vector of centers:

$$\Sigma_1 = \begin{bmatrix} 400 & 0 \\ 0 & 400 \end{bmatrix}, [v_1] = [100 \ -600].$$

In this case, the number of patterns (4320) is the same in the synthetic cloud and the pollutant concentration.

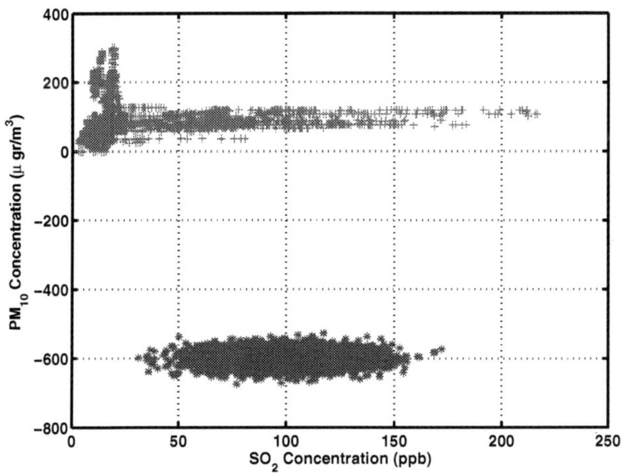

Fig. 2. Air pollution and synthetic cloud patterns.

Fig. 2 shows clearly the synthetic cloud (located in the lower part) and the pollutant concentration patterns (located in the superior part). Once the groups are identified, we apply the PFCM clustering algorithm.

3.5 Correlation coefficient

The correlation coefficient r (also called Pearson's product moment correlation after Karl Pearson Pérez et al. (2000)) is used to determine the strength and direction of the relationship between two variables. This form of correlation requires that both variables are normally distributed, interval or ratio variables. The correlation coefficient is calculated by eq.(16):

$$r = \frac{n\sum x_i y_i - (\sum x_i)(\sum y_i)}{\sqrt{n(\sum x_i{}^2) - (\sum x_i)^2}\sqrt{n(\sum y_i{}^2) - (\sum y_i)^2} \tag{16}$$

where n is the number of data points. The numerical values of correlation coefficient range from +1 to -1. If two variables move exactly together, the value of the correlation coefficient is 1. This indicates perfect positive correlation. If two variables move exactly opposite to each other, the value of the correlation coefficient is -1. Low numerical values indicate little relationship between two variables, such as -0.10 or +0.15 indicate little relationship between on two variable.

4. Results

Fig. 3 shows the distribution of pollutant patterns $[SO_2;PM_{10}]$ at the three monitoring stations (CR, DF and NA). The mesh in Fig. 3 corresponds to the thresholds established by the program to improve the air quality in Salamanca (*ProAire*) INE (2004). Thresholds are Pre-contingency, Phase-I contingency and Phase-II contingency. For example, for SO_2 concentrations equal to or bigger than 145 ppb and smaller than 225 ppb (average per day), a level of environmental pre-contingency is declared. Therefore the spaces between lines in the mesh represent the levels of environmental contingency for SO_2 and PM_{10} concentrations.
In Fig. 3 each symbol (*, • and $\bigtriangledown$) represent the pollutant patterns at each monitoring station. At Nativitas monitoring station we observe that the highest PM_{10} and SO_2 pollutant concentrations are not present at the same time. On other hand, at the Cruz Roja monitoring

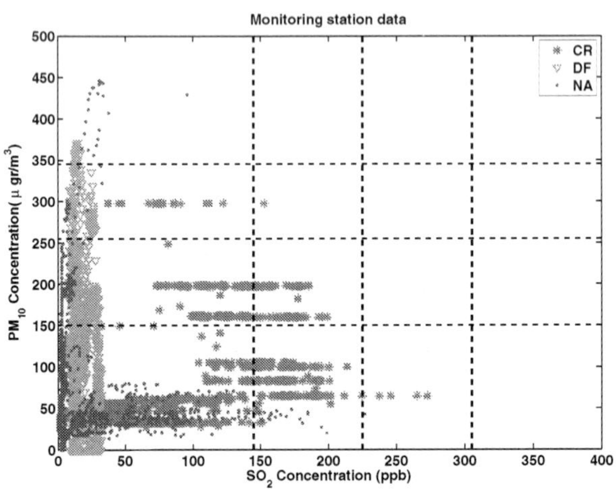

Fig. 3. Monitoring Network per minute.

station we observe that either SO_2 or PM_{10} pollutant concentrations are highest. At the DIF monitoring station we observe the highest PM_{10} concentrations in the AEMN network.
The main proposal in this work is to apply the PFCM clustering algorithm to the AEMN in Salamanca as well to integrate the pollutant measures from the three monitoring stations.
The PFCM initial parameters (a, b, m and η) are very important in order to reduce the outlier effects in the pattern prototypes. Pal *et al*, in Pal et al. (2005) recommend of b parameter value larger than the a parameter value in order to reduce the mentioned effects. On the other hand, a small value for η and a value greater than 1 for m are recommended. nevertheless, choosing a too high of a value of m reduces the effect of membership of data to the clusters, and the algorithm behaves as a simple PCM.
Taking into account the previous recommendations, the initial parameters for the PFCM clustering algorithm were set as follows: $a = 1, b = 5, m = 2$ and $\eta = 2$. The found prototypes (a and b) are shown in Fig. 4.
In Fig. 4(a) the daily averages of SO_2 concentrations are presented for each monitoring station together with the corresponding prototypes. It is observed also that Cruz Roja monitoring station receives the highest emissions of SO_2 concentrations: this is due to its location near to the refinery. The prototypes in this case were very low in comparison with the observed SO_2 concentrations, because only one station observed high SO_2 concentrations (Cruz Roja). According with the analyzed patterns the emitted pollutant is only measured by the Cruz Roja monitoring station (see Fig. 4).
Fig. 4(b) shows the daily averages of PM_{10} concentrations and result prototypes. In this case, the observed averages are very similar at the three monitoring stations. The PM_{10} pollutant dispersion is more uniform then the SO_2 pollutant dispersion in the city.
Table 2 shows the correlation results among SO_2 and PM_{10} pollutants and the meteorological variables. The database used in the correlation analysis correspond to year 2004 of Nativitas. This period was taking because contains more meteorological registrations. The obtained results of the SO_2 correlation coefficient show a high positive correlation between SO_2 pollutant and Wind Speed, also a high and negative correlation between SO_2 pollutant and Wind Direction is observed. The other meteorological variables have not impact. For the PM_{10} pollutant, the meteorological variable with more impact is the Relative Humidity. We observe, when the Relative Humidity increases the pollutant concentration decreases. The PM_{10} particles are caught and fall to the ground during rain.

	SO_2	PM_{10}
SO_2	1	0.0731
PM_{10}	0.0731	1
WS	0.4756	-0.1385
WD	-0.6151	0.1478
T	-0.0329	-0.0007
RH	-0.0322	-0.4416
BP	0.1462	0.1806
SR	-0.021	-0.1207

Table 2. Correlation Coefficient between pollutant concentration and meteorological variables.

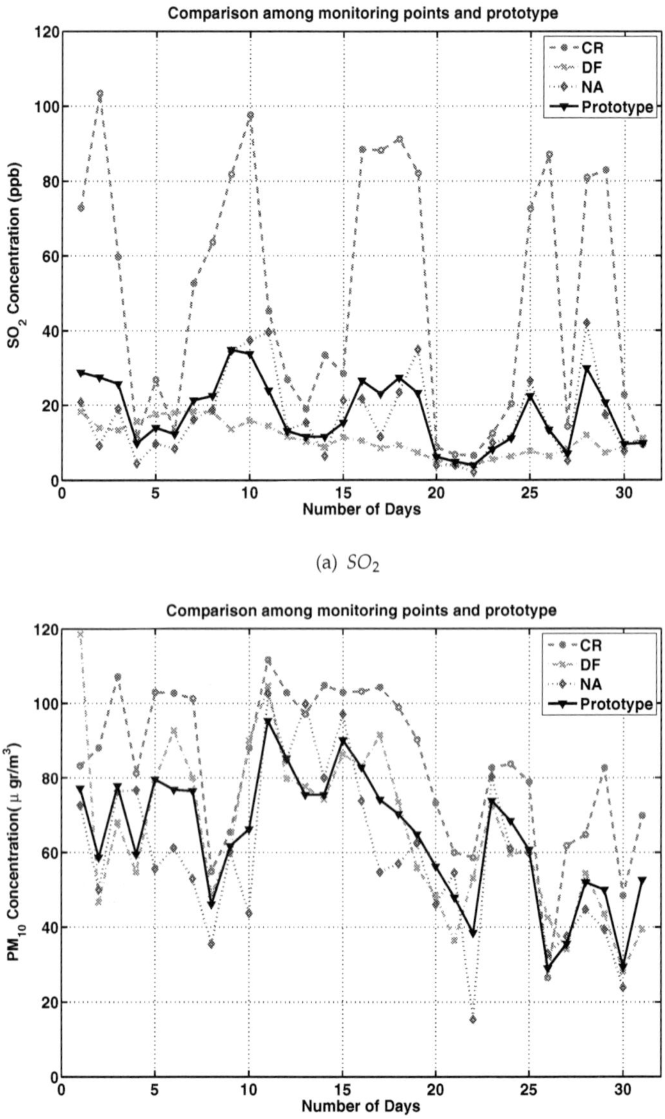

(a) SO_2

(b) PM_{10}

Fig. 4. Comparison between air pollutant averages and estimated prototypes.

5. Conclusions

Nowadays, there is a program to improve the air quality in the city of Salamanca, Mexico. Besides, this program has established thresholds for several levels of contingencies depending on the SO_2 and PM_{10} pollutant concentrations. However, a particular level of contingency for the city is declared taking into account the highest pollutant concentration provided by one of the three monitoring stations. For example, if a pollutant concentration exceeds a given threshold in a single monitoring station, the alarm of contingency applies to the whole city. This value is normally provided by the Cruz Roja station, due to its proximity to the refinery and power generation industries.

Looking for local and general contingency levels in the city, we have proposed to estimate a set of prototypes such that they can represent a calculated measure of pollutant concentrations according to the values measured in the three fixed stations. In such a way, a local alarm of contingency can be activated in the area of impact of the pollution depending on each station, and a general alarm of contingency according to the values provided by the prototypes. Nevertheless, the last case requires adjusting the thresholds, as the actual values would be only used for local contingency because they depend on the measured values of pollutant concentrations, and the general contingency requires thresholds as a function of calculated values.

6. References

Andina, D. & Pham, D. T. (2007). *Computational Intelligence*, Springer.

Barron-Adame, J. M., Herrera-Delgado, J. A., Cortina-Januchs, M. G., Andina, D. & Vega-Corona, A. (2007). Air pollutant level estimation applying a self-organizing neural network, *Proceedings of the 2nd international work-conference on Nature Inspired Problem-Solving Methods in Knowledge Engineering. IWINAC-07*, pp. 599–607.

Bezdek, J. C. (1981). *Pattern Recognition With Fuzzy Objective Function Algorithms*, Kluwer Academic.

Bezdek, J. C., Keller, J., Krishnapuram, R. & Pal, N. R. (1999). *Fuzzy Models and Algorithms for Pattern Recognition and Image Processing*, first edn, Boston, London.

Celik, M. B. & Kadi, I. (2007). The relation between meteorological factors and pollutants concentration in karabuk city, *G.U. Journal of science* 20(4): 87–95.

Cortina-Januchs, M. G., Barron-Adame, J. M., Vega-Corona, A. & Andina, D. (2009). Prevision of industrial so2 pollutant concentration applying anns, *Proceedings of The 7th IEEE International Conference on Industrial Informatics (INDIN 09)*, pp. 510–515.

Dunn, J. (1973). A fuzzy relative of the isodata process and its use in detecting compact well-separated clusters, *Journal of Cybernetics* 3(3): 32–57.

EPA (2008). Air quality and health, chapter Environmental Protection Agency, National Ambient Air Quality Standards (NAAQS).

Fenger, J. (2009). Air pollution in the last 50 years - from local to global, *Journal of Atmospheric Environment* 43(1): 13–22.

Hoppener, F., Klawonn, F., Kruse, R. & Runkler, T. (2000). *Fuzzy Cluster Analysis, Methods for classification, data analysis and image recognition*, Chistester, United Kingdom.

INE (2004). *Programa para mejorar la calidad del aire en Salamanca*, 2 edn, Instituto de Ecología del Estado de Guanajuato, Calle Aldana N.12, Col. Pueblito de Rocha, 36040 Guanajuato, Gto.

INEGI (2005). *National Population and Housing Census 2*, National Institute of Geography and Statistics. www.inegi.org.mx.

Krishnapuram, R. & Keller, J. (1993). A possibilistic approach to clustering, *International Conference on Fuzzy Systems* 1(2): 98–110.

Krishnapuram, R. & Keller, J. (1996). The possibilistic c-means algorithm: Insights and recommendations, *International Conference on Fuzzy Systems* 4, no 3: 385–393.

Ojeda-Magaña, B., Quintanílla-Dominguez, J., Ruelas, R. & Andina, D. (2009b). Images sub-segmentation with the pfcm clustering algorithm, *Proceedings of The 7th IEEE International Conference on Industrial Informatics (INDIN 09)*, pp. 499–503.

Ojeda-Magaña, B., Ruelas, R., Buendía-Buendía, F. & Andina, D. (2009a). A greater knowledge extraction coded as fuzzy rules and based on the fuzzy and typicality degrees of the GKPFCM clustering algorithm, *In Intelligent Automation and Soft Computing* 15(4): 555–571.

Pal, N. R., Pal, S. K. & Bezdek, J. C. (1997). A mixed c-means clustering model, *IEEE International Conference on Fuzzy Systems, Spain*, pp. 11–21.

Pal, N. R., Pal, S. K., Keller, J. M. & Bezdek, J. C. (2004). A new hybrid c-means clustering model., *Proceedings of the IEEE International Conference on Fuzzy Systems, FUZZ-IEEE04, I. Press, Ed.*

Pal, N. R., Pal, S. K., Keller, J. M. & Bezdek, J. C. (2005). A possibilitic fuzzy c-means clustering algorithm, *IEEE Transactions on Fuzzy Systems* 13(4): 517–530.

Pérez, P., Trier, A. & Reyes, J. (2000). Prediction of pm 2.5 concentrations several hours in advance using neural networks in santiago, chile, *Atmospheric Environment* 34(8): 1189–1196.

Ruspini, E. (1970). Numerical method for fuzzz clustering, *Information Sciences* 2(3): 319–350.

Timm, H., Borgelt, C., Döring, C. & Kruse, R. (2004). An extension to possibilistic fuzzy cluster analysis, *Fuzzy Sets and systems* 147, no 1: 3–16.

Timm, H. & Kruse., R. (2002). A modification to improve possibilistic fuzzy cluster analysis., *Conference Fuzzy Systems, FUZZ-IEEE, Honolulu, HI, USA.*

WHO (2008). *Air quality and health*, chapter World Health Organization.

Zamarripa, A. & Sainez, A. (2007). *Medio Ambiente: Caso Salamanca*, Instituto de Investigación Legistativa, H. Congreso del Estado de Guanajuato, LX legislatura.

Zuk, M., Cervantes, M. G. T. & Bracho, L. R. (2007). Tercer almanaque de datos y tendencias de la calidad del aire en nueve ciudades mexicanas, *Technical report*, Secretaría de Medio Ambiente, Recursos Naturales Instituto Nacional de Ecología, México, D.F.

5

Real-Time In Situ Measurements of Industrial Hazardous Gas Concentrations and Their Emission Gross

F.Z. Dong et al.*
Anhui Institute of Optics and Fine Mechanics, Chinese Academy of Sciences, Science Island, Hefei, P. R. China

1. Introduction

Over the past few decades environmental protection has been of greatly worldwide concerns due to the fact of global warming and air quality deterioration particularly in the fast developing countries like China and India (Platt, 1980; Edner, 1991; Sigrist, 1995; Culshaw, 1998; Fried, 1998; Linnerud, 1998; Weibring, 1998; Nelson, 2002; Liu, 2002; Christian, 2003 & 2004; Taslakov, 2006; de Gouw, 2007; Karl, 2007 & 2009; http://www.cnemc.cn). These have resulted in large demands and tremendous efforts for new technology developments to monitor and control industrial gas pollution (Lindinger, 1998; Dong, 2005; Kan, 2006 & 2007; Wang Y.J., 2009; Wang F., 2010; Xia, 2010; Zhang, 2011). CO_2, CO, NH_3, H_2S, HF, HCI, and volatile organic compounds (VOCs) are very important gases generated in many industrial processes; therefore to implement on-line monitoring of these industrial emitted gases is a key factor for industrial process control. Furthermore if one can simultaneously measure the gas flow path-averaged velocity and gas concentrations in a smokestack, all the industrial emissions from the targeted smokestack would be real-time obtained. This could be much beneficial to the administrative implementation of global environmental protection policy on reduction of gas pollution and environmental management.

Tunable diode laser absorption spectroscopy (TDLAS) is a kind of technology with advantages of high sensitivity, high selectivity and fast responsibility. It has been widely used in the applications of green-house measurements (Feher, 1995; Nadezhdinskii, 1999; Kan, 2006), hazardous gas leakage detection (May, 1989; Uehara, 1992; Iseki, 2000 & 2004), industry process control (Linnerud, 1998; Deguchi,2002) and combustion gas measurements (Zhou, 2005; Rieker, 2009). Proton transfer reaction—mass spectrometry (PTR-MS) is a relatively new technology firstly developed at the University of Innsbruck, Austria, in the 1990s (Hansel, 1995). PTR-MS has been found being an extremely powerful and promising technology for on-line detection of VOCs at trace level (Smith, 2005; Jordan, 2009). Optical flow sensor (OFS-2000) based on the concept of optical scintillation to measure airflow velocity (Wang T.I., 1981;

---

* W.Q. Liu, Y.N. Chu, J.Q. Li, Z.R. Zhang, Y. Wang, T. Pang, B. Wu, G.J. Tu, H. Xia, Y. Yang, C.Y. Shen, Y.J. Wang, Z.B. Ni and J.G. Liu
Anhui Institute of Optics and Fine Mechanics, Chinese Academy of Sciences, Science Island, Hefei, P. R. China.

http://www.opticalscientific.com), which is first developed by Optical Scientific INC., has been widely used in the market. OFS-2000 utilizes the high frequency signal of optical scintillation cross-correlation (OSCC) which is from the fluctuations of temperature or refractive index. However, OFS-2000 is not applicable when the temperature fluctuation within the measurement area is small or even ignorable. Recently we have developed a new kind of optical flow sensor which is based on the low frequency signal of OSCC resulting from the particle concentration fluctuations. Therefore the newly developed optical flow sensor could also measure the particle concentration in the stack.

The content of this chapter will first briefly describe the operational principles based on TDLAS, PTR-MS and OSCC technologies for industrial pollution on-line monitoring. Then the instruments developed by our group to measure the emission gross will be introduced. In the third section some experimental results from the field test will be presented. Finally the discussions and conclusions will be given.

2. Basic operational principles of the instruments

2.1 TDLAS technique

For detecting low concentration gases at atmospheric pressure with TDLAS technique, two-tone modulations and harmonic detection method are commonly adopted. The diode laser is modulated with the homemade current and temperature controllers to the wavelength of 1.567μm which precisely locates at the selected absorption line central of target gas CO. The laser wavelength is scanned through the selected absorption line by a saw-tooth signal at low frequency of 147Hz and simultaneously modulated by a sinusoidal signal at frequency of 20 KHz. The modulated laser beam is divided into two parts with a 1×2 fiber splitter. One arm (20%) is used to go through a 10cm calibration cell as a reference signal, while the other arm (80%) is used to measure the flue gas concentrations. Two transmitted laser beams are collimated and then collected by two coincident InGaAs photodiodes after passing through absorption gases, respectively. These two current signals are then transmitted into the digital control module (DCM) to gain the harmonic signals. At last, these signals are sent to computer for processing and harmonic signal detection technique is used for calculation of the target gas concentration. The schematic diagram of the online TDLAS experimental setup is shown in Figure 1.

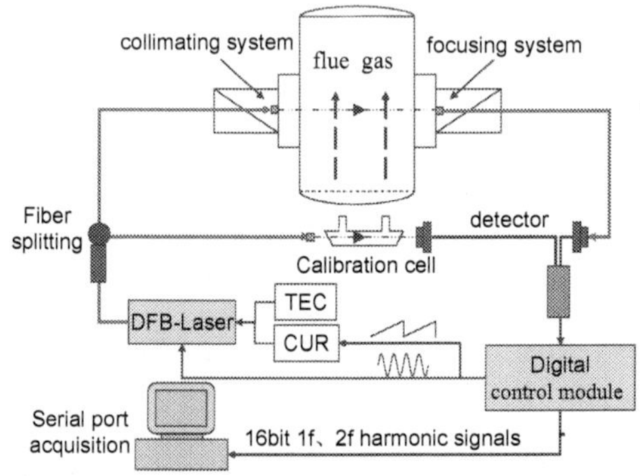

TEC: thermo-electronic cooler; CUR: current controller.

Fig. 1. On-line experimental apparatus for TDLAS system.

When the light passes through flue gases, lots of factors can reduce the light intensity, like dust scattering and absorption in transmission medium. Considering about the intensity reduction by gas absorption, Beer-Lambert law is used. The responses can be described as:

$$\frac{I}{I_0} = \exp(-kL) \tag{1}$$

Where I represents the light intensity after passing the absorption gas, and I_0 represents the light intensity before passing the absorption gas, k is a reducing coefficient and L denotes the path length. When the gas absorption is very small, i.e., $kL \leq 0.05$ (Reid, 1981; Cassidy, 1982), equation (1) can be simplified as:

$$\frac{I}{I_0} = 1 - kL = 1 - \sigma(\nu)CL \tag{2}$$

Where $\sigma(\nu)$ is the absorption coefficient. C and L stand for gas concentration and total optical length. The intensity of second harmonic (2f) signal can be expressed as below (Reid, 1981; Kan, 2006):

$$I_{2f} \propto I_0 \sigma_0 CL \tag{3}$$

Where I_{2f} is proportional to the incident laser intensity I_0 and absorption coefficient σ_0 at the central wavelength of the absorption line.. Nonlinear least square multiplication method is used to fit the 2f signal with reference signal for gaining the calibration coefficient a (Kan, 2007):

$$I_{01} C_{Mea} L_{01} = a\, I_{02}\, C_{Ref}\, L_{02} \tag{4}$$

Where C_{Mea} and C_{Ref} are the concentrations of the target gas to be measured and reference gas in the calibration cell, respectively; I_{01}, I_{02} are the initial intensities of the two laser beams; L_{01} and L_{02} are the length of measurement optical path and the calibration cell, respectively. From equation (4), we could obtain:

$$C_{Mea} = a\, I_{02}\, C_{Ref}\, L_{02} \,/\, I_{01} L_{01} \tag{5}$$

While a saw-tooth current is added on the DFB diode laser, the light wavelength will scan in a certain region, then the gas can be detected if there is a gas absorption line in that region. For detection of high concentration gas, direct absorption method is often used. This method is very simple but the sensitivity is suffered from massive random noises, which is mainly the 1/f noise from the diode laser and the photon detector. However, for low concentration gas detection, in order to eliminate serious noises in the system and enhance the sensitivity, another high frequency sine modulation current is added on the ramp signal. The gas absorption signal can be then achieved with high SNR by monitoring the second harmonic signal of absorption in a very narrow frequency band using a lock-in amplifier (LIA). If one does not pay enough attention, there will be so many factors like dust scattering and imperfect performance of laser source itself affecting the measurement accuracy. In addition, for a practical TDLAS system there are always various noises inevitably existed resulting from predictable or unpredictable sources. For instance, quickly changing random noise affects the sensitivity, and slow signal distortion limits long-term stability of the system because of its large amplitude. It has been reported that a lot of reasons like wavelength drifts and etalon fringe structure change because of thermal effect can result to slow 2f signal distortion (Werle, 1996). Few technologies had been reported to eliminate those distortions like rapid background subtraction (Cassidy & Reid, 1982) and digital signal processing (Reid, 1980), but there are some limits of those ideas when the condition is changed. In fact it is inconvenient to get the background structure in real time for a in situ gas analytical system, particularly when the interference or distortion has similar frequency with the absorption signal in which the digital method could not work well.

Over the past decades many advanced digital signal processing methods for TDLAS system development have been reported. Peter Werle et al (Werle, 1996 & 2004) have demonstrated a method to avoid the effects of noise disturbances and laser wavelength drifts during integration and background changes. To decrease high frequency noise and enhance the stability of a practical TDLAS system, except of optimizing hardware, advanced signal processing algorithm is also needed and have been explored by our group (Xia, 2010 ; Zhang, 2010). One of the novel features in our research is the use of digital signal processing for harmonic signals for which the laser output wavelength can be locked at the absorption line center and fit with reference harmonic signal by utilizing nonlinear least squares routine. The signal-correlation must be computed rapidly. The Fast Fourier Transform (FFT), low-pass filter and Inverse Fast Fourier Transform (IFFT) algorithm are adopted. The correlation version for an N-point spectrum signal is:

$$C(S,S)_i = S^{Ref} \otimes S^{Mea} = \sum_{j=0}^{N} S_j^{Ref} \cdot S_{i+j}^{Mea} \tag{6}$$

Where S^{Ref} and S^{Mea} are the reference and measurement signals acquired during calibration and subsequent measured harmonic signal, respectively with the lag represented by i. Using the discrete FFT the correlation signal $C(S, S)_i$ can be written as:

$$C(S,S)_i = F_j(S^{Ref})F_j^*(S^{Mea}) \tag{7}$$

where $F_j(S)$ stands for the FFT of S. The low pass filter is used to remove high frequency noise simultaneously in the process. The IFFT result between measured signal and reference signal in the above process is used to get the correlation data. Then using the peak-find routine the drift MAX-value position is obtained. At last, the corrected signal position is translated getting the proper data to decrease effects caused by the temperature, current and other external uncertain factors.

2.2 PTR-MS

Proton transfer reaction mass spectrometry (PTR-MS) was first developed at the Institute of Ion Physics of Innsbruck University in the 1990's. Nowadays PTR-MS has been a well-developed and commercially available technique for the on-line monitoring of trace volatile organic compounds (VOCs) down to parts per trillion by volume (ppt) level. PTR-MS has some advantages such as rapid response, soft chemical ionization (CI), absolute quantification and high sensitivity. In general, a standard PTR-MS instrument consists of external ion source, drift tube and mass analysis detection system. Fig. 2 illustrates the basic composition of the PTR-MS instrument constructed in our laboratory using a quadrupole mass spectrometer as the detection system.

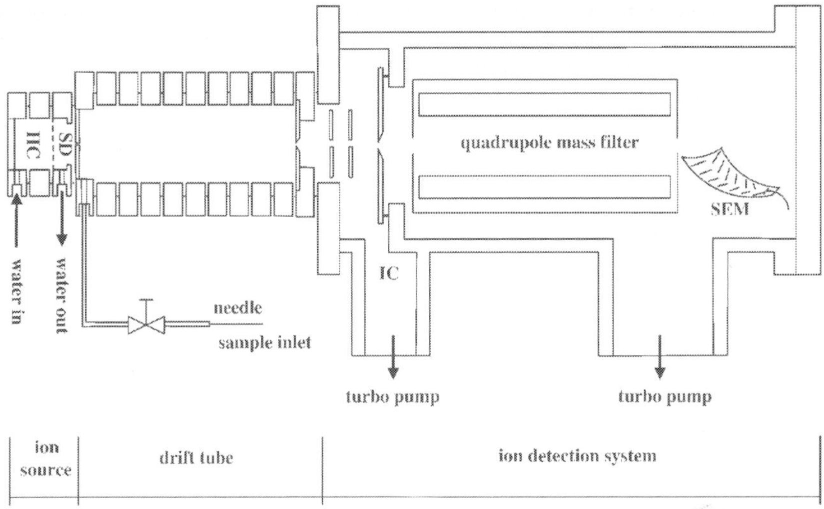

Fig. 2. Schematic diagram of the PTR-MS instrument that contains a hollow cathode (HC), a source drift (SD) region, an intermediate chamber(IC) and a secondary electron multiplier (SEM).

Perhaps the most remarkable feature of PTR-MS is the special chemical ionization (CI) mode through well-controlled proton transfer reaction, in which the neutral molecule M may be converted to a nearly unique protonated molecular ion MH^+. This ionization mode is completely different from the traditional MS where electron impact (EI) with energy of 70 eV is often used to ionize chemicals like VOCs. Although the EI source has been widely used with the commercial MS instruments most coupled with a variety of chromatography techniques, these MS platforms have a major deficiency: in the course of ionization the molecule will be dissociated to many fragment ions. This extensive fragmentation may result in complex mass spectra pertain especially when a mixture is measured. If a chromatographic separation method is not used prior to MS, then the resulting mass spectra from EI may be so complicated that identification and quantification of the compounds can be very difficult. In PTR-MS instrument, the hollow cathode discharge is served as a typical ion source [Blake, 2009], although plane electrode dc discharge [Inomata, 2006] and radioactive ionization sources [Hanson, 2003] recently have been reported. All of the ion sources are used to generate clean and intense primary reagent ions like H_3O^+. Water vapor is a regular gas in the hollow cathode discharge where H_2O molecule can be ionized according to the following ways (Hansel, A.,1995).

$$e + H_2O \rightarrow H_2^+ + O + 2e \tag{8}$$

$$e + H_2O \rightarrow H^+ + OH + 2e \tag{9}$$

$$e + H_2O \rightarrow O^+ + H_2 + 2e \tag{10}$$

$$e + H_2O \rightarrow H_2O^+ + 2e \tag{11}$$

The above ions are injected into a short source drift region and further react with H_2O ultimately leading to the formation of H_3O^+ via ion-molecule reactions:

$$H_2^+ + H_2O \rightarrow H_2O^+ + H_2 \quad (12a)$$

$$\rightarrow H_3O^+ + H \quad (12b)$$

$$H^+ + H_2O \rightarrow H_2O^+ + H \quad (13)$$

$$O^+ + H_2O \rightarrow H_2O^+ + O \quad (14)$$

$$OH^+ + H_2O \rightarrow H_3O^+ + O \quad (15a)$$

$$\rightarrow H_2O^+ + OH \quad (15b)$$

$$H_2O^+ + H_2O \rightarrow H_3O^+ + OH \quad (16)$$

Unfortunately, the water vapor in the source drift region can inevitably form a few of cluster ions $H_3O^+(H_2O)_n$ via the three-body combination process

$$H_3O^+(H_2O)_{n-1} + H_2O + A \rightarrow H_3O^+(H_2O)_n + A \quad (n \geq 1) \quad (17)$$

where A is a third body. In addition there are small amounts of NO^+ and O_2^+ ions occurred due to sample air diffusion into the source region from the downstream drift tube. Thus an inlet of venturi-type has been employed on some PTR-MS systems to prevent air from entering the source drift region (Duperat, 1982; Lindinger, 1998). At last the H_3O^+ ions produced in the ion source can have the purity up to > 99.5%. Thus, unlike SIFT-MS technique (Smith, 2005), the mass filter of the primary ionic selection is not needed and the H_3O^+ ions can be directly injected into the drift tube. In some of PTR-MS, the ion intensity of H_3O^+ is available at 10^6~10^7 counts per second on a mass spectrometer installed in the vacuum chamber at the end of the drift tube. Eventually the limitation of detection of PTR-MS can reach low ppt level.

Instead of H_3O^+, other primary reagent ions, such as NH_4^+, NO^+ and O_2^+, have been investigated in PTR-MS instrument (Wiche, 2005; Blake, 2006; Jordan, 2009). Because the ion chemistry for these ions is not only proton transfer reaction, the technique sometimes is called chemical ionization reaction mass spectrometry. However, the potential benefits of using these alternative reagents usually are minimal, and to our knowledge, H_3O^+ is still the dominant reagent ion employed in PTR-MS research (Blake, 2009; Lindinger, 1998; de Gouw, 2007; Jin, 2007).

The drift tube consists of a number of metal rings that are equally separated from each other by insulated rings. Between the adjacent metal rings a series of resistors is connected. A high voltage power supplier produces a voltage gradient and establishes a homogeneous electric field along the axis of the ion reaction drift tube.

The primary H_3O^+ ions are extracted into the ion reaction region and can react with analyte M in the sample air, which through the inlet is added to the upstream of the ion reaction drift tube. According to the values of proton affinity (PA) (see Table 1), the reagent ion H_3O^+ does not react with the main components in air like N_2, O_2 and CO_2. In contrast, the reagent ion can undergo proton transfer reaction with M as long as the PA of M exceeds that of H_2O (Lindinger, 1998).

$$M+H_3O^+ \rightarrow MH^{+}+H_2O \quad (18)$$

Thus, the ambient air can be directly introduced to achieve an on-line measurement in the PTR-MS operation. Due to the presence of electric field, in the reaction region the ion energy is closely related to the reduced-field *E/N*, where *E* is the electric field and *N* is the number density of gas in the drift tube. In a typical PTR-MS measurement, *E/N* is required to set to an appropriate value normally in the range of 120~160 Td ($1Td=10^{-17}$ Vcm^2 $molecule^{-1}$) which may restrain the formation of the water cluster ions $H_3O^+(H_2O)_n$ (n=1-3) to avoid the ligand switch reaction with analyte M (Lindinger, 1998):

$$H_3O^+(H_2O)_n+M \rightarrow H_3O^+(H_2O)_{n-1}M+H_2O \quad (19)$$

However, a higher reduced-field *E/N* can cause the collision-induced dissociation of the protonated products, thereby complicating the identification of detected analytes.

Compound	Molecular formula	Molecular weight	Proton affinity (NIST database) (kJ mol^{-1})
Helium	He	4	177.8
Neon	Ne	20	198.8
Argon	Ar	40	369.2
Oxygen	O_2	32	421
Nitrogen	N_2	28	493.8
Carbon dioxide	CO_2	44	540.5
methane	CH_4	16	543.5
Carbon monoxide	CO	28	594
Ethane	C_2H_6	30	596.3
Ethylene	C_2H_4	28	680.5
Water	H_2O	18	691
Hydrogen sulphide	H_2S	34	705
Hydrogen cyanide	HCN	27	712.9
Formic acid	HCOOH	46	742
Benzene	C_6H_6	78	750.4
Propene	C_3H_6	42	751.6
Methanol	CH_3OH	32	754.3
Acetaldehyde	CH_3COH	44	768.5
Ethanol	C_2H_5OH	46	776.4
Acetonitrile	CH_3CN	41	779.2
Acetic acid	CH_3COOH	60	783.7
Toluene	C_7H_8	92	784
Propanal	CH_3CH_2COH	58	786
O-xylene	C_8H_{10}	106	796
Acetone	CH_3COCH_3	58	812
Isoprene	$CH_2C(CH_3)CHCH_2$	68	826.4
Ammonia	NH_3	17	853.6
Aniline	C_6H_7N	93	882.5

Table 1. Proton affinities of some compounds

At the end of the drift tube there is an intermediate chamber in which most of the air from the drift tube through a small orifice is pumped away. The ions in the drift tube are extracted and focused by the ion optical lens and finally in a high vacuum chamber are detected by a quadrupole mass spectrometer with ion pulse counting system. The ionic count rates $I(H_3O^+)$ and $I(MH^+)$ are measured in counts per second (CPS), which are proportional to the respective densities of these ions. Although quadrupole mass filter is a traditional analyzer in the current PTR-MS instrument, other MS analyzers have been investigated including time-of-flight (TOF) (Blake, 2004; Ennis, 2005; Jordan, 2009), ion trap (Prazeller, 2003) and linear ion trap mass spectrometer (Mielke, 2008).
Normally, PTR-MS can determine the absolute concentrations of trace VOCs according to well-established ion-molecular reaction kinetics. If trace analyte M reacts with H_3O^+, then the H_3O^+ signal does not decline significantly and can be deemed to be a constant. Thus, the density of product ions $[MH^+]$ at the end of the drift tube is given in Eq.20 (Lindinger, 1998).

$$[MH^+] = [H_3O^+]_0 (1 - e^{-k[M]t}) \tag{20}$$

Where $[H_3O^+]_0$ is the density of reagent ions at the end of the drift tube in absence of analyte M, k is the reaction rate constant of reaction (18) and t is the average reaction time the ions spending in the drift tube. In the trace analysis case, $k[M]t << 1$, Eq.(20) can be further deduced to the following form.

$$[M] = \frac{[MH^+]}{[H_3O^+]_0} \frac{1}{kt} \tag{21}$$

Eq.21 is often used in a conventional PTR-MS measurement. However, when the concentration of analyte M is rather high, the intensity change of reagent ions H_3O^+ is not ignorable. In this case, the relation $k[M]t << 1$ is not tenable, therefore the regular Eq.21 is no longer suitable for concentration determination. For a more reliable measurement, the following Eq.22, deduced from Eq.20, can be used to determine the concentration of analyte M. For instance, the concentrations of gaseous cyclohexanone inside the packaging bags of infusion sets were found to be rather high, and its concentrations at several tens of ppm level could be detected according to Eq.22 (Wang Y.J., 2009).

$$[M] = \ln \frac{[H_3O^+]_0}{[H_3O^+]_0 - [MH^+]} \frac{1}{kt} \tag{22}$$

In PTR-MS instrument, the signal intensities of primary and product ions can be measured. And the reaction time can be derived from the instrument parameters and the reaction rate constant can be found in literatures for most substances or calculated by the theoretical trajectory model (Chesnavich, 1980; Su, 1982) using dipole moment and polarizability. Thus the absolute concentration of trace component can be easily obtained without calibration.

2.3 Optical scintillation

The industrial stack gas is one of the major sources of particulate matter and pollution in the atmosphere. With the high speed development of economy, this situation will exist for a long time. It plays an important role in the environmental management and pollution control to monitor exhaust gas continuously. Using optical scintillation caused by stack gas flow to measure velocity has greater advantage than some traditional velocity measurement techniques, such as Pitot tube, hot wire anemometry and laser Doppler velocimeter (LDV). However, the corresponding theory is not consummate yet.

A light beam passes through the stack gas flow in an industrial setup, the light intensity will fluctuate due to a variety of reasons. First of all, particles move in or out the view of sight in random will induce optical intensity fluctuations (Chen, 1999 & 2000; Yuan, 2003). This optical scintillation made by particle concentration statistical fluctuations can only be observed when the view of sight is small, the optical path is short, the particle diameter is large and the concentration is low. Commonly, large size apertures of transmitter and receiver are used to measure optical scintillation in the large stack of factory, this kind of scintillation signals is rarely used for measurements of gas flow velocity. Secondly, in high temperature stack gas flow, the refractive index is affected by the turbulence, and it will fluctuate in both the temporal and spatial domains. The characteristic frequency of scintillation caused by the above two reasons can be expressed as (Ishimaru, 1986; Andrews, 2000):

$$f \approx \frac{\upsilon}{D_r} \tag{23}$$

Where υ is the mean velocity, D_r is the diameter of the receiver's aperture. If υ =10m/s, D_r <1mm, the characteristic frequency is above 10^4 Hz. The frequency of optical scintillation caused by turbulence is higher and reaches hundreds or thousands Hz. There has been a technique (Wang, T.I., 2003) which uses the scintillation signals of high frequency caused by refractive index fluctuations to measure velocity of stack gas flow, and the refractive index fluctuations is determined by temperature field gradient. It would be difficult to measure velocity when temperature field distributes uniformly.

The fluctuations of particle concentration field can also cause optical scintillation in low frequency range which is commonly below than tens Hz. In the low frequency part of optical scintillation spectrums, the scintillation intensity shows good linearity with particle concentration. This linearity has been used to measure particle concentration (Клименко, 1984). The low frequency of optical scintillation that caused by stack gas flow is relative to the particle concentration fluctuations at random, and it is an experiential knowledge, but this problem still need further investigations in theory.

The scintillation signals of low frequency caused by particle concentration fluctuations are employed in this research work, and parallel double transceiver technique is adapted to measure the velocity and particle concentration of stack gas flow. In this case, even if the temperature field distributes uniformly and refractive index fluctuation is weak, the velocity and particle concentration could still be measured at the same time. The received optical scintillation signal is analyzed and the result illustrates that the power ratio of optical scintillation spectrum in part of low frequency is -8/3.

The signals are received in manner of Fig.3. The emitted light beams are divergent spherical waves, and both beams propagate along x -axis and their origin are both at $x=0$. The diameter of transmitter aperture is D_t and the diameter of the two receivers is D_r. The distance between transmitter and receiver is L, and the distance between the two receivers is l. The direction of stack gas flow is y axis, the mean velocity is υ. The system with two point source trasmitters and two point receivers is discussed here.

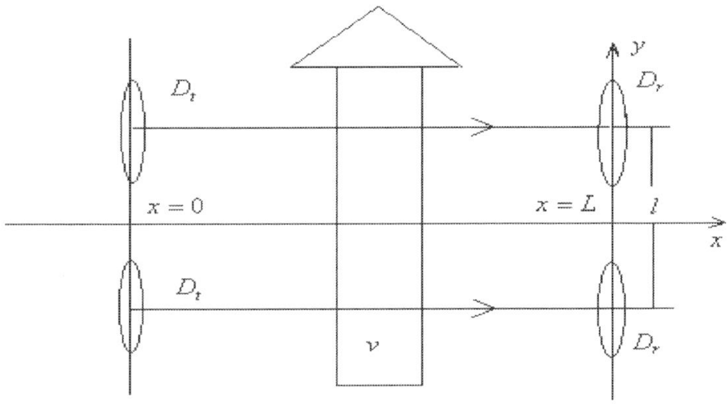

Fig. 3. The layout of optical scintillation measurement

Let the extinction coefficient of stack flow be $\alpha(r,t)$, according to the law of Beer-Lambert, the received logarithmic light intensity is

$$\ln I(t) = \ln < I > - \int_0^L \alpha'(r,t)dx \tag{24}$$

where $<\cdot>$ is the assemble average, $\alpha'(r,t)$ is the perturbation part.

The cross-correlation function of the two scintillation signals received by two independent receivers can be written as :

$$C_{\ln I}(r,\rho,t,\tau) = < \int_0^L \alpha'(r_1, t-\tau)dx_1 \int_0^L \alpha'(r_2,t)dx_2 > \tag{25}$$

where τ is time delay. For homogeneous isotropic time stationary turbulence, the correlation function is only relative to the distance of the two receivers and time delay, then the cross-correlation function is

$$C_{\ln I}(\rho,\tau) = \int_0^L \int_0^L R_\alpha(r_1 - r_2, \tau)dx_1 dx_2 \,, \tag{26}$$

where $R_\alpha(r_1 - r_2, \tau)$ is the correlation function of extinction coefficient. Because of the movement of the stack gas along y-axis, according to Taylor frozen turbulence hypothesis, and by the geometric relations shown as Fig.3, we obtain:

$$R_\alpha(r_1 - r_2, \tau) = R_\alpha(x_1 - x_2, l - v\tau, 0)\,, \tag{27}$$

Inserting Eq. (27) into Eq. (26), Eq. (26) reduces to

$$C_{\ln I}(l,\tau) = 2\int_0^L (L-x)R_\alpha(x, l - v\tau)dx \tag{28}$$

And

$$R_\alpha(x, l - v\tau) = \int\int\int_0^\infty \cos(\kappa_2(l - v\tau) + \kappa_1 x)\varphi_\alpha(\kappa)d\kappa_1 d\kappa_2 d\kappa_3 \tag{29}$$

where $\varphi_\alpha(\kappa)$ is the three-dimensional power spectrum of extinction coefficient fluctuations . For a stationary random process, correlation function can be expressed as

$$C_f(\tau) = \frac{1}{2}[D_f(\infty) - D_f(\tau)] \tag{30}$$

where f is a stationary random function, $D_f(\tau)$ is structure function .

The low frequency of optical scintillation caused by stack gas flow is relative to the particle concentration random fluctuations, meanwhile extinction coefficient is linear with particle concentration,

$$\alpha = K_m m\ , \tag{31}$$

where K_m is the relative extinction coefficient and it is concerned with the particle scale distribution and refractive index, m is particle concentration, the extinction coefficient fluctuations can be expressed as

$$\alpha' = K_m m'\ , \tag{32}$$

So we can start from the extinction coefficient fluctuations to discuss the spectrum characteristics of optical scintillation in stack gas flow. Suppose that the particle concentration obeys conservation law and passive scalar quantity, for sufficiently developed turbulence, the extinction coefficient structure function is

$$D_\alpha(r) = C_\alpha^2 r^{2/3}\ , \quad (l_0 << r << L_0) \tag{33}$$

where C_α^2 is the structure constant of extinction coefficient and r is the distance of two arbitrary points in turbulence field, l_0 and L_0 are the inner-scale and out-scale of turbulence, respectively.

Replacing ∞ with the out-scale of turbulence L_0 in Eq. (30), and insert Eq. (33) into Eq. (30), we then obtain：

$$R_\alpha(x, l - v\tau) = \frac{1}{2}C_\alpha^2(L_0^{2/3} - r^{2/3})\ , \tag{34}$$

where $r = \sqrt{x^2 + (l - v\tau)^2}$, while $r > L_0$, $R_\alpha = 0$.

Inserting Eq. (34) into Eq. (28),

$$C_{\ln I}(l, \tau) = C_\alpha^2 \int_0^L (L - x)(L_0^{2/3} - r^{2/3})dx\ , \ (r = \sqrt{x^2 + (l - v\tau)^2}\ ) . \tag{35}$$

Fig.4 shows the numerical simulation results of Eq. (35).

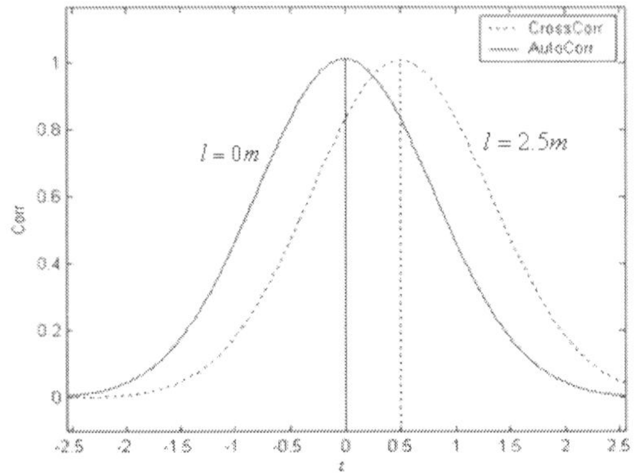

Fig. 4. The numerical computer simulations of Eq. (35). Here v =5m/s, L=2m and L_0=10m.

In Fig. 4, the time delay at the peak of the cross-correlation function is 0.5s, which is equal to l / v. So when we know the time delay at the peak of the cross-correlation function, the mean velocity of stack gas flow could be then easily obtained.

The reasons of optical scintillation caused by a light beam passing through stack gas flow are very complex. Particles move in or out the view of sight in random will induce optical intensity fluctuations, but it is hard to obtain the scintillation signals in industrial environment. Turbulence causes the optical scintillation, the frequency of this kind of scintillation commonly reaches hundreds or thousands Hz. Particle concentration fluctuations at random will also induce optical scintillation, and its frequency is commonly lower than hundreds Hz. As demonstrated above, the low frequency part of optical scintillation can be used to measure gas flow velocity and particle concentration simultaneously.

3. Brief description of the instruments

3.1 TDLAS instrument developed for hazard gas online monitoring

With the features of tunability and narrow line-width of distributed feedback (DFB) laser and by precisely tuning its wavelength to a single isolated absorption line of the target gas, TDLAS technique can be utilized to accurately perform online gas concentration monitoring with very high sensitivity. However, to develop a real practical TDLAS system with high sensitivity and reliability there are many works needed to be done. For instance, signal measurements with a sensitive device inevitably suffer from the predictable or unpredictable sources such as various noises, light intensity fluctuations and laser output wavelength dithers. In order to eliminate or at least reduce the measurement uncertainty and gain better reliability, a close-circle digital-control module (DCM) with functions of digital signal generator, digital lock-in-amplifier (D-LIA), data acquisition and processing have been developed.

The single-board DCM is tailored dedicatedly and specially designed for TDLAS applications in which several functions like digital lock-in amplifier, signal generator, data acquisition and processing are all included. In addition, a high precision temperature / current controller board and display board based on ARM 9 are also constructed. With the newly developed DCM, the total amount of PCB needed for a whole TDLAS system has been decreased from the previous 7 independent cards to 3. Moreover, DCM could set

TEC's parameters through software and a digital interface communicating DCM with TEC. In addition, DCM provides a serial port connecting with a host CPU. The host CPU (MCU or PC) transmits data to DCM setting the parameters, such as frequency, gain, time constant, phase, *1f* or *2f* selection. The host also receives harmonic signal data from DCM. Since the DCM has synchronized the data acquisition and signal generation, the received data are also packaged in onboard memory with 1024 points each period. Fig.5 is the picture of the developed TDLAS system.

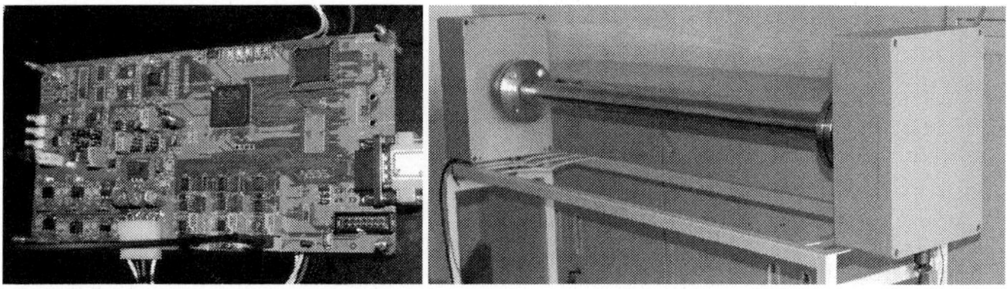

Fig. 5. Developed DCM and TDLAS system for online monitoring of industrial emitted hazard gases.

Though gas analysis based on tunable diode laser absorption spectroscopy (TDLAS) provides features of high sensitivity, fast response and high selectivity. However, many gaseous pollutants with generally low and variable concentrations and large local differences bring challenging requirements to analytical techniques. For example, when the target gas is CO and its concentration is below a few parts-per-million, the TDLAS system becomes more and more sensitive to noise, interference, drift effects and background changes associated with low level signals. Fig.6 shows typical second-harmonic absorption signals in detecting low concentration gas of CO under several noises in a practical TDLAS system. In this case it is very necessary to select proper signal processing and digital filtering technique to remove the effects of noise and distortion, and thus to improve the system performance (Xia, 2010). Fig.7 and Fig.8 show the effective signal improvement by employing wavelet transform method choosing proper wavelet basis and decomposition scale.

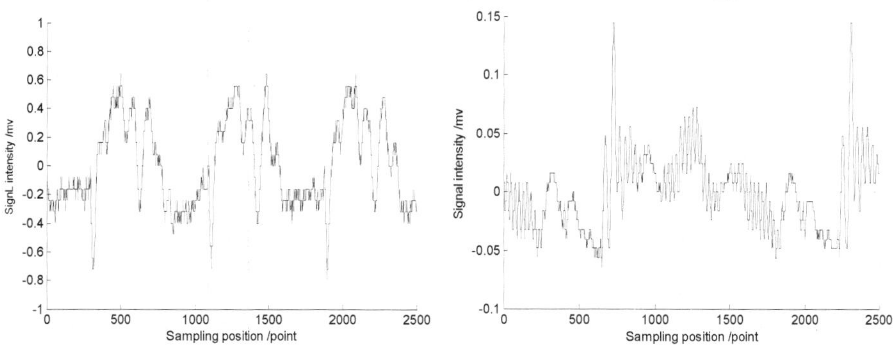

Fig. 6. Typical absorption signals under several noises in a practical TDLAS system.

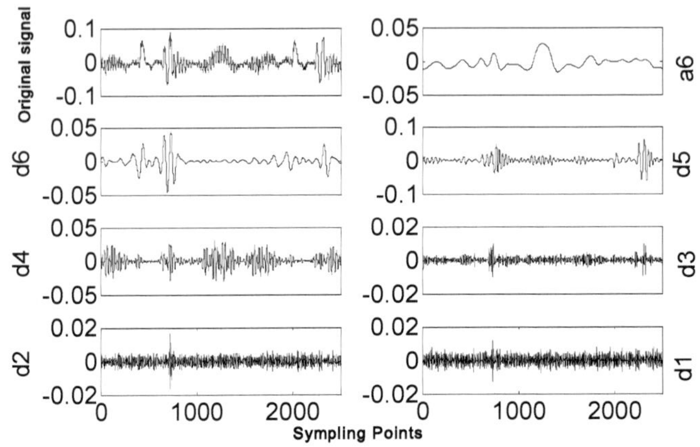

Fig. 7. Signal decompositions at different scales based on wavelet transform.

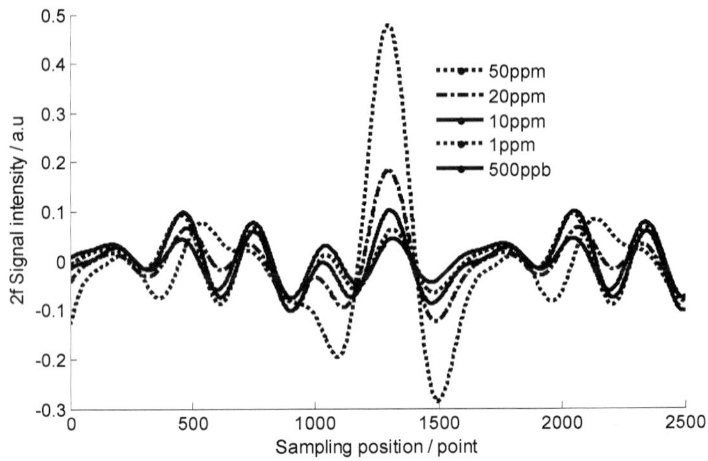

Fig. 8. The treated 2f signals after wavelet transform with proper wavelet basis and decomposition scale.

3.2 PTR-MS instrument constructed for VOCs monitoring

The widest application of PTR-MS is in the field of atmospheric monitoring. In air, VOCs originate from diverse sources but primarily from biogenic origin. Many VOCs have effects on the sources and sinks of ozone, aerosol formation and climate change. In addition, some VOCs are also toxic to human beings [Monks, 2005], so it is important to monitor their concentrations in a wider environments. Nowadays PTR-MS has been used to detect VOCs from plants, forest, human activities and industry processes. Fig.9 is the picture of the PTR-MS instruments we have developed recently for on-line monitoring of industrial emitted VOCs. The left is the standard PTR-MS instrument with detection sensitivity of ppb level.

The right one is the high sensitive PTR-MS instrument with detection sensitivity of ppt level. Fig.10 shows the experimental results for monitoring of acetone, benzene, acetaldehyde and toluene in laboratory with the developed PTR-MS instrument

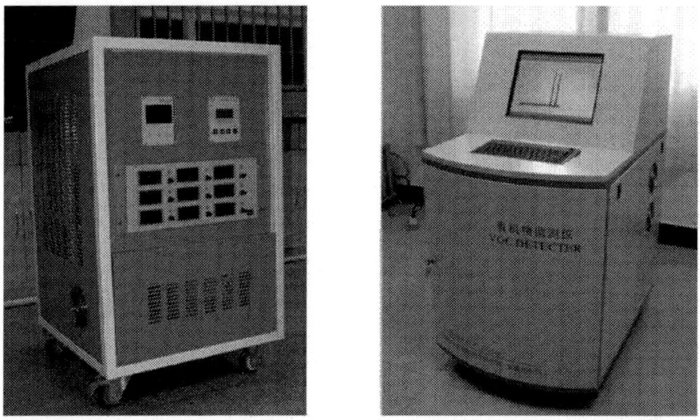

Fig. 9. A series of PTR-MS instruments developed for different requirements.

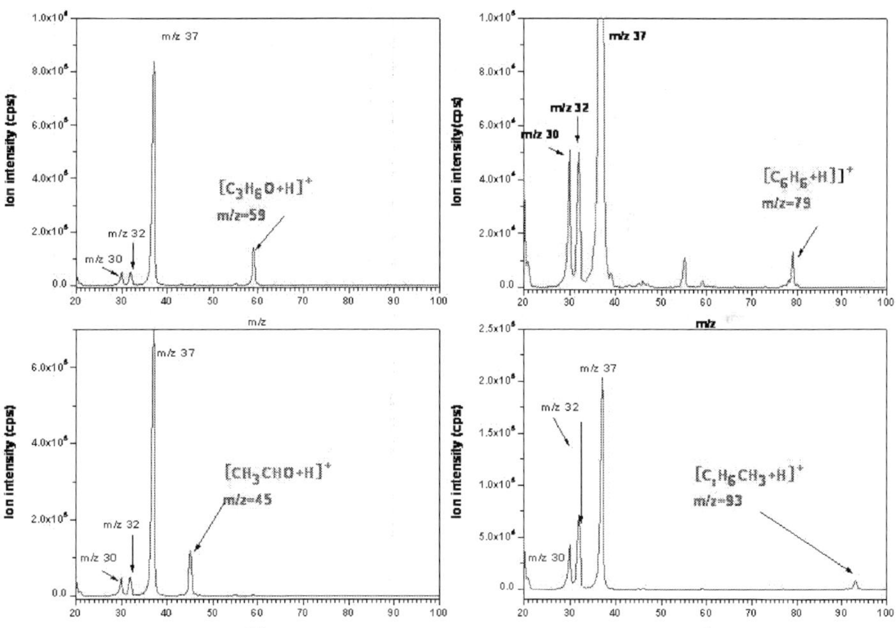

Fig. 10. The measured mass spectra of acetone, benzene, acetaldehyde and toluene in laboratory with the developed PTR-MS instrument.

3.3 OSCC instrument developed for gas flow velocity measurement

In order to measure gas flow velocity in stack, a gas flow velocity sensor was constructed based on the low frequency part of the double-path optical scintillation cross correlation. The schematic diagram of velocity and particle concentration measuring system is shown in Fig.11. Both processed LED light sources emit ideal Gauss spherical waves, the wavelength is 630 nm and the output power is 1 w. The receivers are silicon photoelectric diodes. The received signals will be magnified and filtered by low pass filter, then collected by a A/D card, and finally an industrial computer gets the data to process. Fig.12 shows the developed instruments. The left is the picture of instruments, and the right one is the picture installed on an industrial emission pipe for testing).

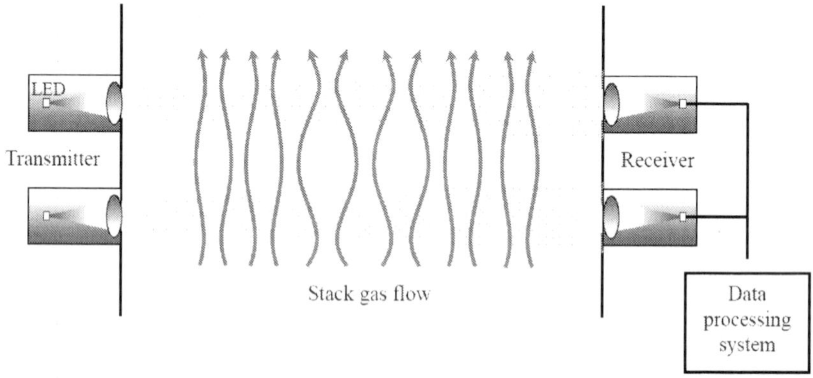

Fig. 11. The schematic diagram of velocity and particle concentration measuring system

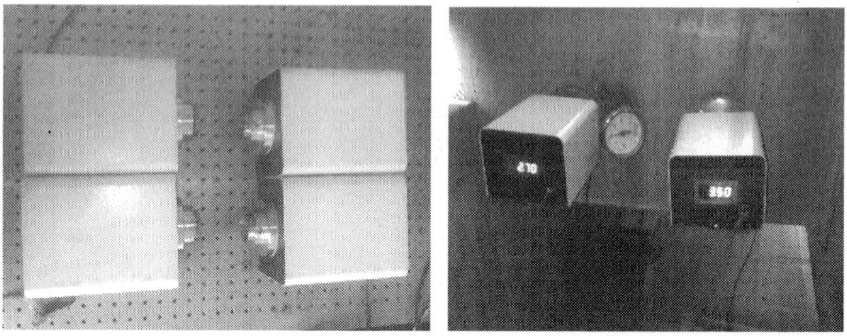

Fig. 12. Developed instrument for online measurements of gas flow velocity.

The field testing measurement was carried out at a chemical factory in Weifang of Shandong province. It is a rectangular stack with the length of 2 m and the width of 0.55 m. The distance between transmitter and receiver is the length of the stack, and the distance between the two receivers is 0.35 m. The diameters of transmitter aperture and receiver aperture are both 30mm. (as shown in the right picture of Fig.12). The velocity of the stack gas flow is about 4 m/s calibrated with a commercial Pitot tube and the temperature is about 150°C. The stack gas is produced from the burning of coal. Fig.13 is the received data plot from both receivers. Fig.14 shows that the power ratio of the optical scintillation spectrums in part of low frequency is -8/3.

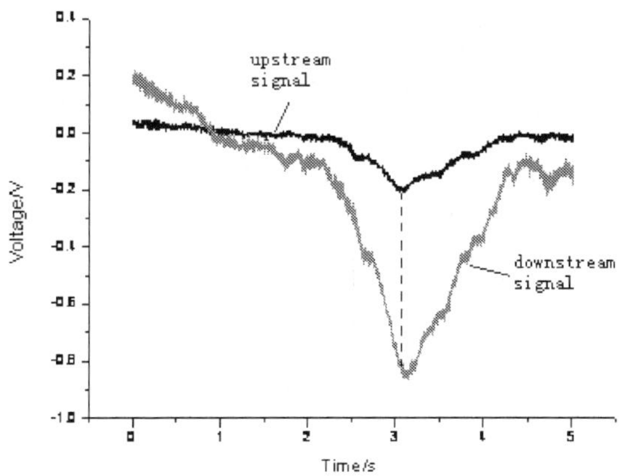

Fig. 13. The received data signal from both receivers

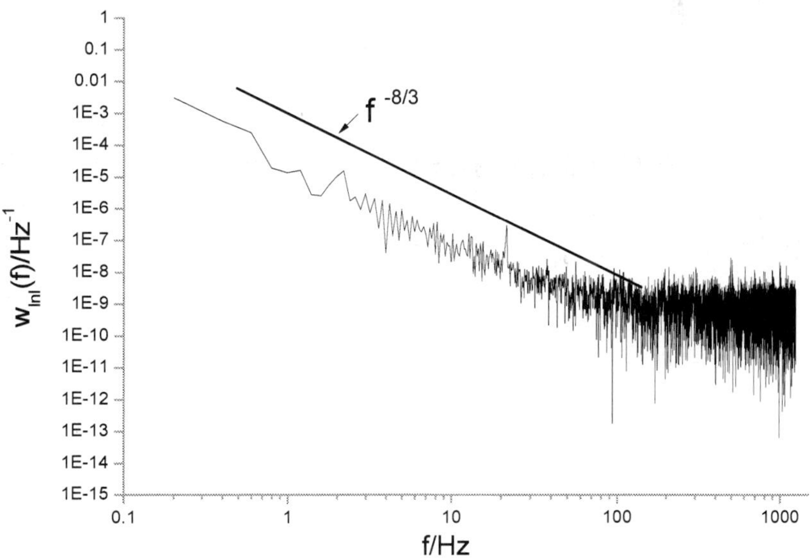

Fig. 14. The low frequency part of optical scintillation spectrum.

The low frequency of optical scintillation caused by stack gas flow is relative to the particle concentration fluctuations at random, the scintillation caused by the fluctuations of particle concentration is analyzed. Fig.15 shows the continuous measurement results of gas flow velocity which shows good agreement comparing with Pitot tube point measurement results.

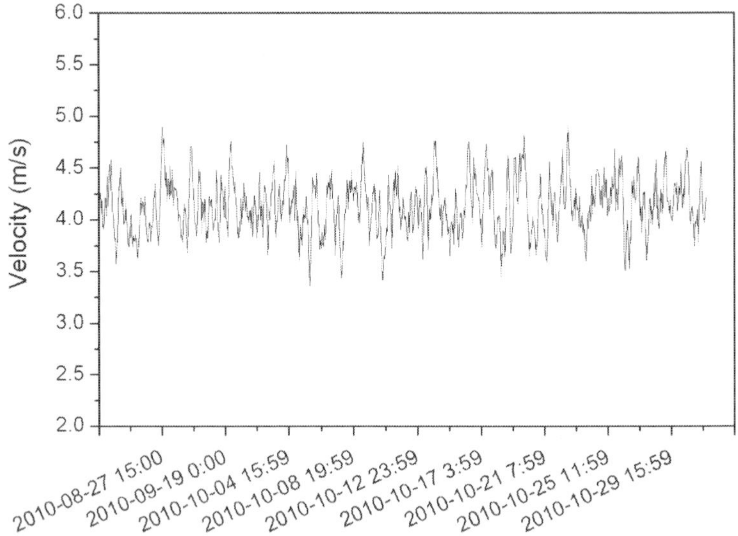

Fig. 15. Continuous measurement results of gas flow velocity with double-path OSCC sensor.

3.4 Online monitoring of industrial emission gross

The aim of the project we have carried out during the past few years is to develop a novel system to realize in site real-time monitoring the industrial emission gross. The idea is through on-line measurements of the targeted gas concentrations and gas flow velocity within a stack before emitted to air and plus with help of the theoretical path-weighted function built based on the configuration of stack cross section to gain *in situ* monitoring of industrial emission gross. Although most works including development of theoretical model and instrument constructions have been fulfilled. However, nowadays it is very hard for us to find a standard commercial instrument to certify the accuracy of our measurements and calculation. Obviously there are still lots of works needed to be further carried out. Here we still use the traditional method to calculate the emission gross presented in the section 4, i.e.,

$$F = M \times p \times (S \times v \times C) / RT \tag{36}$$

Where F is the emission gross, M is the molecular weight, S is the cross section of the stack at the measurement path, C and v are the measured gas concentration and mean gas flow velocity, respectively.

4. Experimental results and discussions

In order to demonstrate the developed instruments, a number of preliminary field trials have been carried out at few sites under different industrial field circumstances as shown in Fig.16 and Fig.17.

Fig. 16. One of the testing circumstances for in situ on-line monitoring of industrial emissions.

Fig. 17. Another testing place for in situ on-line monitoring of industrial emissions.

Fig.18-21 are the industrial field testing results employing the instruments described in this chapter.

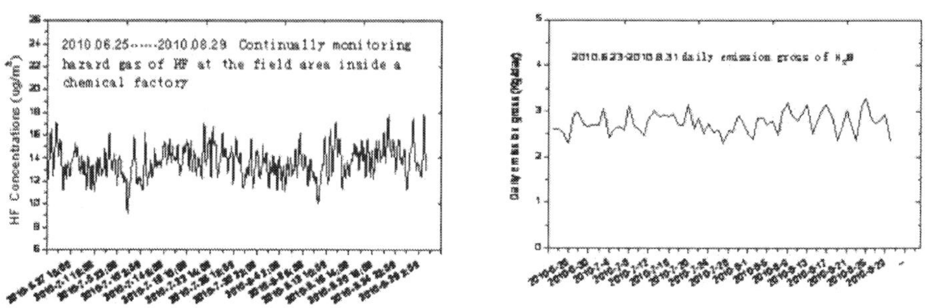

Fig. 18. The continually measurement results of industrial emitted HF and HCI with TDLAS instrument.

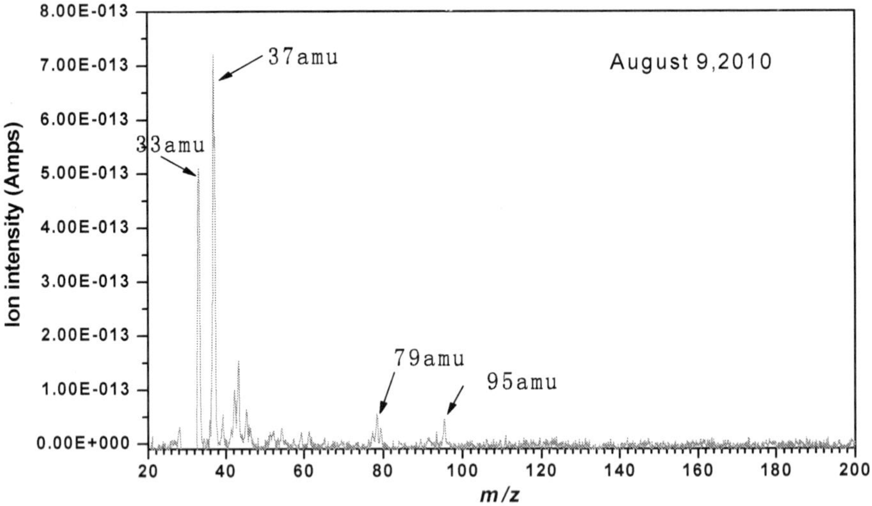

Fig. 19. The total mass scans of VOCs inside the stack measured with PTR-MS

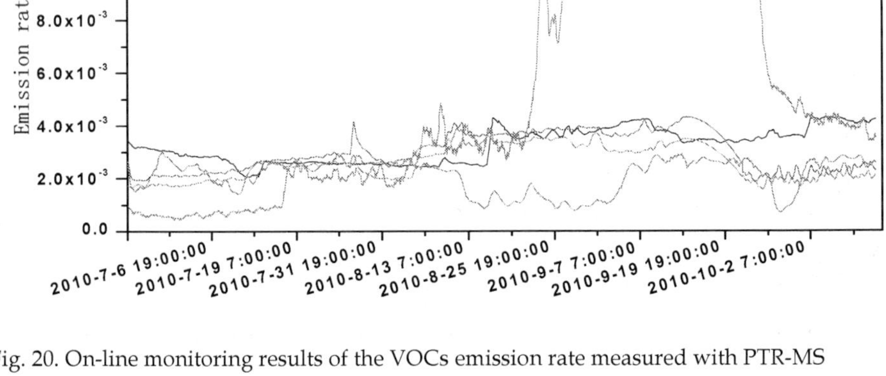

Fig. 20. On-line monitoring results of the VOCs emission rate measured with PTR-MS

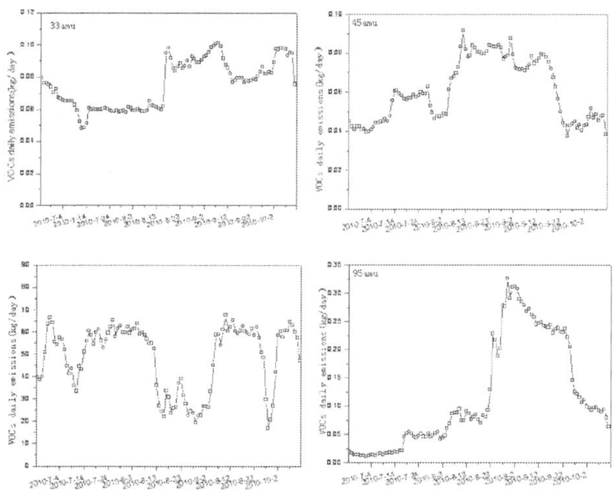

Fig. 21. The measurement results of daily emission gross for VOCs.

5. Conclusions

In conclusion, based on TDLAS, PTR-MS and OSCC techniques we have developed a system to monitor a number of industrial hazard gas emissions. However, it should point out here, the measurement results for on-line monitoring the gross of industrial emissions reported here are still in the early stage. Many further works need to be done and will be published in the future. For instance, we have developed a new complex theory based on path-weighted function and the averaged gas flow velocity to calculate the total emissions with the help of gas concentration measurements, however since nowadays there are no any instruments available to certify the measurement accuracy. Therefore in this chapter the common method for calculation of total emissions is still used.

6. Acknowledgement

The authors acknowledge the financial support from the National High-tech Research and Development Program of China (Grant No. 2007AA06Z420).

7. References

Andrews, L.C.; Phillips, R.L & Hopen, C.Y. (2000). Aperture averaging of optical scintillations: power fluctuations and the temporal spectrum , *Wave Random Media*, Vol.10, pp53-70.

Blake, R.S.; Whyte, C.; Hughes, C.O.; Ellis, A.M. & Monks, P.S. (2004). Demonstration of proton-transfer reaction time-of-flight mass spectrometry for real-time analysis of trace volatile organic compounds. *Analytical Chemistry*, Vol.*76*, pp3841-3845.

Blake, R.S.; Wyche, K.P.; Ellis, A.M. & Monks P.S. (2006). Chemical ionization reaction time-of-flight mass spectrometry: Multi-reagent analysis for determination of trace gas composition. *International Journal of Mass Spectrometry*, Vol.*254*, pp85-93.

Blake, R.S.; Monks, P.S. & Ellis, A.M. (2009). Proton-transfer reaction mass spectrometry. *Chemical Reviews*, Vol.*109*, pp861-896.

Burakov, V.S.; Tarasenko, N.V.; Nedelko, M.I. ; Kononov, V.A.; Vasilev N.N. & Isakov, S.N. (2009). Analysis of lead and sulfur in environmental samples by double pulse laser induced breakdown spectroscopy, *Spectrochimica Acta Part B: Atomic Spectroscopy*, Vol.64, No.2, pp141-146.

Cassidy D. T. & Reid, J. (1982). Harmonic detection with tunable diode lasers two-tone modulation", *Appl.Phys.B*, Vol.29, pp279-285.

Chen A.S.; Hao, J.M.; Zhou Z.P.; et al (1999). Theoretical solutions for particulate scintillation monitors，*Opt. Comm.*，Vol. 166，pp15-20.

Chen A.S.; Hao, J.M.; Zhou Z.P.; et al (2000). Particulate concentration measured from scattered light fluctuations，*Opt. Lett.*，Vol.25, No.10, pp689-691.

Chen A.S.; Hao, J.M.; Zhou Z.P.; et al (2000). Measuring particulate concentration by means of scattered light scintillation，*Proc. Of SPIE*, Vol.4222, pp71-75.

Chesnavich, W.J.; Su T. & Bowers, M.T. (1980). Collisions in a noncentral field: A variational and trajectory investigation of ion--dipole capture. *The Journal of Chemical Physics*, Vol.*72*, pp2641-2655.

Christian, T.J.; Kleiss, B.; Yokelson, R.J.; Holzinger, R. ; Crutzen, P.J.; Hao, W.M.; Saharjo, B.H. & Ward, D.E. (2003) Comprehensive laboratory measurements of biomass-burning emissions: 1. Emissions from Indonesian, African, and other fuels, *J. Geophys. Res. Atmos.*, Vol.108, pp4719.

Christian, T.J. ; Kleiss, B. ; Yokelson, R.J. ; Holzinger,R. ; Crutzen, P.J. ; Hao, W.M. ; Shirai,T. & Blake,D.R. (2004) Comprehensive laboratory measurements of biomass-burning emissions: First intercomparison of open-path FTIR, PTR-MS, and GC-MS/FID/ECD. *J. Geophys. Res. Atmos.*, Vol.109, pp2311-2319.

Claudia, S.G.; John P.T. & Dennis, F.R. (2003). Tunable diode laser absorption spectrometer measurements of ambient nitrogen dioxide, nitric · acid, formaldehyde, and hydrogen peroxide in Parlier, California. *Atmospheric Environment*, Vol.37, pp1583-1591.

Culshaw, B.; Stewart, G.; Dong, F. Z.; Tandy, C. & Moodie, D. (1998). Fiber optic technique for remote spectroscope methane detection from concept to system realization, *Sensors and Actuators B*, Vol.51, pp25-37.

D'Amato F.; Mazzinghi, P. & Castagnoli, F. (2002). Methane analyzer based on TDL's for measurements in the lower stratosphere: design and laboratory tests. *Appl.Phys. B*, Vol. 75, pp195-202.

de Gouw, J. A. & Warneke, C. (2007). Measurements of volatile organic compounds in the earth's atmosphere using proton-transfer-reaction mass spectrometry. *Mass Spectrometry Reviews*, *26*, 223-257.

Deguchi, Y.; Noda, M.; Fukuda, Y.; Ichinose, Y.; Endo, Y.; Inada, M.; Abe, Y. & Iwasaki, S. (2002). Industrial applications of temperature and species concentration monitoring using laser diagonostics. *Meas. Sci. Technol.*, Vol.13, R103-R115.

Dong F.Z.; Liu W.Q.; Liu J.G.; et al (2005). Online roadside vehicle emissions monitoring (Part 1). *Journal of test and Measurement Technology,* Vol.19, No.2, pp119-127 (in Chinese).

Dong F.Z.; Liu W.Q.; Liu J.G.; et al (2005). Online roadside vehicle emissions monitoring (Part 2). *Journal of test and Measurement Technology,* Vol.19, No.3, pp237-244 (in Chinese).

Dupeyrat, G.; Rowe, B.R.; Fahey, D.W. & Albritton, D.L. (1982). Diagnostic studies of Venturi inlets for flow reactors. *International Journal of Mass Spectrometry and Ion Processes, 44,* 1-18.

Edner, H. & Svanberg, S. (1991). Lidar measurements of atmospheric mercury, *Water, Air, & Soil Pollution,* 1991(1): pp131-139.

Ennis, C.J.; Reynolds, J.C.; Keely, B.J. & Carpenter, L.J. (2005). A hollow cathode proton transfer reaction time of flight mass spectrometer. *International Journal of Mass Spectrometry, 247,* 72-80.

Feher, M. & Martin T.A. (1995). Tunable diode laser monitoring of atmospheric trace gas constituents. *Spectrochim. Acta Part A,* Vol.51, pp1579-1599.

Fried, A.; Henry, B. & Wert, B. (1998). Laboratory, ground-based, and airborne tunable diode laser systems: performance characteristics and applications in atmospheric studies. *Appl. Phys. B,* Vol.67, pp317-330.

Hansel A.; Jordan A.; Holzinger, R.; Prazeller P.; Vogel W. & Lindinger W. (1995). Proton-transfer reaction mass-spectrometry - online trace gas-analysis at the ppb level. *International J. of Mass Spectrometry and Ion Processes, 149/150,* 609-619.

Hanson, D.R.; Greenberg, J.; Henry, B.E. & Kosciuch, E. (2003). Proton transfer reaction mass spectrometry at high drift tube pressure. *International J. of Mass Spectrometry, 223,* 507-518.

http://www.opticalscientific.com

http://www.cnemc.cn

Inomata, S.;Tanimoto, H.; Aoki, N.; Hirokawa, J. & Sadanaga Y. (2006). A novel discharge source of hydronium ions for proton transfer reaction ionization: design, characterization, and performance. *Rapid Communications in Mass Spectrometry, 20,* 1025-1029.

Iseki, T.; Tai H. & Kimura K. (2000). A portable remote methane sensor using a tunable diode laser. *Meas. Sci. Technol.* Vol.11, pp594-602.

Ishimaru A. (1986). Wave propagation and scatter in Random medium (Beijing: Science Press), pp674 (in Chinese).

Jin, S.P.; Li, J.Q.; Han, H.Y.; Wang, H.M.; Chu Y.N. & Zhou S.K. (2007). Proton transfer reaction mass Spectrometry for Online detection of trace volatile organic compounds. *Progress in Chemistry, 19,* 996-1006.

Jordan, A.; Haidacher, S.; Hanel, G.; Hartungen, E., Herbig, J., Mark, L., Schottkowsky, R., Seehauser, H., Sulzer, P.,Mark, T.D. (2009). An online ultra-high sensitivity proton-transfer-reaction mass-spectrometer combined with switchable reagent ion capability (PTR+SRI-MS). *International Journal of Mass Spectrometry,* Vol.*286*,pp 32-38.

Jordan, A.; Haidacher, S.; Hanel, G.; Hartungen, E.; Mark, L.; Seehauser, H.; Schottkowsky, R.; Sulzer, P. & Mark, T.D. (2009). A high resolution and high sensitivity proton-transfer-reaction time-of-flight mass spectrometer (PTR-TOF-MS). *International Journal of Mass Spectrometry*, Vol.*286*, pp122-128.

Kan, R.F.; Liu, W.Q.; Zhang, Y.J.; et al (2006). Large scale gas monitoring with tunable diode laser absorption spectroscopy. *Chin. Opt. Lett.*, Vol.4, No.2, pp116-118.

Kan, R.F.; Liu, W.Q.; Zhang, Y.J.; et al (2007). A high sensitivity spectrometer with tunable diode laser for ambient methane monitoring, *Chin. Opt. Lett.*, Vol.5, No.1, pp54-57.

Karl, T.G.; Christian, T.J.; Yokelson, R.J.; Artaxo, P.; Hao, W.M. & Guenther, A. (2007). The tropical forest and fire emissions experiment: method evaluation of volatile organic compound emissions measured by PTR-MS, FTIR, and GC from tropical biomass burning. *Atmospheric Chemistry and Physics*, Vol.7, 5883-5897.

Karl, T.; Apel, E.; Hodzic, A.; Riemer, D.D.; Blake, D.R. & Wiedinmyer, C. (2009). Emissions of volatile organic compounds inferred from airborne flux measurements over a megacity. *Atmospheric Chemistry and Physics*, 9, 271-285.

Клименко А.П. (1984). *Continuous Monitoring of Dust Concentration* (Beijing: China National Defence Industry Press) pp31-41, pp120～121 (in Chinese).

Lindinger, W.; Hansel, A. & Jordan A. (1998). On-line monitoring of volatile organic compounds at pptv level by means of proton-transfer-reaction mass spectrometry (PTR-MS) - Medical applications, food control and environmental research. *International Journal of Mass Spectrometry*, Vol.*173*, pp191-241.

Linnerud, I.; Kaspersen, P. & Jæger, T. (1998). Gas monitoring in the process industry using diode laser spectroscopy, *Appl. Phys. B*, Vol. 67, pp297-305.

Liu, W.Q.; Cui, Z.C. & Dong, F.Z. (2002). Optical and spectroscopic techniques for environmental pollution monitoring. *Optoelectronic technology & information*. Vol.15, No.5, pp1-12.

Liu, W.Q.; Liu, H.L.; Zeng, Z.Y. & Jiang, Y. (2008). Analysis of spectrum characteristics of optical scintillation in stack gas flow，*Chin. Phys.*, Vol.15, No.8, pp1777-1782.

May, R.D. & Webster, C.R. (1989). In situ stratospheric measurement of HNO_3 and HC1 using the balloon-borne laser in situ sensor tunable diode laser spectrometer. *J. Geophys. Res.*, Vol.94, pp16343-16350.

Mielke, L.H.; Erickson, D.E.; McLuckey, S.A.; Muller, M.; Wisthaler, A.; Hansel, A. & Shepson, P.B. (2008). Development of a proton-transfer reaction-linear ion trap mass spectrometer for quantitative determination of volatile organic compounds. *Analytical Chemistry*, Vol.*80*, pp8171-8177.

Mihalcea, R.M.; Webber, M.E.; Baer, D.S.; Hanson, R.K.; Feller, G.S. & Chapman, W.B. (1998). Diode-laser absorption measurements of CO_2, H_2O, N_2O, and NH_3 near 2.0 um. *Appl. Phys. B*, Vol.67, pp283-288.

Nadezhdinskii, A.; Berezin, A.; Chemin, S.; et a1 (1999). High sensitivity methane analyzer based on tuned near infrared diode laser. *Spectrochimica Acta Part A*, Vol.55, pp2083-2089.

Nelson, D. D.; Shorter, J. H.; Mcmanus, J. B. et al. (2002). Sub-part-per-billion detection of nitric oxide in air using a thermoelectrically cooled mid-infrared quantum cascade laser spectrometer. *Appl. Phys. B*, Vol.75, pp343-350.

NIST Standard Reference Database Number 69, NIST Chemistry webbook, http://webbook.nist.gov/chemistry/.

Platt, U. & Pemer, D. (1980). Direct measurements of atmospheric CH_2O, O_3, NO_2 and SO_2 by different optical absorption in the near UV. *J. Geophys. Res.*, 85(C10): 7453~7458.

Prazeller, P.; Palmer, P.T.; Boscaini, E.; Jobson, T. & Alexander, M. (2003). Proton transfer reaction ion trap mass spectrometer. *Rapid Communications in Mass Spectrometry, 17,* 1593-1599.

Reid J.; El-Sherbiny, M. & Garside, B. K. (1980). Sensitivity limits of a tunable diode spectrometer with application to the detection of NO2 at the 100-ppt level. *Appl. Opt.*, Vol.19, No.19,

Reid J. & Labrie D. (1981). Second-harmonic detection with tunable diode lasers--Comparison of experiment and theory. *Appl. Phys. B*, Vol.26, pp203-210.

Rieker, G.B.; Jeffries, J. B. & Hanson, R.K. (2009). Calibration-free wavelength-modulation spectroscopy for measurements of gas temperature and concentration in harsh environments", *Appl. Opt.*, Vol.48, No.29, pp5546-5560.

Riise H.; Carlisle C.B.; Carr. L.W.; Cooper D. E.; Martinelli R. U. & Menna R.J. (1994). Design of an open path near infrared diode laser sensor: application to oxygen, water and carbon monoxide. *Appl. Opt.*, Vol.33, pp7059-7066.

Rocco, A.; Natale, G. De; Natale, P. De; et a1 (2004). A diode laser based spectrometer for in situ measurements of volcanic gases. *Appl. Phys. B*, Vol.78, No.2, pp235-240.

Smith, D. & Spanel, P. (2005). Selected ion flow tube mass spectrometry (SIFT-MS) for on-line trace gas analysis. *Mass Spectrometry Reviews, 24,* 661-700.

Somesfalean, G.; Alnis, J.; Gustafsson, U.; Edner, H. & Svanberg, S. (2005). Long-path monitoring of NO_2 with a 635nm diode laser using frequency-modulation spectroscopy, *Appl. Opt.*, Vol.24, pp5184-5188.

Sigrist, M.W., (1994). *Air monitoring by spectroscopic Techniques*, Published by JOHN WILEY &SONS, INC.

Somesfalean, G.; Alnis, J.; Gustafsson, U.; Edner, H. & Svanberg, S. (2005). Long-path monitoring of NO_2 with a 635nm diode laser using frequency-modulation spectroscopy. *Appl. Opt.*, Vol.24, pp5184-5188.

Su, T. & Chesnavich, W.J. (1982). Parametrization of the ion-polar molecule collision rate constant by trajectory calculations. *The Journal of Chemical Physics, 76,* 5183-5185.

Taatarskii, V. I. (1978). Wave Propagation in a Turbulent Medium (Beijing: Science Press), pp23 (in Chinese)

Taslakov, M.; Simeonov, V.; Froidevaux, M. & Van den Bergh, H. (2006). Open-path ozone detection by quantum-cascade laser. *Appl. Phys. B,* Vol.82, No.3, pp501-506.

Uehara, K. & Tai, H. (1992). Remote detection of methane with 1.66um diode laser. Appl. Opt., Vol.31, No.6, pp809-814.

Wang, F.; Cen, K.F.; Li, N.; Huang, Q. X.; Chao, X.; Yan, J. H. & Chi. Y. (2010). Simultaneous measurement on gas concentration and particle mass concentration by tunable diode laser, *Flow Measurement and Instrumentation*, Vol.21, No.3, pp382-387.

Wang J.; Maiorov M.; Baer D.S.; Garbuzov D.Z.; Connolly J.C. & Hanson R.K. (2000). In situ combustion measurements of CO with diode-laser absorption near 2.3 um. *Appl. Opt.*, 39(30), pp5579-5589.

Wang, T.I.; Ochs, G.R.& Lawrence, R.S. (1981). Wind measurements by the temporal cross-correlation of the optical scintillations , *Appl. Opt.*, Vol.20, No.23, pp4073-4081.

Wang ,Ting-I (2003). , United States Patent 6,661,319 B2.

Wang, Y.J.; Han, H.Y.; Shen, C.Y.; Li J.Q.; Wang, H.M. & Chu, Y.N. (2009). Control of solvent use in medical devices by proton transfer reaction mass spectrometry and ion molecule reaction mass spectrometry. *Journal of Pharmaceutical and Biomedical Analysis, 50*, 252-256.

Webber, M.E.; Wang, J.; Sanders, S.T.;. Baer, D.S. & Hanson R.K. (2000).In situ combustion measurements of CO, CO2, H_2O and temperature using diode laser absorption sensors. *Proc. Comb. Inst.*,Vol.28, pp407.

Weibring, P.; Edner, H. & Svanberg, S. (1998). Monitoring of volcanic sulphur dioxide emissions using differential absorption lidar (DIAL), differential optical absorption spectroscopy (DOAS), and correlation spectroscopy (COSPEC). *Appl. Phys. B*, Vol.67, No.4, pp 419-426.

Werle P. & Lechner S. (1996). Recent findings and approaches for suppression of fluctuation and background drifts in tunable diode laser spectroscopy. *Proc. Of SPIE*, Vol.2834, pp68-78.

Werle, P. (1998). A review of recent advances in semiconductor laser based gas monitors", *Spectrochimica Acta Part A*, Vol.54, No.2, pp197-236.

Werle P.; Mazzinghi P.; Amato F.D.; et al, (2004). Signal processing and calibration procedures for in situ diode-laser absorption spectroscopy. *Spectrochim. Acta Part A*, Vol. 60, pp1685-1705.

Wyche, K.P.; Blake, R.S.; Willis, K.A.; Monks, P.S. & Ellis A.M. (2005). Differentiation of isobaric compounds using chemical ionization reaction mass spectrometry. *Rapid Communications in Mass Spectrometry, 19*, 3356-3362.

Xia, H.; Dong, F.Z.; Tu, G.J.; et al (2010). High sensitive detection of carbon monoxide based on novel multipass cell , *Acta Optica Sinica*, Vol.30, No.9, pp2596-2601(in Chinese).

Yuan, Z. F.; Wang, X. D.; Zhou, J.; Pu, X. G. & Cen, K. F. (2003). Experimental studies on measurement of particle flow velocity using optical scintillation cross-correlations. *Thermal Power Generation*, Vol.3, pp46-50 (in Chinese).

Zeninari, V.; Parvitte, B.; Joly, L.; Le Barbu, T.; Amarouche, N. & Durry, G. (2006). Laboratory spectroscopic calibration of infrared tunable laser spectrometers for the in situ sensing of the earth and martian atmospheres. *Appl. Phys. B*, Vol. 85, pp265-272.

Zhang Z.R.; Dong F.Z.; Tu G.J.; et al. (2010). Selection of digital filtering technique in trace gas concentration measurements with tunable diode laser absorption spectroscopy. *Journal of Optoelectronics·Laser*, 2010, 11(21), pp1672-1676(in Chinese).

Zhang Z.R.; Dong F.Z.;Wang Y.; et al (2011). Online monitoring of industrial toxic gases with a digital control module. *Acta Optica Sinica*, Vol.31, Supplement, s100304.1-6 (in Chinese).

Zhou, X. (2005). Diode-laser absorption sensors for combustion control. Stanford University (Ph. D. thesis).

6

Geochemical Application for Environmental Monitoring and Metal Mining Management

Chakkaphan Sutthirat[1, 2]
[1]*Department of Geology, Faculty of Science, Chulalongkorn University, Bangkok,*
[2]*Center of Excellence for Environmental and Hazardous Waste Management (NCE-EHWM), Chulalongkorn University, Bangkok,*
Thailand

1. Introduction

Metal mines have been increasing continuously due to high growth rate of population and rapid development of industry throughout the world. Various kinds of metal ores are supplied into industries. Precious metals such as gold, platinum and silver have been utilized for ornamental purposes due to their beauty, rarity and durability whereas industrial ores are demanded by many sectors. These ores usually occur in different geological conditions which lead to diversity of depositional characteristics. Multiple elements occur naturally in the same mineral deposits; some of these elements, particularly heavy metals, may in turn have potential impact to the environment. Therefore, heavy metals are the most crucial aspects for toxicity. However, these metals have several chemical binding forms which just a few forms appear to contaminate environment.

Moreover, Acid Mine Drainage (AMD) is another environmental concern. AMD appears to have been accelerated during mining processes when metal sulfides in mineralized rocks and solid wastes are exposed to oxygen and water allowing rapid oxidizing reaction. Oxidation of metal sulfide has potential to produce sulfate which may turn into sulfuric acid. Subsequently, it may be dissolved by rain and leading to acidity drainage. AMD can also cause heavy metal leaching from waste rock and tailings; consequently, some toxic metals (e.g., lead, zinc, copper, arsenic, selenium, mercury and cadmium) may contaminate runoff and groundwater. AMD with high metal concentrations may in turn yield severe toxicological effects on aquatic ecosystems. Biota will be affected primarily and subsequently toxic levels would be increased through food chain. Although, some heavy metals such as copper and zinc are required with small quantities for normal metabolism, their high concentrations become toxic and can cause malfunctioning of human organs.

Geochemical exploration has been carried out by mining geologists for investigation and evaluation of mineral deposit, metal ore in particular. It can also be applied to environmental impact assessments and monitoring. Moreover, mineralogical and chemical characteristics are one among many scientific tools that will lead to identification of potential sources of such problems. Appropriate prevention and mining plans can be designed based on these data. Unfortunately, most mining geologists always apply geochemistry for exploration and mining without concern of environmental impacts; on the other hand, most environmental scientists have little knowledge on geology and mining materials, mining designs and processes. Environmental protection should be carefully planned in order to eliminate and/or minimize any short- and long-term environmental impacts that may occur. Otherwise, serious problems may occur that may be very difficult to remediate and extremely cost enormously.

This chapter will review standard procedures for evaluation of AMD potential of rock waste and tailing generated from mining activity. Digestion techniques for analysis of heavy metals are also considered to give basic knowledge for environmental monitoring and impact assessment. Some cases studies in Thailand will be given for better understanding.

2. Mining wastes

Mine operations may include 3 principle activities which are mining, mineral processing/dressing and metallurgical extraction/refining. All of these activities usually produce wastes that are unwanted and non-economic value. Solid mining wastes and other related wastes may be generated from each activity were summarized by Lakkopo (2002) as shown in Table 1.

Dusts, ashes and other atmospheric emissions may be routinely monitored by environmental scientists who have experienced in other industrial plants. Slag and waste water can also be tested before suitable processes of treatment and disposal will be designed by environmental engineers. Therefore, heterogeneous geological materials including overburden soils and rocks appear to be the most crucial solid wastes due to lack of geological knowledge of both environmental scientists and engineers. Geologist and mining engineer should share their opinion for environmental plans. However, top soils may have been utilized as construction materials during mining activities and reclamation at the mining end. Although, these top soils may contain some natural contaminants, particularly heavy metals in this case, they would have low impact to the environment. This is because they have been undertaken naturally erosion and weathering processes for several hundreds or thousands of years then transportation of contaminant have been taken place slowly ever since. Moreover, quantities of these top soils are usually much lower than waste rocks and tailings. Rocks usually have stable chemical forms of minerals but mining processes such as blasting, grinding and milling will reduce their sizes and increase surface of reaction. Consequently, chemical reactions would be activated rapidly leading to metal leach out form these rocks. On the other hand, tailings are the other solid waste left after ore and metal extractions which usually involve with chemical additives as well as alteration of the natural chemical bonding. This waste type should then be concerned for environmental monitoring plan.

Activities	Mining Wastes
Open pit and underground mining	Waste rocks, overburden soils, mining water, atmospheric emissions
Mineral processing, coal washing, mineral fuel processing	Tailing, sludge, mill water, atmospheric emissions
Pyrometallurgy, hydrometallurgy, electrometallurgy	Slag, roasted ores, flue dusts, ashes, leached ores, process water, atmospheric emission

Table 1. Summary of mining activities and their solid, gaseous and liquid wastes (modified after Lakkopo, 2002)

Waste Rocks: Large amount of waste rocks may have been removed from mining site, particularly for quarrying and excavation, to access to the ore body. These waste rocks are eventually remained in the site and surrounding areas after the mining end (see Fig. 1). Subsequently, they may become sources of environmental impacts. Although, mining design can reduce quantity of waste rocks; for example, mining excavation generates very less amount of waste rocks in comparison with open-pit mining. Geologic setting and ore formation are however the main factor for the mine planning; the open-pit mine may be economically more suitable in many cases. Besides, some waste rocks can be used for construction within the mining site; however, they must be tested prior to appropriate utilization. Otherwise, unexpected threats may occur.

Various types of waste rocks situated within ore deposits usually have different compositions that would be characterized for both mineralogical and geochemical constituents. Apart from heavy metals contained in these rocks, Acid Mine Drainage (AMD), a potential threat, may be activated and lasted for long period of time. AMD actually lowers pH of water; subsequently, the low pH drainage may flow over waste dumps including waste rocks and tailings and may in turn leach some heavy metals and contaminate surrounding area. Surface water and ground water would be crucial pathways of such contamination to ecosystem and food chain. However, most of these threats can be protected and prevented by good environmental management and monitoring plans.

Tailings: During the mineral processing (dressing), ore minerals and their host rocks have to be ground and milled prior to separation; besides, chemical additives may be added during the processes. Although, most of these chemicals are usually recovered and reused in the process, some of them may still remain in these tailings. Some chemical additives can be decomposed naturally within short period but many of them may be bound strongly and long lasted within the tailings. Moreover, these tailings may contain concentrate non-economic minerals such as silicates, oxides, hydroxides, carbonates and sulfide that have never been collected throughout the dressing process. Therefore, these modified ingredients may partly be toxic and harm ecosystem. Tailings are similar to slurry, a mixture of fine-grained sediment and water that have been disposed into tailing pond (see Fig. 2).

Fig. 1. Huge amount of rock waste generated from a gold mine in Thailand: left photo is waste dumping site; right photo shows placing process based on geochemical properties of each type of waste rock

Fig. 2. Tailing pond (left photo) for disposal of slurry-like waste produced from gold dressing and sample collection (right photo), a routine monitoring program which has to be carried out regularly

Due to tailings comprise both solid wastes mixing with water during the operation period of mine and they will become drying after the mining end, redox reaction would be taking place and may in turn change stability of some elements which can be leached out to the environment by accidence. Moreover, their property may also cause AMD. Therefore, routine monitoring plans for both water and tailing must be designed and continuously followed up. Monitoring data should be used for protection at the end and may be very useful for development of the mineral processing.

3. Acid mine drainage

Acid Mine Drainage (AMD) is the problem of acid drainage, traditionally referred in Australia and North America as Acid Rock Drainage (ARD). It seems to be a significant environmental impact of mining activities especially in opencast mines. It may damage long after the operation has ended because process and reaction have taken time. Runoff passing through the sourcing area can then give rise to severe threat. Moreover, AMD potentially dissolves and leaches out some toxic metals from the heap, mining waste dump and even natural soil and rock prior to contamination of surface water and groundwater. AMD is usually generated by the oxidation of sulfides in mining wastes; consequently, water supply from the area would be sulfide-rich drainage with acidic leaching property that may lead to mobilization of metals. Sulfides bound up in the waste rocks and tailings usually have various forms. Mineral sulfides are crystalline substances that contain sulfur combined with metal or semi-metal without oxygen. The most general form is "pyrite" (FeS_2), moreover, other forms also include Fe1-xSx, Fe_3S_4. FeS, $CuFeS_4$, ZnS, PbS, HgS, CoAsS etc. After these sulfide minerals are exposed to the air and water, the sulfide ions are oxidized into soluble sulfates as well as toxic metal ions and hydrogen ions may in turn be released into the environment. Initial factors for acid generation are: 1) sulfide minerals in the solid wastes (e.g., rocks and tailings); 2) water or a humid atmosphere; 3) an oxidant (usually oxygen in the form of O2). Therefore, processes of acid generation and metal release would be taken place together during the formation of AMD which are closely related to oxidation of pyrite and precipitation of Fe hydroxides. There are four common chemical reactions represent AMD formed from pyrite. The first equation shows that an important oxidant of pyrite is oxygen. Ferrous iron is released and sulfur is oxidized and changed to sulfate. This equation shown 2 moles of acidity generated from each mole of pyrite. The second equation is the conversion of ferrous iron to ferric iron. It consumes one mole of acidity. The third equation is a hydrolysis reaction which splits the water molecule; consequently, moles of acidity are generated as by-product. The fourth reaction is the oxidation of additional pyrite by ferric iron. The ferric irons generated in reaction steps 1 and 2 are cycle and propagation of the overall reaction. They take place very rapidly and continue until either the ferric iron or pyrite are depleted. In this reaction, iron is the main oxidizing agent instead of oxygen.

$$\underset{\text{pyrite}}{2FeS_2} + \underset{\text{oxygen}}{7O_2} + \underset{\text{water}}{2H_2O} \rightarrow \underset{\text{ferrous iron}}{2Fe^{2+}} + \underset{\text{sulfate}}{2SO_4^{2-}} + \underset{\text{acidity}}{4H^+} \quad (1)$$

$$\underset{\text{ferrous iron}}{4Fe^{2+}} + \underset{\text{oxygen}}{O_2} + \underset{\text{acidity}}{4H^+} \rightarrow \underset{\text{ferric iron}}{4Fe^{3+}} + \underset{\text{water}}{2H_2O} \quad (2)$$

$$\underset{\text{ferric iron}}{4Fe^{3+}} + \underset{\text{Water}}{12H_2O} \rightarrow \underset{\text{ferrichydroxide}}{4Fe(OH)_3^-} + \underset{\text{acidity}}{12H^+} \quad (3)$$

$$\underset{\text{pyrite}}{FeS_2} + \underset{\text{ferric iron}}{14Fe^{3+}} + \underset{\text{water}}{8H_2O} \rightarrow \underset{\text{ferrous iron}}{15Fe^{2+}} + \underset{\text{sulfate}}{2SO_4^{2-}} + \underset{\text{acidity}}{16H^+} \quad (4)$$

Many procedures have been developed to assess the acid forming characteristics of mine waste materials. The most widely used methods are Acid-Base Accounting (ABA) test and the Net Acid Generation (NAG) test. These procedures are described below.

3.1 Acid-base accounting

Characterization of rock types and geologic setting in the mine should be initially concerned prior to determination of capacity acid drainage generation of these rocks (Environment Australia, 1997). Acid-Base Accounting (ABA) is the most commonly-used static procedure that has been used for estimation/qualification of the acid generation potential of mine wastes (Furguson & Erickson, 1988). This procedure was developed at West Virginia University in late 1960s. ABA tests are designed to measure the balance between potentially acid-generating potential, particularly oxidation of sulfide materials and acid neutralizing potential in sample such as dissolution of alkaline, carbonates, displacement of exchangeable bases and weathering of silicate. The values arising from ABA are referred to the Maximum Potential Acidic (MPA) and the Acid Neutralizing Capacity (ANC), respectively. After MPA and NAC have been determined for a sample, both values are compared with set criteria. Two methods of combination commonly used are: 1) The difference in value between MPA and ANC or Net Acid Producing Potential (NAPP) where NAPP = MPA-ANC; 2) The ratio of ANC to MPA (ANC/MPA). NAPP is a theoretical calculation commonly used to indicate where a waste material has potential to produce acidic drainage. NAPP values represent balance between capacity of acid generation and capacity of acid neutralization. Unit of NAPP is also expressed as kg H_2SO_4/t in MPA and ANC. In addition, ANC/MPA ratio is also considered for assessment of acid generation from mine waste material. The main purpose of ANC/MPA ratio is to indicate relatively safety margin of material. Safe values for prevention of acid generation are reported with different ANC/MPA values ranging from 1 to 3. The higher ANC/MPA value indicates high probability of the material that may remain circum-neutral in pH and should not be problematic by acid rock drainage. Both NAPP value and ANC/MPA ratio are usually used together for placement planning of rock waste and other overburdens (Skousen et al., 1987). Sulfur and ANC data are often used in combination with ANC/MPA ratio as presented in Fig. 3.

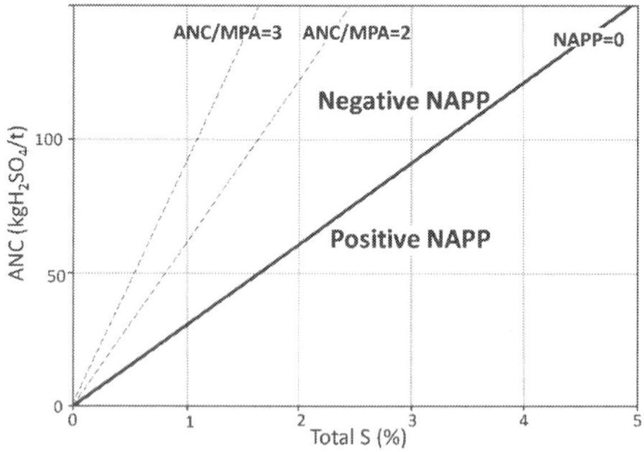

Fig. 3. Plots of all parameters considered in Acid-Base Accounting (ABA)

Maximum Potential Acidic: MPA is the maximum amount of acid that can be produced from the oxidation of sulfur-containing minerals in the rock material. It can be measured and calculated from the sulfur content. Total sulfur content of a sample is commonly determined by the LECO high temperature combustion method or other appropriate methods. For instant, it is assumed that all sulfurs occur as iron-sulfide (or pyrite; FeS_2) and this iron-sulfide reacts under oxidizing condition to generate acid according to the following reaction:

$$FeS_2 + 15/4\,O_2 + 7/2\,H_2O \rightarrow Fe(OH)_3 + 2\,H_2SO_4$$

According to the stoichiometry, the maximum amount of acid that could be produced by a sample containing 1%S as pyrite would be 30.6 kilograms of H_2SO_4 per ton of material. The MPA is calculated from the total sulfur content as:

$$\text{MPA (kg } H_2SO_4/\text{t)} = \text{(Total \%S) X } 30.6$$

Acid Neutralizing Capacity: ANC is calculated from the amount of acid neutralizer in the sample and it is expressed in metric tons/1000 metric tons of material. Acid generated from pyrite oxidation will be partly reacted by acid neutralizing minerals contained within the sample. This inherent acid buffering is resulted in term of the ANC. Most of the minerals which contribute the acid neutralizing capacity usually are carbonates such as calcite and dolomite. The modified Sobek method is the most common method used to determine ANC. This method is determined experimentally by reaction of a known amount of standardized acid (hydrochloric acid, HCL) with a known amount of sample and then the mixed solution sample is back-titrated by sodium hydroxide (NaOH). The amount of acid consumed represents the inherent acid neutralizing capacity of the sample. Calculation will be carried out and expressed in terms of kg H_2SO_4/t.

3.2 Net acid generation

Net Acid Generation (NAG) test was developed as an assessment tool for acid producing potential of sample for longer than 20 years ago. The NAG test is usually used in association with NAPP. It is direct method to measure ability of sample to produce acid via sulfide oxidation. Hydrogen peroxide (H_2O_2) is used to activate and complete oxidation process of the sulfide minerals contained in the sample. H_2O_2 added during the NAG test leads to simultaneous reactions of acid generation and acid neutralization. Then pH measurement of solution has to be carried out after the completion of reaction. The acidity of solution under the NAG is a direct measurement of net acid generation of sample. Shu et al. (2001) studied the effect of lead/zinc mine acidity on heavy metal mobility using both NAG test and ABA method. They concluded, based on their results that NAG test, direct measurements of ANC from acid produced from oxidized sulfide, yields more accurate than that of ABA method. This is because prediction of acid forming potential from the total pyritic sulfur content as done for ABA method may overestimate amount of acid generation due to uncompleted acidification of pyritic sulfur.

However, classifications of waste rock have generally used NAPP estimation based on ABA method in combination of NAG pH testing. Schematic classification is present in Fig. 4. Three types of west rocks from mining activity can be grouped as No Net Acid Forming (NAF), Potentially Net Acid Forming (PAF), and Uncertainly Net Acid Forming (UC). Definitions of these groups are given below.

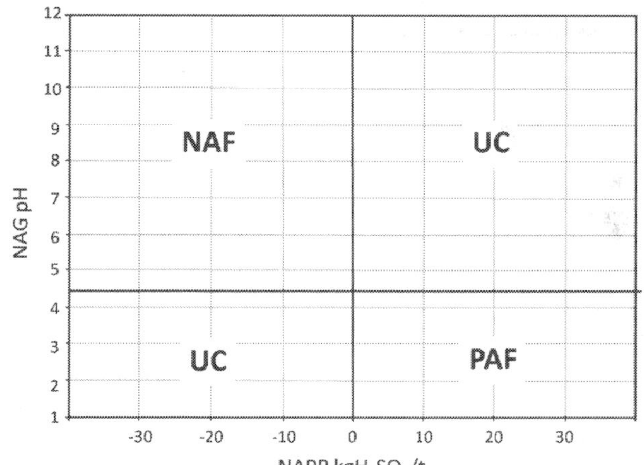

Fig. 4. NAG pH plot against NAPP for classification potential of net acid formation of waste rock

No Net Acid Forming (NAF): either there is minimal or no sulfides present or the neutralization potential exceeds the acid potential. This type of waste rock gives a negative NAPP and NAG pH greater than or equal to 4.5.

Potentially Net Acid Forming (PAF): the acid potential exceeds the neutralization potential. These rocks are described as potentially acid forming. They may generate AMD if they are exposed to sufficient oxygen to allow sulfide oxidations. Geochemical tests usually yield positive NAPP and NAG pH below 4.5.

Uuncertain Net Acid Forming (UC): uncertain classification is obtained when there is an apparent conflict between the NAPP result and NAG pH; for example, NAPP is negative but NAG pH lower than 4.5 or NAPP is positive but NAG pH higher than 4.5. However, further testing work would be performed for such rock types to determine proportion between NAF and PAF if they occur.

Recently, this classification has been using widely for geochemical study of waste rock and assessment of acid forming potential. Tran et al. (2003), for an example, also used NAG together with NAPP tests to figure out key criteria for construction design of waste rock dumps to avoid AMD. They collected samples from 2 sites in which have different temperatures. NAG and NAPP tests were applied to classify PAF, NAF and UC materials prior to placement control of waste rocks within the dumps. They succeeded to have reduced AMD load that may be generated from both dumps.

4. Heavy metals

As mentioned earlier, heavy metals contained in mine wastes, particularly rocks and tailings, may in turn become contamination to water systems around the dumping site. Analyses of these solid wastes must be very crucially considered for environmental protection plan during the mining operation. In fact, these heavy metals usually have different forms appeared in these rocks and tailings. Some forms are quite stable and durable to natural reactions such as weathering and erosion; however, some forms may be leached and available to contamination. Moreover, their stable-chemical bonds may have been destroyed during the mining process, mineral dressing and metal extraction. Therefore, placement and dumping of these solid wastes should concern about these geochemical characteristics. Several standard procedures have been proposed for analyses of heavy metals contained in geological materials such as soils, stream sediments and rocks. These methods were initially engaged for geochemical exploration searching for potential area of mineral deposits. Although, they can also be applied for environmental purpose, some assumption must be taken into consideration as well as limitation of selected method must be understood clearly before interpretation will be carried out. Some methods are designed for total concentrations of element contained in the samples; on the other hand, some of them are planned for partial portions of these elements reliable for specific concern. However, some methods have been developed for environmental impact assessment. In this section, some selective standard procedures are described for suitable application of mining waste and related fields.

4.1 Total digestion

Whole-Rock Geochemical Analyses: this method is designed for analysis of total chemical concentrations contained in the rock materials. This method may not be suitable to the environmental concern because major and minor compositions of these rocks are usually non toxic and they are quite stable. However, their trace compositions may have partly impact after accumulation and transportation have taken place for some periods of time, particularly due to AMD. Moreover, these whole-rock analyses are very useful for geological classification as well as mining operation. Placement and disposal may be designed based on this classification in cooperation with other testing methods. Rock powdering using appropriate crusher and miller must be done prior to further analyses. Subsequently, the powdered rock samples may be fused to glass beads or pressed as pellet for X-ray Fluorescence (XRF) analyses of 9 major oxides (i.e., SiO_2, TiO_2, FeO_t, MnO, MgO, CaO, Na_2O, K_2O and P_2O_5) and perhaps some trace elements (e.g., Ba, Zn, Sr, Rb, Zr, Co, Cr, Ni, Y and V). Rock standards should be used for calibration at the same analytical condition. Moreover, loss on ignition (LOI) should also be measured by weighting rock powders before and after ignition at 900° C for 3 hrs in an electric furnace. Trace and rare earth elements may be additionally analyzed using advanced instruments such as Inductively Coupled Plasma (ICP) Spectrometer, Atomic Absorption Spectrometer (AAS) and other spectrometric techniques. Rock samples have to be digested totally without remaining of rock powders. About 0.1000 g (±0.0001 g) of powdered samples are weighted and then dissolved in a concentrate HF-HNO_3-$HClO_4$ acid mixture in sealed Teflon beakers. The digested samples were diluted immediately and added mixed standard solution to all samples. Proportion of these concentrate acids is usually adapted in laboratory as well as time of digestion. Hotplate has been engaged traditionally but it may take long time. Alternatively, microwave has been applied to shorten the digestion time. This method is total digestion which most elements including toxic elements and non toxic ones are dissolved for analyses. However, these contents do not clearly reflect environmental impact. Microwave-assisted acid solubilization has been proved to be the most suitable method for the digestion of complex matrices such as sediments and soil. This method shortens the digestion time, reduces the risk of external contamination and uses smaller quantities of acid (Wang et al., 2004). However, there are different procedures required for appropriate sample types. Some standard digestion techniques are usually used for soil, sediment and sludge; for example, EPA 3052, EPA 3050B and EPA 3051 are described below.

EPA 3052: This method is an acid digestion of siliceous matrices, and organic matrices and other complex matrices (e.g., ashes, biological tissues, oils, oil contaminated soils, sediments, sludges and soils) which they may be totally decomposed for analysis. Powdered sample of up to 0.5 g is added into 9 ml of concentrated nitric acid and usually 3 ml hydrofluoric acid for 15 minutes using microwave. Several additional alternative acids and reagents have been applied for the digestion. These reagents include hydrochloric acid and hydrogen peroxide. A maximum sample of 1.0 g can be prepared by this method. Mixed acids and sample are placed in an inert polymeric microwave vessel then sealed prior to heating in the microwave system. Temperature may be set for specific reactions and incorporates reaching 180 ± 5 °C in approximately shorter than 5.5 minutes and remaining at 180 ± 5 °C for 9.5 minutes to complete specific reactions. Solution may be filtered before appropriate volume is made by dilution. Finally, the solution is now ready for analyses (e.g., AAS or ICP). More details should be obtained from EPA (1996).

EPA 3050: Two separate procedures have been proposed for digestion of sediment, sludge and soil etc. The first procedure is preparation for analysis of Flame Atomic Absorption Spectrometry (FLAA) or Inductively Coupled Plasma-Atomic Emission Spectrometry (ICP-AES) whereas the other is for Graphite Furnace AA (GFAA) or Inductively Coupled Plasma Mass Spectrometry (ICP-MS). Appropriate elements and their detection limits must be concerned and designed for selection of both methods (EPA, 2009). Alternative determination techniques may also be modified as far as scientific validity is proven. This method can also be applied to other elements and matrices but performance need to be tested. It should be notified that this method is not a total digestion for most types of sample. However, it is a very strong acid digestion that may dissolve most elements that could cause environmental impact. In particular, silicate-bonding elements are unlikely to be dissolved by this procedure. About 1-2 g (wet weight) or 1 g (dry weight) sample is dissolved by repeated additions of nitric acid and hydrogen peroxide. For GFAA or ICP-MS analysis, the digested solution is reduced in volume while heating then the final volume is made to 100 ml. This method may refer to EPA 3050B. On the other hand, for ICP-AES or FLAA analyses, hydrochloric acid (HCl) is additionally poured into the previous digested solution; consequently, the solubility of some metals may be increased which may refer to EPA 3050A. After filtering, filter paper and residue are dissolved by additional HCl and then filtered again. Final digested solution is diluted to 100 ml (EPA, 2009).

A simplified procedure of EPA 3050B has been suggested as following detail. Powdered sample (e.g., soil, sediment and sludge) is mixed in 10 ml of 1:1 HNO_3, then sample is covered with a watch glass. Subsequently, the sample is heated to 95±5 °C and refluxed for 10 to 15 minutes without boiling. When the sample is allowed to cool, 5 ml of concentrate HNO_3 is added and covered and refluxed for 30 minutes. If brown flumes are generated, indicating oxidation of the sample by HNO_3, repeat this step (addition of 5 ml of HNO_3conc.) over and over until no brown flame will be given off by the sample indicating the complete reaction with HNO_3. The solution has to be evaporated to approximately 5 ml without boiling or heating at 95±5 °C for 2 hrs. After the sample had been cooled, 2 ml of water and 30 ml of 30% H_2O_2 are added into the sample. In addition, 1 ml of 30% H_2O_2 has been continuously added with warming until the generated sample appears to have no further change. The sample has to be heated until the volume reduces to about 5 ml. Finally, the sample is then diluted to 100 ml with D.I. water after cooling. Particulates in the solution must be removed by filter (Wattman No.41). The sample is now ready for analyses of ICP or AAS.

EPA 3051: is an alternative to EPA 3050 procedure which is a rapid acid digestion of multielement for analysis. Leaching levels must be designed. In case, hydrochloric acid is required for digestion of certain elements; therefore EPA 3050A would be applied. Otherwise, EPA 3051 may be considered. After 0.5 g of sample is placed in a digestion vessel, 5 ml of 65% HNO_3 is added and the vessel is closed with a Teflon cover. Then, the sample will be heated at 170±5°C for approximately 5.5 minutes and remained at 170-180°C for 10 minutes to accelerate the leaching process by microwave digestion system. Heating temperature and time may be adjusted as appropriate to each microwave system produced by various manufacturers. After cooling, the solution must be filtered by membrane filter of 0.45 µm pore diameter. Finally, the filtered solution is further diluted in 50 ml volumetric flask. The sample is now ready to be analyzed by ICP and AAS.

It has to be notified that EPA 3050 and 3051 methods usually are not total digestions; undigested materials will be remained after acid is added into the sample. However, most of the chemical bonding forms potentially environmental impact appear to have been dissolved. Silicate bonding in particular is a stable form and unlikely to be removed; it actually has no impact. Both methods are suitable for mining wastes that can be used for environmental monitoring and protection plans. In addition, Aqua Regia, mixture of hydrochloric acid and nitric acid, may also be applied for digestion. It is quite similar to EPA 3050A method. Gold can be dissolved in this mixed acid which the method is usually applied for stream sediment collected for mineral exploration.

4.2 Sequential extraction

In the environmental field, determination of total metal concentrations in mining wastes does not give sufficient information about the mobility of metals. Metals may be bound to particulate matter by several mechanisms such as particle surfaces absorption, ion exchange, co-precipitation and complexation with organic substances. For example, not all of heavy metals in soil are available for plant uptaking, only the dissolved metals content in soil solution is moveable enough for plant to absorb. Therefore, heavy metals speciation in form of water soluble fraction and free weak acid soluble fraction out of total heavy metal content are the maximum amount of heavy metals possibly uptaken by plant. However, actual bioavailability of heavy metals by each species of plant must be determined from the plant itself. This will lead to protection and reclamation plans after the mine close. Chemical extraction is played an important role to define metal fractions, which can be related to chemical species, as well as to potentially mobile, bioavailable, or ecotoxic phase of sample. The mobile fraction is defined as the sum of amount dissolved in the liquid phase and an amount which can be transferred into the liquid phase. It has generally accepted that ecological effects of metals are related to such mobile fractions rather than the total concentration.

Sequential extraction procedures are operationally defined as methodologies that are widely applied for assessing heavy metal mobility in sediment. It is also used for the fractionation of trace metal within sediment (Quevauviller et. al., 1993 ; Ure et. al., 1993) and allows for the study of the bioavailability and behavior of metals fixed to the sediment (Pazos-Capeáns et al. 2005). BCR has been applied to characterize the metal fraction of a variety of matrices, including sediment with distinct origin, sewage sludge, amended soils and different industrial soil (Mossop & Davidson, 2003).

There are many methods to determine the different forms of metals. BCR three-step sequential extraction procedure is one of them, which was proposed by the Standards, Measurements and Test Programme (SM&T-formerly Community Bureau of Reference, BCR) of the European Union. It has been applied for the determination of trace metals (e.g., Cd, Co, Cr, Cu, Fe, Mn, Ni, Pb, and Zn) binding various forms. It is strongly recommended to quantify the fractions of metal characterized by the highest mobility and availability applied for sample which the total concentration is high enough. This procedure provides a measurement of extractable metals from a reagent such as acetic acid (0.11 mol/l), hydroxlyammonium chloride (0.1 mol/l) and hydrogen peroxide (8.8 mol/l), plus ammonium acetate (1 mol/l), which are exchangeable, reducible and oxidizable metals, respectively. There are many researchers have studied about this procedure and results indicated that this procedure gave excellent recoveries for all six elements (e.g., Cu, Cr, Cd, Zn, Ni and Pb). The concentration of metal extracted by the various reagents above gave a good reproducibility on species bonded to carbonates, Fe/Mn-oxides, and the residual fraction. Characters of each fraction are simplified and shown in Fig. 5 which summary of these fractions are given below and details were described by Serife et al. (2003).

BCR 1: is an exchangeable, water and acid-soluble fraction. This fraction represents amounts of elements that may be released into the environment if the condition becomes more acidic. Acetic acid is applied for this extraction. The extracted solution includes water-soluble form, easily exchangeable (non-specifically adsorbed) form and carbonate bonding form which are vulnerable to change of pH and sorption-desorption processes. In addition, plants can uptake this fraction easily; consequently, this metal form may in turn contaminate into food chain. It is therefore the most dangerous form for the environment concern.

BCR 2: is a reducible fraction. It theoretically represents contents of metals bond to iron and manganese oxides/hydroxides. These oxides/hydroxides are excellent cleaners of some trace metals that have been weathered and transported from the initial sources. They are thermodynamically unstable under anoxic conditions (Panda et al., 1995). Hydroxylamine hydrochloride is used for this extraction. Levels of extraction in this step should be effected by efficiency and selectivity of reagents used in the previous BCR 1. Therefore, this fraction may be too high if the carbonates have not been completely dissolved or too low if parts of the iron and manganese hydroxides have already been extracted.
BCR 3: is an oxidisable fraction or organic bound. Hydrogen peroxide and ammonium acetate are applied for this extraction. Metals can bond to various forms of organic matter. The complexities of natural organic matter are well recognized, as the phenomenon of bioaccumulation in certain living organisms. These organic matters can be degraded naturally under oxidizing conditions in waters leading to release of soluble metals. An oxidizing condition may have occurred during exposure to the atmosphere either by natural or artificial processes.
BCR 4: is defined as final residue. The final fraction can be calculated as the difference between metal contents extractable from Aqua Regia method (using nitric and hydrochloric acids) and metal contents released from the previous sequential extractions. Metal contents of all three previous fractions are considerable as more mobile and bioavailable than the residual fractions (Tack & Verloo, 1995; Ma & Rao, 1997). The residual metals appear to have relation with mineral structures that are the most difficult to be extracted (Kersten & Förstner, 1991).

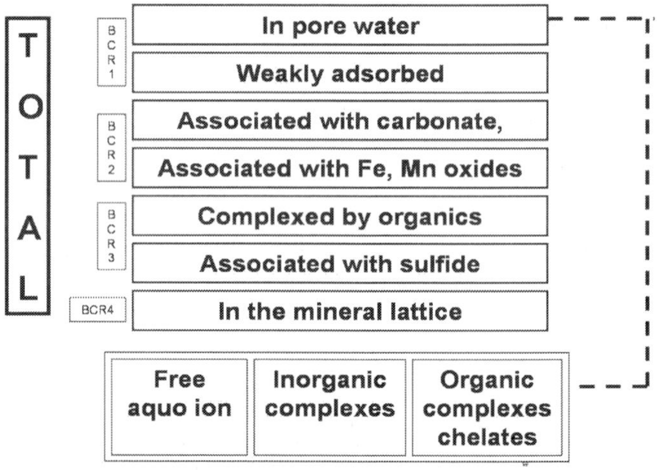

Fig. 5. Chemical fractions of metals in sediments and their characters.

5. Case study in Thailand

Geochemical investigations as mentioned above were applied to the environmental aspects of Akara Gold mine in Pichit Province of Thailand (i.e., Changul et al., 2010 a and b;

Sutthirat et al., 2011). Although, obvious environmental impacts have never been directly evidenced, some concerns have been raised by some sectors. Waste rocks from particular mining pit and tailings from tailing pond were characterized based on their geochemistry. Apart from AMD assessment, investigation of the geochemical characteristics, including their heavy metal contents and the potential of each of these metals to leach, is the first step to develop the best practice for environmental protection. Results of these studies are summarized below.

5.1 Waste rocks

Six types of waste rocks including volcanic clastic, porphyritic andesite, andesite, silicified tuff, silicified lapilli tuff and sheared tuff were collected under supervision of mining geologists. Whole-rock geochemistry, particularly their major compositions (rock powders analyzed by XRF), can be used to differentiate these rocks clearly as shown in Fig. 6; moreover, some trace elements and rare earth elements, using EPA 3052 digestion and analyzed by ICP-OES, were applied for determination of their geneses and evolutions (Sutthirat et al., 2011). Although, these may not be related to environmental aspect they should be initial investigation, at least to distinguish types of waste rock clearly before further testing program will be designed.

Subsequently, nitric leaching of these rocks was experimented following the EPA 3051 method. Amounts of leachable elements were then compared with the total digestion. Almost linear relationship between both forms of at least eight heavy metals was observed (Fig. 7). Except for As, the nitric recoverable levels of the heavy metals were slightly lower than the total concentrations. In conclusion, the maximal leaching potential (%) of these heavy metals were calculated as 30.5 - 63.2% for As, 80.4 - 81.9% for Ag, 0 - 92. 8% for Cd, 63.6 - 87.6% for Co, 91.1 - 100% for Cu, 87.9 - 99.7% for Mn, 85.3 - 93.5% for Ni and 0 - 82.8% for Pb, respectively. Three of the six rock types, i.e., porphyritic andesite, silicified tuff and silicified lapilli tuff, are of the greatest concern because they contain a high heavy metal load (proportional concentration) each with a high maximal acid leaching potential. In the worst case scenario, over 50% of the total heavy metal load would be leached by a very strong acid passing through these rocks and impacting the environment, consequently; however, this case is unrealistic and unlikely to happen.

Acid Base Accounting (ABA) and Net Acid Generation (NAG) tests were applied for evaluation of acid generation potential of these waste rocks (Changul et al., 2010a). Experimental results reveal silicified lapilli tuff and shear tuff are potentially acid forming materials (PAF); on the other hand, the other rocks, i.e., volcanic clastic, porphyritic andesite, andesite and silicified tuff are potentially non-acid-forming (NAF). Among these west rocks, shear tuffs appear to be the most impact to the environment, based on their highest potential of acid forming. Therefore, great care must be taken and focused on this rock type. Finally, they also finally concluded that AMD generation from some waste rocks may be occur a long time after mine closure due to the lag time of the dissolution of acid-neutralizing sources. In addition, environmental conditions, particularly the oxidation of sulphides which is usually activated by oxygen and water, are the crucial factor. Consequently, waste rock dumping and storage must be planned and designed very well that will lead to minimization of risk from AMD generation in the future. Surface management system and addition storage pound should be installed to control the over flood and runoff direction away from the rock waste dump. Environmental monitoring plan including water quality should be also put in place.

5.2 Tailings

Tailing samples were also systematically collected and analyses for chemical composition and mineral assemblages (Changul et al., 2010b). Consequently, these tailings have little differences of chemical compositions quantitatively from place to place but their mineral assemblages could not be clearly distinguished. They suggested that these end-processed tailings were mixed between high and low grade ores which may have the same mineral assemblages. Variation of chemical composition appeared to have been modified slightly by the refining processes that may be somehow varied in proportion of alkali cyanide and quick lime in particular. Moreover, content of clay within the ore-bearing layers may also cause alumina content in these tailings, accordingly. Total heavy metals in the tailing samples were analyzed using solution digested following the EPA 3052 method. Toxic elements including Co, Cu, Cd, Cr, Pb, Ni, Zn etc. range within the Soil Quality Standards for Habitat and Agriculture of Thailand. Only Mn contents are higher than the standard.

Potential of acid generation of these tailings was tested on the basis of Acid-Base Accounting (ABA) and Net Acid Generation (NAG) tests. Tailing samples appear to have high sulfur content but they also gave high acid neutralization capacity; therefore, they were generally classified as a non-acid forming (NAF) material. However, they still suggested that oxidizing process and dissolution should be protected with great care. Clay layer may be placed over the pound prior to topping with topsoil for re-vegetation after the closure of the mining operation. Native grass is suitable for stabilization of the surface and reduction of natural erosion. In addition, water quality should also be monitored annually.

Mining and environmental management programs usually require considerable data for best practice of mining operation and environmental monitoring. The management techniques include the sampling and classification of waste rock types.

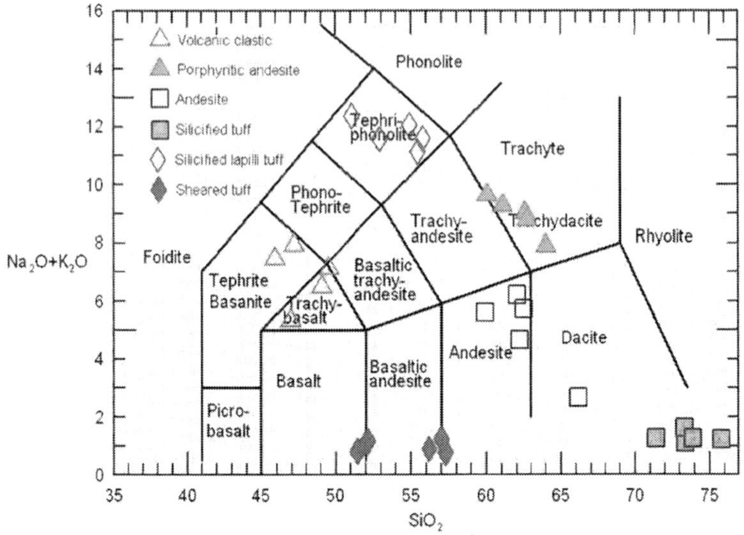

Fig. 6. Alkali-silica discrimination diagram of Le Bas et al. (1986) applied for whole-rock geochemical analyses of waste rocks from the Akara Gold Mine, Thailand

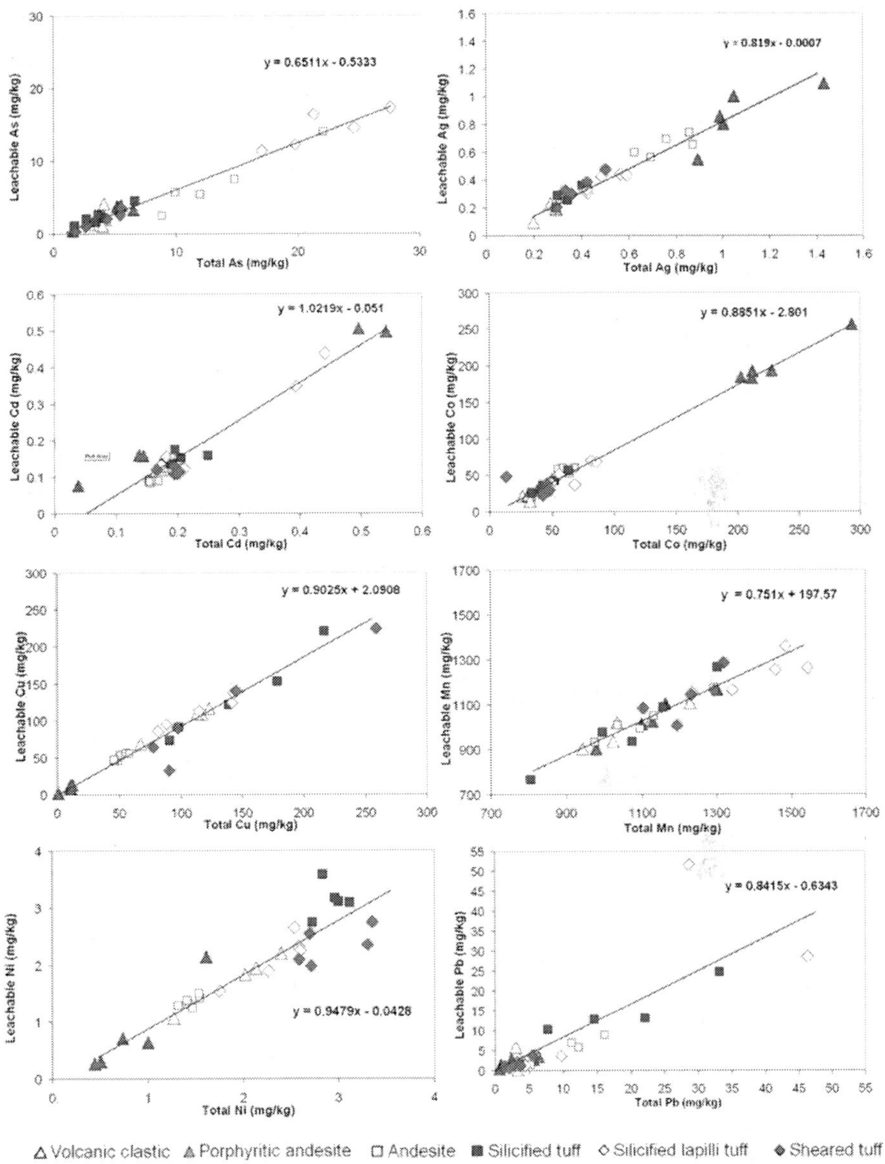

Fig. 7. Correlations between the total and nitric-leachable concentrations of eight heavy metals from various waste rocks from Akara Gold Mine, Thailand, showing linear regression relation

6. Conclusions

Solid mining wastes including host rocks and tailings must be managed during the whole period of operation. Some of them can be utilized for construction and other activities; however, some of them may also cause severe environmental impacts. Moreover, unexpected occasions can be happened individually even routine monitoring program has been carried out during all time of the operation. Therefore, all concerns must be taken into account since mining plan is developed initially. All mining wastes generated from each step of operation should be tested and put into the long term monitoring plans. Besides, all types of top soil and host rock must be sampled systematically for analyses of AMD and heavy metals prior to waste categorization and placement design. Dealing with natural materials, both rock and top soil in this case, variety of chemical composition may lead to complexity. Many of these chemicals are stable and unable to leaching out; however, just in case of some leachable form exiting, it may turn to harmfulness and difficulty of operation. Protection and prevention should therefore be planned well to keep mining operation moving smoothly and clearly to be inspected.

Regarding to rock waste and top soil, both AMD and heavy metal have become the most concerns for mining and environmental management. Some materials are unlikely to cause AMD but they contain high amounts of heavy metals that seem to be well leachable. These materials must be placed away from AMD potential wastes. Otherwise, mixing up of both types can threaten the surrounding area leading to widely land contamination. Neutralizer should be provided during the placement process. Limestone has been used as natural neutralizer which is easy to find and quite cheap. Liners should also be provided particularly for waste materials trending to have potentials of acid generation and/or heavy metal contaminants. Both natural and artificial materials can be used in individual cases, based on nature of the site and characteristics of mining waste. Cares must be taken very well during operation as well as monitoring program must be carried out regularly. It would also be notified that unexpected events can occur all the time; therefore, detailed investigations have to be initiated anytime whenever unusual signature is reveled either by regular monitoring or accident finding.

7. Acknowledgements

The author would like to thank all staff member of Geology Department, Faculty of Science, Chulalongkorn University for their support. Dr. Chulalak Changul had been helping and providing information earned from her PhD thesis research. This book chapter is a part of work initiated by a research group named as Risk Assessment and Site Remediation (RASR) which has been supported by the Center of Excellence for Environmental and Hazardous Waste Management (NCE-EHWM), Chulalongkorn University. Moreover, this work was partly supported by the Higher Education Research Promotion and National Research University Project of Thailand, Office of the Higher Education Commission (project code CC1000A).

8. References

Changul C.; Sutthirat C.; Padmanahban G. & Tongcumpou C. (2010a) Assessing the acidic potential of waste rock in the Akara gold mining, Thailand, *Journal of Environmental Earth Science*, Vol. 60, pp.1065-1071

Changul, C.; Sutthirat, C.; Padmanahban, G. & Tongcumpou, C. (2010b). Chemical characteristics and acid drainage assessment of mine tailings from Akara Gold mine in Thailand, *Journal of Environmental Earth Sciences*, Vol. 60, pp. 1583-1595

Environment Australia (1997). Managing sulphidic mine wastes and acid drainage: Chapter 4, *Best Practice Environmental Management in Mining*, ISBN 0 642 19449 1

EPA (1995). Applicability of the toxicity characteristic leaching procedure to mineral processing waste, *Technical Background Document Supporting the Supplement Proposed Rule Applying Phase IV Land Disposal Restrictions to Newly Identified Mineral Processing Wastes*, Office of Solid Waste, Available via DIALOG http://www.environ-develop.ntua.gr/htdocs/pantazidou/tcremand.pdf

EPA (1996). *Microwave Assisted Acid Digestion of Siliceous and Organically Based Matrices*, Available via DIALOG http://www.epa.gov/waste/hazard/testmethods/sw846/pdfs/3052.pdf

EPA (2009). *Accelerated Acid Digestion*, Available via DIALOG http://cfpub.epa.gov/ncer_abstracts/index.cfm/fuseaction/display.abstractDetail/abstract/1390

Ferguson, K.D. & Erickson, P.M. (1988). Pre-mine prediction of acid mine drainage, In *Dredged Material and Mine Tailings*, Edited by Salomons W. & Forstner, U., Copyright by Springer-Verlag Berlin Heidelberg.

Kersten, M. & Förstner, U. (1991). Speciation of trace elements in sediments. In:, *Trace Element Speciation: Analytical Methods and Problems*, Batley, G.E., Editor, CRC Press, Boca Raton (1991), pp. 245–317

Lapakko, K. (2002). *Metal Mine Rock and Waste Characterization Tools: An Overviews*, International Institute for Environmental and Development.

Le Bas, M.J. ; Le Maitre, R.W.; Streckeisen, A. & Zanettin, B. (1986). A chemical classification of volcanic rocks based on the total alkali-silica diagram, *Journal of Petrology*, Vol. 27, pp.745–750

Ma, L. Q. & Rao, G. N. (1997). Chemical fractionation of cadmium, nickel, and zinc in contaminated soils. *Journal of Environmental Quality*, Vol. 26, pp. 259-264

Mossop, K.F. & Davidson, C.M. (2003). Comparison of original and modified BCR sequential extraction procedures for the fractionation of copper, iron, lead, manganese and zinc in soils and sediments, *Analytica Chimica Acta* ,Vol. 478, Iss. 1, pp. 111-118

Panda, D.; Subramanian, V. & Panigrahy, R.C. (1995). Geochemical fraction of heavy metals in Chilka Lake (east coast of India)- A tropical coastal lagoon, *Environmental Geology*, Vol.26, pp.199-210

Pazos-Capeáns, P.; Barciela-Alonso, M.C.; Bermejo-Barrera, A. & Bermejo-Barrera, P. (2005). Chromium available fractions in arousa sediments using a modified microwave BCR protocol based on microwave assisted extraction.

Quevauviller, P.; Imbert, J. & Olle, M. (1993). Evaluation of the use of microwave oven systems for the digestion of environmental samples. *Mikrochim. Acta*, Vol. 112, pp.147–154

Serife, T.; Senol, K. & Gokhan, B. (2003). Application of a three-stage sequential extraction procedure for the determination of extraction metal contents in highway soils, *Turkish Journal of Chemistry*, Vol.27, pp. 333-346

Shu, W.S.; Ye, Z.H.; Lan, C.Y.; Zhang, Z.Q. & Wong, M.H. (2001). Acidification of Pb/Zn mine tailings and its effect on heavy metal mobility, *Environ Int.*, Vol. 26, pp. 389–394

Skousen, J.G.; Sencindiver, J.C. & Smith, R.M. (1987). *A Review of Procedures for Surface Mining and Reclamation in Areas with Acid-Producing Materials*, National Research Center for Coal and Energy, National Mine Land Reclamation Center, Morgantown, WV.

Sutthirat, C. ; Changul, C. & Tongcumpou C. (2011). Geochemical characteristics of waste rocks from the Akara Gold Mine, Pichit Province, Thailand, *A manuscript submitted to The Arabian Journal for Science and Engineering.*

Tack, F.M.G. & Verloo, M.G. (1995). Chemical speciation and fractionation in soil and sediment analysis, *International Journal of Environment and Analytical Chemistry,* Vol. 59, pp. 225–238

Tran, A.B.; Miller, S.; Williams, D.J.; Fines, P. & Wilson, G.W. (2003). Geochemical and mineralogical characterisation of two contrasting waste rock dumps-the INAP waste rock dump characterization project, *6th ICARD (International Conference Acid Rock Drainage)*, Caims, Australia, 12-18 July 2003.

Ure, A.M.; Quevauviller, Ph.; Muntau, H. & Griepink, B. (1993). Specition of heavy metal in solids and sediment: An account of the improvement and harmonization of extraction techniques under taken under the auspices of the BCR of the commission of the European Communities, *International Journal of Environment and Analytical Chemistry*, Vol. 51, pp. 135-151

Wang, J.; Nakazato, T.; Sakanishi, K.; Yanada, O.; Tao, H. & Saito, I. (2004). Microwave digestion with HNO3/H2O2 mixture at high temperatures for determination of trace elements in coal by ICP-OES and IPC-MS, *Analytica Chimica Acta*, Vol. 514, pp. 115-124.

7

Determination of Fluoride and Chloride Contents in Drinking Water by Ion Selective Electrode

Amra Bratovcic and Amra Odobasic
University of Tuzla, Faculty of Technology,
Bosnia and Herzegovina

1. Introduction

The fluoride element is found in the environment and constitutes 0.06 – 0.09 % of the earth's crust. Fluoride is not found naturally in the air in large quantities. Average concentration of fluoride in air are in the magnitude of 0.5 ng/m^3.[1] Fluoride is found more frequently in different sources of water but with higher concentrations in groundwater due to the presence of fluoride-bearing minerals. Average fluoride concentrations in see water are approximately 1.3 mgL^{-1}. Water is vitally important to every aspect of our lives. Water is a risk because of the possible input and transmission of infectious pathogens and parasitic diseases. We use clean water to drink, grow crops for food and operate factories. The most common pollutants in water are chemicals (pesticides, phenols, heavy metals and bacteria). [2] According to the US Environmental Protection Agency, there are 6 groups which cause contamination of drinking water: microorganisms, disinfectants, disinfection byproducts, inorganic chemicals, organic chemicals, radioactive substances. This chapter concerns the importance of continuously monitoring of fluoride and chloride in drinking water by using a fluoride (F-ISE) and chloride (Cl-ISE) ion-selective electrodes.

Disinfectants that are added to reduce the number of microorganisms, as well as disinfection byproducts can cause a series of disorders in body (anaemia, impaired function of liver, kidneys, nervous system). Chemical disinfection is economically most favourable when it comes to processing large amounts of water, for the preparation of drinking water and wastewater treatment. That is why this type of disinfection is used almost exclusively in Bosnia and Herzegovina. Chlorine is one of the most widely used disinfectants. Water monitoring information helps us to control pollution level. In this context, our work concerns the determination of fluoride in spring waters from different villages in Tuzla's Canton in Bosnia and Herzegovina, and chloride in drinking tap water from Tuzla and Gradacac as well as one sample of bottled water. Spring water sample from "Tarevcica" is designed by SW1, from "Zatoca" by SW2, from "Sedam vrela" by SW3 and "Toplica" by SW4 while a tap water from Tuzla by TW and tap water from Gradacac by GW and bottled water by FW.

The development of potentiometric ion-selective electrode has a wide range of applications in determining ions in water and other mediums. These electrodes are relatively free from interferences and provide a rapid, convenient and non-destructive means of quantitatively determining numerous important anions and cations. [3] The use of ion-selective electrodes enables the determination of very low concentrations of desired ions (to 10^{-6} mol L^{-1}). The amount of fluoride present naturally in non-fluoridated drinking water is highly variable, being dependent upon the individual geological environment from which the water is obtained. It is well known that fluoridation of drinking water is an important tool in the prevention of tooth decay. Adequate fluoride ingestion is helpful to avoid caries, but over ingestion induces dental and skeletal fluorosis, which may result in malfunction of the bone and joint system. [4, 5]. The severity depends upon the amounts ingested and the duration on intake. Dental fluorosis is a condition where excessive fluoride can cause yellowing of teeth, white spots and pitting or mottling of enamel. Skeletal fluorosis is a bone disease exclusively caused by excessive consumption of fluoride.

The procedures of determination of fluoride and chloride will be described in detail. Moreover, it will be discussed advantages and disadvantages of this method. These spring waters are in used for tap water supply. The average fluoride concentration in 4 different fresh spring waters was in a range of 0.04 to 0.12 mg L^{-1}. The fluoride concentrations obtained from the analyses of samples were compared with the permissible values given by the Environmental Protection Agency, World Health Organization, American Dental Association as well as Agency for safety food of Bosnia and Herzegovina who defined maximum amount that is allowed in drinking water. The average chlorine concentration in examined tap water was in a range of 4.55 mg L^{-1}.

2. Importance of fluoride and chloride content in water

Chlorine and fluor are very reactive elements and because of that they easily bind to the other elements. They belong to the group of halogens. Fluoride (F^-) is an important anion, present in water, air and food. Fluorides come naturally into water by dissolving minerals that contain fluor, such as fluorite (CaF_2), cryolite (Na_3AlF_6) and fluorapatite ($Ca_5(PO_4)_3F$). Rocks rich in alkali metals have a larger content of fluoride than other volcanic rocks. Small amounts of fluoride are vital for the human organism, but it's toxic in larger amounts. Fluoride levels in surface waters vary according to geographical location and proximity to emission sources. Surface water concentrations generally range from 0.01 to 0.3 mg L^{-1} (ATSDR, 1993). Fluoride in drinking water is generally bioavailable. It has been shown, that with all the human exposure to fluoride that varies from region to region, drinking water is the largest single contributor to daily fluoride intake.[6] Due to this fact, daily fluoride intakes (mg/kg of body weight are based on fluoride levels in the water and water consumption per day per litter). There are maximum guiding values for fluoride in drinking water. There are no minimum imposed limits, however there are recommended values to ensure no potential health risks from lack of fluoride within the drinking water. World Health Organisation (WHO) places international standards on drinking water that should be adhered to for health purposes, however is not enforceable and each individual nation may places its own standards and conditions on drinking water. This can be seen in the United States, where the Environmental Protection Agency (EPA) places more lenient drinking water standards than that of the WHO. This can be seen in the table 1.

Primary drinking water standards are those that must be enforced. Secondary drinking water standards are non-enforceable guidelines regulating contaminants that may cause cosmetic effects (such as skin or tooth discoloration) or aesthetic effects (such as taste, odour or colour) in drinking water.[7] The WHO maximum guideline value of 1.5 is higher than the recommended value for artificial fluoridation of water supplies, which is usually 0.5 – 1.0 mgL^{-1}. [1]

Fluoride guideline value drinking water standards	Recommended minimum value (mgL^{-1})	Maximum Value (mgL^{-1})	Reference
WHO	0.5	1.5	WHO, 1993
USA			
Primary	0.5	4.0	US EPA, 1985
Secondary	0.5	2.0	
ADA	0.7	1.2	
Agency for Safety Food, B&H	-	1.5	Statute, 2007

Table 1. International and national drinking water standards of fluoride contents

Determination of chloride ions is important in many different fields such as clinical diagnosis [8, 9] environmental monitoring [10, 11, 12] and various industrial applications [13, 14]. Considering the fact that chloride channels play crucial role in physiological processes it is not surprising that missregulation of chloride ions transport by these channels can cause serious disorders. One of disease is cystic fibrosis. [15]
Chloride ions in large quantities are present in sea water and sediments of the Earth's crust where it is associated with ions Na^+, K^+; Mg^{2+}. Chlorides are widely distributed in nature as salts of sodium (NaCl), potassium (KCl), and calcium ($CaCl_2$). Chlorides are leached from various rocks into soil and water by weathering. Exposure to chloride in air has been reported to be negligible. [16] The taste threshold of the chloride anion in water is dependent on the associated cation. Taste thresholds for sodium chloride and calcium chloride in water are in the range 200–300 mg/litre [17]. Sodium chloride is widely used in the production of industrial chemicals such as caustic soda, chlorine, sodium chlorite, and sodium hypochlorite. In the human body it is also found in the form of chloride. In humans, 88% of chloride is extracellular and contributes to the osmotic activity of body fluids. The electrolyte balance in the body is maintained by adjusting total dietary intake and by excretion via the kidneys and gastrointestinal tract. A normal adult human body contains approximately 81.7 g chloride. On the basis of a total obligatory loss of chloride of approximately 530 mg/day, a dietary intake for adults of 9 mg of chloride per kg of body weight has been recommended (equivalent to slightly more than 1 g of table salt per person per day). For children up to 18 years of age, a daily dietary intake of 45 mg of chloride should be sufficient. [16] A dose of 1 g of sodium chloride per kg of body weight was reported to have been lethal in a 9-week-old child [18] Daily requirements for intake of chloride are up to the age range, from newborn to 500 mg and to 2000 mg for adults. Chlorination as a method of water purification is used in 99% cases of the disinfection of municipal water. The chlorine can be added directly into the water. The taste of chlorinated water could be slightly acidic and it is probably because of the presence of chlorine is in the form of hypochloric acid. Permissible concentration of chlorine as a means of disinfections is up to 3 mg/L. Numerous analytical methods for chloride ions in a variety of samples have been developed, such as ion chromatography [19, 20] near-infrared spectrometry [21] spectroscopy [22] light scattering [23] ionselective electrode method [13, 24, 25] turbidimetric method [26] and flow based methods coupled with different detectors [27, 28, 29].

3. Potentiometric analysis

The potentiometric method is based upon measurements of the potential that measures electromotive force of a galvanic element. Direct potentiometric determinations are almost always performed using ion selective electrodes (ISEs), which are capable of rapid and

selective measurements of analyte concentration. Ion-selective potentiometry (ISP) is a non-destructive method, which means that the sample can be used for further analysis. Ion-selective electrode (ISE) such as chloride or fluoride, which is used in our investigation, as detector provides a range of possibilities in the analysis of samples of biological material. [30]
Work of ion-selective electrode is based on the fact that there is a linear relationship between the electrical potential established between the ISE and reference electrode and the logarithm of activity (or effective concentration) of ions in the solution. This relationship is described by Nernst equation:

$$E = E^{\circ} + \frac{2,303RT}{zF}\log(a) \tag{1}$$

where E is the total potential in mV developed between the sensing and reference electrode, z is the ion charge which is negative for anions, log(a) is the logarithm of the activity of the measured ion. The factor 2,303 RT/F has a theoretical value of 59 mV at 25 °C. The equation is valid for very dilute solutions or for solutions were the ion strength is constant. The activity is equivalent to the concentration in dilute solutions but becomes increasingly lower as the ionic strength increases. The activity (a) represents the effective concentration, while the total fluoride ion concentration may include some bound ions as well. The electrode responds only to free ions so it is important to avoid the formation of complexes that are meant to be measured. In this case, the complexation would lower the activity and therefore the electrode response. This is effectively the equation of a straight line:

$$y = mx + c \tag{2}$$

where y = E = the measured electrode response in mV, x = log (a), c = E° = the intercept on the y axis, m = - 0,0592/z = the electrode slope.
Ion selective electrodes are available for measuring more than 20 different cations for instance Ag^+, Na^+, K^+, Ca^{2+}, and anions such as F^-, Cl^-, S^{2-}, CN^-.
The function of ion-selective electrode is based on selective leakage of positively charged specie from one phase to another, creating a difference in potential. Working principle is based on measuring the electrode potential (mV) depending on the concentration of tested ions in the solution. The reference electrode has a constant potential, and potential of ISE is changing with the concentration of certain ions.

3.1 Ion selective electrode as an efficient tool for monitoring of desired ion

An ion selective electrode is sensitive to analyte concentration due to the properties of the ion-selective membrane that provides the interface between the ion-selective electrode and the sample solution. The ability of the ion selective membrane to conduct current depends in some manner on the presence of analyte in the solutions on both sides of the membrane. The mechanism of this dependence varies but usually depends on some reaction of analyte at the surface of the membrane. Analysis were carried out using a MICROPROCESSOR pH/ION METER pMX 3000 WTW equipped with a reference electrode WTW R 500 and the F 500 and Cl 500 as an ion-selective electrode. In Figure 1 is schematically shown reference electrode and an ion selective electrode, where 1 indicate the filling opening for the bridge electrolyte, fluid level of the bridge electrolyte, 3 the inner junction which must be covered with bridge electrolyte and 4 the ground junction which indicate the minimum depth of immersion.

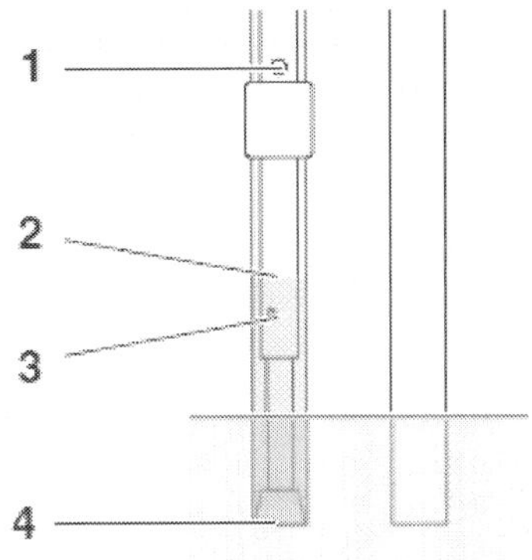

Fig. 1. Schematic representation of reference and an ion-selective electrode. In this picture 1 - indicate the filling opening for the bridge electrolyte, 2 - fluid level of the bridge electrolyte, 3 - the inner junction which must be covered with bridge electrolyte and 4 - the ground junction.

For measurements with the F 500 fluoride electrode and Cl 500 chloride electrode, a reference electrode is required. In our investigation has been used R 500 as a reference electrode. The two electrodes together form a double rod combination electrode. Ion selective electrodes have been storage into diluted aqueous standard solution. Measuring range for fluoride electrode is 0.02 mg L^{-1} or 10^{-6} mol L^{-1} and for chloride electrode from 2 to 35000 mg L^{-1} or from 10^{-5} to 1 mol L^{-1}.

There are many advantages to use an ion-selective electrode as means of analysis, including its efficiency, selectivity, ease of sample preparation and lack of interference and reactivity with sample itself.

3.1.1 Fluoride electrode

One of the most significant of the solid - state electrode is the lanthanum fluoride electrode. The membrane consists of a slice of a single crystal of lanthanum fluoride that has been doped with europium (II) fluoride to improve its conductivity. The membrane, supported between a reference solution and the solution to be measured, shows a theoretical response to changes in fluoride ion activity from 0 to 10^{-6} mol dm^{-3}. The electrode is selective for fluoride ion, only hydroxide ion appears to offer serious interference.[31] The unique property of a europium-doped lanthanum fluoride crystal to form a membrane apparently permeable to fluoride ion and virtually no other anion or cation, provided the first specific ion - selective fluoride electrode. This electrode gives Nerstian response to fluoride ion concentrations from above 1M to below 10^{-5} M, and only OH^- seems to interfere with this response.

Srinivasan and Rechnitz [32] noted that stirring sometimes had a substantial effect on the observed potential. In 10^{-3} M NaF solution, the potential changed from - 61.5 mV in a quiescent solution to - 55.5 mV in a rapidly stirred solution. This shift was less at high concentrations and negligible in the presence of 0.1 M $NaNO_3$ supporting electrolyte, even at fluoride concentration as low as 5 x 10^{-5} M. They recommended that readings be taken with slow stirring (by a Teflon - coated magnetic bar), and that under these conditions reproducibility was excellent: "The potentials were found to be quite stable, changing not more than 0.1 mV even after an hour. The reproducibility on the same day for two different solutions of the same concentration was within 0.1 mV".

The kinetic response of the electrode is almost instantaneous [32, 33], limited by the recorder response time of 0.5 sec., at least in the solutions containing fluoride concentrations greater than millimolar. In very dilute solutions, the response time is has been reported to be very long.

3.1.2 Chloride electrode

The chloride ion-selective electrode is a polycrystalline solid-state electrode that contains a membrane. The membrane consists of a solid salt of silver sulfide/silver chloride. The membrane must be insoluble in the analyte solution and contain the analyte ion of interest. The membrane is placed at the end of a solid plastic tube. This membrane is in contact with the analyte solution during the measurement. Inside of the tube is a reference solution, which contains a known and fixed concentration of analyte (Cl^-) solution. The concentration difference between this inner solution and analyte solution causes the migration of charged species across the membrane. This ion exchange process at the surface of the membrane causes a potential to develop. Since the potential of both the reference electrode and the inner reference (immersed in the standard solution) are constant, any change in measured potential is caused only by a change in potential across the membrane and is a function of the analyte chloride ion activity (or concentration).

The electrode is designed to detect chloride ions in aqueous and viscous solutions and is suitable for use in laboratory investigations. The method allows the determination of chloride in treated water, natural water, drinking water and most waste water with high accuracy and sensitivity. The method is applicable only to samples containing more than 10 000 mgL^{-1} dissolved substances.

All reagents used were of analytical reagent grade and were used without further purification.

4. Results

In the experimental work ISP as a choice method was used, and Mohr's method as a standard was the control method for the determination of chloride ions in drinking tap water. As a comparative method could be use the UV/vis spectrophotometric method with zirconium (IV) ion oxychloride and alizarin S for analysis of Fluoride contents. For the determination of chloride and fluoride ions in represented drinking water has not been required previously sample pre-treatment. Quantitative analyses were performed with calibration curves obtained with standard solutions. The calibration curve has been constructed by plotting obtained electrode potential vs. logarithm of concentrations of standard chloride and fluoride solutions. In our experiments, several standard solutions with different concentrations were prepared. Then, we measured the cell potential for each individual standard solution and plot E_{cell} vs. log C_{F^-}. This curve is our calibration curve and has been used to determine the concentration of the unknown. The F-ISE method for the fluoride determination can be applied either without pretreatment technique, namely conventional potentiometric method, or with pretreatment technique, such as co-precipitation and steam distillation. Frant and Ross [34] pointed out that there were changes in potential as the pH of fluoride solutions was changed.

Since ion-selective electrode responds to activity of the analyte, it is extremely important ionic strength solution. From the literature it is known that the OH^-ions are only interfering ions for fluoride electrode, at pH greater than eight. However, at pH lower than five, the hydrogen ions also interfere, but the pH can not be too low due to the formation of HF,

which is a weak acid and whose salt with water gives alkaline reaction. The interference for this fluoride electrode is pH less than 5 and higher than 7.
In this work has been used the electrode without the addition of any ionic buffer for the determination of F^- in examined water. The composition of the water and the total ionic strength were analysed and were not over allowable limit for this methods in a range of allowed concentration. The interference on the fluoride electrode from hydroxyl ion (OH^-) is eliminated by ensuring that pH is kept below 8. Consequently, there was no necessity to add TISAB buffer to ensure constant ionic strength.

5. Experimental part

5.1 Potentiometric determination of fluoride

A 1000 mg L^{-1} sodium fluoride stock solution was prepared by dissolving 2,21 g NaF in a 1000 mL polystyrene volumetric flask with deionised water. Sodium fluoride has been previously oven-dried at 105 °C for 1 hour and stored in a dessicator. The concentration of this stock solution is 1000 mgL^{-1}. Standards at the required concentration were prepared by appropriate dilution of the stock solution.
Calibration diagrams were obtained by measuring of potential of six different sets of fluorid standard solutions ordered from low to high concentration. The concentration range is from 0.07 to 1.0 mgL^{-1}. The meter reading was taken after a constant value has been attained that is drift < 0,1 mV/min. The results are given in Table 2.

Concentration of F^- (mgL^{-1})	0.07	0.1	0.3	0.5	0.7	1.0
Log C_{F^-}	-1,154	-1,0	-0,522	-0,301	-0,154	0.0
Potential (mV)	33,6	21,3	-0,1	-12,2	-24,7	-31,5

Table 2. Potentiometric responses of the membrane towards different concentrations of fluoride ion.

On the basis of these results has been constructed diagram 1.

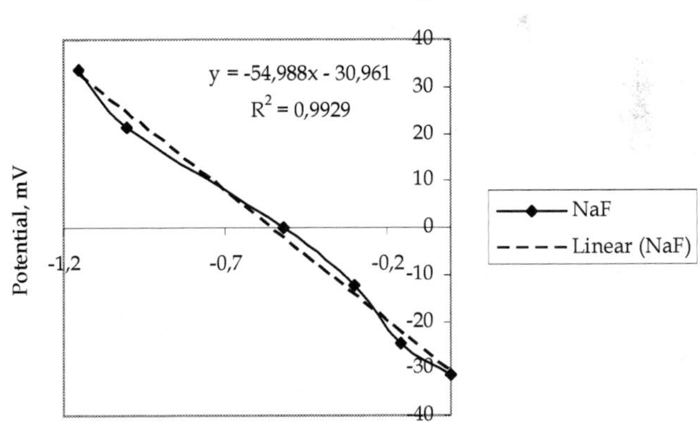

Diagram 1. Calibration curve for Fluoride ISE obtained for fluoride standard solutions in range of concentration from 0,07 to 1 mol L^{-1}. This calibration graph has been used for determination of the samples marked as SW1, SW2, SW3 and SW4.

For determining the concentration of F-ions, the samples were placed in a clean, dry glass in quantities of 50 ml. [35] First of all, has been determined the pH of the sample. The measured pH value was in the range from 5 to 7, and then the sample has been stirred by using a magnetic stirrer for 5 minutes. After that, it has been measured the concentration of fluoride ions, by immersion of the reference and fluoride ion-selective electrode connected to the ion-meter. After a few minutes were read values of the potential. Each sample was measured three times in order to reduce experimental error. Based on the measured potential, was calculated the concentration of fluoride for each individual measurement, and then, calculated the average value of concentration.

In Table 3 are represented the obtained concentrations in samples marked as SW1, SW2, SW3 and SW4.

Sample	SW1	SW2	SW3	SW4
Potential (mV)	44,5	47,1	36,5	18,7
Conc. F^- (mgL^{-1})	0,042	0.038	0.059	0.12

Table 3. Concentrations of fluoride obtained for samples: SW1, SW2, SW3 and SW4.

For these samples also have been determined the concentration of chloride by mercurimetric titration. The results are shown in table 4.

Sample	Concentration of Chloride (mgL^{-1})
SW1	2,81
SW2	4,80
SW3	3,60
SW4	7,25

Table 4. Concentrations of chloride obtained by mercurimetric titration method.

5.2 Potentiometric determination of chloride

Specific ion electrodes measure activity and not concentration, a large amount of an inert strong electrolyte (e.g. nitrate ion) can be added to fix the ionic strength to a constant value. When the ionic strength is constant, the activity is constant and concentration can be accurately measured. To determine the concentration of chloride ions, samples were prepared as follows: in a glass flask of 100 ml was measured 2 ml of 5% $NaNO_3$, and diluted to mark with water that is being analyzed (5% $NaNO_3$ concentrations in all samples was 0.1 mol L^{-1}). Then, 5 mL of prepared sample was transferred in clean, dry glass and stirred using a magnetic stirrer for 5 min with immersed electrodes. After 5 minutes of stirring, the magnetic stirrer has been stopped and then red the potential. Response time for all samples was in a range from 1 to 5 minutes. The samples marked by FW, TW and GW have been analyzed on chloride concentration using a chloride selective electrode. The sample designed as FW was analyzed using a calibration curve represented in diagram 2, while the samples marked as TW and GW by using a calibration curve shown in diagram 3. In Table 5 are given the concentration of chloride solutions for KK1 calibration curve.

Conc. Cl^- (mgL^{-1})	60	120	180	230	280
Potential (mV)	155.1	148.6	143.7	140.2	135.3

Table 5. Electrode response on prepared chloride standard solutions.

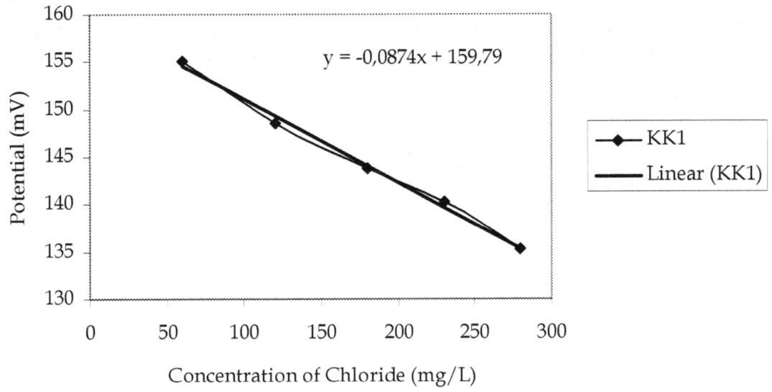

Diagram 2. Shows the obtained calibration curve KK1.

In Table 6 are given the concentration of chloride solutions for KK2 calibration curve.

Conc. Cl^- (mgL^{-1})	1	3	5	10	15	20
Potential (mV)	183.3	179.6	177.8	171.5	166.8	161.3

Table 6. Electrode response on prepared chloride standard solutions.

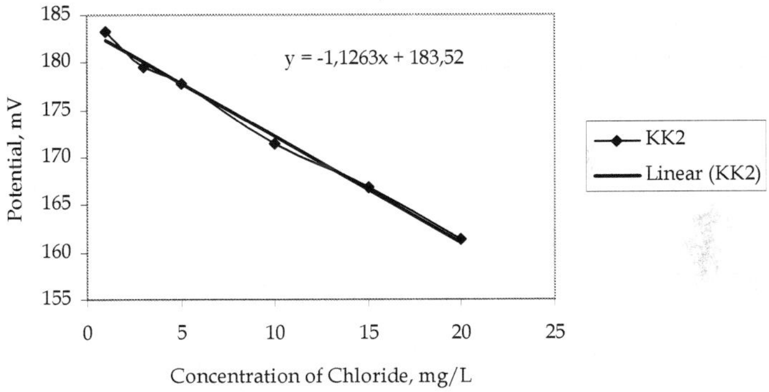

Diagram 3. Shows the obtained calibration curve KK2.

In Table 7 are shown the average concentrations of chloride ion determined in our tested samples.

Sample	FW	TW	GW
Potential, mV	129.9	171,9	178,4
Concentration, mg/L	341.99	10,12	4,55

Table 7. Concentrations of chloride obtained for samples: FW, TW and GW determined using appropriate calibration curve.

5.3 Mohr's method

Cations and anions are systematized according to the analytical groups to make it easier to prove. When the sample contains a lot of cations and anions is difficult or even impossible to prove, because they interfere with each other. Ions belonging to a different groups defined by their relationship to reagent with which the ion is deposited in hard soluble salt. Chloride ion belongs to the fourth group of anions that precipitate reagent $AgNO_3$. Mohr's method is used for volumetric determination of chloride by titration with $AgNO_3$ solution in neutral or slightly alkaline solution and using of potassium or sodium chromate as indicator. It is based on the reactions of the formation of hardly soluble precipitates with the condition that the reaction of precipitation is fast and that there is a true indicator that shows the end of the titration. To determine the concentration of chloride by Mohr, samples were prepared as follows: the sample has been transferred by pipette of 25 mL into Erlenmayer flask and diluted by distilled water (about 100 mL) and added 2 mL of 5% K_2CrO_4. Thus, titration of the sample prepared in this way has been done with standard solution of 0,0984 mol/L $AgNO_3$. The standardization of $AgNO_3$ has been done previously. Titration was completed when appeared a reddish solution.

The amount of chloride was calculated using the equation:

$$m_{Cl^-} = C_{AgNO_3} \cdot V_{AgNO_3} \cdot M_{Cl} \cdot R$$

where:

m_{Cl} - the amount of chloride in water (g)

C_{AgNO3} - concentration of solution (mol L^{-1})

V_{AgNO3} – volume of $AgNO_3$ used for titration (L)

R- dilution

Calculated values of chloride concentration by Mohr method is 14.8 mg L^{-1} for TW sample. TW sample shows a significant discrepancy in values between the two methods used. The difference is caused by problems that can occur when working with a chloride electrode. Interference can cause:

- Complexes with Bi^{3+}, Cd^{2+}, Mn^{2+},Pb^{2+},Sn^{2+}, Tl^{2+}
- Reducing agents
- Interfering ions: 10 % error with the following concentration ratio.

(concentration ratio = interfering ion/measured ion):

In the table are given values of concentration relations for some interfering ions:

OH^-	Br^-	J^-	S^{2-}	CN^-	NH_3	$S_2O_3^{2-}$
80	3x10^{-3}	5x10^{-3}	1x10^{-6}	2x10^{-7}	0.12	0.01

To determine accurately interfering ion present and its concentration in the sample, TW, require long and detailed chemical and bacteriological analysis of water. The results obtained for the GW and TW indicate that the chloride content is in the range of permissible limits prescribed by WHO.

Results for the FW sample show that the concentration of chloride ions is extremely high and exceeds the maximum limit. According to the Regulations of the Republic of Serbia, given that Bosnia and Herzegovina has no defined Rules on allowable concentrations of cations and anions in water, for the chloride limit is 200 mgL^{-1} (Official Gazette of SFRY 42/98). The limit in drinking water is 250 mgL^{-1}. European Economic Community

Directive 80/777/EEC provides that in case of bottled natural mineral waters, chloride concentrations exceed 200 mgL^{-1}, and then the water is declared on the label as chlorinated.

6. Conclusion

Electroanalytical methods based on potentiometry with ion-selective electrodes seem to be the most popular and convenient methods of fluoride and chloride ion determination. Fluoride and chloride selective electrodes can be used to determine fluoride and chloride concentrations in drinking water due to its high selectivity, specificity and low detection limits. The advantages of this study include a short analysis time, elimination of sample pretreatment, simplicity of the measuring system and relatively low instrument cost. The concentration of fluoride ion was determined in 4 drinking water samples, while the concentrations of chloride have been determined in 3 samples (FW, TW and GW) by a chloride selective electrode as well as by Mohr's method. All these samples were analyzed with use direct reading method. By our experimental data we can conclude that the concentration of fluoride in samples marked as SW1, SW2, SW3 and SW4 is within allowed concentration according to World Health Organisation. On the basis of the results of analysis carried out on the water content chloride ions can be concluded that the applied electrochemical measurements and analytical shown that the content is the same within the limits of permissible concentration laid down by WHO. Method ISP when it proved more effective, fast and reliable enough to determine chloride ions in the water and the concentration in the range of 10^{-4} mol L^{-1} to 10^{-5}mol L^{-1}. Additionally, it has an advantage over any other analytical method because it is non-destructive and allows the use of samples for other types of analysis. Based on the results obtained it can be concluded that there are many advantages of using ion-selective potentiometry (ISP) in reference to standard spectrophotometric and Mohr's methods, because measurements with the ISP are faster, efficient and reliable. It does not require the use of many different chemicals, and does not require any preparation of samples before analysis, which directly affects the economic availability. Our experimental data give in evidence that the concentration in these samples are within the allowed concentration according to World Health Organisation except the concentration of chloride in tested bottled water. Therefore, determining of Fluoride and Chloride in drinking water is of great significance for human health because of daily consumption of certain amounts.

7. References

[1] Fluoride in Drinking - water, WHO, 2004.

[2] Rowell, R. M.; Removal of metal ions from contaminated water using agricultural residues, 2nd International Conference on Environmentally - Compatible Forest Products, Portugal (2006), 241-250.

[3] Hutchins, R. S.; Bachas, L. G.; In: Handbook of Instrumental Techniques for Analytical Chemistry, (Ed.), Chapter 38, 727-748, Upper Saddle River, NJ: Prentice-Hall, 1997.

[4] Institute of Medicine, (1997), Fluoride. In "Dietary reference intakes for calcium, phosphorus, magnesium, vitamin D, and fluoride", 288-313. National Academy Press. Washington, D.C., U.S.A.

[5] World Health Organisation (WHO) 2002, Fluorides, World Health Organization (Environmental Health Criteria 227).

[6] Appropriate use of fluorides for human health, J. J. Murray, 1986.
[7] United States Environmental Protection Agency (US EPA), 1985.
[8] Jiang, Q.S.; Mak, D.; Devidas, S.; Schwiebert, E.M.; Bragin, A.; Zhang, Y.L.; Skach, W.R.; Guggino, W.B.; Foskett, J.K.; Engelhardt, J.F., J. Cell Biol. 1998, 143, 645-657.
[9] Huber, C.; Werner, T.; Krause, C.; Klimant, I.; Wolfbeis, O.S., Anal. Chim. Acta 1998, 364, 143-151.
[10] Montemor, M.F.; Alves, J.H.; Simoes, A.M.; Fernandes, J.C.S.; Lourenco, Z.; Costa, A.J.S.; Appleton, A.J.; Ferreira, M.G.S., Cem. Concr. Compos. 2006, 28, 233-236.
[11] Huber, C.; Klimant, I.; Krause, C.; Werner, T.; Mayr, T.; Wolfbeis, O.S., Fresenius J. Anal. Chem. 2000, 368, 196-202.
[12] Martin, A.; Narayanaswamy, R., Sens. Actuator B-Chem. 1997, 39, 330-333.
[13] Babu, J.N.; Bhalla, V.; Kumar, M.; Mahajan, R.K.; Puri, R.K., Tetrahedron Lett. 2008, 49, 2772-2775.
[14] Badr, I.H.A.; Diaz, M.; Hawthorne, M.F.; Bachas, L.G., Anal. Chem. 1999, 71, 1371-1377.
[15] Ratjen, F.; Doring, G., Lancet 2003, 361, 681-689.
[16] Department of National Health and Welfare (Canada). Guidelines for Canadian drinking water quality. Supporting documentation. Ottawa, 1978.
[17] RC Weast, ed. CRC handbook of chemistry and physics, 67th ed. Boca Raton, FL, CRC Press, 1986.
[18] Sodium, chlorides, and conductivity in drinking water: a report on a WHO working group. Copenhagen, WHO Regional Office for Europe, 1978 (EURO Reports and Studies 2).
[19] Mesquita, R.B.R.; Fernandes, S.M.V.; Rangel, A., J. Environ. Monit. 2002, 4, 458-461.
[20] Pimenta, A.M.; Araujo, A.N.; Conceicao, M.; Montenegro, B.S.M.; Pasquini, C.; Rohwedder, J.J.R.; Raimundo, I.M., J. Pharm. Biomed. Anal. 2004, 36, 49-55.
[21] Wu, R.H.; Shao, X.G., Spectrosc. Spectr. Anal. 2006, 26, 617-619.
[22] Philippi, M.; dos Santos, H.S.; Martins, A.O.; Azevedo, C.M.N.; Pires, M., Anal. Chim. Acta 2007, 585, 361-365.
[23] Cao, H.; Dong, H.W., J. Autom. Methods Manag. Chem. 2008, Article No 745636, 5.
[24] Kumar, K.G.; John, K.S.; Indira, C.J., Indian J. Chem. Technol. 2006, 13, 13-16.
[25] Shishkanova, T.V.; Sykora, D.; Sessler, J.L.; Kral, V., Anal. Chim. Acta 2007, 587, 247-253.
[26] Mesquita, R.B.R.; Fernandes, S.M.V.; Rangel, A., J. Environ. Monit. 2002, 4, 458-461.
[27] Junsomboon, J.; Jakmunee, J., Talanta 2008, 76, 365-368.
[28] Pimenta, A.M.; Araujo, A.N.; Conceicao, M.; Montenegro, B.S.M.; Pasquini, C.; Rohwedder, J.J.R.; Raimundo, I.M., J. Pharm. Biomed. Anal. 2004, 36, 49-55.
[29] Bonifacio, V.G.; Figueiredo-Filho, L.C.; Marcolino, L.H.; Fatibello-Filho, O., Talanta 2007, 72, 663-667.
[30] Mentus S., Electrochemistry (Belgrade, 2001).
[31] Douglas A. Skoog, Donald M. West, F. James Holler, Stanley R. Crouch, Fundamentals of Analytical Chemistry, 8th edition, pag. 607.
[32] Srinivasan K., Rechnitz G. A., Anal. Chem. 40, 509 (1968).
[33] Srinivasan K., Rechnitz G. A., Anal. Chem. 40, 1818 (1968).
[34] Frant, M., Ross, J. W., Jr., Science 154, 1553 (1966).
[35] Bratovcic A., Master thesis, Determining of Fluoride contents in waters by application of contemporary of electrochemical methods, 2008, Tuzla, Bosnia and Herzegovina.

8

Environmental Background Radiation Monitoring Utilizing Passive Solid Sate Dosimeters

Hidehito Nanto[1,2], Yoshinori Takei[1,2] and Yuka Miyamoto[2,3]
[1]*Advanced Materials Science R&D Center,*
[2]*Research Laboratory for Integrated Technological Systems,*
Kanazawa Institute of Technology, Hakusan, Ishikawa,
[3]*Oarai Research Center, Chiyoda Technol Corporation, Oarai-machi, Higashi Ibaragi,*
Japan

1. Introduction

Natural environmental background radiation is radiation that is constantly present in the environment and is emitted from a variety of natural and artificial sources. Primary contribution comes from sources in the earth, from space and in the atmosphere. Naturally occurring sources are responsible for the vast majority of radiation exposure. However, not including direct exposure from radiological imaging or therapy, about 3% of background radiation comes from man-made sources such as self-luminous dials and signs, global radioactive contamination due to historical nuclear weapons testing, nuclear power station or nuclear fuel reprocessing accidents, normal operation of facilities used for nuclear power and scientific research, emission from burning fossil fuels and emission from nuclear medicine facilities and patients.

We are all exposed to ionizing radiation every day. In fact, the environmental background radiation contributes about two-thirds of our radiation exposure. Therefore, it is important to determine the exact environmental background radiation dose. Active dosimeters have been formally appropriate for monitoring dose equivalent rates of environmental background radiation. On 2001 in Japan, not only dose equivalent rate but also dose equivalent can be applied to environmental background radiation monitoring, which is based on the Japanese law modification concerned with radiation protection. Thus, there is the possibility that passive solid state dosimeters are also appropriate for environmental background radiation monitoring.

So far, some types of solid state dosimeter have been developed not only for personal monitoring but also for environmental background radiation monitoring. For instance, a thermoluminescence (TL) dosimeter has been studied to monitor the environmental background radiation (Nanto, 2011). Recently newly passive solid state dosimeters utilizing optically stimulated luminescence (OSL), direct ion storage (DIS) and radiophotoluminescence (RPL) phenomena have been developed to monitor the personal and environmental radiation (Ranogajec-Komor, 2008; Koyama, 2010).

In the following, the basic principle of the passive solid state dosimeters utilizing TL, OSL, DIS and RPL phenomenon are reviewed and the results on environmental background monitoring using these passive dosimeters, especially personal dosimeter utilizing RPL penopmenon, are shown and discussed.

2. Passive solid state dosimeters

Active dosimeters have been formally appropriate for monitoring dose equivalent rates of environmental natural radiation. In 2001, not only dose equivalent rate but also dose equivalent can become applied to environmental natural radiation monitoring; the dose equivalents at the boundary of the controlled area and the area is limited to be less than or equals to 1.3 mSv/3 months. Thus, there is the possibility that passive dosimeters are also appropriate for the environmental natural radiation monitoring. An application of various kinds of passive dosimeters, especially the dosimeter utilizing TL phenomenon (TL dosimeter), has been studied to monitoring the environmental natural radiation (Saez-Vergara, 1999). Recently, new passive dosimeters such as the dosimeter utilizing OSL phenomenon (OSL dosimeter), the dosimeter utilizing DIS phenomenon (DIS dosimeter) and the glass dosimeter utilizing RPL phenomenon (RPL glass dosimeter) have been developed as the personal dosimeter. In this study, the RPL glass dosimeters as well as the OSL dosimeter and the DIS dosimeter has been applied to monitor the environmental natural background radiation .

2.1 Operation principle of the solid state dosimeters

In this section, the operation principle of the solid state dosimeters, such as RPL glass dosimeter, the OSL dosimeter and the DIS dosimeter are discussed.

2.1.1 Glass dosimeters

Ag^{+}-doped phosphate glass after exposure to ionizing radiation has an intense luminescence by the excitation with ultraviolet light. This phenomenon is called radiophotoluminescence (RPL). When Ag^{+}-doped phosphate glass is exposed to ionizing radiation, electron and hole pairs are produced. The electrons are captured by the Ag^{+} ions in the glass structure, and

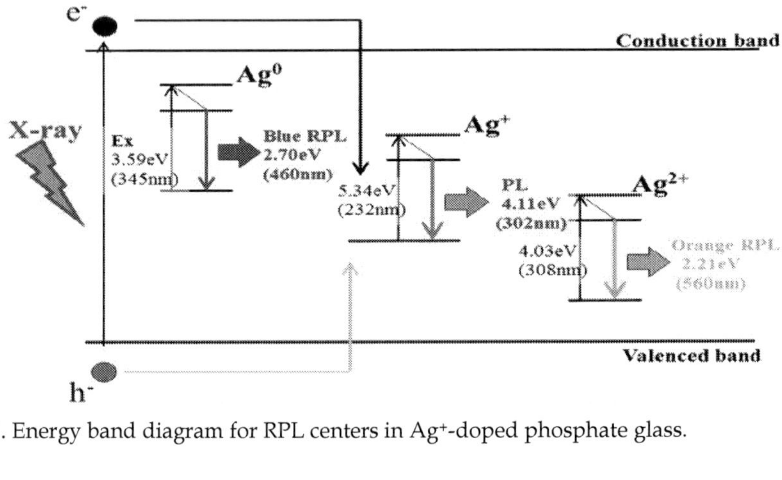

Fig. 1. Energy band diagram for RPL centers in Ag^{+}-doped phosphate glass.

then Ag^+ ions change to Ag^0 ions. On the other hand, the holes are captured initially by PO_4 tetrahedra and then migrate to produce Ag^{2+} ions. It has been reported (Miyamoto, 2011) that both Ag^0 and Ag^{2+} ions can be the centers of luminescence in the phosphate glass as shown in Fig.1. Morever, once trapped, luminescence centers are stable unless the glasses are annealed at high temperature at about 400℃. Figure 2 shows photograph of orange RPL from the glass dosimeter which was exposed to x-ray. As the RPL intensity is proportional to the amount of irradiation, the Ag^+- doped phosphate glass can be used in individual monitoring of ionizing radiation.

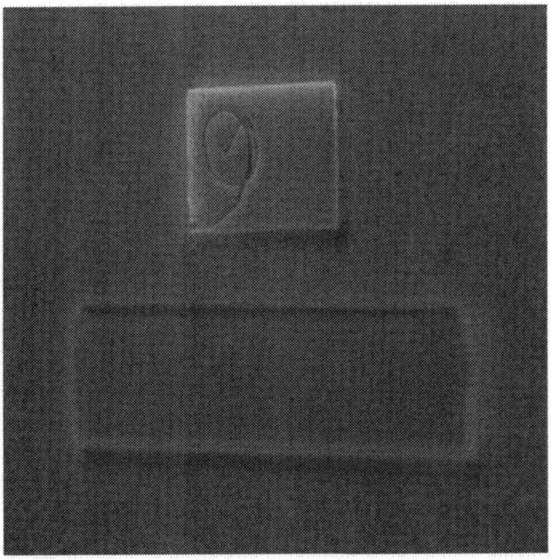

Fig. 2. Photographs of emitted RPL from the glass dosimeter which was exposed to x-ray (upper photograph) and without x-ray irradiation (down photograph).

2.1.2 OSL dosimeters

The OSL process as well as TL process is based on the presence of electron and/or hole traps and luminescence centers in storage phosphor materials (Nanto, 1998). Figure 3 shows the energy band diagram of Eu doped BaFBrI (BaFBrI:Eu) photostimulable storage phosphor which is used as the storage phosphor material of the imaging plate (IP) (Nanto, H. 2006) for the computed radiography. Upon irradiation with ionizing radiation such as x-ray to storage phosphor materials, free electrons in conduction band (C.B.) and holes in valenced band (V.B.) are promoted via band-to-band excitation. The free electrons are, then, trapped at anion vacancies such as F, I and Br vacancies to produce the F centers as the electron trap centers. While the free holes are trapped at the Eu^{2+} impurity centers to produce the Eu^{3+} impurity centers. Detrapping of these carriers requires energy.

In OSL process, the energy is provided by stimulating the phosphor materials with visible or near infrared light after irradiation. During a detrapping transition, free electrons stimulated from the F centers into the conduction band recombine with the luminescence centers of the Eu^{3+} ions, whereby visible photons (OSL) are emitted as shown in Fig.3.

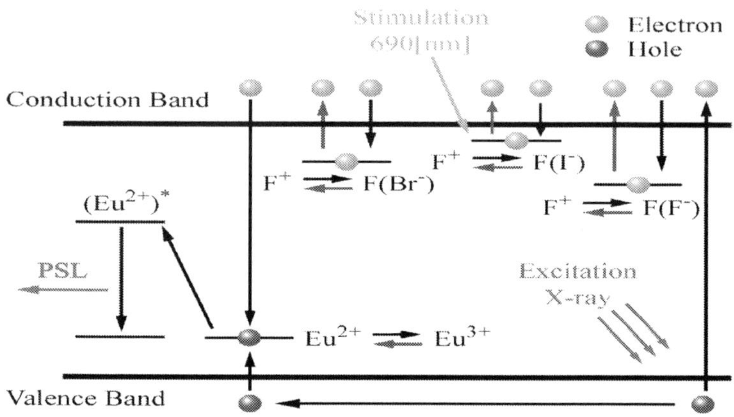

Fig. 3. Energy band diagram for OSL process in BaFIBr:Eu phosphor.

In TL process the energy is provided by heating the phosphor materials. Since the OSL intensity as well as TL intensity is proportional to x-ray dose, the phosphor materials which exhibit the OSL or TL phenomenon offer an alternative to conventional x-ray film (Nanto, 1999). Figue 4 shows typical OSL spectra and their stimulation spectra of various IPs. In all IPs, the OSL peaked at about 400 - 450 nm is observed by stimulating with about 550-650 nm light. The OSL phenomena can, therefore, be applied to the computed radiography using IP with BaFBrI:Eu phosphor materials as well as to individual radiation monitoring and environmental monitoring using LiF:Mg (Saez-Vergara, 1999) TL dosimeters or Al_2O_3:C OSL dosimeter (Sarai, 2004). The OSL of the Luxel badge using Al_2O_3:C photostimulabule phosphor can be observed at about 420 nm by stimulating with about 520 nm light.

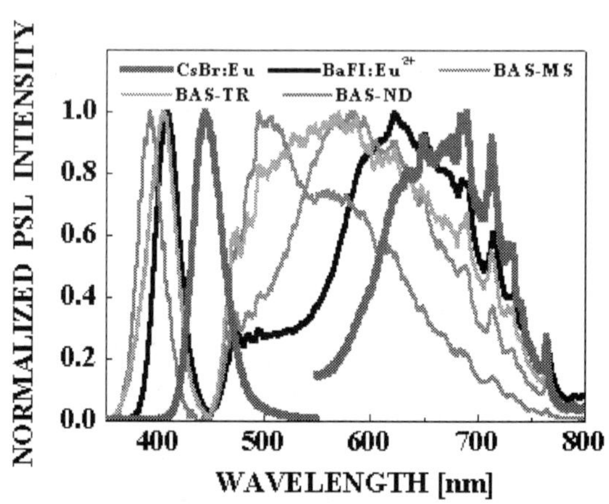

Fig. 4. OSL spectra and its stimulation (excitation) spectra of various imagig plates. Here, the IP, type-BAS-MS using BaFBrI:Eu photostimulable phosphor is commercially available from Fuji Film Corp.

2.1.3 DIS dosimeters

The DIS dosimeter is composed of metal-oxide-semiconductor field effect transistor (MOSFET) with ionizing chamber (Wernli, 1998) as shown in Fig.5. The basic principle of the DIS dosimeter is as follows; a nonvolatile solid state memory cell is stored in the form of electric charge being trapped on the floating gate of a MOSFET in air or gas space surrounded by a conductive wall. The DIS dosimeter (Type DIS-1) which responds to X, γ and β-rays (Kobayashi, 2004) can widely detect a radiation dose within the range from 1to 40 [μSv].

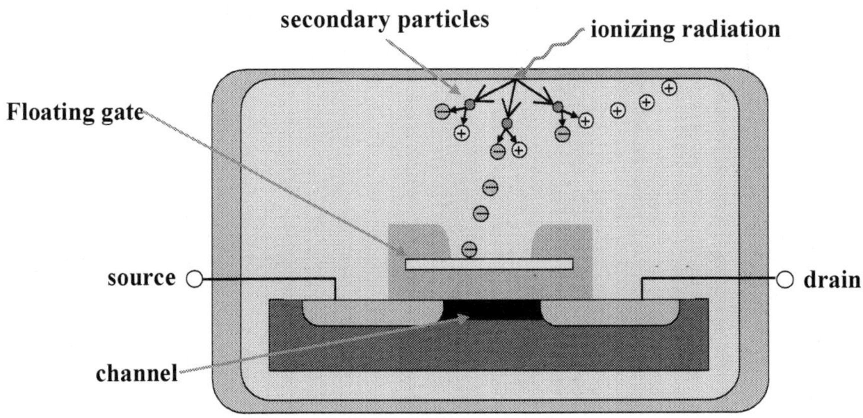

Fig. 5. Schematic diagram of the DIS dosimeter

2.2 Comparison of features of each solid state dosimeter

The comparison of various basic characteristics, such as the readout process, the sensitivity for x-ray, β-ray and neutron, the energy dependence, of each solid state dosimeter is shown in Table 1. We would like to emphasize here is that the RPL glass dosimeter has good fading characteristics which means the luminescence centers are stable at room temperature unless the glasses are annealed at high temperature at about 400℃.

Phenomenon (Materials)	Readout	Sensitivity	Energy Dependence	Fading
RPL (Phosphate glass)	Excitation with UV light	0.1 - 10,000 [mSv]	10keV～10MeV(x-ray, γray), 300keV～3MeV（βray） 0025eV-15MeV(Neutron)	Excllent
OSL (Al_2O_3:C, BaFIBr:Eu, KCl:Eu)	Excitation with visible light	0.01 [mSv] - 10 [Sv] (X・γ-ray) 0.01 [mSv] - 10 [Sv] (β-ray) 0.1 [mSv] - 6 [mSv] (Neutron)	5keV～10MeV(X・γ-ray) 150keV～10MeV(β-ray) 0.025eV～0.5eV(Neutron)	OK
Ionization of air (Si -MOSFET)	Electrical signal	1～40 [μSv]	6keV - 9 MeV (X・γ-ray) 0.06 - 0.8 MeV (β-ray)	Good

Table 1. Basic characteristics of each solid state dosimeter

3. Experimental

The RPL glass dosimeter , the OSL dosimeter and the DIS dosimeter developed as the passive dosimeter were used in the environmental natural radiation monitoring. Figure 6 shows photographs of a personal glass dosimeter of type GD-450 used in this study. The GD-450 is made of Ag^{+}-doped phosphate glass (AGC Techno Glass Co., Ltd.), supplied by Chiyoda Technol Corp.

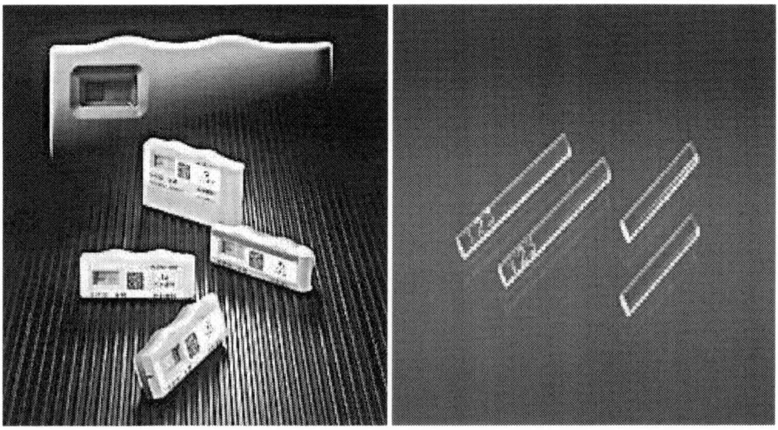

Fig. 6. Photographs of the GD-450 (left) and of glass used in GD-450 (right).

The OSL dosimeters (Luxel badge: S-type) used in this study as shown in Fig.7 (left) were supplied by Nagase Landauer Co., Ltd. (Kobayashi, 2004). In this study, the OSL dosimeter, which was made of an Al_2O_3:C phosphor material, was used for environmental natural radiation monitoring. The Al_2O_3:C phosphor emit 420 nm OSL emission with intensity in proportion to the exposure dose under optical stimulation with the wavelength of 523 nm.

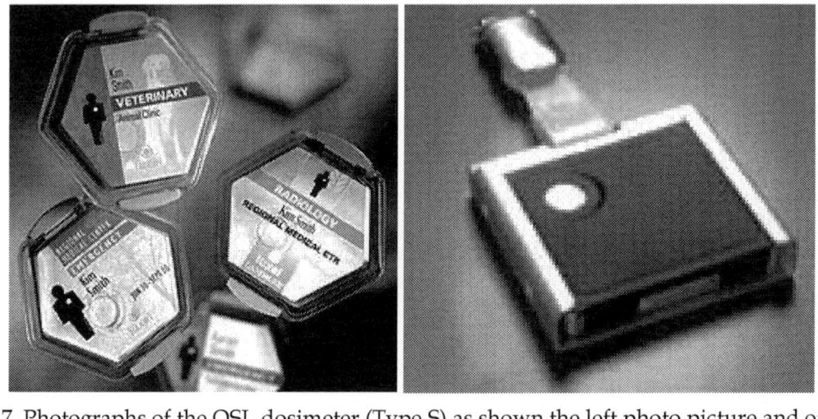

Fig. 7. Photographs of the OSL dosimeter (Type S) as shown the left photo picture and of the DIS dosimetr (Type DIS-1) as shown in the right photo picture.

The DIS as shown in Fig.7 (right) was supplied from RADOS Technology, Finland. The basic principle of the DIS is as follows; a nonvolatile solid-state memory cell is stored in the form of electric charge being trapped on the floating gate of a MOSFET transistor in air or gas space surrounded by a conductive wall. The DIS dosimeter is based on Analog-EEPROM (Analog Electrically Erasable Programmable Read Only Memory). The DIS responds to X, γ, β-rays and neutron. This dosimeter has an excellent energy characteristic and can be read repeatable without quenching of the data. The DIS dosimeter can widely detect a radiation dose within the range from 1 μSv to 40 Sv

The personal dosimeters GD-450, Luxel badge (Type S) and DIS-1 were set on 7 points in Ishikawa prefecture as shown in Fig.8.

Each of 7 points are represented as alphabet of from A to G. (A: Tsurugi-machi, B: Tatsunokuchi, C: Inside of house of Mt. Shishiku, D: Outside of Mt. shishiku, E: Inside of Ogoya Mines, F: Outside of Ogoya Mines, G: Public Health and Environmental Science). Each data was obtained monthly. Photographs of local points in which the dosimeters were set up are shown in Fig.9. Data were obrained monthly. The each accumulated monthly data was divided into daily data and multiplied 30 days. The each data was compensated appropriately with the each formula for the dosimeters (Sarai, 2004). The same point data were averaged and the standard deviations were calculated. The data of GD-450 were compared with the data of the other dosimeters.

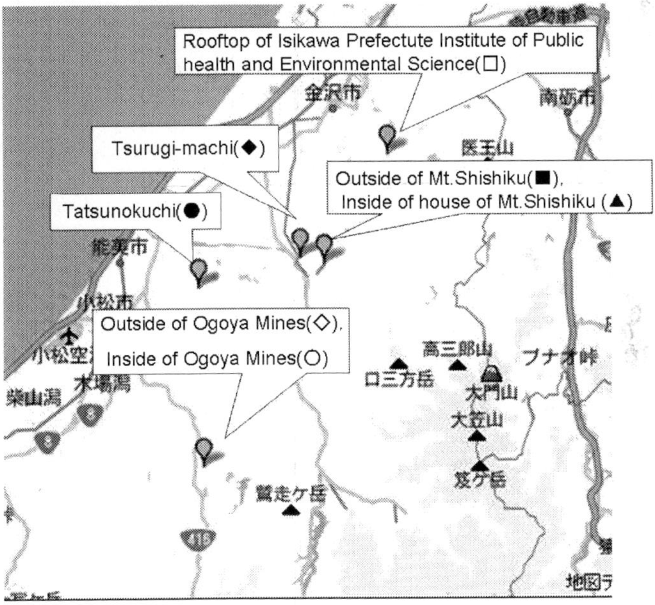

Fig. 8. Map of seven points such as Tsurugi-machi (◆), Tatsunokuchi (●), outside of Mt.Shishiku (■), inside of house in Mt.Shishiku, (▲), outside of Ogoya Mines (◇), Inside of Ogoya Mines (○) and rooftop of Ishikawa Prefecture Institute of Public health and Environmental Science (□). in Ishikawa prefecture, in which the environmental radiation dose using the glass dosimeter were measured.

(a) (b) (c) (d)

Fig. 9. Photographs of 4 points such as (a) point D, (b) point E, (c) point Fand (d) point G in Ishikawa prefecture, in which the dosimeters were set up.

4. Results and discussion

4.1 Basic characteristics of RPL glass dosimeter

Typical RPL emission and its excitation spectra of x-ray irradiated Ag^{+}-doped phosphate glass are shown in Fig.10. It can be seen that the RPL emission spectrum consists of two emission bands peaked at about 2.70 eV (460 nm) and 2.21 eV (560 nm). On the other hand, the RPL excitation spectrum consists of two excitation bands peaked at about 3.93 eV (315 nm) and 3.32 eV (373 nm). The fact that RPL emission spectrum consists of two emission bands such as yellow color emission and blue color emission has been reported in previous report (Miyamoto, 2010). The radiative lifetime of yellow and blue RPL peaks are estimated. The lifetime is 2~4 μs for yellow RPL and 2~10 ns for blue RPL, respectively, which was dependent on the irradiation dose (Kurobori, 2010). The RPL emission mechanism is explained using Fig.11 as follows; when the Ag^{+}-doped phosphate glass is exposed to ionizing radiation such as x-ray, the electron-hole pair will be produced. The electrons are captured into Ag^{+} ions in the glass structure and then the Ag^{+} ions change to Ag^{0} ions. On the other hand, the holes are captured by the PO_4 tetrahedron at the begining of migration and then produce Ag^{2+} ions owing to interaction with Ag+ ions over time. It has been reported that both Ag^{0} and Ag^{2+} ions can be played in role as luminescence centers for blue and yellow RPL, respectively (Miyamoto, 2010).

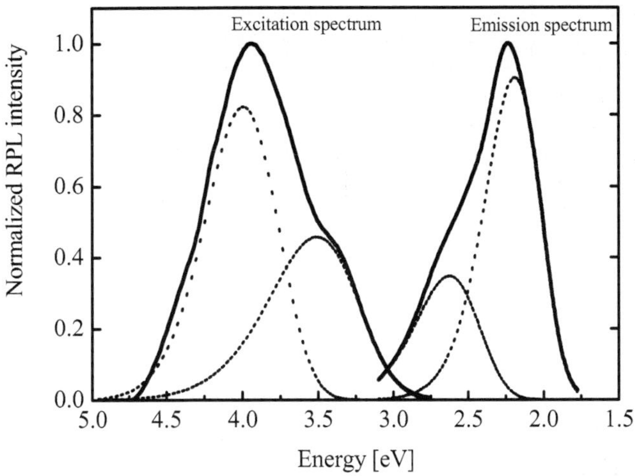

Fig. 10. Typical RPL emission and excitation spectra of Ag^{+}-doped phosphate glass after x-ray irradiation. The peak separation of the excitation and emission spectra of RPL indicated using dashed lines were carried out using the component separation of Gaussian bands(dashed lines).

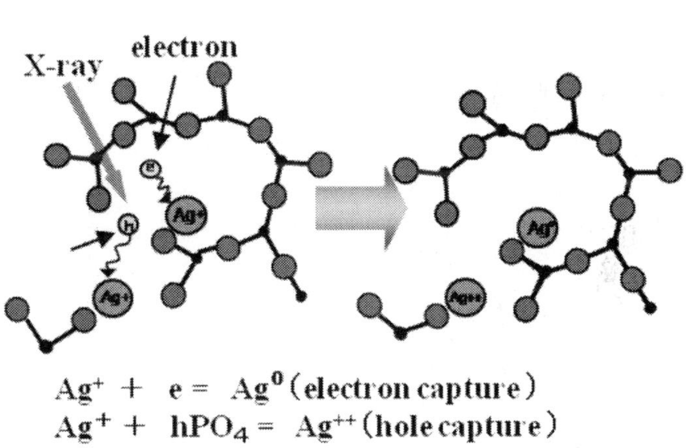

Fig. 11. Formation of RPL luminescence centers such as Ag^{0} and Ag^{2+} ions in x-ray irradiated Ag^{+}-doped phosphate glass.

The RPL emission images as a function of x-ray absorbed dose, when the x-ray irradiated Ag^{+}-doped phosphate glass is excited using UV light, are shown in Fig. 12, where it is seen that the intensity of yellow color emission increases with the absorbed dose. This result coincides with that of previous report (Shih-Ming Hsu, 2007), in which RPL intensity was almost linearly increased with x-ray absorption dose up to 10 Gy.

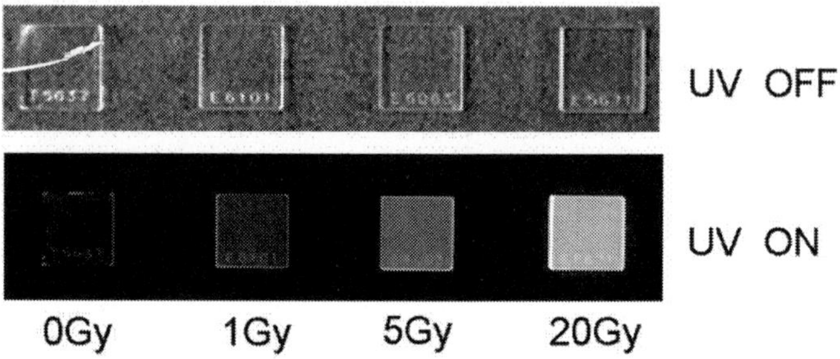

Fig. 12. RPL emission images of Ag^+-doped phosphate glass as a function of x-ray absorbed dose : (a) Ag^+-doped phosphate glass under visible light, (b) Ag^+-doped phosphate glass under UV light.

4.2 Results of environmental natural background radiation monitoring

Before the environmental background radiation monitoring was carried out, the self dose measurement and radioactive nuclide identification were made in an extremely low level background field of the tunnel of Ogoya Copper Mine (Ogoya underground laboratory of Kanazawa University), where muon intensity of cosmic ray is reduced to two orders of magnitude in comparison with the ground (Murata, 2002). The Luxel badge and DIS dosimeter were set in a shielding box of an ancient lead which contains few ^{210}Pb isotope (half-life 22.3 years). Five units of the Luxel badge and the DIS dosimeter were prepared to measure the self-doses. Self-doses were measured by the month during three months.

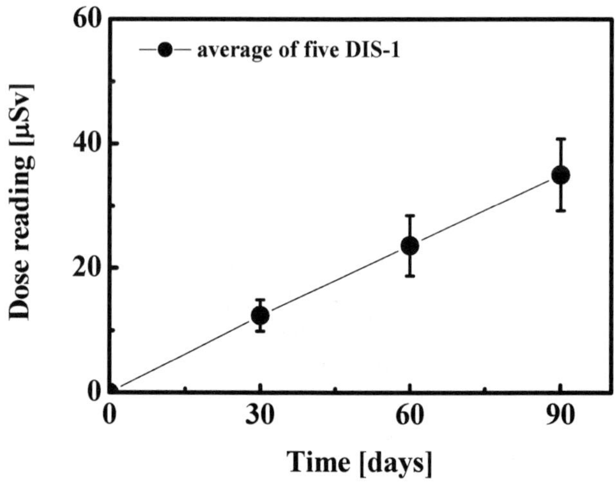

Fig. 13. Self-dose of the DIS-1 dosimeter. Each data point is averaged over doses of five DIS units

Figure 13 shows self-dose of the DIS dosimeter. The average self-dose acumulated in the DIS dosimeter increases almost linearly with increasing time. The average self-dose of the DIS dosimeter within a month was estimated tobe about 12μSv. On the other hand, the averaged self-doses accumulated in the Luxel badge also increases lienarly with increasing the time except for the bigining of measurement as shown in Fig.14. The value at the bigining of measurement is different from othe two values. This deviation may be caused by the exposure to the natural radiation during the transportation of dosimeters to Nagase Landauer in Tokyo by air. Except for the data point at the bigining of measurement, the averaged self dose of the Luxel Badge is estimated to be about 9μSv.

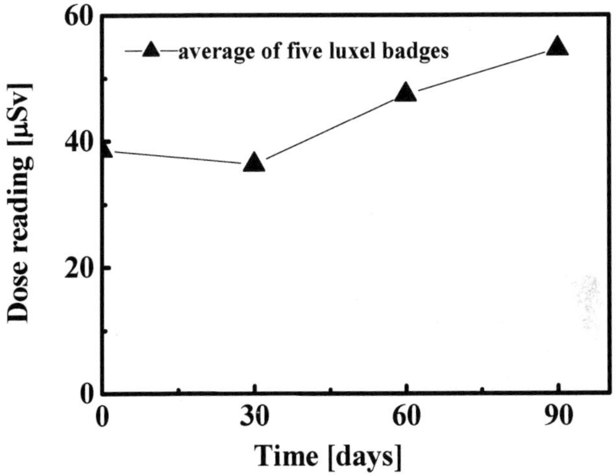

Fig. 14. Self-dose of the luxel badge dosimeter. Each data point is averaged over doses of three Luxel badge units.

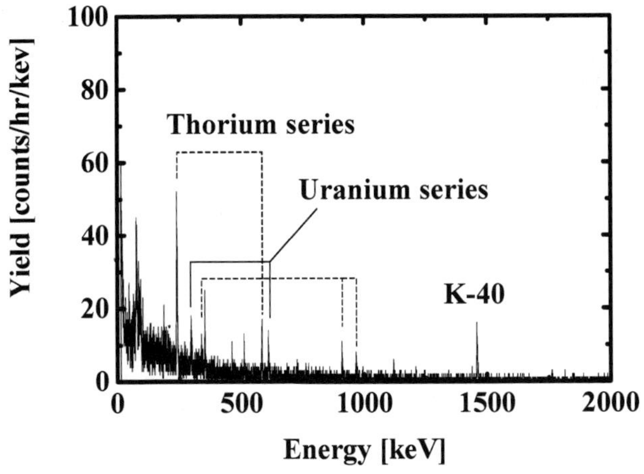

Fig. 15. Typical γ-ray spectrum obtained from the DIS dosimeter.

The origin of the self-dose was identified using high pure Ge semiconductor detector in the Ogoya underground laboratory. Typical gamma-ray spectrum obtained from the DIS dosimeter is shown in Fig.15.

dosimeter	parts	^{238}U (dpm)	^{210}Pb (dpm)	^{232}Th (dpm)	^{40}K (dpm)
DIS	Whole DIS	1.40		2.00	22.0
	Label				0.38
	Spring	0.88			
	Al frame	0.10			0.75
	IC long	1.30		1.50	
	IC fat	0.83		0.85	
	Battery	0.10			
Luxel	Al_2O_3 crystal	2.00		1.50	
	Ag filter				
	Sn filter	0.07	1.70	0.04	5.53

Table 2. Identificated radioactive nuclides contained in each personal dosimeters.

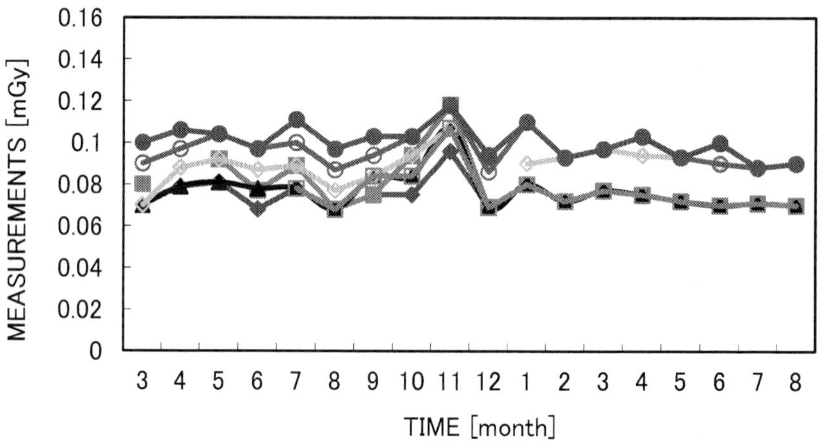

Fig. 16. Measured environmental radiation dose using the GD-450 glass dosimeter in seven points such as Tsurugi-machi (◆), Tatsunokuchi (•), outside of Mt.Shishiku (■), inside of house in Mt.Shishiku, (▲), outside of Ogoya Mines (◇), Inside of Ogoya Mines (○) and rooftop of Ishikawa Prefecture Institute of Public health and Environmental Science (□). in Ishikawa prefecture. The measurements of environmental radiation dose were carried out from March in 2008 to August 2009.

The sveral peaks under 1000 keV correspond to nuclides of ^{232}Th and ^{238}U series. The ^{40}K peak with the energy of 1460 eV has been also detected. Measured parts and identified radioactive nuclies are listed in Table 2. The ^{40}K, ^{232}Th and ^{238}U have been contained in almost all dosimeters. So, it is difined that the self-dose of each dosimeter for a month is about 10-15 μSv. Data was, therefore, compensated for each dosimeter which based on the sel-dose rate of about 12 μSv/month.

The environmental backgroung radiation dose at 7 points for one month were monitored using the glass dosimeter (GD-450) as well as the Luxel badge and the DIS dosimeters. The monitoring results of typical environmental background radiation dose in gray (Gy) as the absorbed dose using the GD-450 from March in 2008 to August 2009 are shown in Fig.16 for 7 points in Ishikawa prefecture.

Although natural background radiation doses with the GD-450 dosimeter at each point in Ishikawa prefecture were significantly different, the standard deviations were very small. Although the values were a little bit different between the GD-450 glass dosimeter and the Luxel badge (OSL dosimeter), the tendencies of the environmental dose at each point were very similar as shown in Fig.17. The higher dose at point B (Tatsunokuchi) than at other points is due to the use of radioisotopes at the Lowere Level Radiation laboratory in Kanazawa University. Morever, the values of the GD-450 dosimeter and the DIS dosimeter were very close and there was no significant difference between them as shown Fig.18. We have made the comparison of different types of RPL glass dosimeters such as Type: GD-450 for personal dosimeter and Type:SC-1 for enviromental monitoring, which were supplied from Chiyoda Technol Corp, as shown in Fig.19. It was found that there is no significant difference at each points.

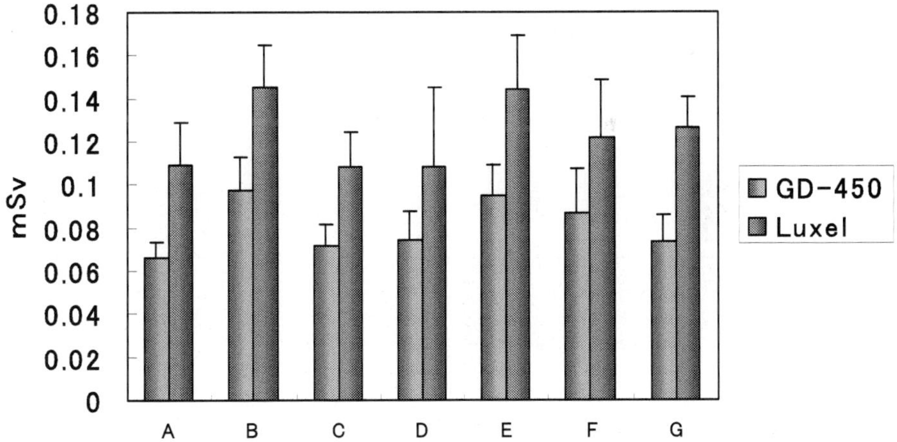

Fig. 17. Dose response at each point in Ishikawa prefecture (A: Tsurugi-machi, B: Tatsunokuchi, C: Inside of house of Mt. Shishiku, D: Outside of Mt. shishiku, E: Inside of Ogoya Mines, F: Outside of Ogoya Mines, G: Public health and Environmental Science) using GD-450 (blue bars) or Luxel badge (orange bars) dosimeters.

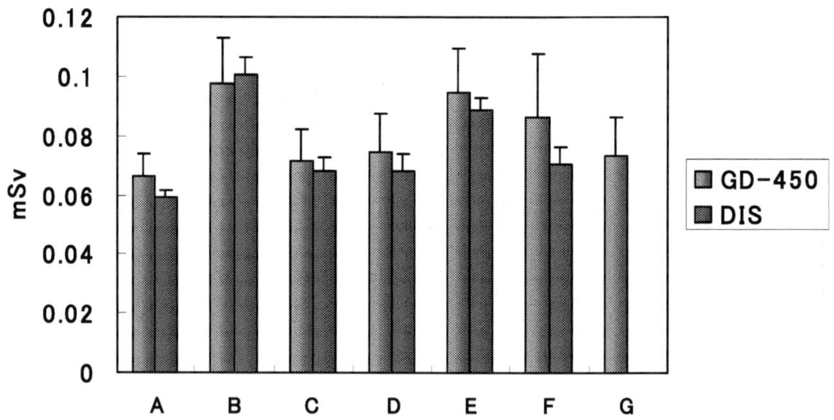

Fig. 18. Dose response at each point in Ishikawa prefecture (A: Tsurugi-machi, B: Tatsunokuchi, C: Inside of house of Mt. Shishiku, D: Outside of Mt. shishiku, E: Inside of Ogoya Mines, F: Outside of Ogoya Mines, G: Public health and Environmental Science) using GD-450 (blue bars) or DIS (purple bars) dosimeters. There is no data at G for DIS.

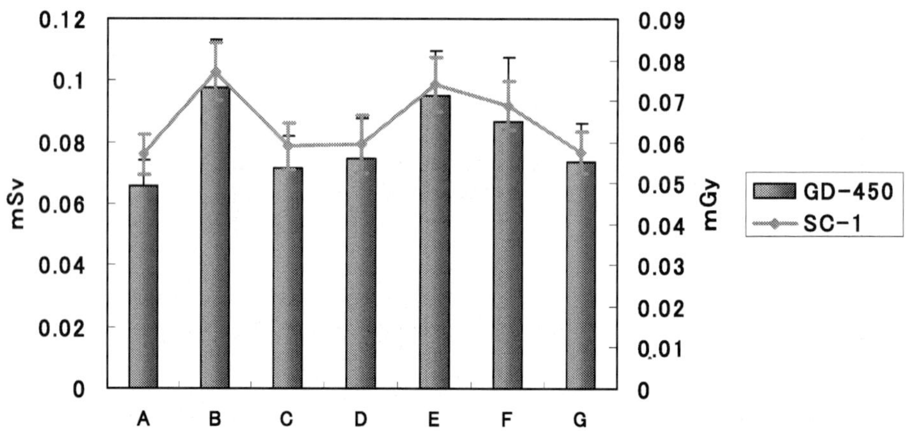

Fig. 19. Dose response at each point in Ishikawa prefecture (A: Tsurugi-machi, B: Tatsunokuchi, C: Inside of house of Mt. Shishiku, D: Outside of Mt. shishiku, E: Inside of Ogoya Mines, F: Outside of Ogoya Mines, G: Public health and Environmental Science) using GD-450 (blue bars) or SC-1 (green line) dosimeters. The unit of the GD-45 and SC-1 are represented by mSv and mGy, respectively.

From the results as described above, Monitoring environmental natural background radiation dose with a personal GD-450 seems to be feasible and consequently, one can say that the GD-450 dosimeter can be suitable for monitoring environmental natural background radiaiton dose.

5. Summary

Environmental natural background radiation dose values at 7 points in Ishikawa prefecture determined using the personal glass dosimeter, type GD-450 were compared with these determined some other personal dosimeters such as DIS dosimeter utilizing a MOSFET with an ioniization chamber and OSL dosimeter, Luxel budge, utilizing OSL phenomenon in Al_2O_3:C phosphor. The actual dose values were different from each other, however, the tendency of each dose at each point were very similar. It can be said that the personal glass dosimeter will be very useful for not only monitoring personal dose but also monitoring natural background radiation dose.

6. Acknowledgements

The author wish to thank Dr.Yamamoto, Directer of the Research Center of Chiyoda Technol Corp. for his fruitful discussion and Dr.Kobayashi of Nagase Landauer Co. Ltd, Dr. Kakimoto of Ishikawa Prefecture Institute of Public health and Environment Science for their excellent assistance.

The work on the environmental natural background radiation monitoring using solid state passive dosimeters was partially supported by the foundation for Open-Research Center Program from the Ministry of Education, Culture, Sport, Science and Technology of Japan and Chiyoda Technol Corp.

7. References

Kobayashi, I, (2004), The detection of the Environmental radiation for DIS and Luxel badge, Ionizing Radiation, Vol.30, pp.33-43.

Koyama, S., Miyamoto, Y., Fujiwara, A., Kobayashi, H., Ajisawa, K., Komori, H., Takei, Y., Nanto, H., Kurobori, T., Kakimoto, H., Sakakura, M., Shimotsuma, Y., Miura, K., Hirao, K. And Yamamoto, T., (2010), Environmental Radiation Monitoring Utilizing Solid State Dosimeters, Sensors and Materials, Vol.22, No.7, 377-385.

Miyamoto, Y., Takei, Y., Nanto, H., Kurobori, T., Konnai, A., Yanagida, T., Yoshikawa, A., Shimotsuma, T., Sakakura, M., Miura, K., Hirao, K., Nagashima, Y. and Yamamoto, T., (2011), Radiophotoluminescence from Silver-Doped phosphate Glass, Radiation Measurements, in press.

Murata, Y., Yamamoto, M. and Komura, K., (2002), Determination of low-level ^{54}Mn in soils by ultra low-background gamma-ray spectrometry after radiochemical separation, J. Radiational Nucl. Chem, Vol.254, No.2, pp.249-257.

Hsu, S.M., Yeh, S.H., Lin,M.S. and Chen, W.L., (2006), Comparison on characteristics of radiophotoluminescent glass dosimeters and thermoluminescent dosimeters, Radiation Protection Dosimetry, 119, 327-331.

Nanto.H, (1998), Photostimulated Luminescence in Insulators and Semiconductors, Radiation Effects & Defects in Solids, Vol.146, pp.311-321.

Nanto, H., (1999), Physics of photosimulable phosphor materials, Ionizing Radiaiton, Vol. 25, No.2, pp.9-24. (in Japanese)

Nanto, H., Takei, Y., Nishimura, A., Nankano, Y., Shouji, T., Yanagida, T., Kasai, S., (2006), Novel X-ray Imaging Sensor Using Cs:Br:Eu Phosphor for Computed Radiography, Proc. of SPIE, Vol. 6142, pp.6142w-1-6142w9.

Nanto, H., (2011), Basic princple of accumulation-type personal dosimeter for ionizing radiation and its application, Ionizing Radiation, Vol.37, No.2, pp.3-9.

Ranogajec-Komor, M., Knezevic, Z., Miljanic, S. And Velic, B., (2008), Characteristics of radiophotoluminescent dosimeters for environmental monitoring, Radiation measurements, Vol.43, 392-396.

Saez-Vergara, J.C., (1999), Practical Aspects on The Implementation of LiF:Mg, Cu, P in Routine Environmental Monitoring Program, Radiation Protection Dosimetry, Vol.1-4, pp.237-244.

Sarai, A., Kurata, N., Kamijo, K., Kubota, N., Takei, Y., Nanto, H., Kobayashi, I., Komori, H., and Komura, K., (2004), Detection of self-dose from an OSL dosimeter and a DIS dosimeter for environmental radiation monitoring, J. Nuclear Science and Technology, Suppl. 4, pp.474-477.

Wernli, C., (1998), Direct ion strage dosimeters for individual monitoring, Radiation Protection Dosimetry, Vol.77, pp.253-259.

9

PILS: Low-Cost Water-Level Monitoring

Samuel Russ, Bret Webb, Jon Holifield and Justin Walker
University of South Alabama
United States of America

1. Introduction

The estuarine environment is important both to global ecology and to human economy. Estuaries are the place where freshwater meets saltwater, and so they typically contain a bounty of marine species, and are essential to the life cycle of many marine organisms. For similar reasons, they often contain sea ports and carry commerce of great value.

In order to study estuaries in more detail, we have developed two sets of low-cost sensors using off-the-shelf technology combined with innovative new low-cost circuits. The first, nicknamed "Jag Ski", is a highly mobile water craft for navigating estuarine and littoral areas and providing real-time data. The second, named "PILS", is a network of stationary sensors for making long-term water-level measurements. This paper describes the construction of both, along with actual measurements.

2. Survey of literature

Sensing the environment can be carried out through remote measurements (e.g. satellites (Villa & Gianietto, 2006)) and through in situ measurements (e.g. wireless sensor networks (O'Flyrm et al., 2007; Thosteson et al., 2009)). Both have been demonstrated successfully as means of measuring characteristics of water.

An example of one real-time water-sensor architecture is the Land/Ocean Biogeochemical Observatory (LOBO) system developed by Satlantic and the Monterrey Bay Aquarium Research Institute (MBARI) (Comeau et al., 2007; Jannasch et al., 2008) and has been installed in the field (Sanibel-Captiva Conservation Foundation, 2009). Others include the Ocean Observation Initiative (OOI) (Frolov et al., 2008; National Research Council, 2003; U.S. Commission on Ocean Policy, 2004), NOAA tide gauges for storm surge (Luther et al., 2007), and sonar-based water-level measurements (Silva et al., 2008). Specific to environmental monitoring in the coastal ocean, mobile field assets typically include profiling floats (Roemmich et al., 2004), autonomous underwater vehicles (AUVs) (Rudnick et al., 2004), and unmanned underwater vehicles (UUVs) (Freitag et al., 1998; Frye et al., 2001).

This work is in line with these earlier systems. We have adapted the mobile sensor platform to a highly maneuverable manned platform to navigate shallow-water areas proficiently. The sensor network is designed for relatively low cost and for unattended measurements. It also contains novel sensors for pressure and salinity.

This work is motivated by the fact that computer models of estuaries need refinement. For example, there is disagreement whether wind forcing or river discharge dominates the dynamics of Mobile Bay (Schroeder & Wiseman, 1986; Kim et al., 2008). Data obtained using the sensors will be used to parameterize a linear approximation of a static momentum balance of the estuary (Van Dorn, 1953) to improve simulation and forecasting accuracy.

3. Real-time monitoring: Jag Ski

The University of South Alabama Jag Ski is a three-person Kawasaki Ultra LX personal watercraft (PWC) equipped with state of the art instrumentation developed by YSI, Incorporated, SonTek, VarTech Systems, and others (Fig. 1). In addition to the PWC, a Kawasaki Mule 3010 four-wheel drive utility vehicle can be used for launching and retrieval when a proper boat launch is not available. The Jag Ski contains an onboard small-form PC running the Windows XP operating system, a foldable waterproof keyboard, a fully submersible touch screen LCD display, and four dry-cell 18 amp hour, 12 volt marine batteries to supply enough dedicated power for twelve to fourteen hours of data collection. The PC, power supply, and other assorted equipment are housed in waterproof cases with internal foam padding. All external cabling and bulkhead connectors are fully submersible. Experience has demonstrated that items labeled water resistant and waterproof offer little protection in the corrosive, marine environment.

Fig. 1. The South Alabama Jag Ski and 4x4 towing vehicle.

The use of PWCs for collecting hydrography is not a new idea. There are numerous examples of PWC systems around the country (and world). Some of the earlier successful applications are discussed in (Dugan et al., 1999; Dugan et al., 2001; MacMahan, 2001; Puleo et al., 2003). The PWC has also successfully been used for larval fish sampling in shallow waters (Strydom, 2007). More recently, however, Hampson et al. (2011) have demonstrated the skill of using a kayak as a surveying platform for still shallower survey applications.

What perhaps makes the Jag Ski so unique in the context of PWC hydrographic data collection systems is its suite of instrumentation. Prior to the Jag Ski, the use of the PWC has been mostly limited to bathymetric surveys in nearshore waters. While it certainly has its limitations, the ability of the PWC to traverse the surfzone in hydrographic surveying cannot be rivaled by most traditional vessels. The addition of a PWC to one's hydrographic surveying deployment provides a very good overlap between land-based surveys and those conducted in deeper waters using traditional watercraft. The Jag Ski, however, was

developed to meet broader goals and objectives in the area of coastal, water resources, and environmental engineering.
The Jag Ski contains a SonTek/YSI RiverSurveyor M9 Acoustic Doppler Current Profiler (ADCP) with an integrated Real Time Kinematic Differential Global Positioning System (RTK DGPS) for georeferenced measurements (Fig. 2). The M9 ADCP has a profiling range of 6 cm to 40 m, and is capable of measuring velocity magnitudes up to 20 m/s. The resolution of the velocity measurements is as low as 0.001 m/s, and vertical bin sizes can be as small as 2 cm, or as large as 4 m. The horizontal resolution of the samples is a function of the reported sample rate (generally 1 Hz) and vessel speed (preferably equal to or less than the water velocity). A nominal speed of 1 - 2 m/s is maintained when using the M9 ADCP on the Jag Ski, so a typical horizontal resolution is, accordingly, 1 - 2 m.

Fig. 2. SonTek/YSI RiverSurveyor M9 ADCP and RTK DGPS base station.

The M9 ADCP contains a dedicated 500 KHz vertical beam for depth measurements and bottom tracking, four slanted 1 MHz beams for sampling in deeper water, and four slanted 3 MHz beams for sampling in shallower waters (Fig. 3). This dual-frequency functionality is unique in the ADCP market, and along with its integrated GPS system for vessel-corrected measurements to account for the moving reference frame, makes it attractive for applications in Mobile Bay (Fig. 4). The bay is a broad, mostly shallow (< 4 m), drowned river mouth estuary that is incised by a navigation channel dredged to a maintenance depth of about 15 m. The depth of the channel in the main entrance to Mobile Bay can reach 20 m or more, and is flanked to the west by a broad, shallow area with depths less than 3 m. The dual frequency M9 ADCP performs well when transitioning between the two extremes.
Aside from the technical capabilities of the RiverSurveyor M9 ADCP, the instrument comes with a well-developed, integrated software package for setup and data collection. The RiverSurveyor Live (RSL) software is loaded on the onboard PC, and is fully interactive using the touch screen LCD display. Some very helpful features of the software include dynamic icons that quickly report the status of various systems, like GPS and bottom

tracking, the ability to see a real-time estimate of discharge, and the integrated GIS shapefile functionality for easy navigation and spatial awareness.

Fig. 3. SonTek/YSI RiverSurveyor M9 ADCP head.

Fig. 4. Terra/MODIS imagery of Mobile Bay taken November 8, 2002. Image courtesy: NASA Visible Earth.

The initial research focus for the Jag Ski was fulfilled with the integration of the RiverSurveyor M9 ADCP. That one piece of equipment provides the capability to perform detailed beach profile surveys, detect and image scour holes near bridge foundations, and measure the spatial variability and magnitude of coastal and nearshore currents, as well as riverine flows. And as preparations were being made in April 2010 for upcoming field experiments in coastal Alabama during the months May - August, the explosion and subsequent sinking of the *Deepwater Horizon* drilling platform later that month unveiled a new, and unexpected, application for the Jag Ski: environmental monitoring.

The National Science Foundation (NSF) issued a number of awards for research, instrument acquisition, and instrument development related to the 2010 Gulf Oil Spill through their RAPID program in the months following the initial explosion and sinking of the platform. The Jag Ski received one such award, issued through the NSF Major Research Instrumentation program. The purpose of the award was to purchase an instrument that could be used to measure near-surface water quality parameters, as well as crude oil and refined fuels, in Alabama's coastal waters. The result is a rather unique piece of equipment produced by YSI, Inc. called a Portable SeaKeeper 1500 (Fig. 5). The Portable SeaKeeper, or PSK, is a scaled-down version of the SeaKeeper 1000 systems that are deployed on nearly 50 different vessels of opportunity around the world. Some vessels are used for research, others are operational ferries, and still others are private yachts. Each of these vessels contributes data and research to the International SeaKeepers Society, and now the Jag Ski does, too (Fig. 6).

Fig. 5. The YSI Portable SeaKeeper 1500 mounted on the stern of the Jag Ski.

Fig. 6. Initial testing of the YSI PSK on a local river.

The PSK contains an YSI 6600v2 sonde, a Turner Designs C3 submersible fluorometer, a Thrane & Thrane Sailor Mini-C vessel monitoring system, a diaphragm pump, and a dedicated small-form PC running the Windows XP operating system (Fig. 7). The PSK continuously draws near-surface water by way of a ram intake and pump, routes it through a manifold, and then to flow chambers attached to the YSI 6600v2 and Turner Designs C3. The YSI sonde measures temperature, specific conductivity (salinity), pH, turbidity, dissolved oxygen, and chlorophyll. The Turner Designs fluorometer measures chromophoric dissolved organic matter (CDOM), crude oil, and refined fuels relative to a calibration standard or deionized water. The Sailor Mini-C contains a 12-channel GPS receiver, and Inmarsat-C antenna and transceiver, which provide vessel positioning and data telemetry to the SeaKeepers online data repository. The PSK currently reports samples at 0.0833 Hz, but this value can be increased or decreased by the user. In the coming months, an R.M. Young meteorological station is being added to the Jag Ski and integrated with the PSK system. The meteorological station will provide continuous underway measurements of wind speed and direction, air temperature, relative humidity, and barometric pressure.

If the suite of sensors and measurement capabilities of the PSK are not impressive enough, then perhaps the ability to collect this data while cruising at 40 knots is! The custom-designed ram intake and diaphragm pump allow for a continuous stream of water to be drawn from the near surface (about 10 cm below the surface) regardless of the speed, and the center-point allows it to track with the vessel when turning at high speed (Fig. 8).

The YSI PSK system is playing an important role in the yearlong BP-funded Gulf Research Initiative program that seeks to evaluate the impacts of the *Deepwater Horizon* events on Alabama's coastal resources. With the YSI PSK system, the first synoptic survey of Mobile Bay's near-surface characteristics will be achieved in the summer of 2011. The ability to map a majority of the bay's surface in less than a quarter tidal cycle provides tremendous opportunities for practical, applied research ranging from coastal and estuarine hydrodynamics to watershed management. In terms of the Gulf Research Initiative, the PSK data will be used in combination with the M9 ADCP data to describe transport pathways that are effective in communicating constituent material from the Alabama shelf, through Mobile Bay, and to the Mobile-Tensaw river delta. A number of field experiments are planned for late summer and early fall of 2011 that will isolate the seasonal (i.e. wet/dry, warm/cool, windy/calm) and tidal (i.e. spring/neap) variability of Mobile Bay's dynamics. Beyond academic research, the ability of the PSK to rapidly measure large spatial distributions of dissolved oxygen, turbidity, chlorophyll, and CDOM make it suitable for a number of environmental applications, from tracking and mapping harmful algal blooms (HAB's) to the measurement and analysis of Total Maximum Daily Loads (TMDL) in the Mobile Bay watershed.

While the YSI PSK 1500 has impressive capabilities, its sampling is limited to one location in the water column for the duration of a survey. It is possible to lower the PSK intake to sample from a different portion of the water column, but this is something that would limit the speed of the vessel. Since an estuary like Mobile Bay can be highly stratified at times, the near-surface PSK data may not necessarily be representative of the entire water column; therefore, CTD casts are performed from the PWC at predetermined locations to evaluate stratification at the time of the survey. The idea of performing CTD casts (conductivity-temperature-depth) from a PWC was not practical until the recent release of the YSI CastAway CTD profiler (Fig. 9).

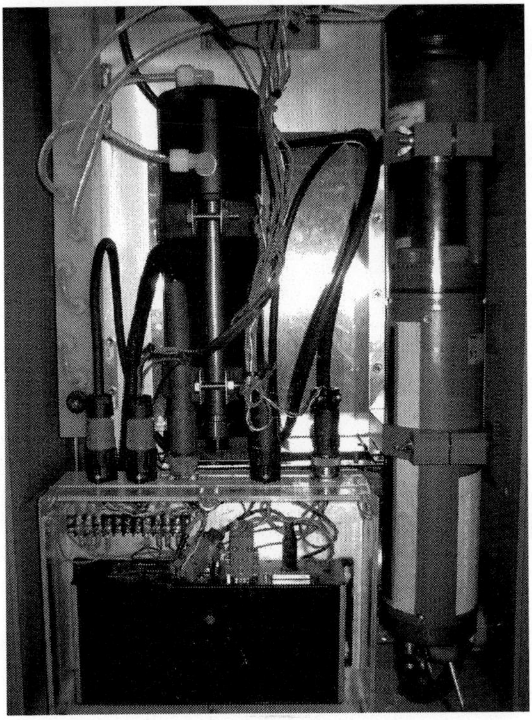

Fig. 7. Internal components of the YSI PSK system. The YSI sonde is on the right, the Turner Designs fluorometer is the black cylinder, the flow manifold is on the left, and the onboard PC is at the bottom. The diaphragm pump is hidden behind the PC.

Fig. 8. The custom-designed center-point swivel and ram intake for the YSI PSK.

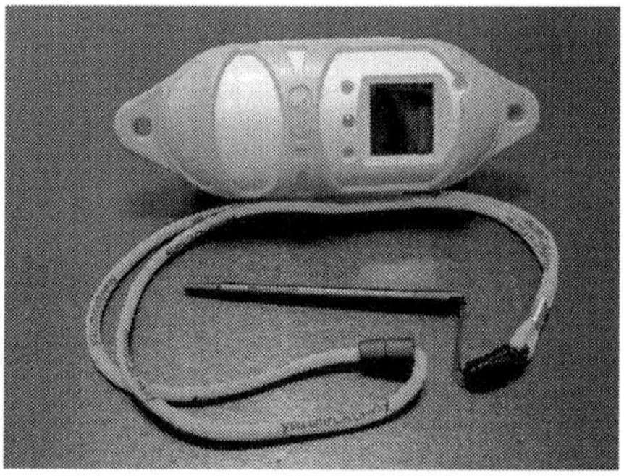

Fig. 9. The YSI CastAway CTD profiler and magnetic stylus.

The CastAway CTD has an internal GPS that logs the time and location of each cast. The user-interface is simple and intuitive, and every operation is controlled using a magnetic stylus. Data offloads are accomplished through a Bluetooth connection between the device and a PC running the CastAway software. The CastAway is ultra-portable, making it suitable for deployment from the Jag Ski.

3.1 Case study – Mobile Bay field experiment

A small field experiment conducted on April 1, 2011 in Mobile Bay (Fig. 10) demonstrates the full capabilities of the Jag Ski described previously. The objective of the experiment was to perform a complete hydrographic survey of the lower portion of Mobile Bay during neap tide conditions. An ADCP transect was collected at each of Mobile Bay's primary connections to surrounding water bodies, continuous underway sampling of near-surface waters was performed, and two CTD casts were obtained.

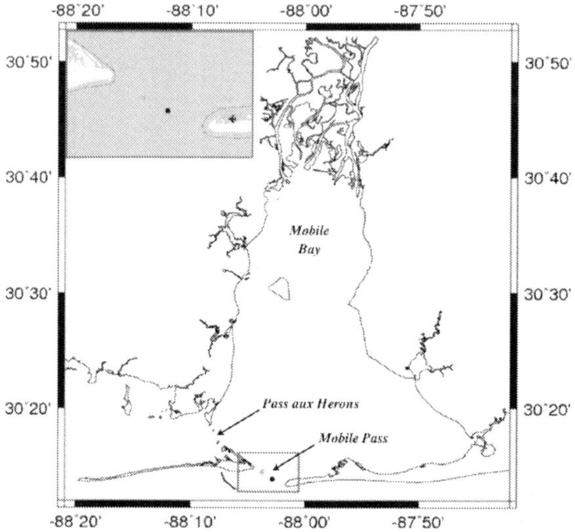

Fig. 10. Overview of study area and locations of CTD profiles at Mobile Pass on April 1, 2011.

The survey took place from 0800 - 1200 hours EDT on Friday, April 1, 2011, beginning and ending at Dauphin Island, Alabama. The tides during the field experiment were in neap, with little variation. Although the survey took place on a falling portion of the tide, the tide was flooding at Mobile Pass and Pass aux Herons throughout the survey, suggesting that the tide propagates into Mobile Bay as a standing wave. A notable departure from the oscillatory tidal signal was evident three days prior to the survey.

Measurements of wind speed and direction, taken from NOAA CO-OPS station number 8735180, for a period four days prior to and during the experiment were analyzed to determine the effects of meteorological forcing on estuarine flows. Conditions during the survey were generally calm, with wind speeds of 3 - 6 m/s out of the west and northwest. Wind speeds were considerably higher three days prior to the survey, and out of the east and southeast. The combination of higher winds and an easterly direction may explain the non-tidal behavior mentioned previously, where Ekman convergence may have produced setup along the Alabama coast. The wind forcing during the study period, however, was weak.

Preliminary (raw) ADCP data at Mobile Pass is shown in Fig. 11. The top panel of Fig. 11 shows the bathymetry between Dauphin Island and Fort Morgan. The middle panel is an overview of the survey location and track, where the green areas denote land. The lower panel of Fig. 11 shows the distribution of velocity magnitude (m/s) across Mobile Pass, where cooler colors denote slow-moving water, and warm colors denote faster-moving water (about 1 m/s). Note that the highest magnitudes occur in the deeper portion of the channel. The total discharge across the pass is nearly 10,400 m³/s.

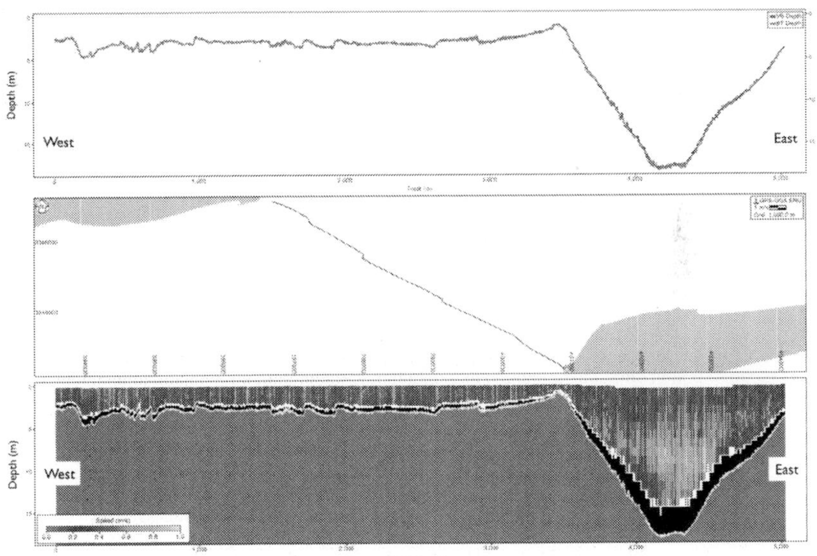

Fig. 11. Bathymetry and velocity magnitude at Mobile Pass for April 1, 2011 during the period 0800 - 0900 hours EDT. The estimated total discharge across the transect was 10,400 m³/s.

Measurements of flow and bathymetry were also collected at Pass aux Herons, to the west of the Dauphin Island Bridge. The preliminary (raw) ADCP data for Pass aux Herons is provided in Fig. 12. The orientation of the plots in Fig. 12 is slightly different than Fig. 11, where north is on the right side of the page in the upper and lower panels. Similar to the flooding tide at Mobile Pass, the strongest flows are confined to the navigation channel and Grant's Pass (just north of the channel), and attain a magnitude of about 1.2 m/s. Unlike Mobile Pass, however, very strong flows are distributed equally over the water column in the channel and pass. The estimated discharge across this transect was 3,300 m^3/s, or about 25% of the total volume flooding into Mobile Bay during the period 0800 - 1100 hours EDT, April 1, 2011, when considering the discharge across Mobile Pass.

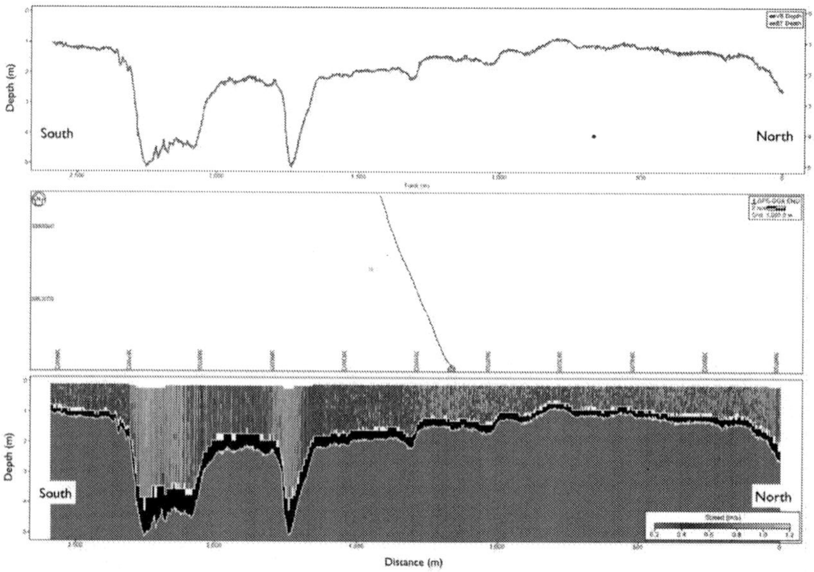

Fig. 12. Bathymetry and velocity magnitude at Pass aux Herons on April 1, 2011 from 1015 - 1100 hours EDT. The estimated total discharge across the transect was 3,300 m^3/s.

An overview of the study area and survey-level view of the CTD locations is shown in Fig. 10. The orange and black dots denote the western and eastern locations of CTD profiles, respectively, provided in Fig. 13. These colors correspond to the orange and black lines in Fig. 13. The vertical profiles of temperature, salinity, and density show only a slight variation over depth near the navigation channel. The CTD cast closest to Dauphin Island suggests a more stratified condition in this portion of the pass, with a notable halocline and pycnocline about 1 to 1.5 m above the bed. Note, however, the very low values of salinity and density at each CTD cast location, even during the flood tide, suggesting the presence of a strong freshwater front.

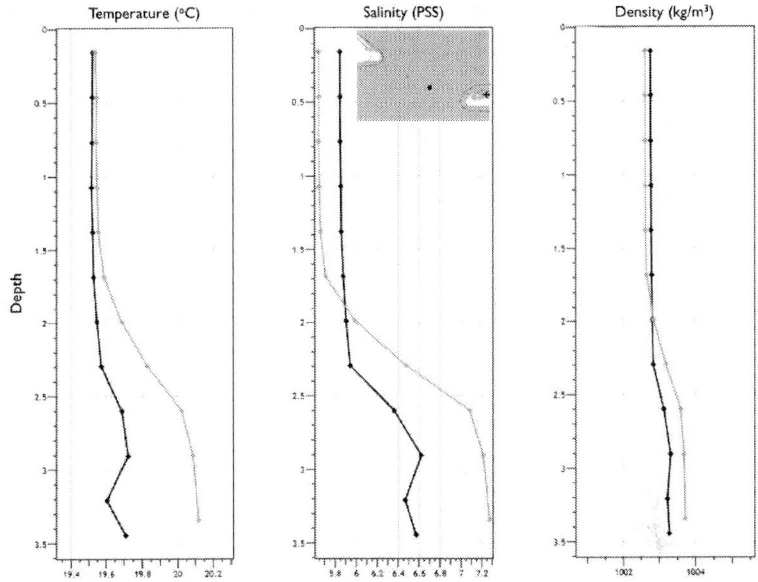

Fig. 13. Vertical profiles of temperature, salinity, and density for two locations at Mobile Pass on April 1, 2011. The orange line represents the western-most CTD cast, while the black line denotes the CTD cast closer to the navigation channel.

Near-surface water characteristics are shown in Fig. 14, where the vessel track is coincident with the spatial distribution of data points. Note the agreement of near-surface temperature and salinity in Fig. 14 with the corresponding values from the CTD profiles shown in Fig. 13. The low salinity environment detected by the CTD profiling is widespread, even on the flooding tide, extending across Mobile Pass and northward into the bay. Values of temperature and salinity entering Mobile Bay from Mississippi Sound across Pass aux Herons, however, were higher. The spatial distributions of near-surface pH, chlorophyll, turbidity, dissolved oxygen, refined fuels, crude oil, and chromophoric dissolved organic matter (CDOM) are also shown in Fig. 14, and their magnitudes and units are specified in each panel. In general, the pH ranged from 7 to 8, the concentration of chlorophyll was low, the turbidity was low, and the dissolved oxygen content was high.

Measurements of refined fuel, crude oil, and CDOM shown in Fig. 14 are made in relative fluorescent units (RFU). For reference, deionized water would have an RFU value of zero, and is commonly used as a calibration standard when the measurement of specific volatile organic compounds cannot be anticipated *a priori*. More simply put, the use of the RFU scale yields a broad-spectrum measurement of the presence of organic compounds in general. In order to measure the volumetric concentration of fuel or crude oil, a corresponding standard would have to be used in the calibration of the instrument. What can be inferred from Fig. 14, though, is that there was a strong return in the measurements of crude oil and CDOM across Mobile Pass and northward into the bay, with much lower values at Pass aux Herons. By comparison, the presence of refined fuels was much weaker, with the exception of one location north of Little Dauphine Island along the centerline of the navigation channel.

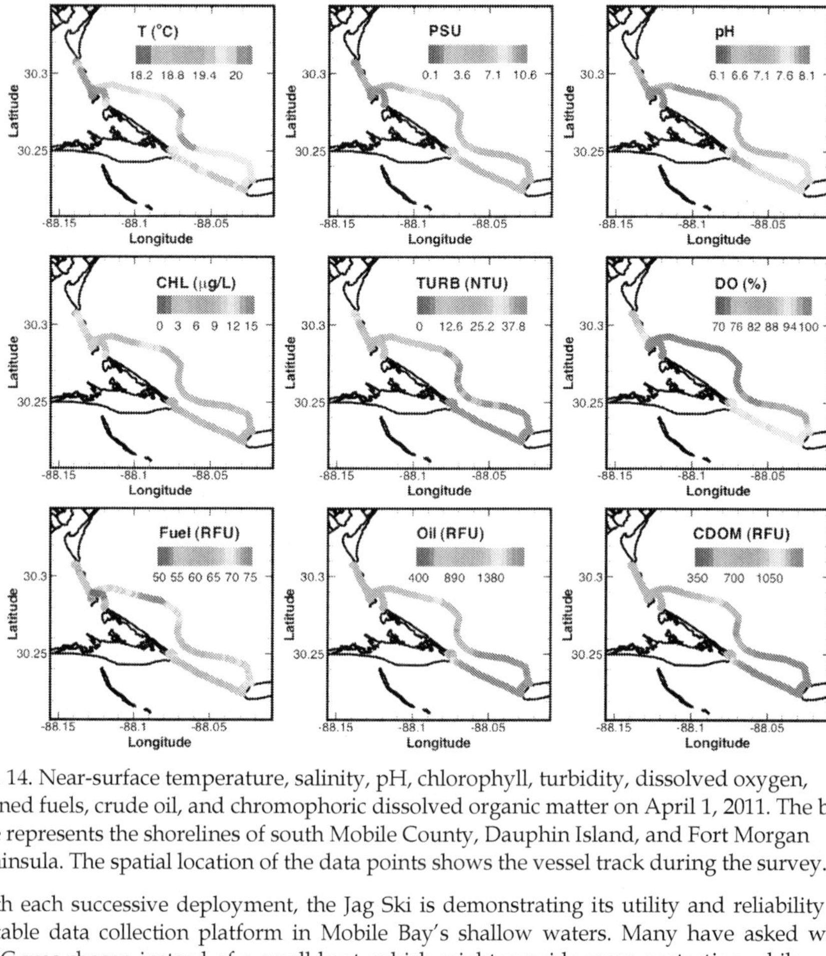

Fig. 14. Near-surface temperature, salinity, pH, chlorophyll, turbidity, dissolved oxygen, refined fuels, crude oil, and chromophoric dissolved organic matter on April 1, 2011. The black line represents the shorelines of south Mobile County, Dauphin Island, and Fort Morgan peninsula. The spatial location of the data points shows the vessel track during the survey.

With each successive deployment, the Jag Ski is demonstrating its utility and reliability as a suitable data collection platform in Mobile Bay's shallow waters. Many have asked why a PWC was chosen instead of a small boat, which might provide more protection while on the water. The simple answer is that in terms of access and ease of use, the PWC cannot be rivaled. The PWC is easy to launch and retrieve, it can be towed by just about any vehicle, and it is much more agile traversing the surfzone than any other craft on the water. In terms of weather conditions, the limitations of the ADCP tend to be more restrictive than the capabilities of the PWC. It is difficult to obtain quality ADCP measurements when the waves are 1 m or greater, but one can still safely operate the PWC in those conditions. Finally, the cost of the PWC is much less than a vessel of any significant size.

4. *In-situ* monitoring: PILS

An effective complement to a mobile platform is a system of low-cost fixed sensors. The goal of the Pressure-Induced Water-Level Sensor (PILS) is to monitor water level over a long period of time, so that it can be correlated to wind, tides, and freshwater flow. In order to be able to deploy a large number of sensors, the PILS unit needs to be low-cost. The units are submerged and estimate water level by measuring water pressure. However, water density varies with temperature and salinity, and so, to measure water depth, temperature and salinity also need to be measured. (The salinity cannot be assumed since, in the brackish estuarine environment, it varies widely.)

Measurement of temperature is straightforward, as integrated temperature sensors are readily commercially available. Since the unit will make intermittent measurements with very low power dissipation, the temperature of the interior of the sensor will be extremely close to that of ambient, and so the temperature sensor will indicate the temperature of the surrounding water. A Maxim DS1621 temperature sensor was chosen; it uses the microprocessor's I^2C bus to communicate.

Measurement of pressure is more complicated because the sensor must be able to register changes in pressure. Thus the pressure sensor must lie outside the waterproof housing. A housing for a commercially available low-cost pressure sensor has been developed and tested, and is described in detail below in section 5.

Measurement of salinity is considerably more complicated because of the ionic nature of seawater. The development of a low-cost pressure sensor is detailed below in section 6.

To make measurements over an extended period of time, the system was designed with flash memory to record readings, a real-time clock to simplify the control of periodic measurements, and a low-cost microcontroller. An Atmel ATMega168 microcontroller was selected along with a serial flash memory and a Maxim DS1337 real-time clock chip. A block diagram of the PILS system is shown below in Fig. 15.

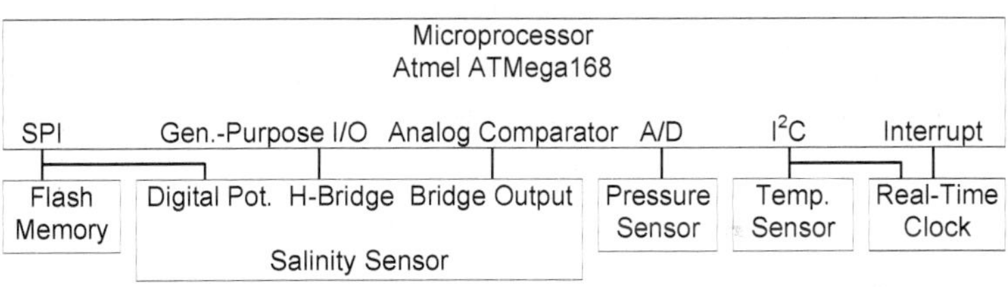

Fig. 15. Block diagram of PILS unit, including its sensor package.

Not counting resistors, capacitors, or a circuit board, the devices listed above have a total cost below $30.

The flash memory is a Winbond W25X80 serial flash. It operates on the microcontroller's SPI bus and has 8 Megabits (1 Megabyte) capacity.

In the process of programming the driver for the flash chip, special considerations were needed to account for the hardware limitations. The problems revolve around the 256 byte page buffer used for programming the flash. If a segment of data was larger than 256 bytes it needed to be broken down into smaller segments. Another, more complicated problem is that the buffer corresponds to a 256-byte page of actual flash (Winbond, 2007). Therefore, if it is necessary to start a segment of data in the middle of a 256 page, it is necessary to end the segment at the end of that page, program the page, and then finish the segment on the next page. These issues were addressed in the design of the flash drivers, and storage of data structures to flash has been tested.

A data structure is needed to store the measurements in flash in an ordered fashion so that they may be retrieved later on. The system must store the time, temperature, pressure, and salinity. The time requires 7 bytes of space for a detailed time stamp. The temperature needs 2 bytes. Sixty pressure measurements are needed (to provide a sample of wave action). With each pressure measurement using 2 bytes, 120 bytes are needed for the wave and water level data. Finally, 2 bytes are needed for the salinity measurements.

A linked list was selected for storage of the data in the flash memory. Each data structure has a 3 byte pointer at the end which gives the address of the next data structure. This allows the software to traverse the list when outputting the data with ease. Additionally, the microprocessor keeps track of where the next set of data must be placed or the tail of the linked list. This allows for quick storing speed without having to read from the flash. A more complicated data structure is not needed because the only time the data is accessed is when the list is parsed at microprocessor start-up. Thus direct access to the data in the middle of the flash is not needed, only the starting address for output of data and the address of the next available slot for storage of new data.

The clock chip was selected to simplify the process of taking periodic measurements and "sleeping" between measurements. The chip uses a 32.768 kHz "tuning fork" crystal, similar to those in wristwatches, to keep time, and has programmable alarms. When the alarm time is reached, the chip asserts an interrupt that "wakes up" the microcontroller. Thus the entire measurement sequence is inside an interrupt service routine.

5. Low-cost pressure sensor

Since the goal of the PILS project is the development of a low-cost deployable sensor, the design proceeded with a low-cost MEMS-based pressure sensor. A Freescale MPXM2010GS sensor was selected. It measures gauge pressure and has a dynamic range of 10,000 kPa (roughly 1 m of water depth). The limited dynamic range was selected for initial tests due to earlier difficulties with sensors having higher dynamic range.

To amplify the signal coming out of the pressure sensor, an op-amp circuit was designed based on an application note from Freescale (Clifford, 2006). Interestingly, the application note explained how to sense water depth in a washing machine. The output of the op-amp circuit was routed into the A/D converter of an Atmel ATMega 168 microcontroller and software was written to obtain samples periodically from the sensor.

The sensor was connected to a piece of tubing with a balloon on the end, so that the prototype unit did not need to be submerged. The balloon was submerged in the wave tank facility at the University of South Alabama, and six seconds of data were obtained. Pictures of the unit under test and of the data are shown below in Figs. 16 and 17.

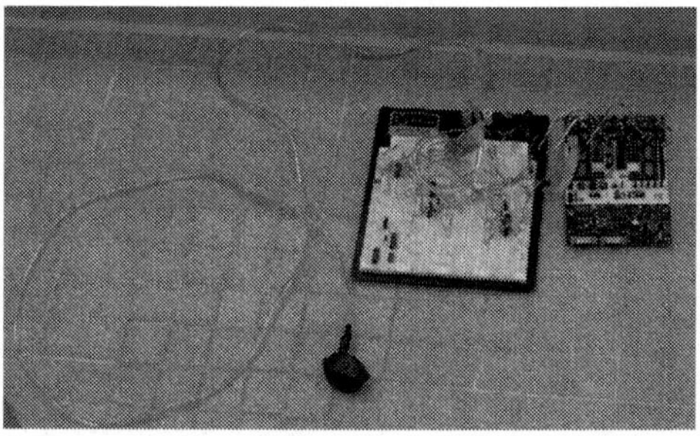

Fig. 16. Pressure sensor. Note balloon and tubing.

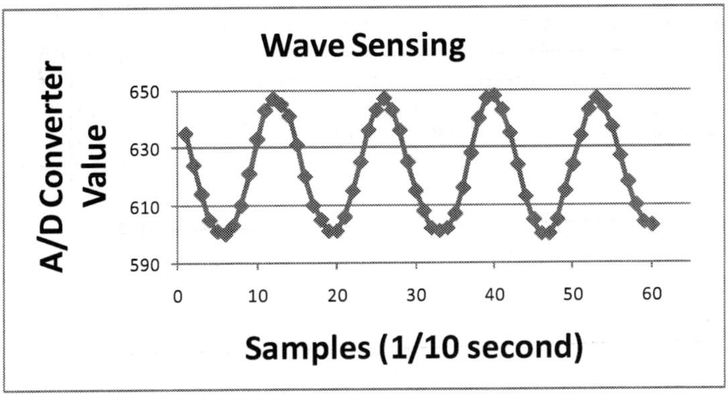

Fig. 17. A/D converter data from the ATMega168. The sample period was 100 ms.

The pressure-sensor data not only measures pressure but also is accurate enough (at the relatively shallow depth of the test) to indicate wave action. Thus the PILS unit will measure not only water level but also wave height.

6. Novel salinity sensor

As noted above, the ability to measure salinity is necessary in order to measure water density and thereby convert a pressure reading to a measurement of water depth. Water salinity can be estimated by measuring the conductivity of a cell of known geometry (that is, the conductance measured between a pair of calibrated electrodes) and then compensating for temperature.

To measure the bulk conductivity of a sample, a set of electrodes of known geometry is used. The set is calibrated ahead of time using solutions of known salinity. The process can be described mathematically as follows.

First, it is well-known that the resistance, R, of a substance can be found as follows

$$R=\rho l/A \tag{1}$$

where ρ is the bulk resistivity of the material, l is the length of the material (in this case, the spacing between the electrodes and therefore the length of the water being measured), and A is the area of the material (in this case, similarly, the area of the electrodes). l/A, then, is the cell constant C which has units of reciprocal-length. ρ is an intrinsic property of the material being measured and C is an intrinsic property of the set of electrodes. (Note that, in this article, we use the terms *resistance* and *conductance* to refer to a measured property of the material being tested and the terms *resistivity* and *conductivity* to refer to the intrinsic property of the material being tested. The actual process will measure resistance and use it to infer conductivity.)

Second, the conductance of a fluid, G, is the reciprocal of resistance (R) and the conductivity of the fluid, σ, is the reciprocal of resistivity R, and so

$$\sigma/G=C \tag{2}$$

Equation (2) can be used to determine the cell constant C by measuring the conductance of a fluid of known conductivity, and can, after being rearranged, be used to determine the conductivity of a fluid by using electrodes of known cell constant C and by measuring conductance.

Third, there are standard equations that are commonly used to estimate the salinity and density of seawater by using conductivity and temperature (Greenberg et al., 1992). Thus the resistance of a seawater sample is measured and converted to conductance, and, using the cell constant C, the conductivity is estimated. The standard equations are then used to estimate seawater density.

Design of a low-cost salinity sensor began with a simple Wheatstone bridge. Its selection was obvious - it permits extremely accurate resistance measurements from imprecise components. For the variable-resistor leg of the bridge, a computer-controlled "digital potentiometer" was used. (An Analog Devices AD8402 was selected.) The selected potentiometer has an eight-bit register that controls the "wiper setting" and so a register value of 0 is minimum resistance and a value of 255 is maximum resistance. A 10kΩ value was selected. (Note that a 100kΩ resistor could be added in parallel for a more accurate reading if so desired.) For the resistor in series with the digital potentiometer, a 20kΩ resistor was selected. For the opposite side of the bridge, the cell (the electrodes to be immersed in seawater) was placed in series with a resistor. The value of the "upper right" resistor is chosen to make the bridge balance across a desired range of salinity, taking into account the geometry of the cell. (The selection process is described in more detail below.) A diagram of the Wheatstone bridge is shown below in Fig. 18.

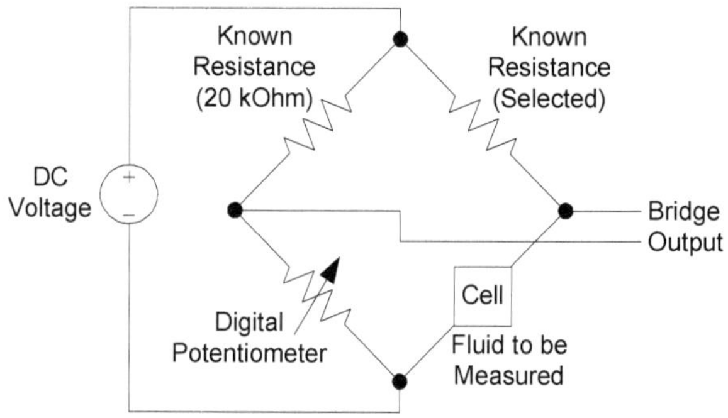

Fig. 18. Wheatstone Bridge used to measure seawater conductance.

The bridge permits an accurate resistance measurement to be made without a precision DC reference, without a current-measuring capability needed, with making only a single measurement (the resistance setting of the potentiometer), and with low-cost components. The measurement process starts by setting the potentiometer to a minimum resistance setting and then increasing its resistance until the polarity of the bridge output reverses. Other algorithms may arrive at a measurement faster, but this algorithm was selected for its simplicity.

To measure salinity, the bridge is first used to measure resistance. Conductance is simply the reciprocal of resistance. From a known, calibrated quantity called "cell constant", the conversion from conductance to conductivity is possible, described in more detail below. The result is a measurement of the bulk conductivity of the seawater.

Initial testing of the Wheatstone bridge was altogether unsuccessful; it never registered a stable resistance measurement.Measurements made with an ohmmeter yielded the same

result. After consultation with a chemical engineering faculty member, it was pointed out that the ionic nature of seawater made a DC measurement impossible. The DC voltages disrupt the ionic distribution of the seawater and resistance measurement is perturbed.
The next step was to replace the DC voltage indicated above in Fig. 18 with an H-bridge. An H-bridge permits the application of a DC voltage in both positive and negative polarity, and is commonly used to control DC electric motors. A Texas Instruments L293D bipolar H-bridge was selected.
During the measurement process, the H-bridge polarity is periodically reversed. More specifically, every time the wiper setting is incremented by one, the polarity is reversed. The software then takes into account that the sign of the bridge output also reverses when the polarity is reversed.
The final circuit is shown below in Fig. 19. Note that the microprocessor's built-in analog comparator was used to lower the cost of the design.
The sensor has an intrinsic limit at the maximum resistance of the potentiometer. Taking into account that fresh water has low conductivity and that conductivity is the reciprocal of resistivity, the result is that the sensor has an intrinsic minimum salinity. The "upper right" resistance in Fig. 19 is selected so that the bridge balances at a high potentiometer setting at the minimum desired salinity reading.
The following process was used to test the circuit over a wide range of salinity.
First, the "upper right" resistance was set so that the sensor produced a reading of decimal 71 (hex 47) at a salinity of 10 parts per thousand (ppt). The resistance value was 38.2 Ohms (56 Ohms in parallel with 120 Ohms).
Second, the salinity was increased in 5 ppt increments, and a resistance measurement made, until a salinity of 40 ppt was reached. (Seawater typically has a salinity of 38 ppt.) The results are tabulated below in Table 1.

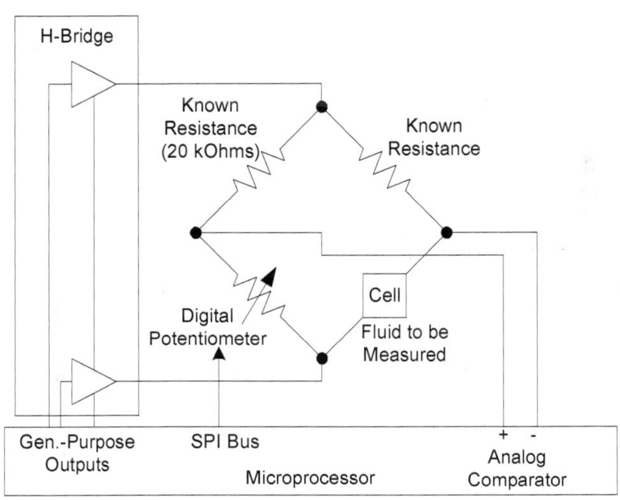

Fig. 19. Final salinity circuit.

The wiper setting is the resistance measurement, where 0 is 0 Ohms and 255 is 10k Ohms. The measured cell resistance is the measured resistance of the cell calculated from the other three bridge resistances. The measured conductance is the reciprocal of the resistance. Finally, the bulk conductivity of water at different salinities is noted from (Weyl, 1964). This last column, then, is the "known" conductivity.

Salt content (ppt)	Digital Pot Wiper Setting	Digital Pot Resistance (Ohms)	Measured Cell Resistance (Ohms)	Measured Cell Conductance (mS)	Bulk Conductivity at 20° C (mS/cm)
10	71	2784	5.32	188.0	15.6
15	51	2000	3.82	261.8	22.4
20	39	1529	2.92	342.3	29
25	33	1294	2.47	404.6	35.4
30	28	1098	2.10	476.8	41.7
35	25	980	1.87	534.0	47.9
40	22	863	1.65	606.9	53.9

Table 1. Measurements used to calibrate the salinity sensor. Bulk conductivity from (Weyl, 1964).

Third, the cell constant of the electrodes had to be estimated from the data. As shown in (2), the cell constant can be estimated by dividing the known conductivity by the measured conductance. The average estimated cell constant over all 7 measurements is 0.0867cm^{-1}. The measured conductivity of the water is plotted against the standard model of the conductivity of seawater using a cell constant of 0.0867 below in Fig. 20.

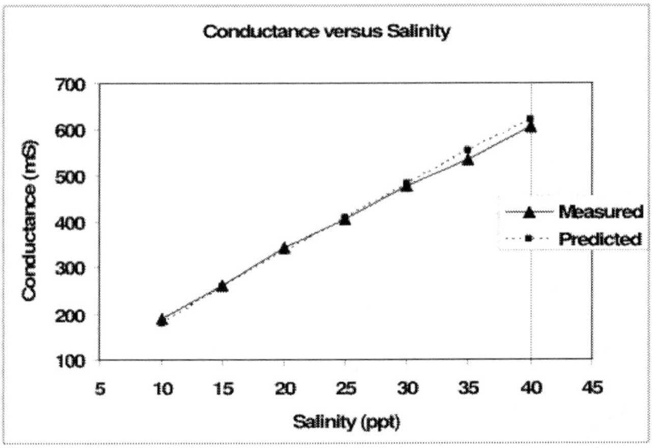

Fig. 20. Correlation of known conductance of seawater (predicted) to actual data (measured).

7. Conclusion

The Jag Ski provides a unique opportunity to collect hydrographic and environmental data in shallow and remote areas typically inaccessible by traditional watercraft. Aside from its

utility as a hydrographic data collection platform, it is small, inexpensive, and relatively easy to maintain. Where a traditional vessel may require two or more people to launch, operate, and recover, the PWC can easily be attended by one person if needed. With the recent addition of the Portable SeaKeeper system, the Jag Ski's capabilities have expanded tremendously. The ability to map large spatial areas in a relatively small amount of time is very helpful in coastal applications, mainly because it reduces the tidal bias of the collected data. The Jag Ski's speed and ease of deployment will also provide opportunities to perform episodic surveys of coastal waters to determine the effects of storms or other events on the near-surface water chemistry of Mobile Bay, Mississippi Sound, and nearby rivers.
The PILS unit combines low-cost components, including a novel low-cost salinity-measuring circuit to provide a powerful and inexpensive environmental-monitoring capability. The sensor package can readily be modified for other, similar missions. For example, development is underway, using the microprocessor, clock, and salinity sensor, to develop a system to control periodic GPS measurements and satellite transmissions to develop a low-cost drifter to measure surface currents in the open ocean.

8. Acknowledgment

The authors wish to acknowledge the support of the following organizations in conducting this work: The University of South Alabama College of Engineering, The University of South Alabama Research Council, and The University of South Alabama University Committee on Undergraduate Research (UCUR) Program. A portion of this material is based upon work supported by the National Science Foundation under Grant No. OCE-1058018.

9. References

Clifford, M. (2006). Water Level Monitoring, In: *Freescale Semiconductor Application Note AN1950*, Rev. 4, Nov. 2006

Comeau, A.; Lewis, M., Cullen, J., Adams, R., Andrea, J., Feener, S., McLean, S., Johnson, K., Coletti, L., Jannasch, H., Fitzwater, S., Moore, C., & Barnard, A. (2007). Monitoring the spring bloom in an ice covered fjord with the Land/Ocean Biogeochemical Observatory (LOBO), *Proceedings of OCEANS 2007*

Dugan, J. P.; Vierra, K. C., Morris, W. D., Farruggia, G. J., Campion, D. C., & Miller, H. C. (1999). Unique vehicles for bathymetric surveys in exposed coastal regions, *Proceedings of the Hydrographic Society of America Conference*, April 27-29, 1999

Dugan, J. P.; Morris, W. D., Vierra, K. C., Piotrowski, C. C., Farruggia, G. J., & Campion, D. C. (2001). Jetski-based nearshore bathymetric and current survey system. *Journal of Coastal Research*, Vol. 17, No. 4, pp. 900-908

Freitag, L.; Johnson, M., & Preisig, J. (1998). Acoustic communications for UUVS. *Sea Technology*, Vol. 39, No. 6, pp. 65–71

Frolov, S .; Baptista, A., & Wilkin, M. (2008). Optimizing fixed observational assets in a coastal observatory. *Continental Shelf Research*, Vol. 28, No. 19, pp. 2644-2658

Frye, D. E.; Kemp, J., Paul, W., & Peters, D. (2001). Mooring developments for autonomous ocean-sampling networks. *IEEE Journal of Oceanic Engineering*, Vol. 26, No. 4, pp. 477-486

Greenberg, A.; Clesceri, L., & Eaton, A. (1992). *Standard Methods for the Examination of Water and Wastewater* 18th Edition, The American Public Health Association, Washington D.C.

Hampson, R.; MacMahan, J., & Kirby, J. T. (2011). A low-cost hydrographic kayak surveying system. *Journal of Coastal Research*, Vol. 27, No. 3, pp. 600-603

Kim, C.; Park, K., Jung, H., & Schroeder, W. (2008) A hydrodynamic modeling study of physical transport in Mobile Bay and Eastern Mississippi Sound, Alabama. Paper submitted to *Estuaries and Coasts*, 2008

Luther, M.; Metz, C. R., Scudder, J., Baig, S. R., Pralgo, L., Thompson, D., Gill, S., & Hovis, G. (2007). Water level observations for storm surge. *Marine Technology Society Journal*, Vol. 41, No. 1, pp. 35-43

MacMahan, J., (2001). Hydrographic surveying from personal watercraft. *Journal of Surveying Engineering*, Vol. 127, No. 1, pp. 12-24

National Research Council (2003). *Enabling Ocean Research in the 21st Century: Implementation of a Network of Ocean Observatories*, National Research Council, Washington D.C.

O'Flyrm, B.; Martinez, R., Cleary, J., Slater, C., Regan, F., Diamond, D., & Murphy, H. (2007). SmartCoast: a wireless sensor network for water quality monitoring, *Proceedings of the 32nd IEEE Conference on Local Computer Networks* (LCN 2007), pp. 815-816

Puleo, J. A.; Farquharson, G., Frasier, S. J., & Holland, K. T. (2003). Comparison of optical and radar measurements of surf and swash zone velocity fields. *Journal of Geophysical Research*, Vol. 108, No. C3, pp. 45-1 – 45-12

Roemmich, D.; Riser, S., Davis, R., & Desaubies, Y. (2004). Autonomous profiling floats: workhorse for broad-scale ocean observations. *Marine Technology Society Journal*, Vol. 38, No. 2, pp. 21-29

Rudnick, D. L.; Davis, R. E., Eriksen, C. C., Fratantoni, D. M., & Perry, M. J. (2004). Underwater gliders for ocean research. *Marine Technology Society Journal*, Vol. 38, No. 2, pp. 73-84

Sanibel-Captiva Conservation Foundation (2009). Accessed 11/24/09, Available from: <http://recon.sccf.org/index.shtml>

Schroeder, W., & Wiseman Jr., W. (1986). Low-frequency shelf-estuarine exchange processes in Mobile Bay and other estuarine systems on the northern Gulf of Mexico, In: *Estuarine Variability*, Ed. D. A. Wolfe, pp. 355-366, Academic Press, New York, NY

Silva, S.; Cunha, S., Matos, A., & Cruz, N. (2008). Shallow water height mapping with interferometric synthetic aperture sonar, *Proceedings of OCEANS 2008*

Strydom, N. A. (2007). Jetski-based plankton towing as a new method of sampling larval fishes in shallow marine habitats. *Environmental Biology of Fishes*, Vol. 78, pp. 299-306

Thosteson, E.; Widder, E., Cimaglia, C., Taylor, J., Burns, B., & Paglen, K. (2009). New technology for ecosystem-based management: marine monitoring with the ORCA Kilroy network, *Proceedings of OCEANS 2009-EUROPE*

U.S. Commission on Ocean Policy (2004). *An ocean blueprint for the 21st century*, Final Report, U.S. Commission on Ocean Policy

Van Dorn, W. (1953). Wind stress on an artificial pond. *Journal of Marine Research*, Vol. 12, No. 3, pp. 249-276.

Villa, P., & Gianinetto, M. (2006). Multispectral transform and spline interpolation for mapping flood damages, *Proceedings of IEEE International Conference on Geoscience and Remote Sensing Symposium (IGARSS 2006)*, pp. 275-278

Weyl, P. (1964). On the change in electrical conductance of seawater with temperature. *Limnology and Oceanography*, Vol. 9, No. 1, (Jan. 1964), pp. 75-78

Winbond. (2007). *W25X80A Datasheet* [Revised 08/09], Available from : <http://www.winbond-usa.com/products/Nexflash/pdfs/datasheets/W25X10_20_40_80g.pdf>

10

An Innovative Approach to Biological Monitoring Using Wildlife

Mariko Mochizuki[1], Chihiro Kaitsuka[1],
Makoto Mori[2], Ryo Hondo[1] and Fukiko Ueda[1]
[1]*Nippon Veterinary and Life Science University, Tokyo,*
[2]*Shizuoka University, Shizuoka,*
Japan

1. Introduction

Biological monitoring using wildlife is a useful and important method that helps us to understand the degree of contamination in the environment. The book "Our Stolen Future" (Colborn et al., 1996) has become an influential bestseller worldwide; the authors of this book have pointed out issues relevant to the monitoring of the state of environmental pollution using wildlife. However, there are also many criticisms of the content of this book. For example, the designation of the control areas as non-contaminated is very difficult in the studies that use wildlife (Krimsky, 2000). In studies that use wildlife, there is a lack of epidemiological information on age, sex, movement range and detailed feeding habits. For example, the content of cadmium (Cd) in animals increases with age (Sakurai, 1997), even when the animals live in non-polluted areas. This is because Cd has a long biological half-life in animals (Friberg et al., 1974). Thus, knowledge of the age of targeted animals is necessary for accurate monitoring. However, obtaining an estimate of age in wildlife is very difficult. Carnivorous animals have been used frequently for biological monitoring (Harding et al., 1998; Helander, et al., 2009; Kenntner, et al., 2007; Meador et al., 1999) because it is well known that various contaminants are bioaccumulated in carnivorous animals as they move up the food chain. However, detailed information on feeding habits is sometimes difficult to obtain. According to bird guides, the greater scaup (*Aythya marila*) is classified as a carnivorous bird. However, its rate of intake of animal food changes between 45 and 97 % depending on the environment (Kaneda, 1996). In such a case, is it correct to categorize the scaup among carnivorous birds?

Despite the lack of epidemiological information, we have been investigating the degree of contamination of wild birds with inorganic elements such as Cd (Mochizuki et al., 2002a, 2011d; Ueda et al., 1998), chromium (Cr) (Mochizuki et al., 2002c), molybdenum (Mo) (Mochizuki et al., 2002c), thallium (Tl) (Mochizuki et al., 2005) and vanadium (V) (Mochizuki et al., 1998, 1999). However, there is also problem in the use of statistical procedures in studies that use wildlife because the distribution of the data is very wide. Normally distributed data are sometimes not obtained from samples of wildlife (Mochizuki et al., 2010b; Ueda et al., 2009a). The effects of toxic elements have also been investigated under experimental conditions using cultured bacteria (Kadoi et al., 2009), cells (Mochizuki et al., 2011b), and various experimental animals (Mochizuki et al., 2000). However, biological monitoring is important for the assessment of risk to human health.

Recently, we developed a solvent for use in biological monitoring using wildlife. This method was established using the significant regression lines obtained from the Cd content of kidney and that of liver (Mochizuki et al., 2008). Given that the data from animals were cited in various studies in which no particular contamination was described, we considered that these lines were indicative of normal metabolism in animals. This theory was supported by some evidence obtained from polluted animals, including humans (Mochizuki et al., 2008; Ueda et al., 2009a). Thus, the degree of contamination of humans (Mochizuki et al., 2008; Ueda et al., 2009a), experimental animals (Mochizuki et al., 2008; Ueda et al., 2009b), domestic animals (Ueda et al., 2011) and wild birds (Mochizuki et al., 2011a,c,d; Ueda et al., 2009a) has been analyzed using those indexes. Further, we developed a similar index for lead(Pb); the basis of this study was presented at an International Conference (Mochizuki et al., 2009), and the modified index has also been submitted to a journal for publication.

However, contamination with multiple elements is also an important problem in environmental science. Recently, we investigated the concentration of various elements in the urine of cats (Mochizuki et al., 2010c). In that study, a significant correlation was obtained among multiple elements in urine obtained from healthy cats, although a similar correlation was not observed in urine obtained from cats with urinary tract disease. A loss of balance and equilibrium among multiple elements had occurred in the urine of the diseased cats. This result suggested that similar indexes involving Cd and Pb can be obtained using measurement of multiple elements.

The new technique for biological monitoring is introduced in the first part of this chapter. Subsequently, we will attempt to establish an index to increase our understanding of the degree of contamination with multiple elements using multivariate analysis.

2. Introduction of CSRL and CEPE

In this section, we explained about Cd standard regression line (CSRL) and Cd equal probability ellipse (CEPE). We selected previous publications that reported the content of Cd in samples of 46 mammals and 55 birds, and we used 101 data points from 27 reports in which the Cd contents were represented as arithmetic means. The 101 data points were plotted on a graph with the Cd content in the liver on the abscissa and the Cd content in the kidney on the ordinate. A significant correlation was obtained, as follows: $Y=0.902X - 1.334$, $Y=\log(y)$, $X=\log(x)$, $R^2=0.944$, $p<0.01$. The regression line obtained after logarithmic transformation was $\log(Y)=0.900\log(X)-0.580$ ($R^2=0.944$, $p<0.01$)(1). When the outliers among the 101 data points were tested by equal probability ellipse, seven data points were identified as outliers as shown in Fig.2. After elimination of these seven points, the regression line obtained was: $\log(Y)= 0.941\log(X)-0.649$, ($R^2=0.965$, $p<0.01$)(2). There were no significant differences between the two lines (1&2) (Ueda et al., 2009a). In mention above, regression line obtained from 101 points and the equal probability ellipse were used as the Cd standard regression line, CSRL, and the Cd equal probability ellipse, CEPE, respectively.

The data from experimental animals to which Cd had been administered were distinct from the CSRL, as shown in Fig. 1.

Similarly, the data from humans who lived in a polluted area and from patients with Itai-itai disease were located outside the CEPE, as shown in Fig. 2. Although the values from humans who lived in non-polluted areas were high, the data were located within the CEPE, as shown in the figure (Fig.2). Detailed information on the references used (Mochizuki et al., 2008), the procedure for calculation of the indexes (Ueda et al., 2009a), and the data from humans and rushes monkeys (Mochizuki et al., 2008) have been described in our previous reports.

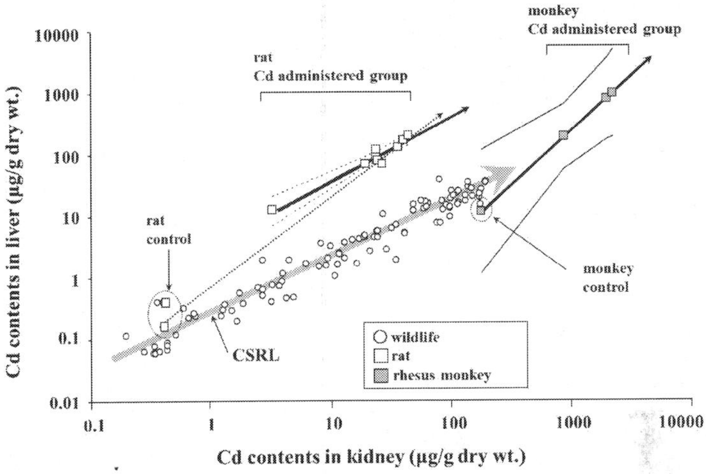

Fig. 1. Comparison of the data from laboratory animals. Original figure from Mochizuki et al. (2008) as modified by Ueda et al. (2009a).

A new development in the research area of biological monitoring has been introduced in this section. In the next section we describe the pilot study for establishment of a similar index using multiple elements.

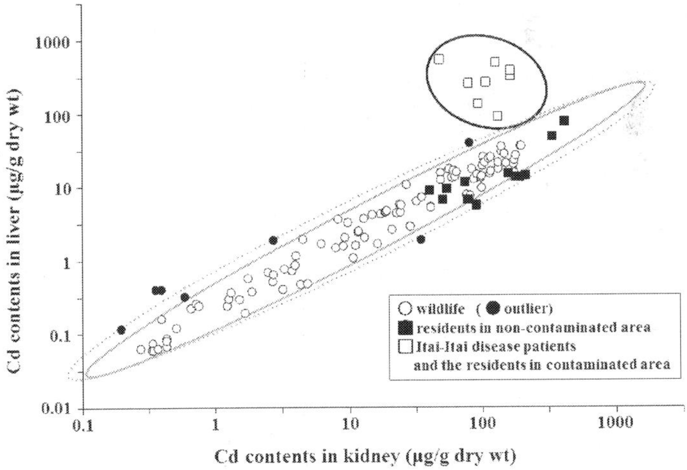

Fig. 2. Comparison of human data. Original figure from Ueda et al. (2009a). Dot-line; equal probability ellipse by 101 data points, solid line; equal probability ellipse by 94 data points.

3. A new index for investigation of contamination by multiple elements

3.1 Materials and methods

3.1.1 The wild birds used in the present study

A total of 127 wild birds, including Anatidae (n=65), seabirds (n=17), common cormorants (*Phalacrocorax carbo*, n=30), Ardeidae (n=10) and others (n=5) was used in the present study. The categories of birds included the following species: the Anatidae included spotbill duck (n=19, *Anas poecilorhyncha*), wigeon (n=15, *Anas penelope*), pintail (n=11, *Anas acuta*), mallard (n=7, *Anas platyrhynchos*), common teal (n=6, *Anas crecca*), gadwall (n=2, *Anas strepera*), common shoveler (n=2, *Anas clypeata*), wood duck (n=1, *Aix sponsa*), garganey (n=1, *Anas querquedula*) and tundra swan (n=1, *Cygnus columbianus*). The seabirds included greater scaup (n=6, *Aythya marila*), tufted duck (n=6, *Aythya fuligula*), Eurasian pochard (n=3, *Aythya ferina*), common scoter (n=1, *Melanitta nigra*) and great crested grebe (n=1, *Podiceps cristatus*). The Ardeidae included black-crowned night heron (n=3, *Nycticorax nycticorax*), little egret (n=3, *Egretta garzetta*), intermediate egret (n=2, *Egretta intermedia*), cattle egret (n=1, Bubulcus ibis) and great egret (n=1, *Egretta alba*). The others included eastern turtle dove (n=1, Streptopelia orientalis), common kestrel, (n=1, *Falco tinnunculus*), sparrowhawk (n=1, *Accipiter nisus*), peregrine falcon (n=1, *Falco peregrinus*) and Eurasian woodcock (n=1, *Scolopax rusticola*). As shown in Fig. 3, the birds were collected from various areas in Japan.

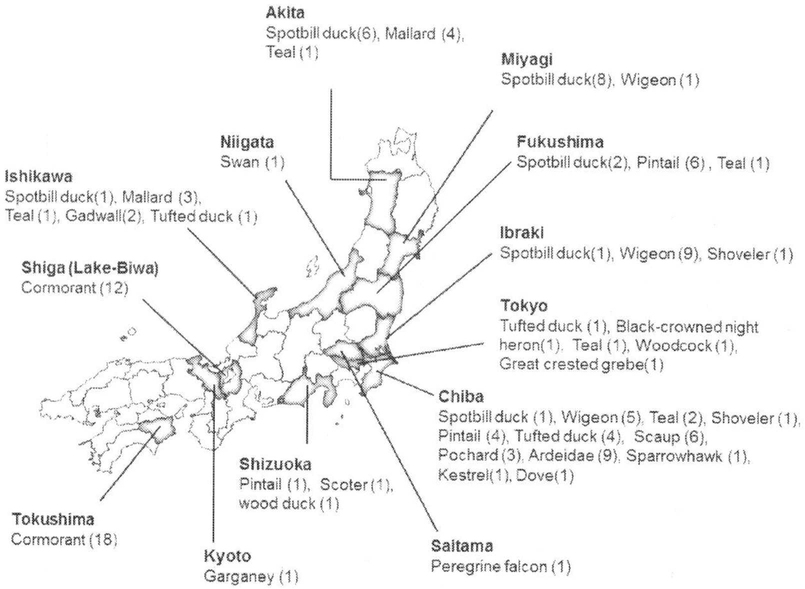

Fig. 3. The wild birds and collection areas used in this study. Number in brackets indicated the number of samples.

Most of the wild birds were collected as part of another National Investigation conducted by the Environment Agency in Japan (the present; Ministry of the Environment in Japan) in

1995. Other birds, which were protected in the Gyoutoku bird observatory in Chiba Prefecture, were transported to our laboratory after death.

3.1.2 Analytical procedure

Samples of kidney were removed from the birds, and about 200 mg of each sample was put into a Pyrex tube (Corning, USA), and dried in an oven at 70℃ to determine the dry weight of the sample. The appropriate volume of HNO_3 : $HClO_4$ (1:1, Wako Pure Chemical, Ltd., Japan) was added to the dried samples, and the samples were digested at 180℃. The contents of various elements in the kidneys of the birds were analyzed using inductively coupled plasma emission spectrometry (ICP-AES, FTP08, Spectro A.I., Germany). The eight target elements were: Cd, Cr, copper (Cu), lithium (Li), Mo, titanium (Ti), Tl and V. The standard additional method was employed for the analysis. The detailed methods of sample preparation and the analytical procedure have been described previously (Mochizuki et al., 2002b).

3.1.3 Statistical methods

The statistical analyses used in the present study were carried out using computer software such as Lotus 2001 (Lotus Development), Excel 2003 (Microsoft Corporation), and JUMP (SAS Institute, Japan) to obtain the regression line, the confidence intervals, and the logarithmic transformation. Factor analysis was carried out using Excel add-in software (Esumi, Japan).

3.2 Factor analysis

3.2.1 Establishment of an index for multiple elements

The contents of the eight elements measured in the kidney are shown in Table 1. The data were recalculated using factor analysis. The multiple variables used in the present study (contents of elements) were merged using factor analysis, a form of multivariate analysis. Thus, a higher factor score was thought to indicate more serious contamination by multiple elements. We obtained three significant factors, as shown in Table 2. No tendency for contamination was observed when the mean values of each category were compared. Thus, it is thought that the comparison using only mean values makes it difficult to understand the degree of contamination by multiple elements.

n	Anatidae 65	Seabird 17	Cormorant 30	Ardeidae 10	Others 5
Cd	8.33±1.48	8.36±4.00	1.97±0.50	4.38±0.98	9.17±6.73
Cr	2.67±0.55	1.69±0.32	1.65±0.37	3.86±0.93	0.33±0.200
Cu	24.21±2.87	40.99±6.78	12.85±0.86	21.12±2.77	19.04±4.19
Li	1.85±0.52	1.24±0.36	1.62±0.43	3.04±0.92	0.32±0.23
Mo	6.88±1.04	3.80±0.94	4.62±0.43	5.35±1.39	12.57±8.98
Ti	2.07±0.56	0.80±0.34	1.32±0.34	2.94±0.88	0.83±0.83
Tl	9.17±2.14	2.87±0.49	3.33±0.49	5.07±1.45	1.97±1.85
V	2.50±0.60	1.10±0.32	2.35±0.48	3.04±0.92	1.75±1.16

Table 1. The contents of the elements in kidneys from birds of each category. The results are represented as mean contents (µg/g dry wt.), and the standard error of the mean.

	n	Factor 1	Factor 2	Factor 3
		Cr, Li, Ti	Mo	Cu
Anatidae	65	0.101±0.155	0.111 ±0.099	-0.026±0.099
Seabird	17	-0.261±0.078	-0.126±0.122	0.656±0.222
Cormorant	30	-0.133±0.103	-0.230±0.063	-0.259±0.019
Ardeidae	10	0.483±0.264	-0.067±0.159	-0.115±0.063
Others	5	-0.600±0.087	0.494±0.848	-0.112±0.206

Table 2. The mean factor score and standard error of the mean for each category of birds.

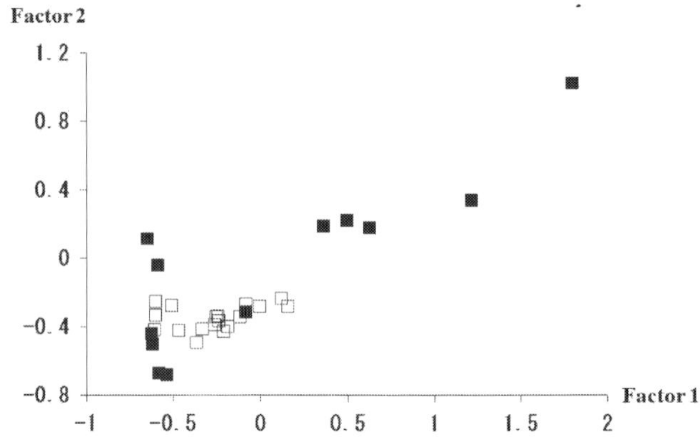

Fig. 4. The relation between the factor score of factor 1 and that of factor 2 for the common cormorant. Filled squares; common cormorants collected in Shiga Prefecture; empty squares; common cormorants collected in Tokushima Prefecture.

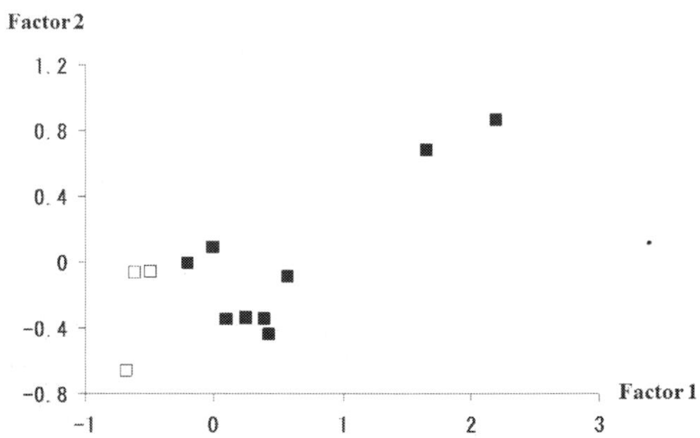

Fig. 5. The relation between the factor score of factor 1 and that of factor 2 for wild birds collected in Chiba Prefecture. Filled squares; Anatidae, empty squares; other birds.

We used previously published data to develop the indexes for Cd and Pb in our previous studies. However, we were unable to use a similar method in this study. Thus, the correlation between the factor score of factor 1 and that of factor 2 was investigated using various methods of classification. The correlation ($Y=0.499X - 0.177$, $R^2=0.656$) was found between the factor score of factor 1 (Cr, Li and Ti) and that of factor 2 (Mo) in the results from common cormorants, as shown in Fig. 4. A similar correlation was obtained using the results from wild birds captured in Chiba Prefecture (Fig. 5).

Further, a correlation was also obtained when Figs 4 and 5 were summarized, as shown in Fig. 6. The regression line obtained was: $Y=0.474X - 0.199$, $R^2=0.698$. When the outliers among the data points were checked using the method of the 95% equal probability ellipse, three data points were identified as outliers. It was thought that correlation between two variables indicated normal equilibration in the target animals investigated using multiple elements. As mentioned above, we decided to use the regression line obtained in Fig. 6 and the equal probability ellipse as the multiple elements standard regression line, MSRL, and the multiple elements equal probability ellipse, MEPE, respectively.

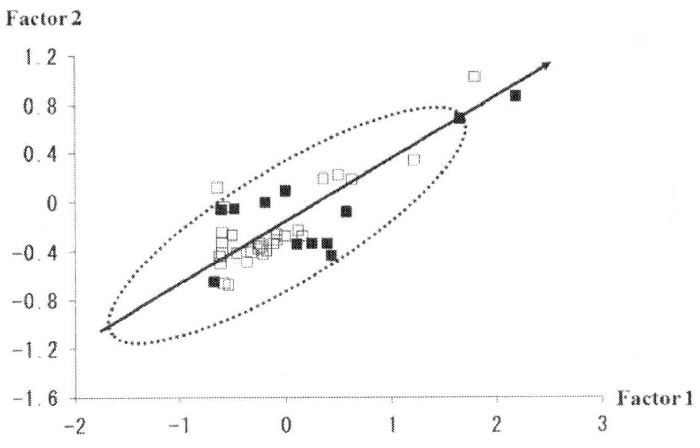

Fig. 6. The relation between the factor score of factor 1 and that of factor 2 in the data used in Figs 4 and 5. Filled squares; birds collected in Chiba Prefecture (Fig. 5), empty squares; common cormorants (Fig. 4). Dotted line; 95% equal probability ellipse, solid line; regression line obtained from wild birds.

3.2.2 Comparison of the degree of contamination of diving and dabbling ducks

The factor scores of diving ducks and dabbling ducks were compared with the index obtained. As shown in Fig. 7, the factor score obtained from diving ducks was observed to fall within the MEPE, except for one data point. Similarly, two of nine data points were observed to fall outside the MEPE when the data from wild birds collected in Chiba Prefecture were compared with the index (Fig. 8).

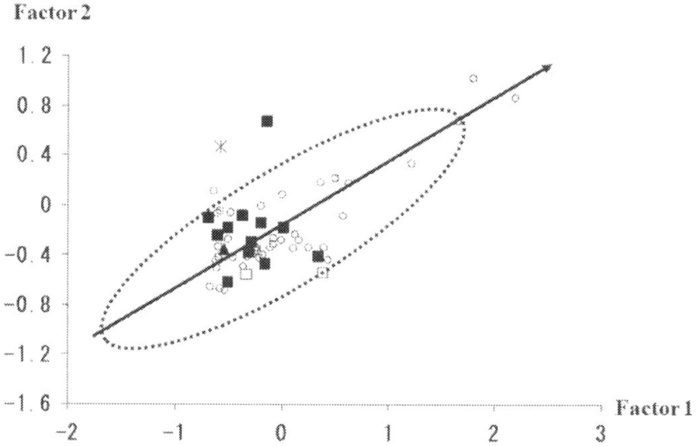

Fig. 7. The comparison between the index and the data obtained from seabirds. Filled squares: diving ducks collected in Chiba Prefecture, empty squares; diving ducks collected in Tokyo, asterisk; seabirds collected in Tokyo, filled triangles; diving ducks collected in Ishikawa Prefecture. Dotted line; 95% equal probability ellipse, solid line; regression line obtained from wild birds (empty circles) used in Fig. 6.

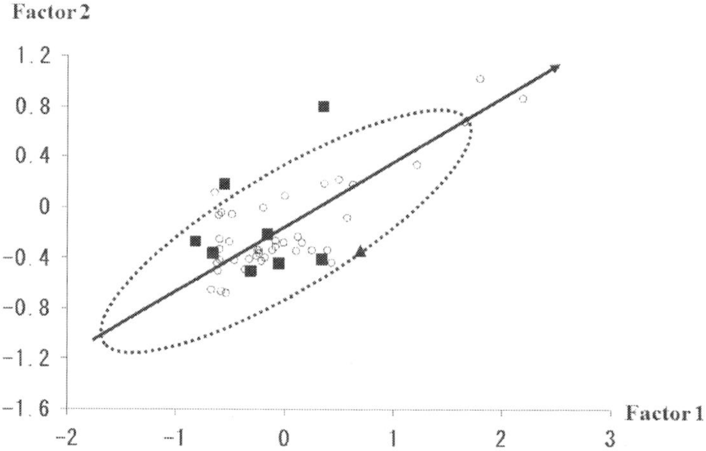

Fig. 8. Comparison between the index and the data obtained from dabbling ducks collected in Chiba Prefecture. Filled squares; wigeon, filled triangles; teal. Dotted line; 95% equal probability ellipse, solid line; regression line obtained from wild birds (empty circles) used in Fig. 6.

On the other hand, the data from dabbling ducks showed a marked tendency to deviate from the MEPE (Figs 9 and 10). Dabbling ducks inhabit inland water environments such as lakes and marshes. Thus, it is thought that the degree of contamination with multiple elements may be more serious in dabbling than in diving ducks. However, two of eight data points from wigeon collected in Chiba Prefecture were observed to fall outside the MEPE, as did seven of nine results from wigeon collected in Ibaraki Prefecture. As mentioned above, the area from which the birds were collected was thought to influence the level of contamination.

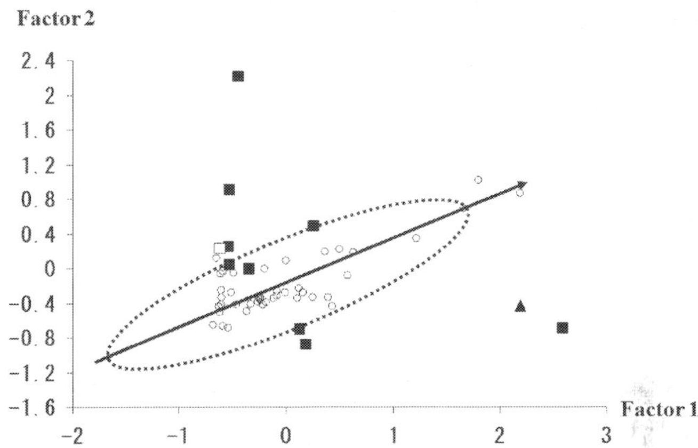

Fig. 9. Comparison between the index and the results from dabbling ducks collected in Ibaraki Prefecture. Filled squares; wigeon, filled triangles; spotbill duck, empty squares; shoveler. Dotted line; 95% equal probability ellipse, solid line; regression line obtained from wild birds (empty circles) used in Fig. 6.

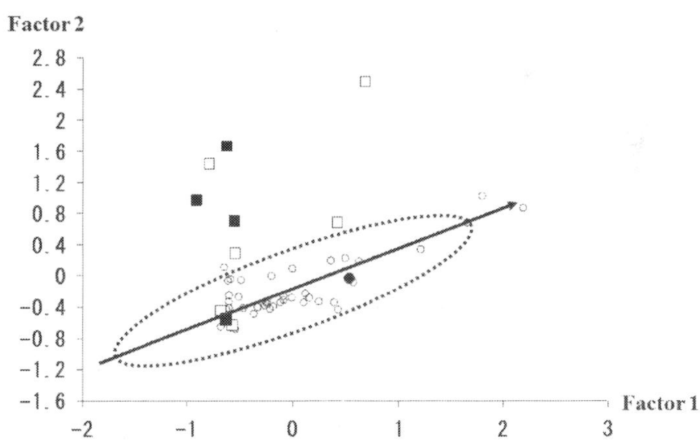

Fig. 10. Comparison between the index and the results from dabbling ducks collected in Akita Prefecture. Filled squares; mallard, empty squares; spotbill duck, filled circles; teal. Dotted line; 95% equal probability ellipse, solid line; regression line obtained from wild birds (empty circles) used in Fig. 6.

4. Conclusion

This study involved the establishment of an index using multiple elements, which is in the early phase of development for use in biological monitoring. Of course, a detailed study using the index is necessary in order to increase our understanding of contamination of wildlife with multiple elements. However, interestingly, the survey revealed that a similar index could be obtained, despite the investigation of multiple elements. Further, the difference between the degree of contamination by multiple elements in dabbling ducks and in diving ducks was clarified using this index. These results suggest that an understanding of the equilibrium among elements in the animal body is important for the investigation of contamination by multiple elements.

5. Acknowledgement

The study of Cd indexes was supported by Grant-in-Aid no. 20580344 from the Ministry of Education, Science, Sports, and Culture, Japan. The study of the index for multiple elements was supported by River Fund in charge of the Foundation of River and Watershed Environment Management (FOREM), Japan (Nr. 22-1215-016). The pilot study of factor analysis was presented at the 35th Annual Meeting of Japanese Avian Endocrinology in Okayama Prefecture. The attendance of students at this meeting was supported by Lion Trading Co., Ltd., Tokyo, Japan.

6. References

Colborn, T., Dumanoski, D. & Myers, J.P. (1996). *Our stolen future: Are we threatening our fertility, intelligence, and survival? -a scientific detective story*, Spieler Agency, New York, USA. (trans. into Japanese, Nagao, T. Syoeisya, Tokyo,). ISBN 978-4881359853

Friberg, L., Piscator, M., Nordberg, G. F. & Kjellström, T. (1974). Cadmium in the environment. CRC press, Ohio, USA, 1974, pp.1-400 (trans. into Japanese, Ishiyaku Publisher)

Harding, L.E., Harris, M.L. & Elliott, J.E. (1998). Heavy and trace metals in wild mink (*Mustela vison*) and river otter (*Lontra canadensis*) captured on rivers receiving metals. *Bulletin of Environmental Contamination and Toxicology*, Vol. 61, No.5, pp.600-607, ISSN 0007-4861

Helander, B., Axelsson, J., Borg, H., Holm, K. & Bignert, A. (2009). Ingestion of lead from ammunition and lead concentrations in white-tailed sea eagles (*Haliaeetus albicilla*) in Sweden. *Science of Total* Environment,Vol. 407, No. 21, pp. 5555-5563, ISSN 0048-9697

Kaneda H (1996). Greater scaup, In: *The encyclopedia of animals in Japan, volume 3, birds 1,* H. Higuchi, H. Morioka & S. Yamagishi S (Eds.), 78 (in Japanese), Heibonsya Limited, Publishers, ISBN4-582-54553-X, Tokyo, Japan

Kadoi, K., Mochizuki, M., Ochiai, Y., Takano, T., Hondo, R. & Ueda, F. (2009). The effects of cadmium on DNA of *Listeria monocytogenes, The 147th meeting of Japanese Society of Veterinary Science*, Utsunomiya, Japan

Krimsky, S. (2000). *Hormonal chaos,* trans. into Japanese: Fujiwara shoten, the translation published by arrangement with the Johns Hopkins University Press through The English Agency Ltd. ISBN 9784894342491

Kenntner, N., Crettenand, Y., Fünfstück, H-J., Janovsky, M. & Tataruch, F. (2007). Lead poisoning and heavy metal exposure of golden eagles (*Aquila chrysaetos*) from the European Alps. *Journal of Ornithology,* Vol. 148, No.2, pp.173-177, ISSN 0021-8375

Meador, J.P., Ernest, D., Hohn, A.A., Tilbury, K., Gorzelany, J., Worthy, G. & Stein, J.E. (1999). Comparison of elements in bottlenose dolphins stranded on the beaches of Texas and Florida in the Gulf of Mexico over a one-year period. *Archives of Environmental Contamination and Toxicology,* Vol.36, No.1, pp. 87-98, ISSN 0090-4341

Mochizuki, M., Hondo, R., Kumon, K., Sasaki, R., Matsuba, H. & Ueda, F. (2002a). Cadmium contamination in wild birds as an indicator of environmental pollution. *Environmental Monitoring and Assessment,* Vol.73, No.3, pp.229-235, ISSN 0167-6369

Mochizuki, M., Hondo, R., & Ueda, F. (2002b). Simultaneous analysis for multiple heavy metals in contaminated biological samples. *Biological Trace Element Research.* Vol. 87, No.1-3, pp. 211-223, ISSN 0163-4984

Mochizuki, M., Kitamura, T., Okutomi, Y., Yamamoto, H., Suzuki, T., Mori, M., Hondo, R., Yumoto, N., Kajigaya, H. & Ueda, F. (2011a). Biological monitoring using new cadmium indexes: cadmium contamination in seabirds, In: *Advances in Medicine and Biology. Volume 33,* L.V. Berhardt, (Ed.), 87-96, Nova Science Publishers, Inc. ISBN 978-1-61761-672-3, New York, USA

Mochizuki, M., Kudo, E., Kikuchi, M., Takano, T., Taniuchi, Y., Kitamura, T., Hondo, R. & Ueda, F. (2011b). A basic study on the biological monitoring for vanadium-effects of vanadium on Vero cells and the evaluation of intracellular vanadium contents. *Biological Trace Element Research,* Vol.142, No.1, pp.117-126, ISSN 0163-4984

Mochizuki, M., Mori, M., Akinaga, M., Yugami, K., Oya, C., Hondo, R. & Ueda, F. (2005). Thallium contamination in wild ducks in Japan. *Journal of Wildlife Diseases,* Vol.41, No.3, pp. 664-668, ISSN 0090-3558

Mochizuki, M., Mori, M., Hondo, R. & Ueda, F. (2008). A new index for evaluation of cadmium pollution in birds and mammals. *Environmental Monitoring and Assessment,* Vol. 137, No.1-3, pp.35-49, ISSN 0167-6369

Mochizuki, M., Mori, M., Hondo, R. & Ueda, F. (2009). A new index for heavy metals in biological monitoring, *Proceedings of 5th international conference on energy, environment, ecosystem and sustainable development,* pp. 185-191, ISSN 1790-5095, ISBN 978-960-474-125-0, Athens, Greece, September 28-30, 2009

Mochizuki, M., Mori, M., Hondo, R. & Ueda, F. (2010a). A cadmium standard regression line: A possible new index for biological monitoring. In: *Impact, monitoring and management of environmental pollution,* Ahmed El Nemr (Ed.), 331-338, Nova Science Publishers, ISBN, 978-1-60876-487-7, New York, USA

Mochizuki, M., Mori, M., Hondo, R. & Ueda, F. (2011c). The biological monitoring of wild birds: Part II - The possibility of a new index for biological monitoring. *International Journal of Energy, Environment, and Economics,* Vol. 19, Issue 6, pp.525-534, ISSN 1054-853X

Mochizuki, M., Mori, M., Kajigaya, H., Hayama, S., Ochiai, Y., Hondo, R. & Ueda, F. (2011d). The biological monitoring of wild birds, Part I : The cadmium content of organs from migratory birds. *International Journal of Energy, Environment, and Economics.* Vol. 19, Issue 6, pp. 535-546, ISSN 1054-853X

Mochizuki, M., Mori, M., Miura, M., Hondo R., Ogawa, T. & Ueda, F. (2010b). A new technique for biological monitoring using wildlife. *International Journal of Energy, Environment, and Economics*, Vol.18, Issue 1-2, pp.285-293, ISSN 1054-853X

Mochizuki, M., Morikawa, M., Yogo, T., Urano, K., Ishioka, K., Kishi, K., Hondo, R., Ueda, F., Sako, T., Sakurai, F., Yumoto, N. & Tagawa, M. (2010c). The distribution of several elements in cat urine and the relation between the content of elements and urolithiasis. *Biological Trace Element Research,* Online First™, 6 November 2010. Accessed 15 Dec 2010, ISSN, 1559-0720

Mochizuki, M., Sasaki, R., Yamashita, Y., Akinaga, M., Anan, N., Sasaki, S., Hondo, R. & Ueda, F. (2002c). The distribution of molybdenum in the tissues of wild ducks. Environmental Monitoring and Assessment. No.77, No.2, pp.155-161, ISSN 0167-6369

Mochizuki, M., Ueda, F., Sasaki, S. & Hondo, R. (1999). Vanadium contamination and the relation between vanadium and other elements in wild birds. *Environmental Pollution* Vol.106, No.2, pp.249-251, ISSN 0269-7491

Mochizuki, M., Ueda, F., Sano, T. & Hondo, R. (2000). Relationship between vanadate induced relaxation and vanadium content in guinea pig taenia coli. *Canadian Journal Physiology and Pharmacology,* 78, No.4, pp. 339-342, ISSN 0008-4212

Mochizuki M, Ueda F, Hondo R (1998). Vanadium contents in organs of wild birds. Journal of Trace Elements in Experimental Medicine Vol 11, No.4, pp.431, ISSN 0896-548X

Sakurai, H. (1997). Genso 111 no shinchisiki, (the new knowledge of 111 elements), Kodansha Ltd., Tokyo Japan (in Japanese), ISBN 978-4062571920

Ueda, F., Mochizuki, M. & Hondo, R. (1998). Cadmium contamination in liver and kidney in Japanese wild birds. *Journal of Trace Elements in Experimental Medicine,* 11(4), pp. 491-492, ISSN 0896-548X

Ueda, F., Mochizuki, M., Mori, M. & Hondo, R. (2009a). A new technique for biological monitoring, *Proceedings of 5th international conference on energy, environment, ecosystem and sustainable development*, pp.176-184, ISSN 1790-5095, ISBN 978-960-474-125-0, Athens, Greece, September 28-30, 2009

Ueda,F., Mori,M., Mochizuki, M. & Hondo, R. (2011). The analysis using new index for cadmium contamination in poultry. *The proceedings of 9th Asia Pacific Poultry Conference*, pp.325, Taipei, Taiwan, March 20-23, 2011

Ueda, F., Mori, M., Takano, T., Ochiai, Y., Hondo, R. & Mochizuki, M. (2009b). Basic investigation for an epidemiological study on cadmium contamination of wildlife - Cadmium distribution in the rat body after intravenous cadmium exposure, *Proceedings of 5th international conference on energy, environment, ecosystem and sustainable development*, pp. 57-63,ISSN 1790-5095, ISBN 978-960-474-125-0, Athens, Greece, September 28-30, 2009

11

Public Involvement as an Element in Designing Environmental Monitoring Programs

William T. Hartwell[1] and David S. Shafer[2]
[1]*Division of Earth and Ecosystem Sciences, Desert Research Institute, Nevada System of Higher Education*
[2]*Office of Legacy Management, United States Department of Energy, USA*

1. Introduction

The monitoring of various environmental parameters may occur for a wide variety of reasons in numerous venues and at scales both large and small. Significant advances in the realms of data collection and communication technologies, as well as advances in remote sensing, have resulted in the ability to collect, transmit, analyze, manage, and disseminate environmental monitoring data at a scale little imagined only a couple of decades ago. These advances have also significantly increased the opportunities and means by which the public can contribute to environmental monitoring.

Some types of environmental monitoring may be targeted at short and long-term observations of changes in ecological systems that are the result of natural processes and their effects, and do not come under significant public scrutiny. However, quite the opposite is true for monitoring of potential effects of various anthropogenic media, especially with regards to their impact on the safety and health of human receptors and associated ecosystems. Members of the public may view the results of such monitoring with suspicion, especially if collected by government agencies or other organizations that could be perceived as having either caused a situation which requires monitoring, or who have a vested interest in the results of the monitoring. Suspicion among the public about radiation monitoring was a major contributing factor to how the "Community Environmental Monitoring Program," discussed later in this chapter, was designed. However, even monitoring of natural phenomena can have critics. Challenges exist in involving the public in environmental monitoring for environmental changes that may be a result of global issues such as climate change (IceWatch Canada and Project BudBurst are described in this chapter if the issues are viewed by some members of the public as being of ideological or political creation. Alternatively, with issues such as climate change, some people feel that the problems are so big that their contributions in measuring the effects of it, or reducing activities that contribute to it, will make no difference (e.g., Norgaard 2006).

Members of the public are often more than willing to participate in environmental monitoring, particularly when they and their own communities have a personal stake in the results or when the monitoring process itself provides tangible benefits. However, sometimes the public does not immediately accept the notion that a monitoring program will have benefits. In fact, there are examples where they have, at least initially, concluded that it would have only resulted in expenses for them. For example, Mori et al. (2005) describes a program of identifying and mitigating landslides in the Republic of Armenia in which it was hoped that the citizens of rural areas could help identify landslide-prone areas too small to be delineated by remote imagery. A key in making the program successful was investing time with people in towns prone to landslides, showing them how recognizing landslide hazards and implementing mitigating measures could help them avoid breakage of waterlines, damage to foundations of homes, and loss of cropland which the people had incurred great cost in time and money in developing. Simply talking about economic impacts of landslides at a national level was of no interest to people at a local level, and even created suspicion that the project was being undertaken to prevent people from freely using their land. The program in Armenia is just one example of how monitoring programs which effectively incorporate significant roles for the public can have a profound effect on the willingness of stakeholders to accept monitoring results, can result in better communication efforts, improve program transparency, and can actually result in a reduction in program costs in some scenarios. However, for these results to come about, the design of the monitoring program must carefully examine how the public perceives the subject, and how they will participate or contribute to the program.

This chapter will discuss the benefits, as well as potential pitfalls, of significant levels of public involvement in environmental monitoring programs. It will highlight mechanisms for designing, implementing, and maintaining viable monitoring programs with significant public components, and provide several real-world examples of programs that are highly inclusive of public stakeholders. Examples will be provided of environmental monitoring that concerns public and ecologic health, emergency response, as well as improved understanding of environmental processes or phenomena. The chapter will also highlight technological advances that have made public participation and transparency much easier to accomplish than in the past.

2. The citizen as scientist

There is a long history of public participation in environmental monitoring and other scientific endeavours. These "citizen scientists" (e.g., Bonney & LaBranche, 2004) have contributed greatly to several scientific bodies of knowledge by providing large, mainly volunteer constituencies, often comprised mostly of individuals without any formal science education or training, who nevertheless are able to carry out various forms of data collection and reporting that might otherwise be difficult or impossible for reasons of funding, time, or geographic distribution, among others. One of the best examples of a long-term monitoring program with significant public involvement is the National Audubon Society's annual Christmas Bird Count, which has been ongoing for 111 years (http://birds.audubon.org/christmas-bird-count, accessed July 2011). From humble beginnings in the year 1900, when twenty-seven individuals took part in the first bird count, the project now includes tens of thousands of participants in more than 15 countries who monitor bird populations and distributions between December 14th and January 5th annually, and enter their results in an online database. Other ornithological research projects have adopted the citizen science model for more regional scale studies (e.g., McCaffrey, 2005). Another area of science that has long embraced citizen scientists is the astronomy community. The 20-millionth observation of a variable star was made by an amateur astronomer in February 2011 as part of a citizen science program that is in its 100th year (http://www.aavso.org/, accessed July 2011). Amateur astronomers also produce a number of regular discoveries of new comets and asteroids that are added to databases of programs (e.g., the Spaceguard Center in the UK: http://spaceguarduk.com/, accessed July 2011) that monitor the skies for near-Earth objects that may one day threaten the planet.

There is a growing recognition amongst scientists and those in environmental communication that the establishment of meaningful partnerships with the public and the identification of significant participatory roles for those who are willing to take on associated responsibilities can help facilitate the communication that occurs between interested, concerned citizens and corporations or agencies and the scientists who perform research or monitoring tasks for them (Groffman et al., 2010; Shneider & Snieder 2011; Shafer & Hartwell, 2011, in press). This is especially true in cases where constituents in the media being monitored are anthropogenic in origin and have the potential, either real or perceived, to inflict harm upon human communities and associated ecosystems.

Willingness and interest on the part of citizens to pursue involvement in environmental monitoring may be driven by simple curiosity or, as mentioned above, by concern or fear surrounding the monitored media's potential to inflict harm and/or distrust of the agency or corporation responsible for conducting the monitoring activity. Regardless of the reason, it behoves the scientific community to take advantage of this interest in the name of cultivating a stronger association with the public whose tax dollars often fund the majority of scientific research that occurs in most countries, and whose sometimes heightened perception of risk of a planned activity can often bring a project to a screeching halt, or at least a significant delay. Providing the public with a greater role than the minimum required by legislative regulation can result in the measurer's recognition as a show of good faith, as well as an opportunity to provide a greater public understanding of monitoring and associated activities, and can produce a network of citizens who not only develop a personal ownership in the project or process, but who also become informal communicators in their communities as we shall see in some later examples.

3. Degree of participation

The degree to which the public may participate most successfully in a project will likely be determined by such factors as public visibility of the project, funding, study length, geographical extent, and especially the willingness of those responsible for the operation of a given project to include and define roles for the public that will be of mutual interest and benefit to everyone involved. For purposes of discussion, we separate public participation into two categories: passive and active. Several brief examples of passive participatory programs are given, with discussion focusing on active public participation.

3.1 Passive participation projects

The arrival over the last decade or so of new information technologies is one of the most significant factors driving greater opportunities for public involvement in scientific monitoring and research endeavours (Kim, 2011; Silvertown, 2009). The realization of personal computers in most homes in developed and developing nations, coupled with the advent of email, the internet, the World Wide Web, and cellular "smart" phones and their associated applications (or "apps") have changed the manner and speed with which data can be gathered, transmitted, accessed, analyzed, and reported. While these innovations have made major contributions to all levels of public involvement, they have leant themselves particularly well to what we refer to as "passive" participation.

By passive participation, we refer essentially to the relatively new phenomenon of allowing one's personal home (or work) computer to be used as a computational resource for studies that require significant computer power which may not be directly available due to funding considerations or due to prior commitments in using resources that are locally available. This essentially free and extensive network of computational power can be an extremely invaluable tool to the researcher who has need of it. This type of participation, while not necessarily providing the participating citizen with physical or intellectual involvement, does give the participant the emotional satisfaction of knowing that he or she is contributing to the understanding or resolution of a problem in which he or she is particularly interested. Aside from installing the software and choosing which projects to support, there is no further participation on the part of the volunteer---all computations run in the background while the user is using the computer for other functions, or when the computer is idle. One benefit to this level of participation is that the home user maintains complete control over which projects to support, the timing of the support, and how much computer processing power to allocate. Several examples are provided below.

3.1.1 SETI@home

SETI@home (http://setiathome.ssl.berkeley.edu/, accessed July 2011) was the first monitoring project to make use of tens of thousands of personal home computers to process data (Anderson et al., 2002). SETI, which stands for Search for Extraterrestrial Intelligence, has a scientific goal of detecting intelligent life outside of the Earth. One part of SETI involves using large radio telescopes to monitor for the presence of narrow-bandwidth radio signals from outer space which, if detected, would likely be indicative of intelligent origin, since such signals are not known to occur naturally. As of July 2011, SETI@home had more than 1.2 million users, with more than 155,000 actually active when it was accessed, representing 204 countries and over 493 TeraFLOPS average floating point operations per second (http://boincstats.com/stats/project_graph.php?pr=sah, accessed July 2011).

3.1.2 BOINC

The Berkeley Open Infrastructure for Network Computing, or BOINC (http://boinc.berkeley.edu/, accessed July 2011) was originally designed to combat the falsification of data by some users of the SETI@home program. BOINC is an open-source software designed for volunteer computing. Since its inception in 2002, it has provided volunteer users worldwide with the opportunity to, among many other things, assist with such endeavours as long-term climate modelling at Oxford University in the UK (http://climateprediction.net, accessed July 2011), help with epidemiological modelling of malaria outbreaks being studied at the Swiss Tropical Institute (http://www.malariacontrol.net/, accessed July 2011), help the Planetary Science Institute monitor and study the hazard posed by near-Earth asteroids (http://orbit.psi.edu/oah/, accessed July 2011), and assist Stanford University in the United States (U.S.) with the monitoring of earthquakes to improve understanding of seismicity in an effort to aid with earthquake preparedness planning (http://qcn.stanford.edu/). The "Quake-Catcher" network, as it is called, is also proactive in involving public schools, providing free educational software designed to help teach about earthquakes and earthquake preparedness (Cochran et al., 2009).

3.2 Active participation projects

Active participation refers to those programs that require participants to take an active role in the collection of and/or observation of data, and to record, enter, or otherwise transmit those data. While internet and phone app technologies are usually components of these projects as well, it is often the citizen scientist who must actively enter the data.

3.2.1 Project BudBurst and related programs in Europe

Global climate change is already resulting in the changes in the timing of leafing, flowering, and fruiting of plants (plant phenophases) with a general lengthening of the growing season. While there have been many local records developed, there remain significant geographic gaps and gaps in the types of plants for which phenological records have been developed (Backlund et al. 2008). Project BudBurst, co-managed by National Ecological Observatory Network (NEON) of the U.S. National Science Foundation (Keller et al. 2008) and the Chicago Botanic Garden (http://neoninc.org/budburst/_AboutBudBurst.php , accessed July 2011) is designed to address these data needs through public participation. The principle objective of NEON is establishing observational and experimental sites in 20 ecoclimatic domains in the contiguous U.S. as well as the states of Alaska and Hawaii. Project BudBurst's contribution is in expanding the number of locations and species for which information on the response to climate change is collected in the U.S. and Canada by using citizen scientists referred to as "Project BudBurst Observers." See Fig. 1.

Fig. 1. On-line banner for Project BudBurst, a collaboration between the NEON program funded by the U.S. National Science Foundation and the Chicago Botanic Garden. The project also aims to integrate phenological observation programs initiated by other organizations, universities, and national laboratories.

Similar to a growing number of programs involving stakeholders in environmental observations, extensive information is available for individuals or groups, including school classes, to participate in the program. A "help site" is also available for assisting in selecting sites, targeting plant species, and interpreting phenological phases. Project BudBurst Observers are encouraged to focus on recording first leaf, full leaf, and first flower, relatively easy phenological observations to make, although data is sought on other events too. For registered users, information is available on the website for interpreting these phases and results can be entered in an on-line journal. Similar to other programs described elsewhere in this chapter, results are available on-line in the form of maps that show the 100 most recent observations for a particular phenomena such as first flower and first pollen in the spring, and 50 percent leaf fall for deciduous plants in the fall. By clicking on the icon for one of the recent observations, information and a photo of the plant of interest and the phenological event observed is provided, and the record number is shown. Particularly for younger participants in the program, these types of on-line results, besides being educational, reinforce that the data they are collecting is contributing to the program.

Although many Project BudBurst participants are making observations on plant species close to where they live or go to school because of the frequency of observations needed at critical times of the year, there are special projects underway. For example, Project BudBurst is teaming with the U.S. Fish and Wildlife Service (USFWS) to have observations made in its refuges that are of particular ecological significance. Also, in different regions, a "most wanted" list of plants is posted and volunteers sought to record data on them.

The U.S. National Park Service (NPS) has been a leader among federal agencies in the U.S. in engaging the public in phenological observations (see "What's Invasive!" later in this chapter). At a workshop in March 2011 lead by the NPS and the USA National Phenology Network (http://www.usanpn.org , accessed August 2011), participants from government organizations, nonprofits, and institutions of higher-education met to explore ways of further engaging the public in phenological observations and standardizing protocols to better compare data from different regions. The workshop included discussion on three ongoing efforts at six NPS pilot parks in California including 1) identifying target species to assess resource response to climate change; 2) testing monitoring protocols; and 3) using different approaches to engage the public in phenological observations and documenting the results of projects in "tool kits" on the Web (Sharron and Mitchell, 2011). Material from the workshop is available at http://www.usanpn.org/nps (accessed August 2011).

Geographically large, phenological observation networks are not limited to the U.S. and Canada. The International Phenological Gardens program, managed by Humboldt University of Berlin, Germany was founded in 1957 (http://www.agrar.hu-berlin.de/struktur/institute/nptw/agrarmet/phaenologie/ipg/ipg_allg-e/, accessed July 2011). Today the network includes gardens in 19 countries in continental Europe as well as Britain and Ireland, ranging from northernmost Finland to sites in southern European countries including Portugal, Spain, Italy, and Macedonia. However, because the natural environment of Europe has been much more extensively altered than those of North America, the International Phenological Gardens restricts its observations to a limited number of plant species common to a large number of gardens in Europe. Locally, a wider range of plant species have been tracked since 2002 by faculty as well as volunteers associated with the Royal Botanic Gardens in Edinburgh, Scotland (http://www.rbge.org.uk/science/plants-and-climate-change/phenology-projects/, accessed July 2011). Although not continuous, phenological research at the Royal Botanic Gardens Edinburgh dates from the 1850s, when curator James McNab began recording the flowering dates of more than 60 species (McNab, 1857).

3.2.2 Citizen scientists and physical phenology

In addition to biological phenology, citizen scientists are contributing to the establishment of records of changes in physical phenology that may be in response to climate change. A good example is "IceWatch Canada." Scientists have found that the freeze-thaw cycles of lakes and rivers in Canada are changing, usually resulting in a longer ice-free period during the year (e.g., Futter, 2003). However, in a country as large (nearly 149 million square kilometers {km^2}) and physiographically diverse as Canada, climate change is not consistent across the country either latitudinally or longitudinally. Observations from citizen scientists are helping provide a greater geographic distribution of freeze-thaw cycle records across the country. IceWatch Canada is one element of "NatureWatch"

Canada, managed by Environment Canada, Nature Canada, and the University of Guelph (http://www.naturewatch.ca , accessed July 2011). Environment Canada is the principle Canadian agency responsible for environmental protection and natural heritage conservation, as well as for providing developing climate data and making weather forecasts across the nation (http://www.ec.gc.ca/default.asp?lang=En&n=BD3CE17D-1/, accessed July 2011). Nature Canada is one of the largest non-profit organizations in Canada supporting protection of rare species of plants and animals, habitat conservation, and environmental education (http://www/naturecanada.ca, accessed July 2011).

IceWatch Canada makes extensive use of the Web for recruiting volunteers, providing training on making freeze-thaw observations, and as a platform for stakeholders to submit observations. As part of quality control, citizen scientists must register with the program where on-line resources guide them through selecting observation points and interpreting freeze-thaw cycles.

Two principle events are the goal of "ice watching" to make observations consistent from one location to another. The first event is the date when ice completely covers a lake, bay, or river and stays intact for the winter. The second is when ice completely disappears from the same body of water. This allows the principle measurement to be determined: the length of ice duration during the year. This calculation is the most common historic measurement made of freeze-thaw cycles in the country, allowing modern records to be combined with historic ones, some of the latter being continuous from the early part of the 20th century. While during the middle of the winter or summer observations are rarely important to make, the on-line training for IceWatch Canada emphasizes the importance of daily observations during the freeze-up or ice break-up period.

In addition to observations that contribute to ice duration, data on other phenomena are sought including:

- The first date that ice completely covers the water body, even when this is a temporary event. For some lakes and rivers, the first ice cover is of short duration and the ice partially or totally melts if temperatures rise. In some cases, this may happen several times before the permanent freeze of the winter occurs.
- Similarly in the spring, ice may disappear, but then partially or completely freeze over again before a permanent ice-free stage is established (Fig. 2).

The web site provides detailed instructions in selecting observation points, with a safety message that there should be no need for a citizen scientist to venture out onto an ice-covered water body to complete the observations. In addition, images are provided of lakes and rivers to show complete or partial stages of freeze and thaw to help participants make similar interpretations at their observation points. Volunteers are also asked to make a detailed description of their observation point, including latitude and longitude, in part so that consistent observations can continue to be made if there is a change in the person or organization responsible for a site so that longer records can be kept for the same location. Participants are cautioned to avoid selecting water bodies in the proximity of anthropogenic processes or features such as dams or water intake facilities which may impact normal freeze-thaw cycles.

What types of results are available on line? One is the pattern of freeze and thaw over time for individual sites that are part of the IceWatch Canada network. The second is a map of Canada showing at any given time the spatial distribution of the stages of freeze and thaw across the country. New citizen scientists are continually being sought for IceWatch Canada, particularly for those parts of the country at higher latitudes where the population of Canada is sparse.

Fig. 2. Although sea ice has partially broken up on this bay, thin ice is beginning to form over the open water areas again. Such observations of episodic ice thaw are one type of observation sought in the IceWatch Canada program.

3.2.3 Citizen scientists, cell phones, and natural resource management

The growing popularity of cellular "smart" phones such as Android™ and iPhone™ with embedded global position system (GPS) capability is allowing citizen scientists to collect and transmit data on environmental phenomena from dispersed locations (Kim, 2011). An example of the use of these tools for natural resource management is the "What's Invasive!" program for tracking the location of invasive plants. "What's Invasive!" is a collaboration of the Center for Embedded Networked Sensing (CENS) at the University of California, Los Angeles; the University of Georgia Center for Invasive Species and Ecosystem Health, and a growing number of local, state, and federal resource management agencies in the U.S. including the NPS, the U.S. Forest Service, and the USFWS (http://whatsinvasive.com, accessed July, 2011). A key element of the app that is downloaded to the smart phone is EDDMaps, or "Early Detection and Distribution Mapping System" that allows for web-based mapping the location of invasive species. EDDMaps was and continues to be developed at the University of Georgia.

The pilot effort of "What's Invasive!" occurred at the Santa Monica Mountains National Recreational Area (NRA) in California. The Santa Monica Mountains protect one of the largest areas of mountainous, Mediterranean-type ecosystem in North America, let alone the world. However, in addition to the NRA protecting habitat of many plant species of limited range, its proximity to the Los Angeles metropolitan area--the second largest in the U.S.--means that the park is a popular getaway for hikers, birdwatchers, and amateur naturalists.

In addition, it is a park that is struggling with the control of invasive plants since so many non-native species have been introduced to southern California.

With "What's Invasive!", the NPS, which manages the Santa Monica Mountains NRA, went from relying on a small staff of federal employees to locate newly established areas of invasives to the much larger community of visitors to the park. From its beginning at the NRA, "What's Invasive!" projects have been established at more than 40 local, state, and federal parks and recreation areas in the U.S., as well as locations in Canada and Denmark.

Application of "What's Invasive!" has common elements at all the locations where it is in use. First, a login is required for a person to provide data to CENS which manages the databases. At a given park, the application provides users with photos and other information to help identify the most common invasive plant or those for which data is most needed. Besides noting the location, the user can send an image of the plant and make qualitative assessments of the population size (one, few, or many). Beyond providing information on correctly identifying the plant, the application also provides educational information such as where the plant is native, characteristics of its growing patterns, and how it is changing the environment where it has become established as an invasive. The citizen scientist can also look at results, such as maps of all the locations in a park where the plant has been observed.

Rather than just relying on periodic observations from visitors to the park, "What's Invasive!" is also being used in a "campaign mode" where an agency collaborates with a school or other organization to rapidly identify where invasive plants have become established in an area. This mode has been the most effective when the application has been used as an educational tool for schools since a large number of results are generated quickly, are visually available in a short amount of time, and it is an opportunity for students to work in teams to collect the information. Finally, use of "What's Invasive!" is not limited to parks with established programs. At any location, the application automatically picks the invasive plants most associated with the nearest location to the user. Those more experienced can also turn off the automated selection list and manually choose from the list of invasive plants.

3.2.4 The Community Environmental Monitoring Program

The Community Environmental Monitoring Program (CEMP) provides a model for embedded public involvement in a monitoring program, and was designed with the specific intent of fostering better communications between participating communities and the federal agency responsible for monitoring through maximizing the involvement of public stakeholders (Hartwell et al., 2006; Shafer & Hartwell, 2011, in press). The CEMP is a network (Fig. 3) of radiation and weather monitoring stations (Fig. 4) surrounding the Nevada National Security Site (NNSS), formerly known as the Nevada Test Site, where the U.S. tested nuclear devices between 1951 and 1992. It has provided a well-defined hands-on role for members of the public since its inception in 1981. Modelled in part after an independent monitoring network that was implemented around the Three Mile Island nuclear power plant in the U.S. after the accident there in 1979 (Gricar & Baratta, 1983), the CEMP seeks to provide maximum transparency of, and accessibility to, monitoring data both through the participation of public stakeholders and by making data available in near real-time on a public web site.

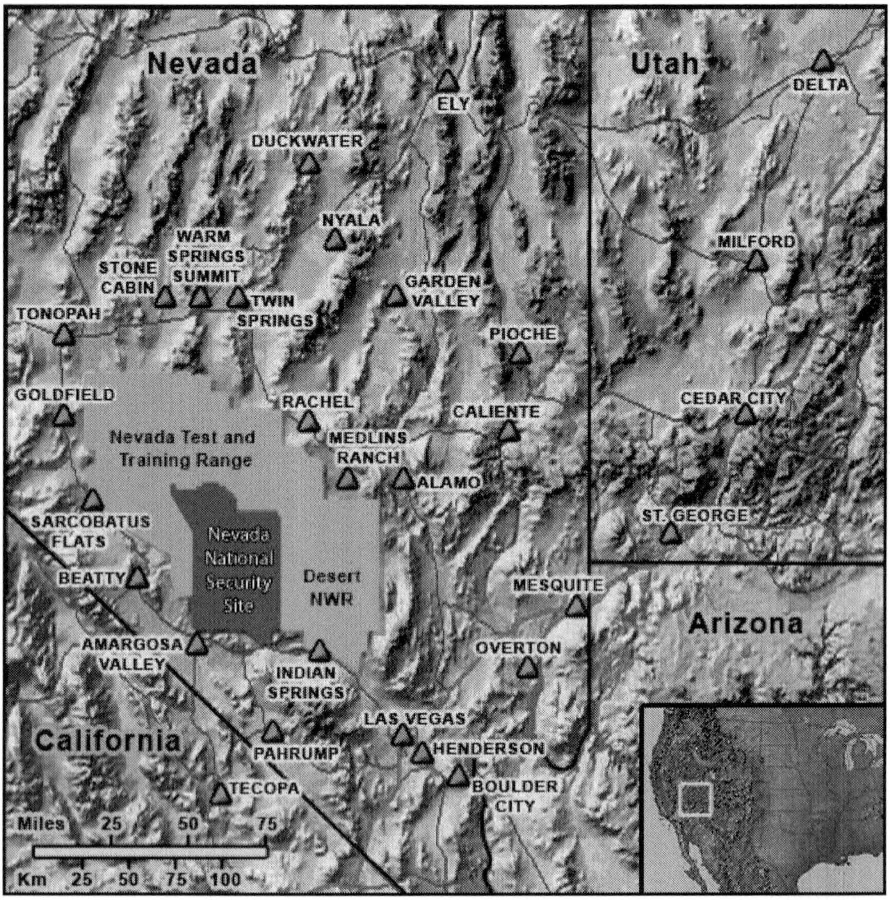

Fig. 3. The monitoring stations that make up the CEMP are located in communities and ranch sites scattered across a 160,000 km^2 area of southern Nevada, south-eastern California, and south-western Utah in the U.S.

The CEMP is funded by the U.S. Department of Energy (DOE), National Nuclear Security Administration, and administered by the Desert Research Institute (DRI), a non-profit environmental research arm of the Nevada System of Higher Education. While the DOE has historically been viewed by many in the region with distrust as the agency responsible for radioactive contamination of downwind areas, particularly during the era of above ground nuclear testing, the administration of the program by a state agency associated with the higher education system helps to improve confidence in the monitoring results. However, it is the participation of residents of the local communities that achieves the greatest benefit for the program in terms of public trust, communication, and education.

Two people per community (Fig. 5) are designated as Community Environmental Monitors (CEMs). Their responsibilities include collection of bi-weekly air filter samples, the posting of monthly summary data at their local monitoring stations, and serving as liaisons between their communities and the DOE and as points-of-contact for local residents who have questions about the monitoring process, results, or ongoing activities at the NNSS.

The original CEMs (a few of whom have participated in the program since its inception in 1981) were nominated by their communities, and largely consisted of school teachers with a general science background. However, many other CEMs are from other walks of life, including clergy, postmasters, volunteer firefighters, and retirees. The only true criteria for selection is that there be a willingness to perform the outlined duties, that they be generally respected members of their communities, and that they have a significant degree of contact with other community members. The tradition of identifying teachers as primary participants has continued throughout the program's existence, with the added benefit of knowledge gained through participation in the program often working its way into the teaching curricula and thus involving the local students.

The public participants in this program are not strictly volunteers, but receive a small monthly stipend for their duties as employees of DRI. The decision to hire them as employees was made in part to stress the importance of their sample collection duties, but also to offer them protection for any injuries that might occur during the discharge of their duties. The amount they are paid (approximately $150 US per month) is small, so as not to create the public perception that they are "in the pocket" of the DOE sponsors, and simply being given messages to parrot to their communities. On the contrary, CEMs are provided with regular training on the basics of ionizing radiation, and become knowledgeable on subjects ranging from radiation detection to local environmental conditions. Through attendance at these training workshops, the CEMs also become effective liaisons between local and federal entities, helping to identify concerns of people in their communities.

Fig. 4. A typical CEMP environmental monitoring station, with a full suite of meteorological sensors, radiation detection equipment, air sampler, and interpretive display with real-time sensor readouts.

CEMs in participating communities are part of the official chain-of-custody for collected samples, and become trained in the basics of ionizing radiation, including detection and potential health effects. They become knowledgeable points-of-contact for other community members. Although the CEMs are the primary means of interacting with and disseminating information to the public, DRI and DOE personnel actively participate in community events (e.g., producing displays and giving presentations for civic organizations and schools).

In 1999, DRI developed a public web site and upgraded communications at the stations so that most could upload their data every ten minutes (http://cemp.dri.edu/, accessed July 2011). In addition, data are archived back to the year 1999 for most stations, and users are able to produce tabular and graphical summary data for multiple parameters in any combination. The advent of the web site ushered in a new era of even greater transparency for the monitoring program, since now the public could access the data in near real-time, and know that they were seeing the data as soon as anyone else, including personnel of the sponsoring federal government. With time, links were developed to multi-level educational information on ionizing radiation, as well as a means to contact and discourse with program personnel.

Fig. 5. A photo of a CEMP station and its CEMs in California, taken when nuclear testing was still ongoing at the Nevada National Security Site, then called the Nevada Test Site.

There are undoubtedly some pitfalls associated with significant public transparency that can be provided by a public web site such as the CEMP. The public sees not only the normal data when it is posted, but also is occasionally privy to "bad" data caused by message mistranslation during communication, power outages, or equipment malfunction. While these incidents can cause significant angst for program personnel for short periods of time (e.g. Hartwell et al., 2008), the overall benefits conveyed by maximum transparency (especially public trust) are much greater than any temporary detriment caused by such an aforementioned incident.

As the CEMP has continued to evolve, it has endeavored to keep pace with the advent of new technologies (you can follow the CEMP on Twitter at @DRICEMP as of May 2011), and has played a significant role in keeping the public informed not only about monitoring results associated with past nuclear weapons testing, but also about other events as well. The CEMP web site reported the program's detection of radionuclides in Nevada in the U.S. resulting from the nuclear accident caused by the earthquake and subsequent tsunami in Japan in March 2011, and responded to hundreds of public inquiries from concerned citizens and requests for media interviews. By reporting data results as they became available, the CEMP was able to keep its network of community citizen scientists (Fig. 6) informed about not only the detection of radioactivity from this incident, but also that levels being measured in the U.S. were not a public health threat. As recognized points-of-contact in their communities, they were able to provide an invaluable service by mitigating much of the concern being expressed by their neighbors over the event.

4. Conclusions

Members of the general public often have a surprising willingness to participate in the process of assisting scientists with the collection of data as well as the dissemination of and communication of results to the public at large. While such public participation is often driven by personal curiosity, in cases where there is either the perception or reality of a potential risk to the personal well-being of an individual, his or her family, or community, many citizens relish the opportunity to become significantly involved when the opportunity is made available.

Most models of public involvement in environmental monitoring or other scientific endeavours have traditionally stopped short of a direct role for public involvement, instead relying solely on practices such as holding public meetings, providing opportunities for written feedback in the form of response to proposals or studies, or the formation of advisory groups to provide input into the decision-making process. While these are all important avenues for public discourse, they are oftentimes regulatory-driven, with little effort or impetus on the part of the agencies or corporations involved to provide additional opportunities for public engagement. The endowment of public stakeholders with a direct role in the process of environmental monitoring (or other scientific research) can convey several potential benefits, both to the stakeholders as well as the entities responsible for conducting monitoring studies. Direct participation by public stakeholders imparts a sense of ownership to those involved as well as to the general community. Careful identification of participating individuals who are in positions of high public trust and who are representative of a broad cross-section of the members of potentially affected communities can be an important contributing factor towards increasing public confidence in monitoring results. A role for direct involvement for the public from the outset (as opposed to in the conduct of damage control following an incident which has caused a loss of public trust) can be seen as a gesture of both good faith and public transparency in the monitoring process. The inclusion of these public stakeholders also helps to engender increased accountability on the part of those conducting the monitoring activities.

Engaging members of the public in a participatory role can actually produce programmatic cost savings, especially in those cases where significant computer resources or data entry is required, or in cases where environmental data must be collected from widely dispersed sites over a large geographic region. Technical tasks that require a minimal amount of training (such as the proper collection and replacement of an air filter sample at a CEMP station) can be accomplished by local residents, often on a voluntary basis, rather than sending technicians out on a regular basis at a significant cost to the monitoring program.

Finally, the process of educating and training citizen participants can create a network of informal communicators who live and work within the communities that may have concern about future or past activities that necessitate environmental monitoring. These citizens can be equipped with the knowledge to become "lay-experts" on related issues of community concern, and can serve both as liaisons between their communities and those conducting monitoring activities, as well as points-of-contact for their neighbours, which can help to identify and defuse rumours or public tensions before they reach unmanageable proportions. While there are invariably some pitfalls that will arise as a result of increased public participation and transparency, the authors believe that the overall benefits conveyed by maximizing public involvement to the greatest extent practical generally far outweigh any detrimental factors.

Fig. 6. Residents of 23 communities in southern Nevada, south-eastern California, and south-western Utah in the U.S., most of whom are schoolteachers, come together for regular workshops that train them to become effective communicators on issues related to the monitoring of ionizing radiation in their communities.

5. Acknowledgments

The authors gratefully acknowledge the Desert Research Institute of the Nevada System of Higher Education for funding for research and production of this manuscript. Work described for the Community Environmental Monitoring Program was accomplished through funding provided through the U.S. DOE, National Nuclear Security Administration Nevada Site Office under contract number DE-AC52-06NA26383. The authors also gratefully acknowledge the citizens whose concern, curiosity, willingness, and volunteerism assist and facilitate scientific endeavours worldwide.

6. References

Anderson, D.P.; Cobb, J.; Korpela, E.; Lebofsky, M. & Werthimer, D. (2002). SETI@home: An experiment in public-resource computing. *Communications of the ACM*, Vol. 45, No. 11, pp. 56-61

Backlund, P.; Janetos, D.; Schimel, J.; Hatfield, J; Boote, K., Fay, P.; Hahn, L, Izaurralde, C.; Kimball, B.A.; Mader, T.; Morgan, J.; Ort, D.; Polley W.; Thomson, A.; Wolf, D.; Ryan, M.G.; Archer, S.R.; Birdsey, R.; Dahm, C.; Heath, L.; Hicke, J.; Hollinger, D.; Huxman, T.; Okin, G.; Oren, R.; Randerson, J.; Schlesinger, W.; Lettenmeier, D.; Major, D.; Poff, L.; Running, S.; Hansen, L.; Inouye, D.; Kelly, P.B.; Meyerson, L.; Peterson, B. & Shaw, R. (2008). *The effects of climate change on agriculture, land resources, water resources, and biodiversity in the United States*, A Report by the U.S. Climate Change Science Program and the Subcommittee on Gobal Change Research, 362pp., Washington, DC.

Bonney, R. & LaBranche, M. (2004). Citizen science: Involving the public in research. *ASTC Dimensions*, May/June 2004, p. 13

Cochran, E.; Lawrence, J.; Christensen, C. & Chung, A. (2009). A novel strong-motion seismic network for community participation in earthquake monitoring. *IEEE Instrumentation & Measurement Magazine*, Vol. 12, No. 6, pp. 8-15

Futter, M. (2003). Patterns in trends in southern Ontario Lake ice phenology. *Environmental Monitoring and Assessment*, Vol. 88, No. 3, pp. 431-444

Gricar, B.G. & Baratta, A.J., (1983). Bridging the information gap at Three Mile Island: radiation monitoring by citizens. *The Journal of Applied Behavioral Science*, Vol. 19, No. 1, pp. 35-49

Groffman, P.M.; Stylinski C.; Nisbet, M.C.; Duarte, C.M.; Jordan, R.; Burgin, A.; Previtali, M.A. & Coloso, J. (2010). Restarting the conversation: challenges at the interface between ecology and society. *Frontiers in Ecology and the Environment* Vol. 8, pp. 284–291

Hartwell, W.T.; Shafer, D.S.; Tappen, J.; McCurdy, G.; Hurley, B. & Farmer, D. (2008). Pitfalls of transparency: Lessons learned from the Milford Flats fire. Proceedings of the WM'08 Conference, Phoenix, Arizona

Hartwell, W.T.; Tappen, J. & Karr, L. (2006). Positive community relations: the keystone to the CEMP. Proceedings of the WM'06 Conference, Tucson, Arizona

Keller, M.; Schimmel, D.S., Hargrove, W.W.; & Hoffman, F.M. (2008). A continental strategy for the National Ecological Observatory Network. *Frontiers in Ecology* Vol. 6: 282-284.

Kim, K.A. (2011). Become a citizen scientist with your cell phone. *Imagine Magazine*, Vol. 18, pp. 10-13, Johns Hopkins University Press

McCaffrey, R.E. (2005). Using citizen science in urban bird studies. *Urban Habitats*, Vol. 3, No. 1, pp. 70-86

McNab, J. 1857. *Transactions of the Botanical Society of Edinburgh* 5: pp. 173, 184.

Mori, M.; Hosoda, T.; Ishikawa, Y.; Tuda, M.; Fujimoto, R. & Iwama, T. (2005). Landslide management by community based approach in the Republic of Armenia. Japan International Cooperation Agency, Government of Armenia

Norgaard, K.M. (2006). "People want to protect themselves a little bit": emotions, denial, and social movement nonparticipation. *Sociological Inquiry*, Vol. 76, No. 3, pp. 372-396

Schneider, J. & Snieder, R. (2011). Putting partnership first: A dialogue model for science and risk communication. *GSA Today*, Vol. 21, No. 1, pp.36-37

Shafer, D.S. & Hartwell, W.T. (2011, in press). Community Environmental Monitoring Program: a case study of public education and involvement in radiological monitoring. *Health Physics*, Vol. 101, No. 5.

Sharron, E. & B. Mitchell. (2011). Tracking global change at local scales: phenology for science, outreach, and conservation. *EOS, Transactions, American Geophysical Union*, Vol. 92, No. 25, pp. 211-212.

Silvertown, J. (2009). A new dawn for citizen science. *Trends in Ecology & Evolution*, Vol. 24, No. 9, pp. 467-471

12

Monitoring Lake Ecosystems Using Integrated Remote Sensing / Gis Techniques: An Assessment in the Region of West Macedonia, Greece

Stefouli Marianthi[1], Charou Eleni[2] and Katsimpra Eleni[3]
[1]*Institute of Geology and Mineral Exploration, Olympic Village, Acharnai,*
[2]*N.C.S.R. "Demokritos", Institute of Informatics & Telecommunications,*
[3]*Geographer, Independent Researcher,*
Greece

1. Introduction

The environment and its land and water systems are put into constant stress through the various human activities, natural and climate processes. Water resource managers have long been incorporating information related to climate in their decisions. They also increasingly recognize that climate is an important source of uncertainty and potential vulnerability in long-term planning for the sustainability of water resources (Hartmann, 2005). These are leading to questions about the relative impacts of shifts in river hydraulics, land use, and climate conditions. Prospects for climate change due to global warming have moved from the realm of speculation to general acceptance. Climate change will have different effects on lakes. Lakes can be extremely sensitive to short- and long- term changes in the weather and so are intrinsically sensitive to climate change through a direct effect, or indirectly by affecting processes that take place in the catchment. Understanding the response of lakes to climate change is of great importance since year-to-year changes in the weather patterns can influence water quality and the ecological status of a lake in the terms of Water Framework Directive.

Characterizing the heterogeneity and temporal change of water quality across surface waters is difficult through conventional sampling methodologies (Tyler et al., 2006). In situ measurements and collection of water samples for subsequent laboratory analyses provide accurate measurements for a point in time and space, but do not give either the spatial or temporal view of water quality needed for accurate assessment or management of water bodies (Schmugge et al., 2002). Traditional monitoring of water quality as well as other environmental parameters involves specialized personnel and both on site and laboratory analysis. Field measurements for monitoring the environment are expensive and difficult to conduct. For example, the water quality monitoring of lakes often includes the monitoring of water clarity using a Secchi disk. Therefore the use of Sechi Disk Transparency (SDT) has been widely adopted in many lake monitoring programs worldwide (Bukata et al. 1988; Wallin and Hakanson 1992; Lee et al., 1995).

Substances in surface water can significantly change the backscattering characteristics of surface water. Remote sensing techniques for monitoring water quality depend on the ability

to measure these changes in the spectral signature and relate these measured changes by empirical or analytical models to water quality parameters. The spectral resolution of most satellite imagery is insufficient to identify (concentrations of) individual components that affect water quality. In most cases, satellite remote sensing is used to investigate the dynamics of sediment loads in reservoirs and lakes (Vrieling, 2006). Many studies found significant linear or nonlinear relationships between in situ determined suspended sediment concentration near the surface of inland water bodies and atmospherically corrected spectral reflectance derived from satellite remote sensing data, such as Landsat (Nellis et al., 1998; Schiebe et al., 1992) and SPOT-HRV (Chacon-Torres et al., 1992). Because sediment characteristics, like texture and color, influence the water reflection, developed empirical relationships are not easily transferable to other regions where erosion entrains different sediment types. Therefore, until a universal equation does not exist, most models of suspended sediment are site-specific (Liu et al., 2003). Thermal infrared (TIR) satellite images can be also used to study transport processes in lakes, such as wind-driven upwelling and surface circulation, providing a measure of spatial variability and horizontal distribution of water temperature that conventional field-based measurements cannot provide, (Steissberg et al., 2006, Zhen-Gang Ji et al., 2006). There still remain many unanswered questions about the effective implementation of integrated remote sensing / GIS techniques into a lake / environmental monitoring program, and these are analyzed in this presentation.
The objective of our research is to better understand the use of integrated application of remote sensing / GIS techniques on monitoring various environmental factors of lake ecosystems.

2. Pilot project area

About 65% of the surface waters of Greece are in its north-western part, in the periphery of West Macedonia. Some of the most valuable lakes of Europe in terms of biodiversity are located in this area, (Figure 1). The analysis of the basins of Macro Prespa and Vegoritis lakes are included in the Chapter.

Fig. 1. Pilot project area

Macro Prespa lake is a transboundary lake that it is shared between FYR of Macedonia, Greece and Albania. The study area is extended to include the catchment of lake Macro & Micro Prespa, as well as that part of the region that is hydro-geologically related to Ohrid lake. The study area has a size of 4769 Km^2 while the Prespa basin covers an area of 1380 Km^2 and is bounded between latitude 40° 38. 3 N to 41° 19.3 N and longitude 20° 33.2 E 21° 18.6 E. Prespa lakes are selected for use as a case study because they have been used in a variety of settings, by multiple agencies, over a long period. Furthermore, Prespa and Ohrid lakes can explicitly accommodate a broad range of resource management concerns (e.g., transnational management, environmental protection / biodiversity concerns, recreation / tourism, water supply, water quality, and power plant support). The study area consists of mountainous ridges, surrounding valleys and the Macro / Micro Prespa and Ohrid lakes. The elevation of the study area lies within ~600 and ~2500 masl with the highest elevations being observed in the central and eastern part of the Prespa basin.
Vegoritis lake basin covers an area of 1894 sq kms and it is bounded between latitude 40° 18. 4 N to 40° 54.2 N and longitude 21° 24.2 E 22° 06.6 E. The lakes range in surface area from 1.8 to 59.7 sq kms with a mean of Secchi depth of 2 m. Emphasis is given on the environmental aspects of Vegoritis lake. Mean depth of the lake is 20 meters. The annual rainfall is about 600mm. There are two main aquifers in Vegoritis hydrologic basin. One of the acquifers is of phreatic type and it is developed in the loose sediments of the basin. Depth of groundwater table varies from about 0 m to more than 40 m. The other one is developed in the karstified limestones and is hydraulically connected directly with the lake.
The criteria for lake monitoring involve complex considerations of meteorological, hydrological, geomorphic and socio-economic factors. The necessary secondary data sources are not always available, or they are out of date. The relevant lake features are either not on the maps or they are inaccurate. Hence the advantages of satellite RS imagery. Both lakes show an abrupt drop of water level during the last decades. Analysis of meteorologic data could not explain the abrupt drop in water level, Figure 2.

3. Data

Optical sensors are widely used for environmental impact monitoring. Satellite images with moderate to high spatial resolution have facilitated scientific research activities at landscape and regional scales. Different sensor properties are important to be considered, when evaluating their possible use for environmental monitoring.
These properties refer to spatial, spectral, radiometric temporal resolution, signal-to-ratio and finally launch date, length of the time series. Multi-temporal Landsat images are the main source of information. LANDSAT-1 was the world's first earth observation satellite, launched by the United States in 1972. Following LANDSAT-1, LANDSAT-2, 3, 4, 5, and 7 were launched. LANDSAT-7 is currently operated as a primary satellite, although an instrument malfunction occurred on May 31, 2003, with the result that all Landsat 7 scenes acquired since July 14, 2003 have been collected in 'SLC-off' mode. Of all remotely sensed data, those acquired by Landsat sensors have played the most pivotal role in spatial and temporal scaling: given the more than 30-year record of Landsat data, mapping land and vegetation cover change and derived surfaces in environmental modeling is becoming commonplace (Cohen and Goward, 2004).

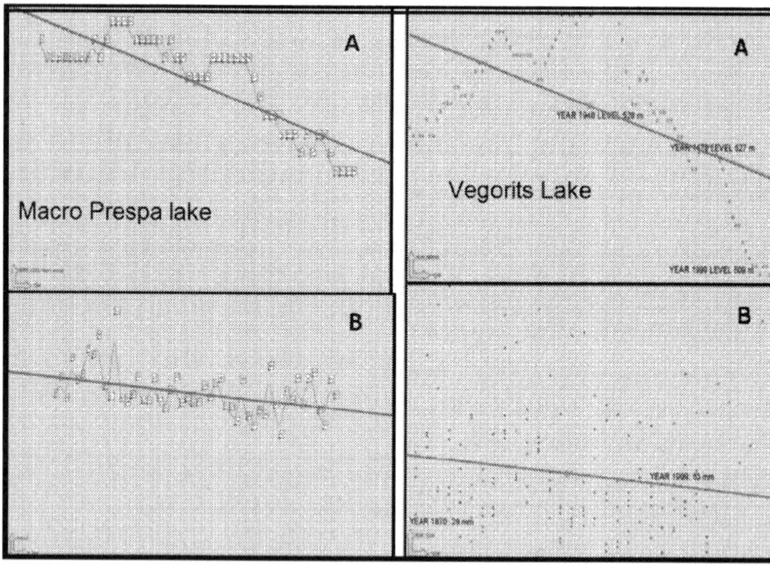

Fig. 2. Climate - rainfall (B) vs hydrology water level(A) measurements of Macro Prespa (A / B left) and Vegoritis lake (A / B right)

The Landsat images that have been used in the current study have been acquired from the USGS / NASA: http://edcsns17.cr.usgs.gov/NewEarthExplorer/ . SRTM DEM data that are available from the USGS server at http://dds.cr.usgs.gov/srtm/ and ASTER-DEM http://www.gdem.aster.ersdac.or.jp/ have been also included in the analysis.
ENVISAT / MERIS, ASTER satellite systems are relatively new systems and their data are also evaluated as far as monitoring the lake water systems is concerned. Much emphasis is given to extract information concerning a variety of parameters like land cover change that influences (indirectly) lake water quality. The effects of anthropogenic land cover modification on lakes are also analyzed by incorporating detailed information about land surface properties derived from Earth Observation data. The data inventory that is prepared and reported in the current submission includes acquisition of land cover maps, geological maps, compilation of hydrogeological maps based on analysis of relevant data, compilation of digital elevation models, analysis of multi-temporal satellite data and land cover maps. The amount of the above information is reviewed and analyzed and the result of the compilation is shown in the form of various maps.

4. Processing techniques

Various data like multi-temporal optical / thermal satellite images of Landsat ETM+, ASTER, ENVISAT systems and image processing / GIS techniques are used for the analysis, Figure 3. To compile the data from the various sources of information, the following problems had to be overcome:

- Different scales of maps, charts and imagery.
- Different coordinate systems.

- Different units of measurement.
- Different types of data.

The interpretation process is set up and activated upon receipt of each image. There are several factors which can influence the quality of RS images and can affect whether or not they are even worth acquiring, e.g.: *Weather, Smoke, Time, Sensors and Sensor performance.* Analysts should be familiar with these factors when interpreting the RS data.

Flawed images, or those having too much cloud cover, were rejected and alternatives sought. Only images with less than 10% cloud cover for the watershed areas or lakes of our pilot study area were used for the analysis. For example 5 out of the 7 collected images for the year 2011 shown in Fig. 4 were used while scenes D and H were rejected. This criterion significantly reduced the pool of images suitable for analysis but most years had at least one winter-spring / one summer-autumn image that met the criterion. From the pool of 38 images we selected 10 Landsat TM images, 2 Landsat ETM and 1 MSS images for reference spanning a ~ 35- year period (1974-2011). Another restriction refers to the scan line problem of the LANDSAT ETM scanner which made practically unusable these images for the analysis of Vegoritis lake. However, Prespa lakes are recorded properly and so Landsat ETM images with acquisition dates after the 2003 have been included in the analysis.

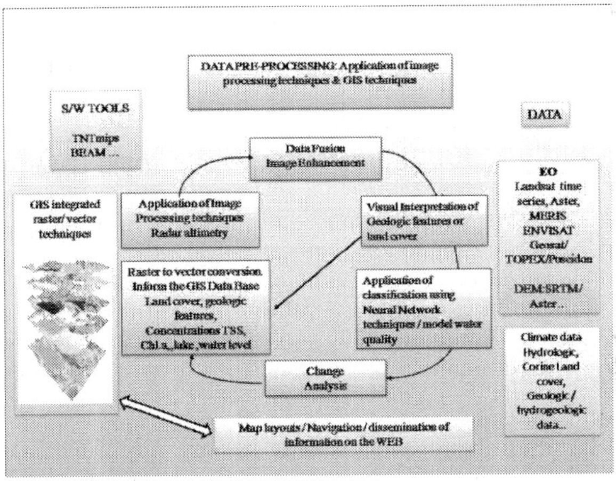

Fig. 3. Data and methods of analysis

All Landsat images were registered to the Greek Geodetic Datum of 1987 (EGSA '87) using the Landsat 17 January 2011 scene as a reference. The root mean square error (RMSE) for positional accuracy was generally less than 0.5 pixels (~ 10 m for Landsat TM). A nearest-neigbour resampling scheme was used to preserve the original brightness values of the images. The most rigorous method of radiometric calibration involves the use of radiative transfer models to produce an absolute correction. However, some data required to perform such a calibration are unavailable for historic images. We tested simple radiometric correction techniques such as dark pixel subtraction, Sun angle correction and normalization of multi temporal images to a single reference scene but found these insufficient as the area should have: (1) similar elevation to the rest of the scene, (2) minimal vegetation, (3) a relatively flat surface, and (4) constant pattern or general appearance over time.

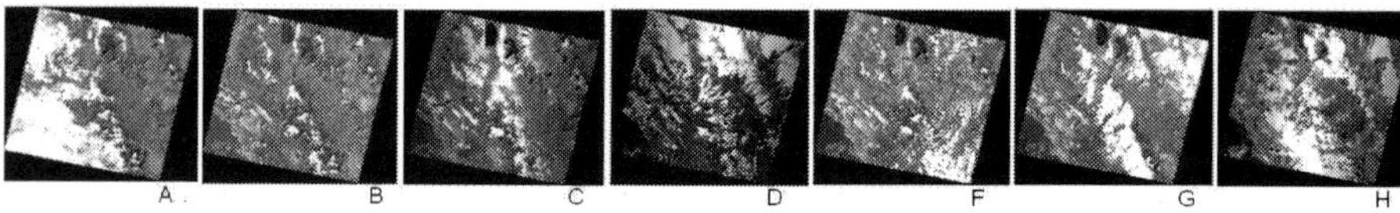

Fig. 4. Available acquisitions of Landsat images for the year 2011: L5:1/January B. L5 17/January C. L7 10/February D. 30/March E. L5 7/April F. 23/ April G.2/June

Data fusion techniques have been used in creating enhanced images at ~ 15 meters resolution for the Landsat ETM images for the watershed areas of the lakes. Digital image processing techniques are applied, plus some necessary image enhancement. The next step in extracting image data for lakes used an unsupervised classification method based on a clustering K Means algorithm with 10 classes and 20 maximum iterations that were then aggregated to land and water classes. Raster to vector conversion techniques were used to outline the polygons of the water surfaces. Due to intense topographic relief of the area shadows were also classified as water surfaces especially in winter scenes. These were eliminated as they have small area extent using vector editing techniques. Auxiliary information was also used to guide AOI selection and correction and this included the use of vector (GIS) layers of map coastlines, bathymetric maps and sampling point locations, Figure 5. Each Lake_AOI polygon was assigned a unique identification number and database fields so as to join the satellite data to the observation database.

The Lake_AOI polygons were used to create the water-only images of the lakes. Multitemporal water-only images of Macro Prespa, Vegoritis and Ohrid lakes have been created and these have been stored as a raster database. Metadata information describing the image acquisition information was also included. Lake surfaces have then been further analyzed using unsupervised classification techniques. Self-organizing Map Classifier - unsupervised classification using neural network techniques proved quite effective in analyzing the lake water surfaces. Available SDT and Cl data are not readily available for the lakes of our region and so some ground measurements are used just for general verification purposes.

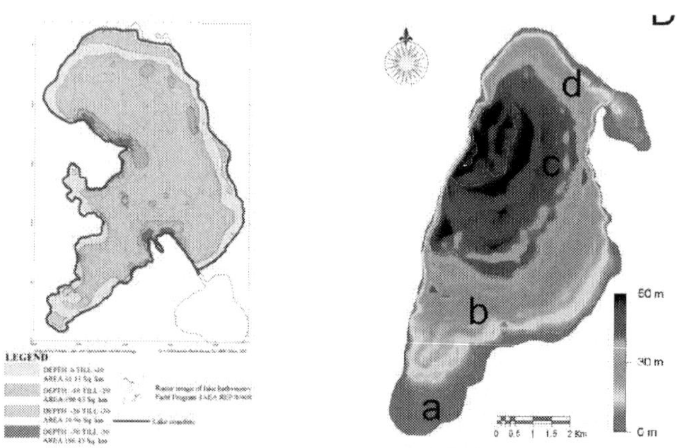

Fig. 5. Auxiliary information: Lake bathymetry of Macro Prespa lake (left), Vegoritis lake (right)

Various ratios of the Landsat bands have been calculated as these are related to SDT measurements. The TM3/TM1 ratio has been tested because previous investigators found it to be a strong predictor of SDT (Cox et al., 1998; Lathrop, 1992), but this was not confirmed by our analysis. All results have been stored to the raster database. Conversion of raster to vector of the lake water surfaces gave the opportunity to identify and store in the database the spatial variability of quantity / quality data of the lakes. GIS techniques have been used to overlay the results obtained from the multi temporal analysis, Figure 14.

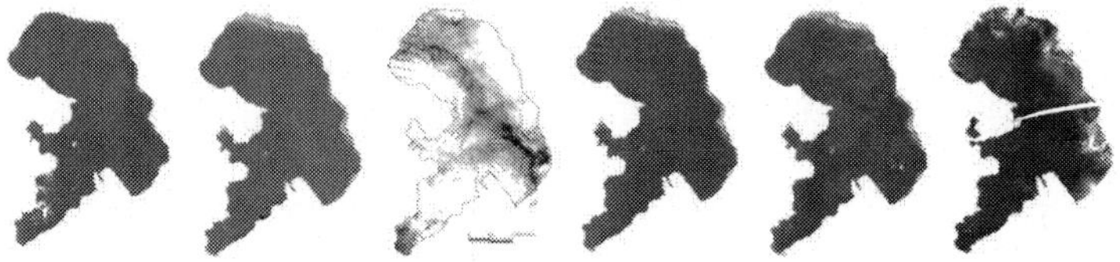

Fig. 6. Extracted surfaces of Macro Prespa lake, using the available map coastline.

High resolution of about 0.5 m ortho-photos available through the WMS service of Greek Cadastral Agency of Greece have been also used to acquire information and verify the results obtained from the analysis of Landsat data. The GIS system gives the opportunity of using the ortho-photos as a background while overlaying any type of GIS data and updating the information. All processing techniques have been applied using the TNTmips Image Processing / GIS S/W system (www.microimages.com). Our case study is intended to give a recent example of the practical applications of RS and GIS to lake monitoring. The RS study is placed first, followed by the GIS study, and finally an integrated interpretation is attempted.

5. Information gathering from remote sensing

Lake physics plays a fundamental role in limnology as temperature structures, circulation patterns and turbulent mixing, all set the environment in which the biology and chemistry within a lake operate. It is also through physics that the initial impact of any changes in climate will be felt within a lake.

5.1 Lakes

5.1.1 Lake inventory

Delineation of water bodies is essential for the estimation of the water balance of the area. Water authorities need to know date, location, extent and variations of these water bodies. The test area covers a broad region while the transnational Prespa lakes basin is included. The problems that are faced are related to:

- The fact that maps are not readily available
- There is lack of updated information
- Digital data are in different scales or coordinate systems
- Accurate measurements of surface areas of Macro - Micro Prespa lakes are lacking.

The 17th of January 2011 Landsat image has been used to make an inventory of all the lakes of the region at a scale of ~ 1:50000, Figure 3. The lake water surfaces have been extracted using classification of infrared bands & conversion of raster to vector techniques. There is up to date information which is readily available in a digital format for the whole of the translational region. Extraction of surface areas / perimeter and spatial context of the location of the lakes is easily obtained. Relationships of the different lake water bodies are also obtained, Figure 7. Accurate mapping of surfaces of the Greek lakes in scales up to ~5000 (Figure 8) is obtained using the WMS - Web service of the Hellenic Cadastre, http://gis.ktimanet.gr/wms/ktbasemap/default.aspx . The acquisition dates of the aerial photography are in the time period of 2007 to 2009.

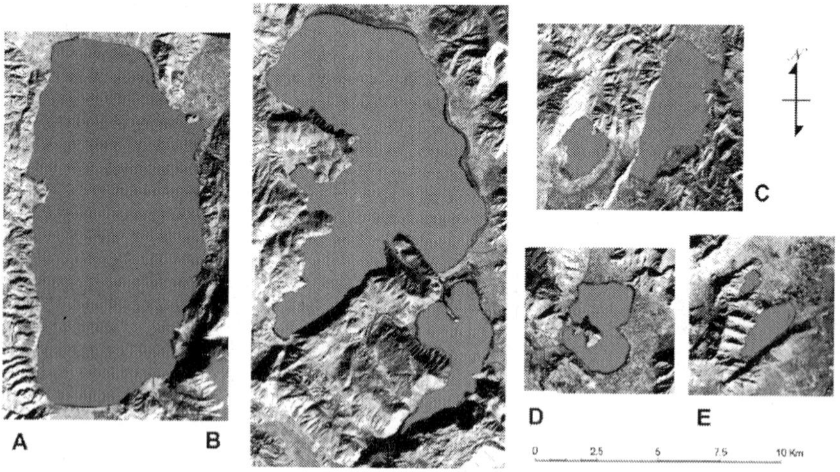

Fig. 7. Lake inventory from the 17th January 2011 image scene. Polygons of the water surface of the lakes have been extracted using classification techniques: A. Ohrid lake B. Macro / Micro Prespa lakes C. Vegoritis / Petron Lake D. kastoria Lake E. Chimaditis / Zazari lakes

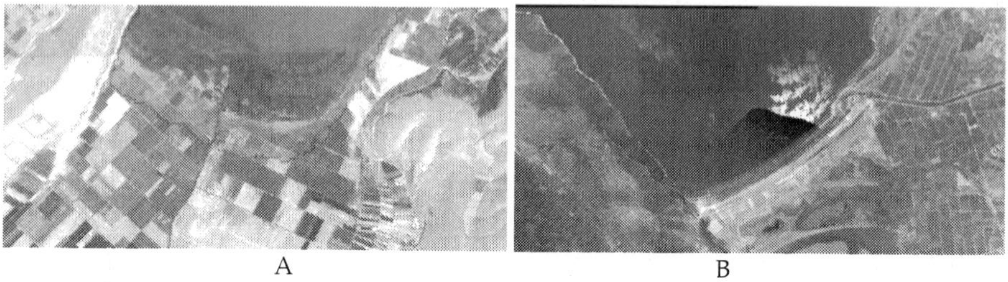

Fig. 8. Overlay of the coastlines extracted from the 17th January 2011 image to the orthophoto of 0.5 m resolution A. south part of Vegoritis lake B. South part of Macro Prespa lake.

5.1.2 Multitemporal analysis of change in surface area / size / shape of lakes

Lakes are sensitive to both climate change and to anthropogenic influence. Drop of water level has been observed in both Macro Prespa and Vegoritis lakes, Figure 2. Time series water level data are available for both lakes even though these measurements are not comparable for Macro Prespa lake as different reference levels are used between the three countries. Water level also does not show the spatial variability of the water surfaces, as changes depend on the bathymetry, the amount of sediment input due to erosion or other factors like geomorphology / geology. Satellite and especially Landsat data can be used to perform multi-temporal studies of lake surfaces.

Data collection included the acquisition of lake coastlines as these are available by the national / local authorities or on the Web. The only readily available data for Vegoritis lake are those of maps provided by the Greek Geographic Service of the Army of 1970s while the boundary of Macro Prespa lake has been made available for a time period on the Web (Traborema EU project). The stored in the GIS database map coastlines have been used to assess changes in water surfaces. These coastlines have the same areal extend as these extracted from the Landsat MSS images of the 1974 and therefore are used as a reference. These lake surfaces / coastlines dated since the 70's have been compared to the ones extracted from the multi-temporal Landsat images and stored as GIS vector layers.

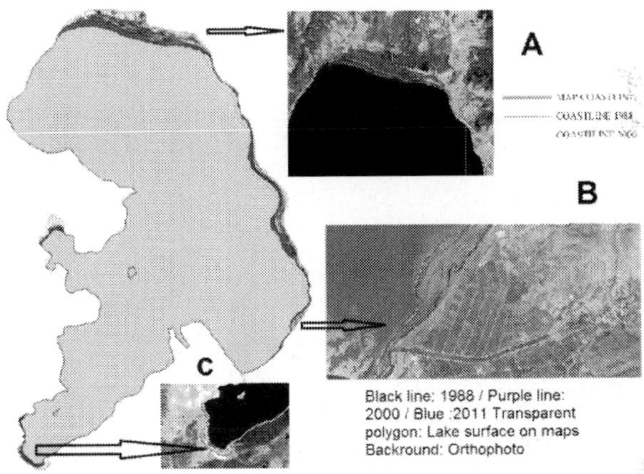

Fig. 9. Incremental changes of Macro Prespa lake for the last ~ 30 years: Changes in the North (A), South East (B)and South West (C)

Both Vegoritis and Macro Prespa lakes have lost their water surface area. A reduction of the surface area of Macro Prespa lake is evident, as estimates of its surface are as following: 20 November 1974 - ~276.5 km^2, August 1988 ~ 273.7 km^2, August 2000~265.2 km^2, 21 August 2008 ~257.2 km^2 and 17 January 2011~ 256.7 km^2. Macro Prespa lake has lost nearly 19.8 km^2 of its surface in the period 1973 to 2011.

Sharp drop of water level of Macro Prespa lake occured in 1975/1977 (1.2 m), 1987 /1990 (3.7m) and 2000/2002 (2.2m.) Figure 9. It is further evident that Macro Prespa lake is still losing its surface, even though the entire Prespa basin has been declared as a trans-boundary protected area, with the establishment of the "Prespa Park" by the Prime Ministers of Albania, Greece and the FYR of Macedonia on 2 February 2000.

Vegoritis lake has lost 30% of its surface (1970: 59.7 km^2 - 2011: 43.8 km^2) in the last ~ 30 years. Changes on its coastline are observed in its southern part, Figure 10. This can be partly explained by its bathymetry as the waters are shallow in the southern part, while its deepest area is in its western part, Figure 5. Comparison with the multitemporal analysis of the other lakes of the area shows that Ohrid, Micro Prespa and Petron lakes have lost only a small part of their surface area. Analysis of the space imagery of the years 1975 and 2011 respectively clearly revealed areas of shore line changes. It is now possible to draw accurate maps which look at the future incremental changes of Vegoritis / Prespa lakes. The modeling of this process is efficiently performed in the GIS.

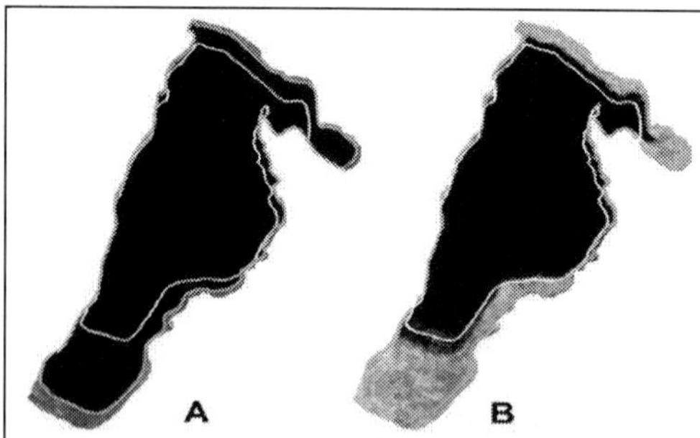	A. & B. The coastlines of 1988 (blue line) as well as that of 2011 (white line) have been plotted on the Landsat 1984 image scene.
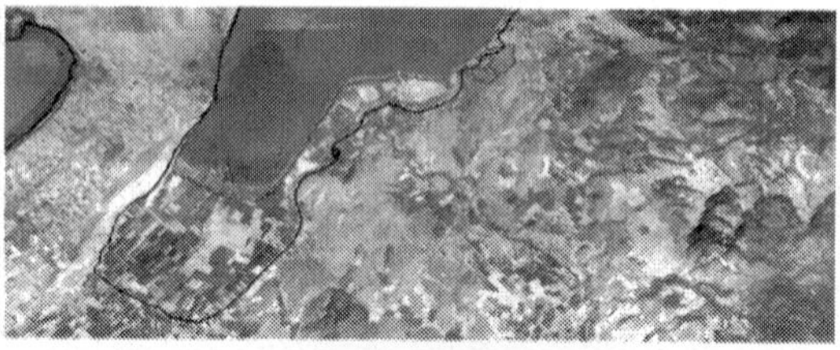 C	C. South part of Vegoritis lake: The black line shows the coastline of the map. The transparent polygon of the lake surface has been outlined from the17th January 2011 Landsat image.
 D	D. The 1988 coastline plotted on the orthophoto: Estimates of the land use change of the lake to agricultural land can be obtained and used by authorities.

Fig. 10. Changes of the Vegoritis lake surface area.

In the framework of the assessment of remote sensing techniques a small scale experiment has been carried out using radar altimetry techniques by Alexei Kouraev (Stefouli et al 2008). Results show that there are annual variations of Ohrid lake water level and these can be measured using radar altimetry. As Macro Prespa lake is hydraulically connected to Ohrid lake and located in higher altitude these could explain its drop of the water level. For some ENVISAT cycles estimates of water level have not been made due to quality control. The difference between the two time series can be up to 15-20 cm, apparently due to land influence in altimetric signal, but in general both in situ and altimetric observation are in good agreement, Figure 11. Time series water level measurements can be obtained through the process of radar altimetry and if it is combined with the estimated surface areas, lake bathymetry can give an indication of the quantitative characteristics of the lakes.

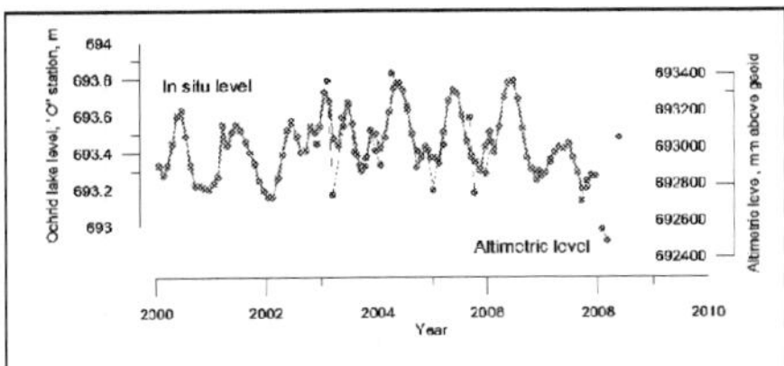

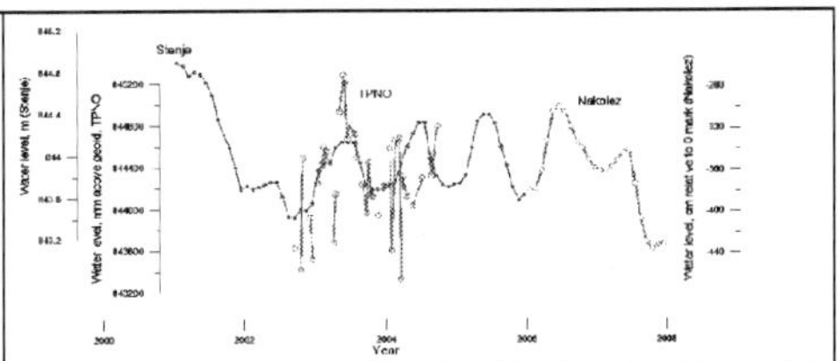

(A) In situ (red line and dots) and altimetric (blue line and dots) water level time series of Ohrid lake. Though absolute values differ for Y axis, vertical scales are identical for both.	**(B)** In situ (blue line and small dots for Stenje station and dark blue line and open circles for Nakolez station) and altimetric TPNO (red line with open circles) water level time series for Macro Prespa lake. Though absolute values differ for Y axis, vertical scales are identical.

Fig. 11. Results of applying test for estimating water level of lakes from radar altimetry data.

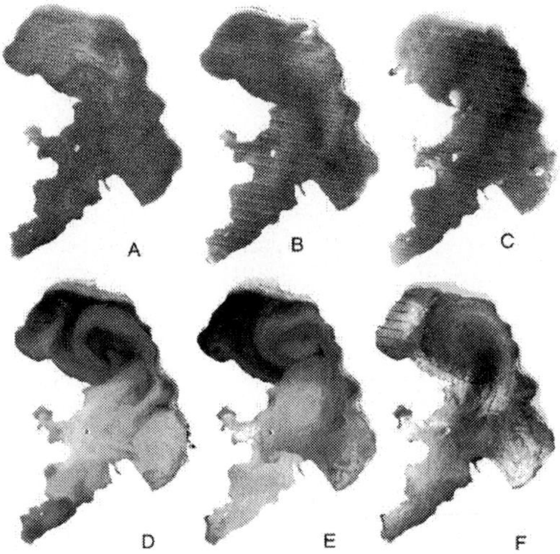

Fig. 12. Seasonal changes of Macro Prespa lake shown on the Landsat images of the year 2010: A. 14 / November B. 4 / April C.7 / June & D./ E./F. 2 / 18 / 26 of August respectively.

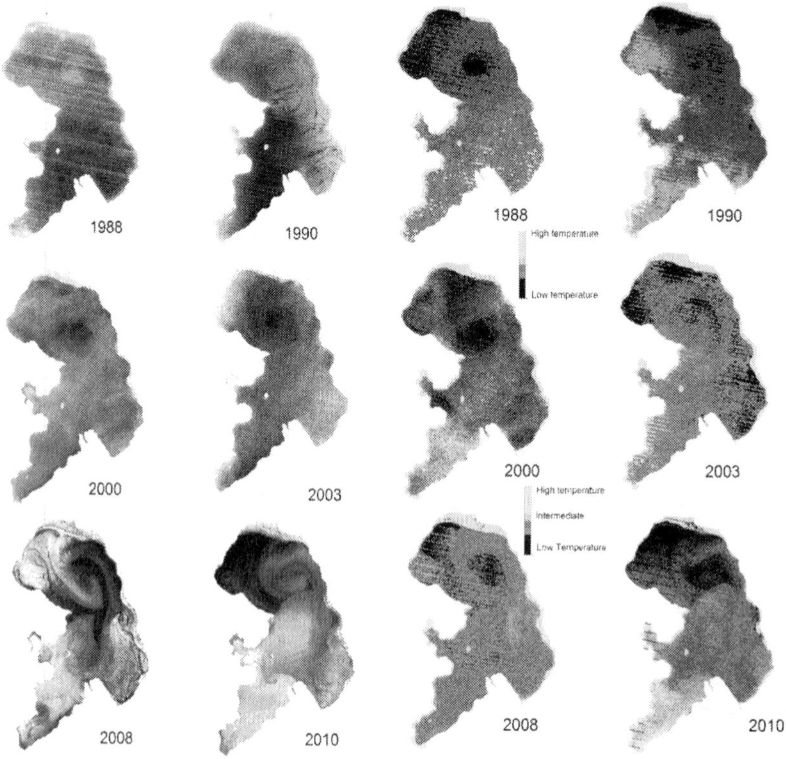

Fig. 13. Surface currents as mapped on using visible part of the spectrum (left image) and the thermal bands (right image) of the summer Landsat images for the period 1988 - 2010.

The images have shown that wind-driven partial upwelling events occur at least throughout the summer stratified period, transporting water from intermediate depths to the surface. These are important events that contribute to the patchiness and heterogeneity that characterize natural aquatic systems. The circulation in Lake Prespa is typically dominated by the northern two-gyre pattern, especially in the summer. The north wind leads to a cyclone (a counterclockwise rotation gyre) in the southwest and an anticyclone (a clockwise rotation gyre) in the northeast.

Analysis shows that a well formed system of gyres is formed during summer D,E,F of the year 2010 while this is not apparent in other seasons of the year i.e. winter / spring or autumn. These results have also been confirmed from the lake surfaces extracted from the ~ 30 years time span. Inter annual changes of the surface currents have been also evaluated. Circular features have been mapped in summer season of every year while some results are shown in Figure 13. These prominent features have been identified in most of the Landsat images. Self organization techniques classification techniques of the visible part of the spectrum proved to be quite effective in mapping lake circulation patterns. Multitemporal data are stored in the GIS database, while synthetic maps can be produced, Figure 14.

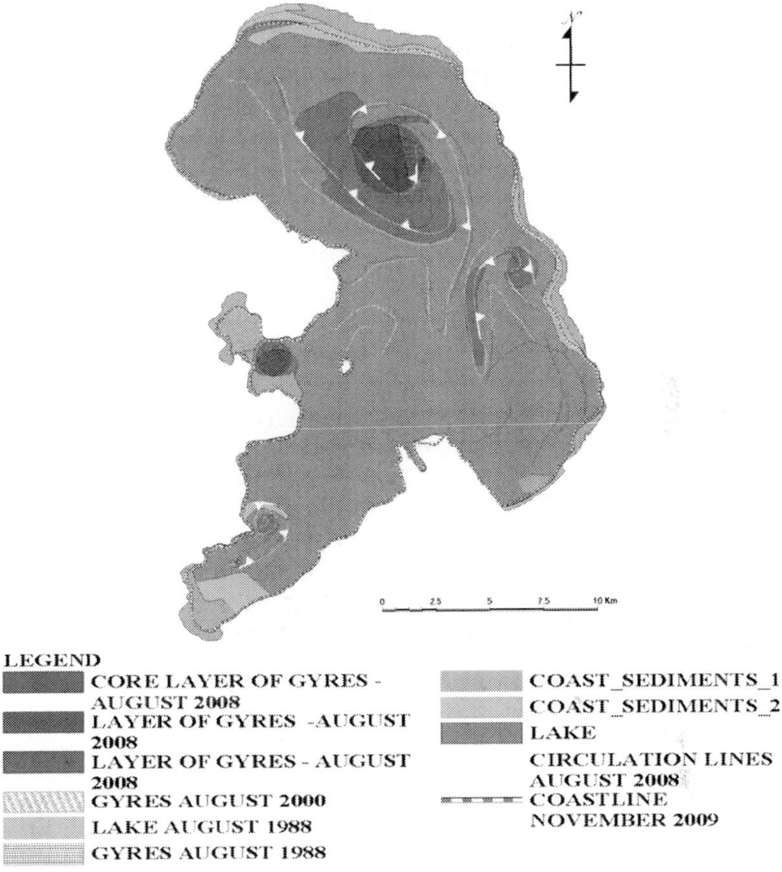

Fig. 14. Synthetic map concerning coastal sediment concentrations surface currents in the form of gyres.

The Landsat and ASTER data have been analyzed for estimating differences of suspended sediment content in Vegoritis lake. The data of band 2 of Landsat images (A, B in Figure 15) and band 1 of the ASTER image (C in Figure 15) have been used in the analysis as they correspond to the same spectral region of 0.52-0.60 μm. The same color palette has been used for displaying the multi-temporal images. Blue-green colors show relatively low sediment content while red - yellow colours high content. The Vegoritis lake thermal regime is also displayed in Figure 15. Inflow patterns of sediments can be interpreted on the satellite imagery in the different acquisition dates. Numbers 1 to 4 show the location of the streams / canals that discharge into the lake.

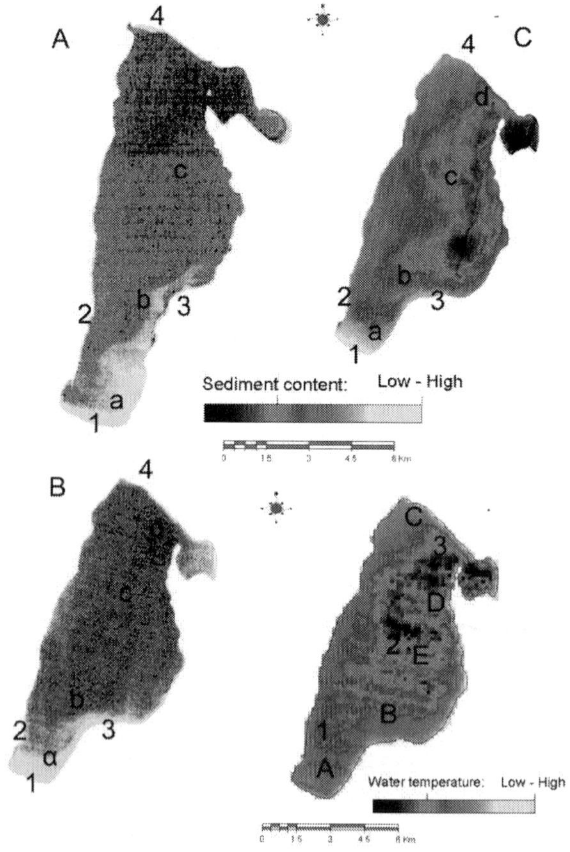

Fig. 15. Circulation patterns of Vegoritis lake a to d: field sampling points.

The high-spatial-resolution TIR images provide a detailed view of fine-scale processes, such as surface jets, that cannot be clearly resolved in moderate-resolution images, and they enable the accurate measurement of surface transport and circulation patterns.

The high spatial resolution of ASTER and ETM+ images allow the surface currents and general circulation in lakes and coastal environments to be accurately delineated. The vector field delineates three gyres as shown in Figure 14, Convergence and divergence zones and inflows can also be clearly resolved in the thermal patterns of the high-resolution TIR satellite images. The analysis enabled the characterization of wind-driven upwelling and the measurement of surface currents and circulation at lakes of West Macedonia. Trends during the last ~25 years of lake hydraulics, concerning surface currents, turbulence charactiristics and transport phenomena are identified.

Dates	(a) Depth 0,5m	(b) Depth 0,5m	(b) Depth 5m	(c) Depth 0,5m	(c) Depth 5m	(d) Depth 0,5m	(d) Depth 5m	Mean value
21-03-2000	1,0	2,1		2,6			2,6	2,1
4-04.2000			2,20		2,80		2,60	2,5
16-04-2000		2,40		2,60		2,60		2,5
7-05-2000	0,50	1,00		2,20		2,40		1,5
22-05 2000	0,7	1,60		1,70		1,50		1,4
5-06-2000	0,5	1,80		2,80		2,40		1,9
21-06-2000	0,8	2,20		2,60		2,40		2,0
10-07-2000	0,5	2,10		2,30		2,20		1,8
16-07-2000	0,6	1,6		1,8		2		1,5

Table 1. Sechi measurements in locations a,b,c,d of Figure 15 with variable depth and in various dates of the year 2000

5.1.3 Suspended sediments – chlorophyll

Optical remote sensing of inland waters has become a task of increasing importance, since the availability of clean fresh water is one of the great environmental challenges. In particular natural lakes and artificial reservoirs have to be monitored on a regular basis to ensure the quality of the water. With its 300 m spatial resolution and 15 spectral bands the imaging spectrometer MERIS on ENVISAT can be used for monitoring of at least larger inland waters. However, the standard algorithms as used for open ocean or even coastal waters are not appropriate because different water constituents occur in particular different phytoplankton blooms with partly extreme high concentrations. To this end the CASE 2 REGIONAL (C2R) processor of the BEAM 4.9 (Envisat/Brockman Consult) has been developed.
A time series of MERIS full-resolution (300 m spatial resolution at nadir) imagery was obtained from ESA's rolling archive at ESRIN https://oa-es.eo.esa.int/ra/mer_frs_l1/index.php and processed using BEAM 4.9. Images were subset to a geographic region bounded by the lat/lon limits of the study area. The BEAM 4.9 C2R processor was applied to data to extract atmospherically corrected radiance and the algal product C2R Chl_conc, according to the methods of Doerffer and Schiller (Doerffer and Schiller, 2008a, b). Default settings were accepted for all processing parameters. The algorithm used for the retrieval of water constituents is based on the Case-2-Water Bio-Optical Model. The input to the algorithm are the water leaving radiance reflectances (i.e. the output of the atmospheric correction) of 8 MERIS bands. The algorithm derives data of the inherent optical properties total scattering of particles (total suspended matter, tsm) b_tsm, the absorption coefficient of phytoplangton pigments a_pig and the absorption of dissolved organic matter a_gelb (gelbbstof) all at 443nm (MERIS band 2). Hence the concentrations of phytoplankton chlorophyll and of total suspended dry weight are determined. The algorithm is based on a neural network which relates the bidirectional water leaving radiance reflectances with these concentration variables. We estimated the concentrations of two parameters: chlorophyll and total suspended matter.
As was already pointed the test area is a cross border area between 3 different countries so it is not easy to establish a classification scheme and find the suitable variables and classification limits for a common water quality classification system. However, a relative classification scheme can be created using MERIS images. According to results shown in

Fig. 16, the quality of water in the lake Ohrid is the highest among all lakes. Then follows Macro Prespa, Micro Prespa and Vegoritis while Petron shows the worst water quality. This MERIS based relative classification of lakes coincides with the classification based on the available in situ data observations.

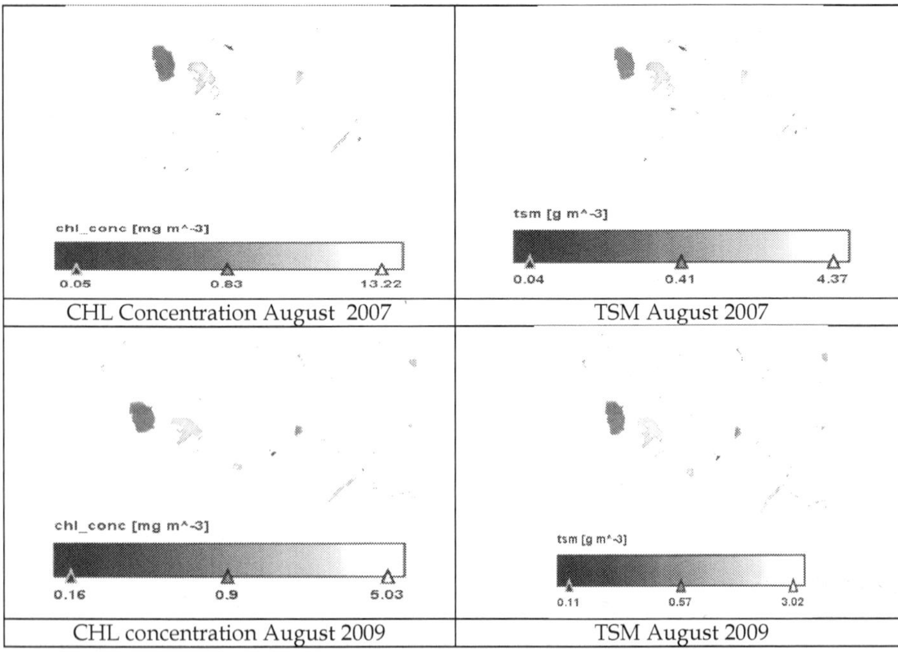

Fig. 16. Chl concentration and tsm

5.2 Catchment areas of lakes

Water authorities need tools to monitor and assess the status and the changes of basins so as to optimize and regulate their usage and to avoid depletion of the water resources. Up to date information about land cover, land use, vegetation status and their changes over time (e.g. seasonally) is important for the understanding and modelling of hydrological processes such as infiltration, runoff rates, evapotranspiration and water needs. Additional EO-derived information such as land cover, DEMs (digital elevation models) or surface water variations can be used to infer properties of surface waters and aquifers, or used in water cycle models (e.g. to calculate evapotranspiration). In order to interpret these discrepancies the water basin status and the changes that are taking place need to be analyzed.

The collected information is reviewed and analyzed and the result of the compilation is shown in the form of various maps. Multi-temporal analysis of Landsat- 5 / 7, Enhanced Thematic Mapper Plus (ETM+) scenes, Envisat MERIS and one ASTER scene have been used in the analysis of the catchment areas. Special emphasis is given on the catchment delineation using DEMs available for the lake basins. The analysis included various types of DEMs like the SRTM (100 m resolution) and ASTER (30 m resolution) DEMs. Catchments of river networks are fundamental to the automation of flow-routing management in distributed hydrologic models and for the morphometric evaluation of river network structure. The analysis of the DEM resulted to the delineation of the hydrographic network of the area of the transnational Prespa basin. The ASTER DEM has been used to delineate the changes of the relief of the Vegoritis lake basin.

Geology plays a role in the region as it allows the interconnections of adjacent river basins, which is the case of Prespa and Ohrid lakes. Ground waters cannot be observed directly by existing EO satellites, however, location, orientation and length of lineaments can be derived from EO and can be used as input for studies of fractured aquifers (e.g. location of sites for water harvesting). Available geologic maps have been scanned, geo referenced, digitized for the whole region within the context of the GIS system, Figure 3. The original maps have been of different scales and information content. A great variety of rocks with varying age and lithology constitute the catchment areas. Available information on location of springs has been also integrated in the GIS database.

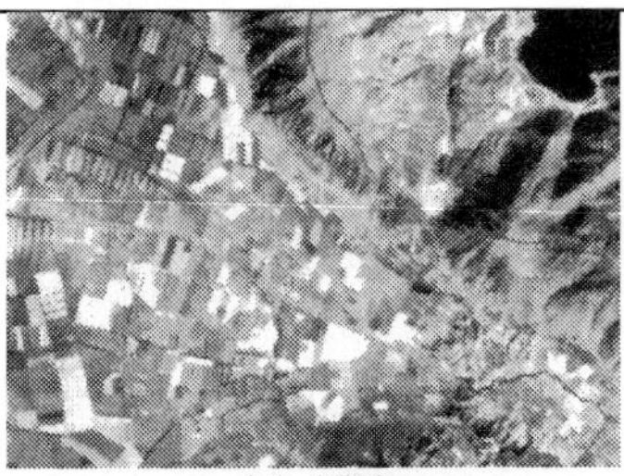

A1. 1988

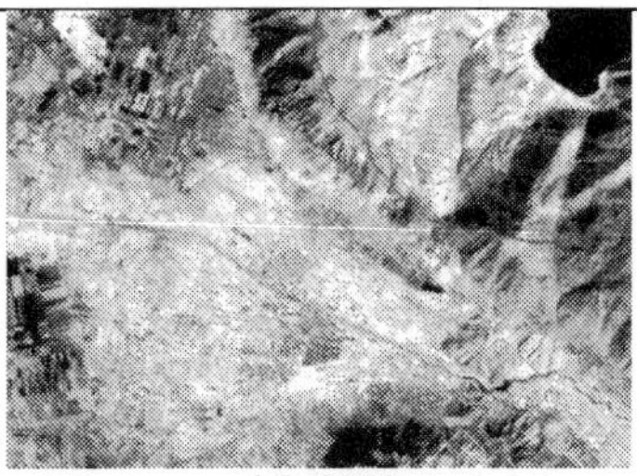

A2. 2000

A1 & A2. Impact of the implementation of Government policies after the 1990's as it shown on the multi temporal images of 1988 to 2000.

B. Emergent vegetation due to siltation

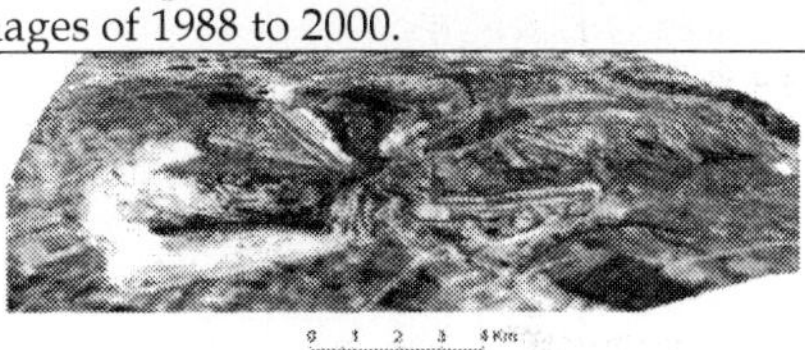

C. Mining activites. 3D representation of relief changes due to surface mining as it is mapped by the ASTER DEM & Landsat image of 2011.

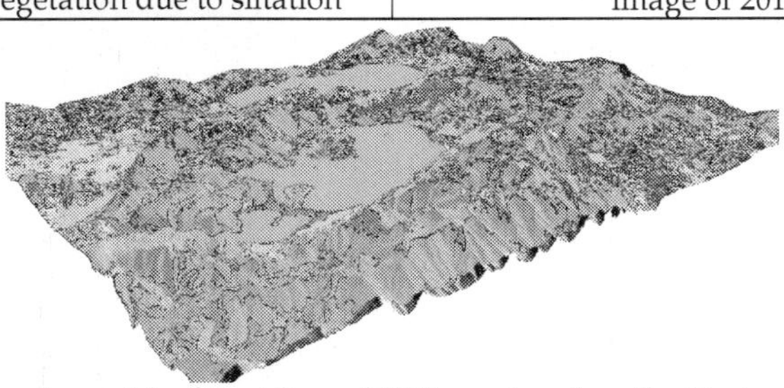

D. Red areas show burned by forest fires of 2007 overlayed on the Corine land cover map.

Fig. 17. Impact of anthropogenic factors to the lakes of the study area.

Natural and anthropogenic processes take place in the basins of Prespa and Vegoritis lakes and these have an impact on the water resources of the basins. The catchments of the three lakes have been described by the GIS based analysis of "Corine Land Cover Classification" Figure 17-D. MERIS data has been used for Corine land cover map updating because of their improved temporal resolution. Burnt areas due to the 2007 forest fires are detected and mapped on the MERIS data.

Surface mining takes place in Vegoritis lake basin with negative impacts of mining on the water resources, both surface and groundwater, which occur at various stages of the life cycle of the mines and even after their closure: 1.From the mining process itself, 2. From dewatering activities which are undertaken to make mining possible. 3. During the flooding of workings after extraction has ceased 4. By discharge of untreated waters after flooding is complete.

Anthropogenic factors seem to play a key role on the deterioration of the water resources of the region. Integrated Earth Observation / GIS techniques help to monitor changes in lake basins and can cover specific water management requirements, Table 2, Figure 17.

Anthropogenic Impact		**Comments**
Transnational treaties	First aggrement 1959- 2nd 2000 Prespa Park 2/2/2010, Petersberg Process (1998), Athens Declaration Process Water Convention 1992, Karipsiadis2008	Implementation is suffering from problems like lack of information, insufficient data.
Infra-structures	Diverson of Aghios Germanos (1936) Diversion of Devolli river (mid-70's) It has deposited about 1.2 million m3 of alluvium in the shores of Micro Prespa Lake. Sluice gates controlling flow of waters from Micro to Macro Prespa lake (2004).	Figure 17_B shows the effect of Devolli river diversion to Micro Prespa lake.
Mining	The environmental effects of the extraction stage: Surface disturbance, and the increased amount of sediments transported to the lake.	Figure 17 C shows the effect of surface mining in the Vegoritis lake basin.
Land cover changes	Multitemporal changes of the surface of lakes 1972-2009 period.	Land cover changes due to forest fires, Figure 17 D
Social changes	After the fall of the Eastern Block regimes the land was redistributed in Albania.	The total 550 agricultural cooperatives were converted to 467,000 small holder farms. These land management practices could have driven or intensified different water usage across Albania that would have influenced hydrologic lake water balances..Figure 17, A1 & A2
Agriculture	Irrigation schemes / pumping stations were created during the period 1950-1980, and occur on mainly flat, or gently sloping and river terrace	Agriculture influence both the quantitative / qualitative characteristics of the lakes

Table 2. Selected natural / anthropogenic impacts on the water resources of lakes

An advantage of using remote sensing is that data for large areas within a single image can be collected quickly and relatively inexpensively, while this can be repeated through selected time intervals. It is clear that in order to make regional assessments, one must develop a means to extrapolate from well-studied areas, as the site of our inter-comparison, to other lakes. Since the strength of satellite imagery for lake monitoring is the regional scale dimension, more than one location has to be taken for reference in order to learn how to separate crucial environmental parameters from all kinds of important interfering phenomena. Deterioration of water quantity and quality parameters is interpreted for Macro Prespa & Vegoritis lakes, while Ohrid lake remains stable.

6. Discussion

Monitoring of the lake ecosystems is of paramount importance for the overall development of a region. Remote sensing provides valuable information concerning different hydrological parameters of interest to a lake assessment project. Monitoring is supported due to the multi-temporal character of the data. Temporal changes for the last 30 years can be analyzed with the use of satellite imagery. Processing techniques that have been applied include integrated image processing / GIS vector data techniques. Satellite data generate GIS database information required for hydrological studies and the application of models. Neural network algorithms are quite effective for the satellite data classification. Generated database can be used to assess changes that are taking place in the lakes and its surrounding environment. The areal extent of the lakes has been mapped accurately in all cases. Using the adopted methodology various parameters concerning the lakes and their basins can be extracted related to the description of catchments, surface area, water-level, hydrogeology and water quality characteristics of the lakes.

Water quality parameters of the lakes can be retrieved from remote sensing. Peristrophic movements (gyres) can be clearly identified in the time series images, both in the optical and thermal bands of the Landsat satellite system for the Macro Prespa lake. Understanding the naturally occurring mixing processes in the lake aids in determining the ultimate fate of pollutants, and supports the application of good management strategies and practice. The high spatial resolution of the satellite images allow the surface currents and general circulation in lakes to be accurately identified using the multi-temporal imagery. This can assist in monitoring the clarity and general water quality of lakes. ENVISAT MERIS satellite data have been used for the assessment of spatio-temporal variability of selected water quality parameters like dispersion of suspended solids and chlorophyll concentration. Deterioration of water quantity and quality parameters is interpreted for both Macro Prespa and Vegoritis lakes. It is indicated that satellite monitoring is a viable alternative for spatio-temporal monitoring purposes of lake ecosystems. However, technology alone is insufficient to resolve conflicts among competing water uses. A more useful approach is to have specialists to support decision makers by making available to them the use of data and techniques.

7. References

Bukata, R. P., Jerome J. H., & Burton J. E. (1988). Relationships among Secchi disk depth, beam attenuation coefficient, and irradiance attenuation coefficient for Great Lakes waters. Journal of Great Lakes Research, 14(3), 347-355.

Chacon-Torres, A., Ross, L., Beveridge, M. & Watson, A., 1992. The application of SPOT multispectral imagery for the assessment of water quality in Lake Patzcuaro, Mexico. International Journal of Remote Sensing, 13(4): 587-603.

Charou E., Katsimpra E., Stefouli M. & Chioni A., Monitoring lake hydraulics in West Macedonia using remote sensing techniques and hydrodynamic simulation (2010) Proceedings of the 6th International symposium on environmental Hydraulics, 22-25 June 2010, pages 887-893.

Cox, R. M., Forsythe, R. D., Vaughan, G. E., & Olmsted, L. L. (1998). Assessing water quality in the Catawba River reservoirs using Landsat Thematic Mapper satellite data. Lake and Reservoir Management, 14, 405– 416.

Doerffer, R. & Schiller, H. (2008a). MERIS lake water algorithm for BEAM ATBD, GKSS Research Center, Geesthacht, Germany. Version 1.0, 10 June 2008.

Doerffer, R. & Schiller, H. (2008b). MERIS regional, coastal and lake case 2 water project – Atmospheric Correction ATBD. GKSS Research Center, Geesthacht, Germany. Version 1.0, 18 May 2008.

Hartmann, H. C. (2005) Use of climate information in water resources management. *In: Encyclopedia of Hydrological Sciences,* M.G. Anderson (Ed.), John Wiley and Sons Ltd., West Sussex, UK, Chapter 202.

Liu, Y., Islam, M. and Gao, J., 2003. Quantification of shallow water quality parameters by means of remote sensing. Progress in Physical Geography, 27(1): 24-43.

Nellis, M., Harrington, J. and Wu, J., 1998. Remote sensing of temporal and spatial variations in pool size, suspended sediment, turbidity, and Secchi depth in Tuttle Creek Reservoir, Kansas. Geomorphology, 21(3-4): 281-293.

Ritchie, J., Schiebe, F. and McHenry, J., 1976. Remote sensing of suspended sediment in surface water. Photogrammetric Engineering and Remote Sensing, 42: 1539-1545.

Schiebe, F., Harrington, J. and Ritchie, J., 1992. Remote sensing of suspended sediments: the Lake Chicot, Arkansas project. International Journal of Remote Sensing, 13(8): 1487 - 1509.

Schmugge, T., Kustas, W., Ritchie, J., Jackson, T. and Rango, A., 2002. Remote sensing in hydrology. Advances in Water Resources, 25: 1367-1385.

Steissberg, T. E.; Hook, S. J.; Schladow, G. American Geophysical Union, Fall Meeting 2006, abstract #H32D-01.

Stefouli M., Charou E., Kouraev A., Stamos A (2011) Integrated remote sensing and GIS techniques for improving trans-boundary water management: The case of Prespa region. In: Selection of papers from IV International Symposium on Transboundary Waters Management, Thessaloniki, Greece, 15th - 18th October 2008 for publication in *Groundwater Series of UNESCO's Technical Documents* , 174-179 pp.

Tyler, A., Svab, E., Preston, T., Présing, M. and Kovács, W., 2006. Remote sensing of the water quality of shallow lakes: a mixture modelling approach to quantifying phytoplankton in water characterized by high-suspended sediment. International Journal of Remote Sensing, 27(8): 1521-1537.

Vrieling, A., 2006. Satellite remote sensing for water erosion assessment: a review. Catena, 65: 2-18.

Wallin, M. L., & Hakanson, L. (1992). Morphometry and sedimentation as regulating factors for nutrient recycling and trophic level in coastal waters. Hydrobiologia, 235, 33-45.

Zhen-Gang Ji and Kang-Ren Jin 2006. Gyres and Seiches in a Large and Shallow Lake, in (Volume 32, No. 4, pp. 764-775) of the Journal of Great Lakes Research, published by the International Association for Great Lakes Research, 2006.

13

Landscape Environmental Monitoring: Sample Based Versus Complete Mapping Approaches in Aerial Photographs

Habib Ramezani[1], Johan Svensson[1] and Per-Anders Esseen[2]
[1]*Department of Forest Resource Management, Swedish University of Agriculture Science, Umeå,*
[2]*Department of Ecology and Environmental Science, Umeå University, Umeå, Sweden*

1. Introduction

Unknown land use premises are to be expected due to changing conditions, e.g. shifting land use priorities, climate change, globalizing natural resource markets or new products in the natural resource sector. As a result the need is obvious for accurate, relevant and applicable landscape data to be used in cause-and-effect analysis concerning changes in environmental conditions (Ståhl et al., 2011).

The current land use strongly influence landscape structure (composition and configuration) and contribute to biodiversity loss (Hanski, 2005; Fischer and Lindenmayer, 2007). In order to consider current status and also to monitor trends within a landscape there is a need for reliable and continuous information as a basis for policy- and strategic - as well as operational decision making (Bunce et al., 2008). For this purpose, many countries have now established or are in the process of establishing monitoring programs that provide information on large spatial scale (e.g., regional and national levels), for instance, the National Inventory of Landscapes in Sweden (NILS) (Ståhl et al., 2011), the Norwegian 3Q (NIJOS, 2001), and similar programs in other countries, e.g., in Hungary (Takács and Molnár, 2009). A major concern in landscape monitoring at national scale is the large complexity and amount of data, and the consequently the labor need in data acquisition, database management as well as data analysis and interpretation.

Description and assessment of landscape conditions and changes require relevant, accurate and applicable landscape metrics, which are defined based on measurable attributes of landscape elements such as patches or boundaries. The suite of metrics must cover both the composition and configuration of the landscape to have potential to detect changes within a given landscape or when comparing different landscapes.

Calculation of landscape metrics is commonly conducted on completely mapped areas based on remotely sensed data. FRAGSTATS (McGarigal and Marks, 1995) is a frequently used software for this purpose. In mapping, homogenous areas are first delineated as polygons. Aerial photo interpretation is usually performed using a manual approach while some automated and computer-assisted approaches have recently become available (e.g., Blaschke, 2004). Important attributes in manual interpretation include tone, pattern, size and

shape (Morgan et al., 2010). The experience of the interpreters is critical and the results from manual interpretation are thus often more accurate than those from automated approaches. However, the manual approach may be time-consuming (Corona et al., 2004), subjective (interpreter-dependent) and considerable variation may occur between photo interpreters. The automated approach is sometimes unreliable, for instance, when land cover classes that are similar in terms of spectral reflectance should be separated (Wulder et al., 2008). In addition, overall time, including delineation and corrections may be large if an inappropriate automated approach is chosen.
Sample based approach is an interesting alternative to extract landscape data compared to complete mapping (Kleinn and Traub, 2003). The argument is that a sample survey takes less time; that it is possible to achieve more accurate result in a well-designed and well-executed sample survey; and that data can be acquired and analyzed more efficiently (Raj, 1968; Cochran, 1977). The efficiency and speed in delivering results is of particular interest in landscape-scale monitoring programs where stakeholders commonly are closely involved and expect outputs within reasonable time. Figure 1 shows examples of complete mapping and sample based approaches (point and line intersect sampling methods) over 1 km × 1 km aerial photo from NILS.

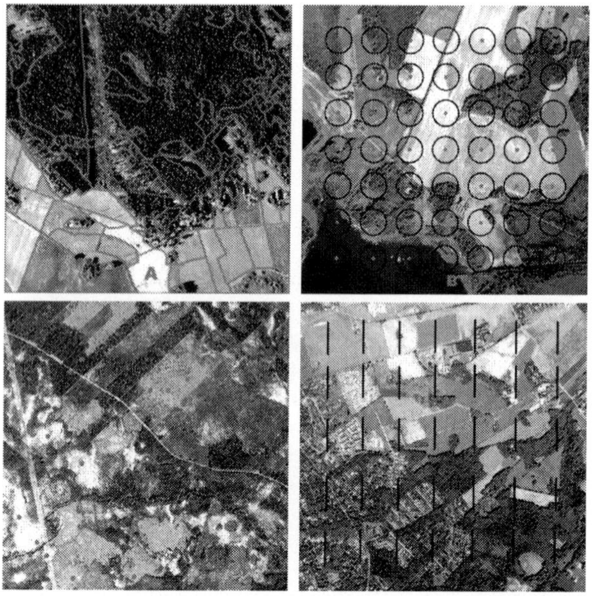

Fig. 1. Examples of complete mapping and sample based approaches to extract landscape metrics in 1 km × 1 km aerial photo. A) Complete mapping, B) systematic point sampling with fixed buffer (40 m), C) point pairs sampling, and D) systematic line intersect sampling.

Since aerial photos are important source of data for many ongoing environmental monitoring programs such as NILS (Ståhl et al., 2011), there is an urgent need to investigate the possibilities and limitations of both mapping and sample based approaches for estimating landscape metrics. The overall objective of this chapter is to compare the advantages and limitations of complete mapping versus sample based approaches for estimating landscape metrics Shannon's diversity, total edge length and contagion from aerial photos. The specific objectives are: (1) to compare point and line intersect sampling for selected metrics in terms of the level of detail and accuracy of data extracted, and the time needed (cost) to extract the data, (2) to compare sample based and complete mapping approaches in terms of time needed, and (3) to investigate statistical properties (bias and RMSE) of estimators of selected metrics using Monte-Carlo sampling simulation.

2. Material and methods

2.1 Study area

The data was collected from aerial photographs and land cover maps from the NILS program (Ståhl et al., 2011), which covers the whole of Sweden. NILS was developed to monitor conditions and trends in land cover classes, land use and biodiversity at multiple spatial scales (point, patch, landscape) as basic input to national and international environmental frameworks and reporting schemes. NILS was launched in 2003 and has developed a monitoring infrastructure that is applicable for many different purposes. The basic outline is to combine 3-D interpretation of CIR (Color Infra Red) aerial photos with field inventory on in total of 631 permanent sample plots (5 km × 5 km) across all terrestrial habitats and the land base of Sweden (see Fig. 2).

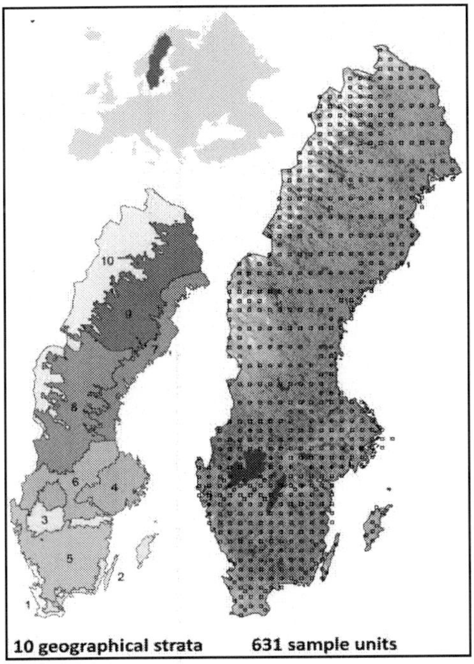

Fig. 2. Illustration of systematic distribution of 631 NILS 1 km × 1 km sample plot across Sweden with ten strata. The density of plots varies among the strata (Ståhl et al., 2011).

The present study is based on a detailed aerial photo interpretation of a central 1 km × 1 km square in the sample plot. Landscape data was extracted from 50 randomly selected NILS 1 km × 1 km sample plots distributed throughout Sweden. The aerial photo interpretation is carried out on aerial photos with a scale of 1:30 000. The aerial photographs in which interpretations were made had a ground resolution of 0.4 m. Polygon delineation is made using the interpretation program Summit Evolution from DAT/EM and ArcGIS from ESRI. According to the NILS' protocol, homogenous area delineated into polygons which are described with regard to land use, land cover class, as well as features related to trees, bushes, ground vegetation, and soils (Jansson et al., 2011; Ståhl et al., 2011).

2.2 Landscape metrics

Landscape metrics are defined based on measurable patch (landscape element) attributes where these attributes first should be estimated. In this study, point (dot grid) and line intersect sampling (LIS) methods were separately applied in (vector-based) land cover map from aerial photos for estimating three landscape metrics: Shannon's diversity, total edge length and contagion. Riitters et al. (1995) demonstrated that these metrics are among the most relevant metrics in landscape pattern analysis. Definition and estimators of the selected metrics are briefly described below.

2.2.1 Shannon's diversity index (H)

This metric refers to both the number of land cover classes and their proportions in a landscape. The index value ranges between 0 and 1. A high value shows that land cover classes present have roughly equal proportion whereas a low value indicates that the landscape is dominated by one land cover class. The index, H , is defined as

$$H = -\frac{\sum_{j=1}^{s} p_j \cdot \ln(p_j)}{\ln(s)} \tag{1}$$

where p_j is the area proportion of the j th land cover class and s is the total number of land cover classes considered (assumed to be known). For $p_j = 0, p_j \cdot \ln(p_j)$ is set to zero. The estimator $\hat{H}$ of H is obtained by letting the estimator $\hat{p}_j$ for land cover class j in Eq. 2 (for point sampling) and in Eq. 3 (for line intersect sampling) take the place of p_j in formula (1). With point sampling, p_j is estimated without bias by

$$\hat{p}_j = \frac{1}{n}\sum_{i=1}^{n} y_i \tag{2}$$

where y_i takes the value 1 if the i th sampling point falls in certain class and 0 otherwise and n is the sample size (total number of points).

With the line intersect sampling (LIS) method (Gregoire and Valentine, 2008), p_j can unbiasedly be estimated by

$$\hat{p}_j = \frac{A}{L} \cdot \sum_{i=1}^{n} l_{ij} \tag{3}$$

where l_{ij} is the intersection length of the j th land cover class with sampling line i, L is the total length of all line transects, and A is the total area.

2.2.2 Total edge length (E)

This metric refers to the amount of edge within landscape. An edge is defined as the border between two different land cover classes. Edge length is a robust metric and can be used as a measure of landscape fragmenattion (Saura and Martinez-Millan, 2001). In a highly fragmented landscape there are more edges and response to those depends on the species under consideration (Ries et al., 2004). The length is relevant for both biodiversity monitoring and sustainable forest magament.

Ramezani et al. (2010) demonstrated that total edge length in the landscape can be estimated using point sampling in aerial photographs without direct length measurement. In such procedure, estimation of the length is based on area proportion of a buffer around patch borders. In Fig. 3 is shown a rectangular buffer around patch border for simulation application. The proportion of sampling points within the buffer can be employed for estimating the buffer area and, hence, the edge length. In practice, however, if a photo interpreter observed a point within distance d from a potential edge, this would be recorded. Figure 2 shows a circular buffer (with fixed radius 40 m) around sampling points on non-delineated aerial photograph for estimating edge length in practice.

According to Ramezani et al. (2010), the buffer area B_j inside the landscape with area A, can be estimated without bias, for a given land cover class by

$$\hat{B}_j = \hat{p}_j \cdot A \tag{4}$$

where $\hat{p}_j$ is the estimator (1) of the buffer area proportion. The length E_j of the edge of the land cover class j is then estimated by

$$\hat{E}_j = \frac{\hat{B}_j}{2d} = \hat{p}_j \cdot \frac{A}{2d} \tag{5}$$

where d is buffer width (m) in one side.

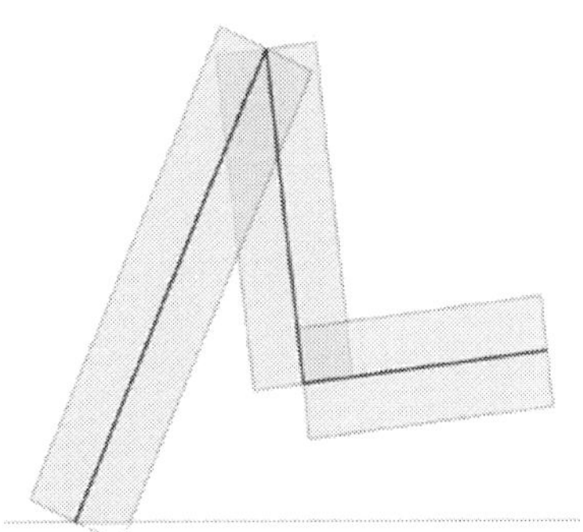

Fig. 3. Illustration of rectangular buffer with fixed width created in both sides of patch border for estimating edge length for simulation application (from Ramezani et al., 2010)

In the LIS method, the estimation of total edge length is based on the method of Matérn (1964). The edge length can unbiasedly be estimated by simply counting the number of intersections between patch border and the line transects. According to Matérn (1964), the total edge length estimator $\hat{E}$ (m ha^{-1}), using multiple sampling lines of equals length, is given by

$$\hat{E} = \frac{10000 \cdot \pi \cdot m}{2 \cdot n \cdot l} \tag{6}$$

where m is the total number of intersections, n is the sample size (number of lines) and l is the length of the sampling line (m).

2.2.3 Contagion (C)

Contagion metric was first proposed by O'Neill et al. (1988) as a measure of clumping of patches. Values for contagion range from 0 to 1. A high contagion value indicates a landscape with few large patches whereas a low value indicates a fragmented landscape with many small patches. Contagion metric is highly related to metrics of diversity and dominance and can also provide information on landscape fragmentation. This metric is originally defined and calculated on raster based map (O'Neill et al., 1988; Li and Reynolds, 1993).
Recently, however, a new (vector-based) contagion metric has been developed by Ramezani and Holm (2011a), which is adapted for point sampling. The new version is distance-dependent and allows estimating contagion metric using point sampling (point pairs).
According to Ramezani and Holm (2011a), for a given distance d the (unconditional) contagion estimator is defined as

$$\hat{C}(d) = 1 + \frac{\sum_{i=1}^{s}\sum_{j=1}^{s} \hat{p}_{ij}(d) \cdot \ln(\hat{p}_{ij}(d))}{2\ln(s)} \tag{7}$$

where the $p_{ij}(d)$ (unconditional probability) is estimated by the relative frequency of points in land cover classes i and j. The estimator $\hat{p}_{ij}(d)$ is then inserted into the Eq. 7 to obtain estimator of $\hat{C}(d)$ the unconditional contagion function and s is the number of observed land cover classes in sampling.
A vector based contagion metric has been developed by Wickham et al (1996), which is defined based on the proportion of edge length between land cover classes i and j to total edge length within landscape. This definition (i.e., Eq. 8) is more adapted to the LIS method. According to Wickham et al (1996), contagion estimator can be written

$$\hat{C} = -\frac{\sum_{i}^{s}\sum_{i \neq j}^{s} \hat{p}_{ij} \cdot \ln(\hat{p}_{ij})}{\ln(0.5(s^2 - s))} \tag{8}$$

Similar to point based contagion (Eq. 7), component $\hat{p}_{ij}$ should be estimated and then inserted into Eq. 8. The estimator $\hat{p}_{ij}$ $(= \hat{E}_{ij}/\hat{E}_t)$ is the proportion of the estimator of edge length between land cover classes i and j ($\hat{E}_{ij}$) to the estimator of total edge length ($\hat{E}_t$)

within landscape. Both $\hat{E}_{ij}$ and $\hat{E}_t$ can unbiasedly be estimated by Eq. 6. In contrast to Eq. 7, a value of 1 from Eq. 8 indicates a fragmented landscape with many small patches.

2.2.4 Monte-Carlo sampling simulation

In this study, Monte-Carlo sampling simulation was used to assess statistical performance (bias and RMSE) of estimators of the selected metric. Bias (or systematic error) is the difference between the expected value of the estimator and the true value. RMSE is the square root of the expected squared deviation between the estimator and the true value.
In point sampling, simulation was conducted for four sample sizes (49, 100, 225, and 400) for both Shannon's diversity and total edge length and five buffer widths (5, 10, 20, 40, and 80 m) for total edge length. In line intersect sampling, simulation was conducted for four sample sizes (16, 25, 49, and 100), three line transect length (37.5, 75, and 150 m), and five transect configurations (Straight line, L, Y, Triangle, and Square shapes). In point pairs sampling (i.e., using Eq.7) simulation was conducted for nine point distances (2, 5, 10, 20, 30, 60, 100, 150, and 250 m) and five sample sizes (25, 49, 100, 225, and 400). Systematic and simple random sampling designs were employed for all cases above.

3. Results

In this study, the statistical properties (RMSE and bias) of the estimators of the selected metrics were investigated for different sampling combinations. But some major results are presented here. In general, a systematic sampling design resulted in smaller RMSE and bias compared to simple random design, for all combinations.

3.1 Shannon's diversity index

In point sampling, both RMSE and bias of Shannon's diversity estimator tended to decrease with increasing sample size in both sampling designs. In Fig. 4 is shown the relationship between bias and sample size of Shannon's diversity estimator in systematic and random sampling designs.

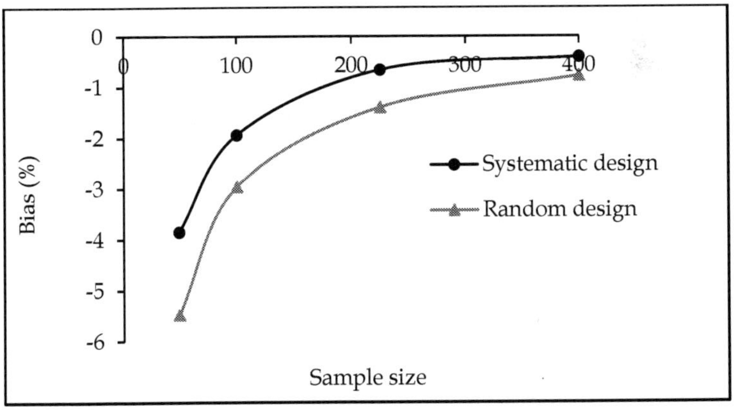

Fig. 4. The relationship between bias and sample size of Shannon's diversity estimator using point sampling method in systematic and random sampling designs (from Ramezani et al., 2010).

In line intersect sampling, similar to point sampling, both RMSE and bias of Shannon's estimator tended to decrease with increasing sample size and line length. The longer line transect (here 150 m) resulted in lower RMSE and bias than shorter one (here 37.5 m), for a given sample size. We found a small and negative bias for the estimator in both point and the LIS methods. The magnitude of bias tended to decrease both with increasing sample size and line transects length. Straight line configuration resulted in lower RMSE and bias than other configurations.

3.2 Total edge length

In point sampling, the magnitude of RMSE of estimator is highly related to buffer width, for a given sample size and a wide buffer resulted in lower RMSE than narrow one. The edge length estimator had bias since parts of buffer close to the map border were outside the map. Bias of estimator tended to increase with increasing buffer width whereas it was independent on sample size. To eliminate or reduce the bias of estimator three corrected methods were suggested which have been discussed in detilas in Ramezani et al. (2010).
In LIS, the magnitude of RMSE of estimator is dependent on the length of the line transect, for a given sample size and the longer transect resulted in lower RMSE than short one. Furthermore, straight line configuration resulted in lower RMSE compared to other configurations (e.g., L and square shape). In Fig. 5 is shown the relationship between relative RMSE and sampling line lengths of total edge length estimator.

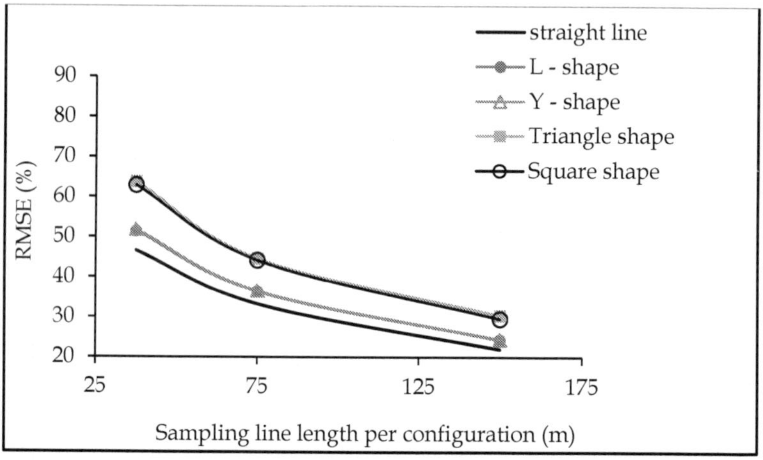

Fig. 5. Relative RMSE of total edge length estimator for different sampling line lengths and configurations of line intersect sampling, for a given sample size (from Ramezani and Holm, 2011c).

3.3 Contagion

Point based contagion (i.e., Eq. 7) is a distance-dependent function that delivers a contagion value that decreased with increasing point distance. The rate of decrease of the contagion value was faster in a fragmented landscape compared to a more homogenous landscape. Examples of such landscapes are shown in Fig. 6. The contagion estimator was biased even if its component (i.e., $p_{ij}(d)$) was estimated without bias. The sources of bias discussed in details in Ramezani and Holm (2011b).

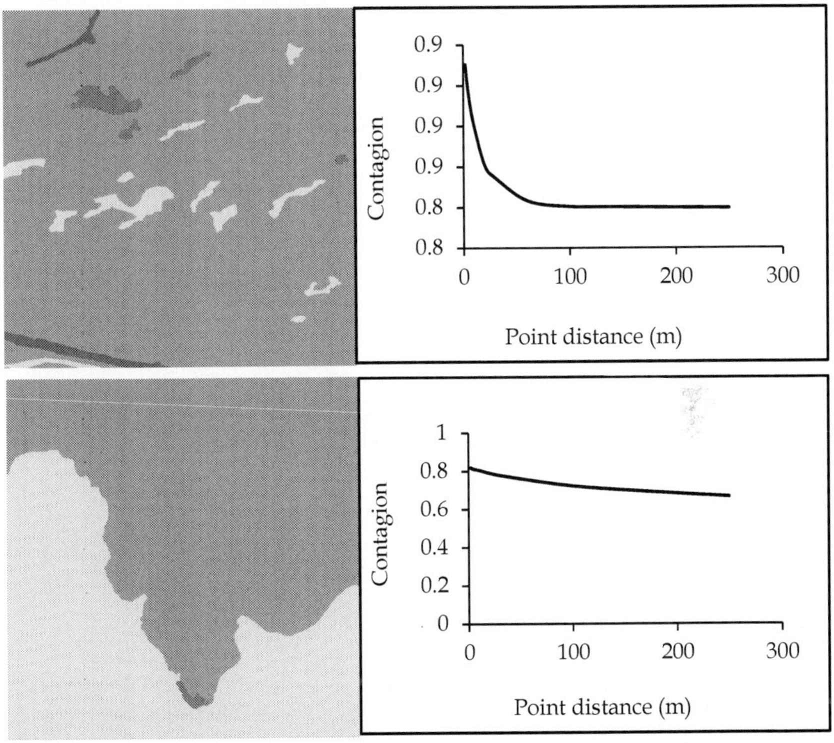

Fig. 6. Example of two landscapes with different degree of fragmentation and their corresponding contagion function (Eq. 7). Top: a high fragmented landscape (four land cover class and nineteen patches) with large rate of decrease of the contagion function. Bottom: a homogenous landscape (three land cover class and three patches) with a small rate of decrease in the contagion function.

In line intersect sampling, both RMSE and bias of the contagion estimator (Eq.8) tended to decrease with increasing sample size and line transects length. Straight line configuration resulted in lower RMSE and bias than other configurations. We found a small and negative bias for the contagion estimator despite its components (i.e., $\hat{E}_{ij}$ and $\hat{E}_t$) can be estimated without bias. The relative RMSE and bias of the contagion estimator through line intersect sampling (LIS) method (Eq.8) is shown in Fig. 7. Note that the two contagion estimators differ as they are based on different equations (i.e., Eqs.7 and 8).

A comparison was also made for variability in terms of range and mean in sample based estimates of Shannon's diversity, edge length and contagion metrics for sample sizes 16 and 100. In Table 1 is provided an example for line intersects sampling method, systematic sampling design, straight line configuration and line length 37.5 m.

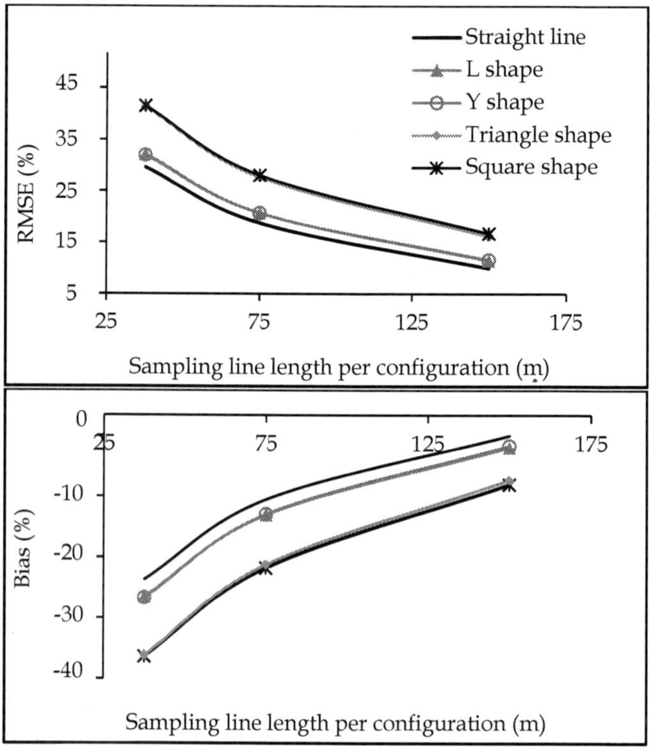

Fig. 7. Relative RMSE (top) and bias (bottom) of contagion estimator (Eq. 8) for different sampling line lengths and configurations, a sample 49 and systematic sampling design

Landscape metrics	Sample size	
	16	100
Shannon' diversity	0.398 (0.019-0.747)	0.423 (0.026-0.784)
Contagion [a]	0.188 (0.006-0.478)	0.407 (0.226-0.758)
Total edge length (m ha^{-1})	92.2 (12.2-197.6)	92.1 (10.5-194.6)

[a] according to Eq.8

Table 1. Variability (mean) in sample based estimates of Shannon's diversity, edge length and contagion in fifty random landscapes (NILS plots) in Sweden for sample sizes 16 and 100. Data collected using line intersects sampling method, systematic sampling design, straight line configuration and 37.5 m length of sampling lines. Ranges are given in parentheses.

3.4 Time study (cost needed for data collection)

A time study was conducted on non-delineated aerial photos from NILS employing an experienced photo interpreter. The results of the time study for Shannon's diversity and total edge length are summarized in Tables 2 and 3.

Method	Time needed (h)
Complete mapping	3.5
Point sampling (number of points)	
9	0.4
100	0.8
225	1.9
400	3.3

Table 2. Average time consumption of data collection on five NILS plots for point sampling and complete mapping for deriving the Shannon's index (from Ramezani et al. (2010))

Sampling method	Time needed (min)	
	Edge length estimator	Shannon' s diversity estimator
Point sampling	25 [a]	28.3
LIS	18.3 [b]	60 [b]

[a] (buffer 40 (m))
[b] (line 150 (m))

Table 3. Average time needed for point and line intersect sampling (LIS) methods for deriving Shannon's diversity and total edge length. For sample size 100 (number of point and lines)

The time needed to collect data was highly related to landscape complexity and the classification system applied. We also found that in a coarse classification system the time needed was less than in a more detailed system. This issue becomes more serious in complete mapping approaches where all potential polygons should be delineated. Furthermore, time was also dependent on sampling method the chosen. With a point sampling method less time was needed for estimating Shannon's diversity compared with other metrics. With line intersect sampling; it was more time efficient to use edge-related metrics. For a given sample size, the time depended on the length of line transect (in LIS) and the buffer width (in point sampling). With the former method it is indicated that the time is independent on line configuration in the aerial photo.

4. Discussion

This study addresses the potential of sampling data for estimating some landscape metrics in remote sensing data (aerial photo). Sample based approach appears to be a very promising alternative to complete mapping approach both in terms of time needed (cost) and data quality (Kleinn and Traub, 2003; Corona et al., 2004; Esseen et al., 2006). However, some metrics may not be estimated from sample data regardless of chosen sampling method since currently used landscape metrics are defined based on mapped data. To describe landscape patterns accurately, a set of landscape metrics is needed since all aspect of landscape composition and configuration cannot be captured through a single metric. On the other hand, all metrics cannot be extracted using a single sampling method. Thus, in a sample based approach a combination of different sampling methods is needed, for instance, a combination of point and line intersect sampling. In such combined design, the start, mid and end points of line transects can be treated as grid of points which is preferred for estimating area proportions of different land cover classes within a landscape and thus Shannon's diversity. It would also be effective in terms of cost if several metrics could simultaneously be derived from a single sampling method.

From a statistical point of view unbiasedness is a desirable property of an estimator. In sample based assessment of landscape metrics, attributes (metrics components) such as the number, size, and edge length of patches must unbiasedly be estimated (Traub and Kleinn, 1999) if an unbiased estimate is needed. However, this is a necessary but not sufficient conditions (Ramezani, 2010). For instance, in the case of Shannon' diversity, there is still bias despite its component i.e., area proportions of land cover classes can be estimated without bias through both point and line intersect sampling methods (Ramezani et al., 2010; Ramezani and Holm, 2011c). The bias is due to non-linear transformation, which also generally is the case for other metrics with non-linear expression such as contagion. Bias of selected metric estimators is very small if the sample size is large and the magnitude of bias depends jointly on type of selected metric, the sampling method, and the complexity of the landscape structure. To achieve an acceptable precision in a complex landscape there is a need for a larger sample size compared to the homogenous landscape.

The landscape metrics used in this study are based on a patch-mosaic model where sharp borders are assumed between patches. In such procedure, as noted by Gustafson (1998) the patch definition is subjective and depends on criterion such as the smallest unit that will be mapped (minimum mapping units, MMU). This becomes more challenging in a highly fragmented landscape where smaller patches than predefined MMU are neglected. Even though these patches constitute a small proportion (area) of the landscape, they contribute significantly to the overall diversity of that landscape; including biodiversity where other type organisms may occupy these patches habitats. However, in sample based approach which can be conducted in non-delineated aerial photos, there is no need to predefine minimum patch size and even very small patches can be included in the monitoring system. Furthermore, point sampling appears to be in consistent with gradient based model of landscape (McGarigal and Cushman, 2005) where landscape properties change gradually and continuously in space and where no subjective sharp border need to be assumed between patches.

Polygon delineation errors are common in manual mapping process. It can be assumed that this error can be eliminated when sampling methods are used for estimating some landscape metrics. As a result, obtained information and subsequent analysis is more reliable than for traditional manual polygon delineation. As an example, for estimating the metrics Shannon's diversity and contagion using point sampling, no mapped data are needed and assessment is only concentrated on sampling locations. This is also true for the LIS, for instance, the total length estimation of linear features within a landscape is to be based on simply counting the interactions between lines transect and a potential patch border. Consequently, assessment is conducted along line transect which, thus, considerable reduce the polygon delineation error.

It is clear, however, that a sample based approach cannot compete with a complete mapping approach, in particular when high quality mapped data is available. With the mapping approach a suite of metrics can be calculated for patch, class, and landscape levels whereas in sample based approach a limited number of metrics on landscape level can often be estimated.

5. Conclusion

A sample based approach can be used complementary to complete mapping approach, and adds a number of advantages, including 1) the possibility to extract metrics at low cost 2) applicable in case of lacking categorical map of entire landscape 3) the possibility in some case to obtain more reliable information and 4) the possibility of estimating some metrics from ongoing field-based inventory such as national forest inventories (NFI). In some cases, there is a need to slightly redefine currently used landscape metrics or develop new metrics to meet sample data. There is obviously plenty of room for further studies into this topic since sample based assessment of landscape metrics is a new approach in landscape ecological surveys.

6. References

Blaschke, T., (2004). Object-based contextual image classification built on image segmentation: IEEE Transactions on Geoscience and Remote Sensing, p. 113-119.

Bunce, R.G.H., Metzger, M.J., Jongman, R.H.G., Brandt, J., de Blust, G., and Elena-Rossello, R., et al., (2008). A standardized procedure for surveillance and monitoring European habitats and provision of spatial data: landscape Ecology, v. 23, p. 11-25.

Cochran, G., (1977). *Sampling techniques*: New York, Wiley, xvi, 428 p.

Corona, P., Chirici, G., and Travaglini, D., (2004). Forest ecotone survey by line intersect sampling: Canadian Journal of Forest Research-Revue Canadienne De Recherche Forestiere, v. 34, p. 1776-1783.

Esseen, P.A., Jansson, K.U., and Nilsson, M., (2006). Forest edge quantification by line intersect sampling in aerial photographs: Forest Ecology and Management, v. 230, p. 32-42.

Fischer, J., and Lindenmayer, D.B., (2007). Landscape modification and habitat fragmentation: a synthesis: Global Ecology and Biogeography, v. 16, p. 265-280.

Gregoire, T.G., and Valentine, H.T., (2008). *Sampling Strategies for Natural Resources and the Environment* Boca Raton, Fla. London, Chapman & Hall/CRC.

Gustafson, J.E., (1998). Quantifying landscape spatial pattern: What is the state of the art?: Ecosystems, v. 1, p. 143-156.

Hanski, I., (2005). Landscape fragmentation, biodiversity loss and the societal response - The longterm consequences of our use of natural resources may be surprising and unpleasant: Embo Reports, v. 6, p. 388-392.

Jansson, K.U., Nilsson, M., and Esseen, P.-A., (2011). Length and classification of natural and created forest edges in boreal landscapes throughout northern Sweden: Forest Ecology and Management.v.262,P.461-469

Kleinn, C., and Traub, B., (2003). Describing landscape pattern by sampling methods, in Corona, P., Köhl, M., and Marchetti, M., eds., *Advances in forest inventory for sustainable forest management and biodiversity monitoring.*, Volume 76, p. 175-189.

Li, H., and Reynolds, J., (1993). A new contagion index to quantify spatial patterns of landscapes: Landscape Ecology, v. 8, p. 155-162.

Matérn, B., (1964). A method of estimating the total length of roads by means of line survey: Studia forestalia Suecica, v. 18, p. 68-70.

McGarigal, K., and Cushman, S.A., (2005). The gradient concept of landscape structure, in Wiens, J., and Moss, M., eds., *Issues and perspectives in landscape ecology*: Cambrideg, Cambrideg University press.

McGarigal, K., and Marks, E.J., (1995). FRAGSTATS: Spatial pattern analysis program for quantifying landscape pattern. General Technical Report 351. U.S. Department of Agriculture, Forest Service, Pacific Northwest Research Station.
Morgan, J., Gergel, S., and Coops, N., (2010). Aerial Photography: A Rapidly Evolving Tool for Ecological Management: BioScience, v. 60, p. 47-59.
NIJOS, (2001). Norwegian 3Q Monitoring Program: Norwegian institute of land inventory.
O'Neill, R.V., Krumme, J.R., Gardner, H.R., Sugihara, G., Jackson, B., DeAngelist, D.L., Milne, B.T., Turner, M., Zygmunt, B., Christensen, S.W., Dale, V.H., and Graham, L.R., (1988). Indices of landscape pattern: Landscape Ecology v. 1, p. 153-162.
Raj, D., (1968). *Sampling theory*: New York, McGraw-Hill, 302pp. p.
Ramezani, H., (2010). Deriving landscape metrics from sample data (PhD thesis): Umeå, Swedish University of Agricultural Sciences (SLU).
Ramezani, H., and Holm, S., (2011a). A distance dependent contagion functions for vector-based data: Environmental and Ecological Statistics (accepted).
–, (2011b). Estimating a distance dependent contagion function using point sample data (in review).
–, (2011c). Sample based estimation of landscape metrics: accuracy of line intersect sampling for estimating edge density and Shannon's diversity . Environmental and Ecological Statistics, v. 18, p. 109-130.
Ramezani, H., Holm, S., Allard, A., and Ståhl, G., (2010). Monitoring landscape metrics by point sampling: accuracy in estimating Shannon's diversity and edge density: Environmental Monitoring and Assessment v. 164, p. 403-421.
Ries, L., Fletcher, R.J., Battin, J., and Sisk, T.D., (2004). Ecological responses to habitat edges: Mechanisms, models, and variability explained: Annual Review of Ecology Evolution and Systematics, v. 35, p. 491-522.
Riitters, K.H., O'Neill, R.V., Hunsaker, C.T., Wickham, J.D., Yankee, D.H., Timmins, S.P., Jones, K.B., and Jackson, B.L., (1995). A factor-analysis of landscape pattern and structure metrics: Landscape Ecology, v. 10, p. 23-39.
Saura, S., and Martinez-Millan, J., (2001). Sensitivity of landscape pattern metrics to map spatial extent: Photogrammetric Engineering and Remote Sensing, v. 67, p. 1027-1036.
Ståhl, G., Allard, A., Esseen, P.-A., Glimskär, A., Ringvall, A., Svensson, J., Sture Sundquist, S., Christensen, P., Gallegos Torell , Å., Högström, M., Lagerqvist, K., Marklund, L., Nilsson, B., and Inghe, O., (2011). National Inventory of Landscapes in Sweden (NILS) - Scope, design, and experiences from establishing a multi-scale biodiversity monitoring system: Environmental Monitoring and Assessment v. 173, p. 579-595.
Takács, G., and Molnár, Z., (2009) National biodiversity monitoring system XI. Habitat mapping (2nd modified ed., p. 54). Ministry of Environment and Water, Budapest.
Traub, B., and Kleinn, C., (1999). Measuring fragmentation and structural diversity: Forstwissenschaftliches Centralblatt, v. 118, p. 39-50.
Wickham, J.D., Riitters, K.H., ONeill, R.V., Jones, K.B., and Wade, T.G., (1996). Landscape 'contagion' in raster and vector environments: International Journal of Geographical Information Systems, v. 10, p. 891-899.
Wulder, M.A., White, J.C., Hay, G.J., and Castilla, G., (2008). Towards automated segmentation of forest inventory polygons on high spatial resolution satellite imagery: Forestry Chronicle, v. 84, p. 221-230.

14

Real-Time Monitoring of Volatile Organic Compounds in Hazardous Sites

Gianfranco Manes[1], Giovanni Collodi[1], Rosanna Fusco[2], Leonardo Gelpi[2], Antonio Manes[3] and Davide Di Palma[3]
[1]*Centre for Technology for Environment Quality & Safety, University of Florence,*
[2]*eni SpA,*
[3]*Netsens Srl,*
Italy

1. Introduction

Volatile Organic Compounds (VOCs) are largely used in many industries as solvents or chemical intermediates. Unfortunately, they include some components, present in the atmosphere, that can represent a risk factor for human health. They are also present as a contaminant or a by-product in many processes, i.e. in combustion gas stacks and groundwater clean-up systems.

Benzene, in particular, shows a high toxicity resulting in a Time-Weighted Average (TWA) limit of 0.5 ppm, as compared, for instance, with TWA for gasoline, in the range of 300 ppm.

Detection of VOCs at sub-ppm levels is, thus, of paramount importance for human safety and industrial hygiene in hazardous environments.

The commonly used field-portable instruments for VOC detection are the hand-held Photo-Ionisation Detectors (PIDs), sometime using pre-filter tubes for specific gas detection. PIDs are accurate to sub-ppm, measurements are fast, in the range of one or two minutes and, thus, compatible with on-field operation. However, they require skilled personnel and cannot provide continuous monitoring.

Wireless connected hand-held PID Detectors start being available on the market, thus overcoming some of the previously described limitations, but suffering for the limited battery life and relatively high cost.

The paper describes the implementation and on-field results of an end-to-end distributed monitoring system integrating VOC detectors, capable of performing real-time analysis of gas concentration in hazardous sites at unprecedented time/space scale.

The system consists of a Wireless Sensor Network (WSN) infrastructure, whose nodes are equipped with distributed meteo-climatic sensors and gas detectors, of TCP/IP over GPRS Gateways forwarding data via Internet to a remote server and of a user interface which provides data rendering in various formats and access to data.

The paper provides a survey of the VOC detector technologies of interest, of the state-of-the-art of the fixed and area wireless technologies available for Gas detection in hazardous areas and a detailed description of the WSN based monitoring system.

2. Regulatory requirements for oil&gas industry

The oil&gas sector is characterised by a high complexity in terms of processes, materials and final products. Consequently, activities related to the oil&gas industry need to be effectively controlled to minimize their impact on the environmental matrices (air, water and soil) and to avoid any potential risks for human health.

Environmental issues related to the oil&gas sector are also strictly dependent on the specific activities performed. In particular, petrochemical and refining sectors are involved in the production of waste materials, such as water and toxic sludge, and atmospheric pollutant emissions, including many VOCs potentially harmful both to the environment and to human health. All these environmental issues are considered areas of high human and environmental risk and therefore subject to stringent international and local environmental regulations.

During the last decade the EU has fixed several Thematic Strategies to improve the management and control on Air Pollution, Soil Protection, Prevention and Recycling of Waste as a follow-up to the Sixth Community Environment Action Programme (Council of 22nd July 2002). In particular, the EU set objectives and regulations on the industrial sector to protect human health and the environment, objectives can be met only with further reductions in emissions arising from industrial activities. The final act of this process was the publication, on 24th November 2010, of the new Directive 75/2010 (IED) on industrial emissions (integrated pollution prevention and control) which recasts together six directives on industrial emissions (IPPC, LCP, VOC, TiOxide).

Based on the principle of the *polluter pays* and also on the *pollution prevention* one, industrial owners should manage their activities in order to protect the environment as a whole, in compliance to the IPPC integrated approach. Furthermore, in accordance with the Århus Convention on access to information and public participation, operators should both improve and promote tools and procedures, such as adopting environmental management system (ISO 14001), increasing the accountability and transparency of the monitoring and reporting data process and contributing to public awareness of environmental issues, and support for the decisions taken.

In order to ensure the prevention and control of pollution, each installation should operate only if it holds a permit, which should include all the measures necessary to achieve a high level of protection of the environment as a whole, and to ensure that the installation is operated in accordance with the general principles governing the basic obligations of the operator. The permit should also include emission limit values for polluting substances or technical measures and monitoring requirements; all conditions should be set on the basis of Best Available Techniques (BAT)[1] applied on each specific installation.

On the other hand, the European Union has issued, in 2008, Directive No 2008/50/EC concerning ambient air quality and cleaner air for Europe.

In order to protect human health and mostly urban environment, the directive addresses the following key points:

---

[1] In the IPPC Directive, BAT are defined as "the most effective technologies available for achieving a high level of environmental protection concerned in an economically feasible and technical view of the costs and benefits". Currently BAT is identified on the basis of an exchange of information organized by the European Commission that occurs between the Member States, industry and non-governmental organisations

- It's very important to prevent and reduce pollutant emissions at source, implementing the best effective reduction measures, both technological and on management. Emissions of air pollutant should be reduced by each member state according to World Health Organisation guidelines.
- The directive establishes the need of a strong monitoring system and the reciprocal exchange of information and data from networks and individual stations measuring ambient air pollution in order to incorporate the latest health and scientific developments and the experience of the Community.
- Each Member State should ensure consistency and representativeness of the information collected on air pollution; standardised measurement techniques and common criteria for the number and location of measuring stations are defined.
- For assessing air quality, information and data collected from fixed measurement stations may be integrated with data from alternative techniques, such as modelling or indicative measurements. The use of measurement methods other than standardised methods allows improving data monitoring and interpretation in some critical areas (such as, for instance, industrial sites) in an economical and feasible way.

Alternative measurement methods may provide indicative results that could be less accurate than those made with the reference method. Indicative measurement techniques based on the use of automatic sensors, mobile laboratories, portable analysers and manual methods of measurement, such as diffusive sampling techniques, are very interesting due to the relatively low cost and simplicity of operation compared with instrumental and operative costs of fixed measuring stations.

3. Volatile Organic Compounds

Volatile Organic Compounds are defined as all compounds containing organic carbon characterized by low vapour pressure at ambient temperature. They are present in the atmosphere mainly in the gas phase.
The number of volatile organic compounds observed in the atmosphere, both in urban and remote areas, is extremely high and includes, in addition to hydrocarbons (compounds containing only carbon and hydrogen), also oxygen species such as ketones, aldehydes, alcohols, acids and esters. Natural emissions of VOCs include the direct emissions from vegetation and the degradation of organic matter; anthropogenic emissions are mainly caused by the incomplete combustion of hydrocarbons, the evaporation of solvents and fuels, and processing industries. On a global scale, natural and anthropogenic emissions of VOCs are of the same order of magnitude.
A lot of volatile organic compounds are highly toxic; this makes them extremely dangerous to human health. In addition, many compounds react with nitrogen oxides and other substances, contributing to the formation of ozone in the lower atmosphere, with impact on climate change and pollution issues (i.e. photochemical smog). Finally, some substances are characterized by a very low odour threshold, resulting in complaints from population and community living around industrial sites.

4. VOC classification

There are many classification systems, based on chemical characteristics, or based on the impact on the environment and human health. The term VOC covers several groups of

organic substances with different chemical and physical characteristics. VOC compounds include in fact compounds containing only atoms of carbon and hydrogen (which include for example aromatic compounds such as benzene). One type of classification used in many states is defined by German regulations (TA Luft - Technical Instructions on Air Quality Control): it identifies three classes of VOCs based on their impact and it defines appropriate prevention and control.
The three classes are:

- extremely hazardous to health - such as benzene, vinyl-chloride and 1,2 dichloroethane
- class A Compounds - that may cause significant harm to the environment (e.g. acetaldehyde, aniline, benzyl chloride)
- class B Compounds - that have lower environmental impact.

Benzene (C6H6) is a volatile organic compound belonging to the family of hydrocarbons and characterized by a monocyclic aromatic structure. It is a natural constituent of petroleum, and it is present in gasoline by virtue of its anti-knock properties (it contributes to increase octane number).

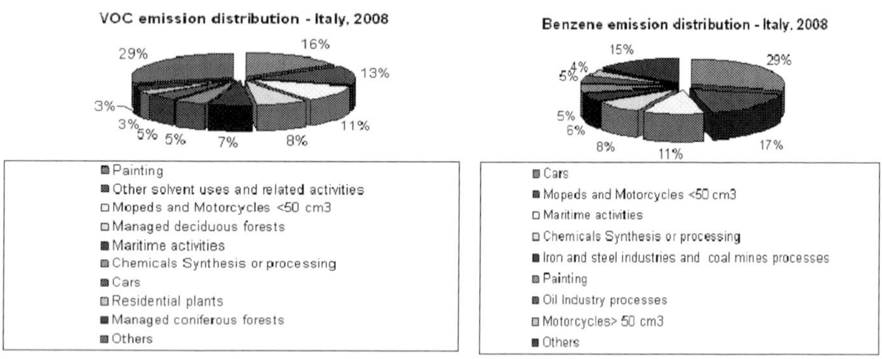

Fig. 1. VOC emission distribution in Italy

In the chemical industry, benzene is a solvent widely used, especially as an intermediate for the synthesis of other products (ethylbenzene, cumene, cyclohexane, etc.) in turn used for the production of plastics, resins, paints, tires, detergents etc.
Benzene exposure is very dangerous to human health; it is classified as a human carcinogen, due to the high toxicity. Among VOCs, benzene is the only compound for which the European directive on air quality has set a limit to 5 $\mu g/m^3$ (about 1.5 ppb), with no margin of tolerance. At work, the TLV-TWA limit is set at 0.5 ppm for prolonged exposure of 8 hours per day and 2.5 ppm for exposures not exceeding 15 minutes (for reference TLW-TWA for gasoline is in the range of 300 ppm).
Benzene emissions related to petroleum activities are about 5% of total emissions, while for the non-methane VOCs the chemical industry appears to be more involved than refining sector.
The graphs in Fig. 1 (2008 VOC and Benzene emission distribution in Italy - data from ISPRA Database) show that motor vehicles are the main pollution sources for benzene, while painting is the main source for non-methane VOCs.

Main VOC sources in petroleum industry

Oil installations, petrochemical plants and refineries are industrial sites that manage several raw materials (crude oil, natural gas, chemical intermediates, etc.), thus having great impact on the environment. Industrial processes may generate VOC emissions to the atmosphere, so prevention and control is becoming a very important issue in the petroleum industry.

The main quantity of VOC releases are due to diffuse and fugitive emission sources. The main sources of VOCs from refineries and petrochemicals are fugitive emission from piping, vents, flares, air blowing, waste water system, storage tanks and handling activities, loading and unloading systems.

Fugitive emissions from piping

Fugitive emissions are defined as emissions of pollutants (gases and dust) in the atmosphere resulting from losses such as pumps, valves, flanges, drains, compressors, sampling points, open ended lines, agitators. The loss of process fluids affects all plant equipment; although the amount emitted from single components may be individually small, the cumulative emissions of the plant can be considerable in some cases.

Fugitive emissions can be considered as the main source of VOCs in the refinery. The application of Best Available Techniques requires industrial facilities to define a Leak Detection and Repair programme (LDAR), which allows the monitoring at defined frequency of the leaks from plant's component, thus providing a swift repair of leaker.

A standard method (EPA 21) is available to define the monitoring criteria. In addition, it is possible to calculate fugitive emissions based on average literature data, but this approach does not provide evidence of improvements and does not allow for leaker repair. For this reason, on-site monitoring is mandatory.

Handling and storage tanks

VOC emissions from storage tanks are due to evaporative loss of the hydrocarbon liquid stored. There are two main types of tanks, fixed roof and floating (internal or external) roof tanks. In the first case, evaporation losses occur mainly from vents and fittings. In floating roof tanks, where the roof is in direct contact with the liquid, emissions may occur from the seals, especially during changes of liquid level.

Emissions depend on the type of product stored and the vapour pressure of the product: higher vapour pressure tends to generate higher VOC emissions.

The emissions are generally estimated by calculation software that takes into account numerous factors such as construction types (type of the roof, seals, colour, etc.), number of loading and unloading cycles, etc.

It is possible to perform monitoring with analytical instrumentation, as long as the requirements of intrinsically safe regulations (ATEX) are met.

During loading, i.e. product stored on vessels, VOC emissions may occur in the vapour phase.

Waste Water Treatment Plants

VOC emissions from Waste Water Treatment Plants are due to evaporation of more volatile compounds from tanks, ponds and sewerage system drains.

Because of contamination of treated water, this type of plant is a major source of odorous emissions, thus causing the need for careful monitoring and control. VOCs are emitted also during air stripping in flotation units and in the biotreaters. Emissions of VOCs and other pollutants into the atmosphere from the treatment ponds and basins can be significantly limited by implementing systems of coverage (almost all industrial sites have this requirement from local authority).

Flare systems

VOC emissions are due to an incomplete combustion of flare gas. However, this type of source does not represent a major cause of VOC emissions.

From a first analysis of the major sources, it is clear that VOC emissions come from widespread areas inside the industrial site. The individual emission sources may have small or large impact, but it is important to consider the overall impact of all sources combined.

Often a regular monitoring at the source may be ineffective, and sometimes the use of methods of monitoring network in the areas close the critical area could be of great help to combat the phenomenon and to achieve a significant reduction of emissions in an economically feasible way.

VOC monitoring systems

Common VOC concentration measurement methods include colorimetric tubes, Infrared Detectors, Photo Ionisation Detectors (PIDs) and Flame Ionisation Detectors (FIDs), portable/transportable Gas Chromatograph (GC) and sampling followed by laboratory analysis. Deployable sensors are of particular relevance, as they are capable to provide on-site monitoring.

Sampling and laboratory analysis

The main sampling technologies for subsequent laboratory analysis are based on the use of active and passive samplers. In the first case, sampling is done by exposing a trap in the site under investigation connected with a pump capable of sucking a steady flow of air. The trap is usually made of absorbent material, e.g. charcoal. The exposure time may vary from a few tens of minutes to hours. The sample is then analysed in the laboratory with gas chromatography techniques (GC).

Passive samplers instead use the diffusive properties of substances dispersed in the atmosphere. They are generally exposed to ambient air for even longer periods (days, weeks), and they are protected in order to prevent damage and contamination caused by weather phenomena (wind, rain). The pollutants are captured at different rates because each of them has different diffusive properties. Sample is then desorbed and analysed in the laboratory (GC). The sampler can be treated with appropriate reagents, in order to obtain selectivity only on a few compound families.

Various passive sampling devices are commercially available. One of the most popular is the sampler Radiello, characterized by radially distributed operation and a better sensitivity due to increased diffusive surface.

The difference between the two types of samplers is linked to the range of compounds they are able to detect; passive sampler are not useful to detect many VOCs (olefins, compounds with less than 5 atoms of carbon, etc.) because they tend not to remain adherent to the passive diffusion sampler, due to prolonged exposure to the atmosphere. The use of one or another depends on the family of VOCs under study.

The main advantage of this sampling technology is the low cost of materials and resources, giving the opportunity to create very dense monitoring networks in an economical feasible way. The disadvantage is the impossibility to continuously collect real-time data, so they are not suitable for emergency management and early warning, but they may be useful for air characterization of an hazardous industrial site, in terms of average concentrations and emission source profiles. Another important application is the use in monitoring networks for checking compliance with the TWA for toxic component (e.g. Benzene).

On-field monitoring

On-field monitoring technologies allow obtaining real-time concentration of pollutants close to a specific source or along the perimeter of the industrial establishment, enabling to manage specific emergency situations in real-time.
The equipment usually yields a response in terms of quantitative concentration levels of VOCs in the atmosphere; in some cases it is even possible to get a specification of the components in the air.
Below an overview of the main methods used on-site, especially at industrial sites, is carried out.

VOC fixed analysers

The use of automatic VOC GC analysers able to collect air samples at regular intervals and analyse them is particularly common when performing monitoring campaigns using fixed stations or a mobile laboratory.
Mobile laboratories (as well as transferable measuring stations) usually combine the advantages of automated measurement methods with the mobility and flexibility.
Many commercially available VOC analysers can be used to perform the task. Unlike active and passive samplers, in this case air sampled is pumped through a sampling probe and is sent directly to the instrument, to run GC analysis by using several detection technologies (photo ionization, flame ionization, thermal conductivity, etc.). The measurement interval is in the range of tens of minutes.
This methodology allows quick answers as well as concentrations for individual compounds to be achieved; however, it does not allow simultaneous monitoring over an industrial site grid, due to the high costs of devices (ten thousand Euros) and operation/ maintenance cost and complexity; furthermore, to cover all the families of compounds of interest - BTEX, C1-C6, sulphur, etc. – more than one analyser is needed.

VOC portable analysers

Portable VOC analysers are instruments of limited size and weight, easily transportable by an operator in the plant and able to provide real-time analysis of gas concentration in hazardous sites.
They are usually equipped with battery life in the range 8-12 hours and allow the storage of data acquired in a time-programmable internal logger. The main application in industry is the detection of gas leaks, leaks from piping, releases in proximity of storage tanks, monitoring of loading and unloading areas, etc.
Based on the sensor technology, they can be classified in the main following typologies.

a. *PID* - Photo Ionization Detectors. These detectors are equipped with a lamp emitting ultraviolet light. The emitted light ionizes targeted VOCs in the air sample so they can be detected and reported as a concentration. Depending on the features of the lamp (there are many on the market able to ionize VOCs depending on ionisation potential), a portable PID can detect a wide range of VOC substances. The analyser is not selective but generally provides a cumulative figure of VOCs; however, knowing emission profiles or mixture composition (in the case of measures directly at the source, such as for fugitive emissions from the plant components), concentration values can be calculated for each substance by applying the response factors. It is usually possible to attach a pre-filter tube to allow detection and selective measurement of a single VOC component (eg. Benzene).

b. *FID* - Flame Ionization Detector. In FIDs sample air is channelled through a chamber where a flame ionizes it, measuring the concentration as a function of electric potential. The fuel gas used is usually hydrogen, contained in a small pressurized cylinder inside the analyser. A charge of hydrogen typically allows for about ten hours undiscontinued operation. This analyser is often used in hazardous industrial sites where there are high concentrations of total organic compound, such as methane.
c. *Portable GC-MS*- This device collects an air sample through a heated probe on a charcoal trap. After sample desorption, the separation is carried out on a chromatographic column and the individual components are analysed by mass spectrometry. It is normally used for on-site monitoring of environmental pollutants (organic substances, sulphur compounds) and to detect oil spills and waste water in the exhaust gases. Unlike PID and FID, this detector is most expensive, weighty and it requires experienced operators.
d. *Colorimetric tubes.* Colorimetric tubes are portable and disposable devices for detecting the concentration of pollutants in ambient air. As for active samplers, a pump draws an air volume within the vial. The sample reagent reacts with the substance causing a colour change proportional to the concentration of the substance to be measured. Devices are disposable and are commercially available for hundreds of pollutants. The advantage is the low cost, rapid response and ease of use; however, measurement accuracy is very low, due to the deterioration of the reagent, contamination and interference with other substances in the sample other than those to be measured.

Optical remote sensing methods

These methods are based on real-time measurement of concentrations of pollutants by taking advantage of the properties of absorption and diffusion of gases in the atmosphere in the visible, ultraviolet and infrared light regions. In fact, the optical path of a light beam of a certain wavelength can be changed by contact with gases and/or dust. The combination with a computerised system allows for automatic management tools, and data processing/presentation to be implemented. Multiple-path optical configurations also allows to measure concentration averaged over a given area .

In the Best Reference Documents (BREF) are mentioned the following optical remote sensing techniques:

i. DIAL (Differential Absorption Infrared Laser): pulsing light is diffused and absorbed by gases in the atmosphere; the analysis of the response time is observed with an optical device that allows the determination of the concentration of the pollutant, and (with modelling support) a generic indication of the origin.
ii. DOAS (Differential Optical Absorption Spectrometry): a continuous light beam is absorbed by pollutants; The receiver, placed at the end of the optical path, directs the beam into an optical fibre and through this to the analyser.
iii. FT-IR (Fourier Transform - Infra Red): absorption in the IR spectrum between a source and a receiver allows the quantitative analysis of numerous substances.
iv. BAGI (Back scatter Absorption Gas Imaging): an infrared laser illuminate the potential source of emission, permitting quantification of gas concentrations by means of the Lambert Beers' law and producing real-time video images of numerous organic vapours of interest.

5. Wireless Sensor Network platform overview

Real-time monitoring of VOCs at unprecedented time/space scale can be performed using Wireless Sensor Networks (WSNs); WSNs have been extensively investigated since the last decade; they consist of spatially distributed autonomous sensors to monitor physical or environmental conditions, such as temperature, sound, vibration and pressure motion or pollutants and to cooperatively pass their data through the network to a sink node.

A WSN generally consists of a base-station (or "gateway") that can communicate with a number of wireless sensors via a radio link. Data are collected at the wireless sensor node, compressed, and transmitted to the gateway directly or, if required, using other wireless sensor nodes to forward data to the gateway. The transmitted data are then presented to the system by the gateway connection and can be accessed worldwide via Internet by authenticated users.

The power of WSNs lies in the ability to deploy large numbers of nodes that assemble and configure themselves, with minimal deployment costs, unlike traditional wired systems, while featuring a high degree of flexibility, re-reconfigurability and scalability.

The most difficult resource constraint to meet is electrical energy, as the WSNs are typically designed as stand-alone systems only relying on autonomous energy sources. Energy budget is shared between the radio/computational unit and the sensor(s), often power hungry and, thus, predominant in power consumption.

The WSN architecture presented here includes both a proprietary hardware platform and communication routines designed specifically to address the needs of an application intended for VOC monitoring in a chemical plant.

The VOC Concentration Precision Monitoring System (VCPMS) based on a Wireless Sensor Network (WSN) has been deployed and tested at the Mantova Petrochemical plant in Italy, starting with May 2011. The lay-out of the installation is represented in Fig. 2.

The system was designed for stand-alone operation, i.e. only relying on autonomous energy and connectivity resources. This is very useful for installation in industrial plant where excavation may be difficult. Internet connectivity is provided via TCP/IP over GPRS using GSM mobile network; wireless connectivity uses the UHF-ISM unlicensed band; electrical power is provided by primary sources (batteries) and secondary sources (photovoltaic cells); highly efficient power saving strategies have been implemented to prolong battery life, as the system is designed to operate undiscontinued and unattended.

The wireless network infrastructure includes base stations operating as Sink Nodes (SNs) exhibiting superior computational capability and energy resources and featuring both TCP/IP over GPRS and wireless connectivity; The SNs are wirelessy connected to distributed wireless units, or End Node (ENs) units.

The SNs are equipped with meteo-climatic sensors thus providing a map of air relative humidity/temperature (RHT) and wind speed/direction (WSD) over the area, while the ENs are equipped with VOC sensors and RHT sensors for accurate VOC sensor read-out compensation. Owing to the extension and complexity of the Mantova plant, covering some 300 acres and featuring complex metallic infrastructures, it was decided to subdivide the area of interest in 7 different sub-areas. Accordingly, each of the sub-areas was equipped with a SN unit and with an appropriate number of EN units. The VCPMS gathers data from the field at minute data rate to produce a real-time VOC concentration map of some key areas in the plant, namely the ST40 chemical plant (eni 6, eni 7), the benzene pipeline (eni 5), the perimeter (eni 1, eni 2, eni 5) and one of the benzene tanks (eni 3). In the ST40 area six

ENs were located, regularly positioned around the plant to detect any VOC emission generated by the plant itself. Two WSD sensors provide information about wind intensity/direction in form of a blue arrow (see Fig. 2). Taking into account for the fact that the wind is turbulent within the plant, the WSD information turns out to be very useful to establish a proper correlation between wind distribution and VOC concentration.

Fig. 2. Lay-out of the installation featuring the SN units (grey) and the EN units (rose)

In the pipeline area two ENs were located in close proximity of the possible sources of fugitive, while three ENs were located along the perimeter. The installation on top of the benzene tank requires ATEX certification which was not yet completed at the time this paper was edited. The units deployed so far consist of a total of seven SNs and ten ENs; owing to the high degree of modularity of the WSN architecture, however, the system is fully scalable by simply deploying additional ENs, with self-configuration capability.

6. System architecture

Various WSN architectures, including mesh and cluster three, have been investigated as potential candidates, each exhibiting advantages and drawbacks; in this application, as it will be explained below, the VOC detectors have to be continuously powered-on and the wireless node has to transmit data at minute data-rate. VOC detector cost, in terms of power consumption, is thus predominant with respect to transmission cost; in that context, the mesh configuration turned-out to be unnecessarily complicate in terms of protocols and less efficient in terms of energy budget; consequently, the much simpler and effective cluster three configuration represented in Fig. 3 was selected. The basic elements of the network are, the SN, the EN and the Router Node (RN). In this application only SNs and ENs were used. The GPRS unit is always connected to the GSM base station and transmits the gathered meteo-climate data down to 1 second rate (e.g wind). The ENs are normally in the low-power sleeping mode; they wake-up for a short time at 1 minute time interval, perform read-out of the VOC sensor and forward the gathered concentration data to the SN unit, along with other climatic and diagnostic information.

7. System requirements

For hazardous and complex industrial sites, it is very important to have a monitoring tool with a whole range of features in an economically feasible way. In particular, when designing a monitoring network is necessary to take into account the following issues:

i. *Data grid*: in the presence of multiple diffuse sources (as for VOC in industrial sites), it is important to implement a grid monitoring network, in order to have simultaneously available data over the whole area of the plant. Correlation with meteorological parameters allows then to better interpret the data and identify major emission sources.
ii. *Real-time acquisition:* the availability of real-time and continuous data is relevant to detect and effectively manage emergencies that may occur within the perimeter of the plant.
iii. *Data rate:* it is important to have high sampling rate (i.e. samping interval of one minute or less) to determine in detail for critical short-term situations and to address the best corrective actions.
iv. *Scalability and reconfigurability:* network scalability and reconfigurability are key issues, in particular in complex industrial sites; in addition to deploy fixed stations (e.g. on the perimeter of the plant), it can be useful to move the monitoring stations in specific areas during critical process phases with potential impact in terms of VOC emissions (eg. stop, start, revamping, etc.).
v. *Data rendering*: depending on the purpose of monitoring (emergency management, monitoring air quality, etc.) it is useful to make available real-time VOC concentration data as well as statistical index or cumulative parameters. This solution can be effective in terms of cost/benefit if specific information for a particular compound is not required.
vi. *Detection threshold*: if the purpose of monitoring is not only the management of emergency situations but also the evaluation of mean VOC concentrations or specific substances (for example, using the fixed monitoring stations), the choice should fall on detectors able to collect data at concentration levels in the order of ppb (as already mentioned, the air quality limit value for benzene in ambient air is about 1.5 ppb).
vii. *Communication*: the use of wireless stations connected to web-based graphical interface allows to significantly reduce operating costs, infrastructure and personnel involved.

8. System implementation

Based on the previous requirements the WSN-based VOC monitoring system prototype was implemented and tested at *Mantova, Italy,* petrochemical plant.

The aim was to test a new distributed instrument for collecting VOC emission data in real-time with a high degree of flexibility and scalability, thus transferable to other monitoring stations as needed, reconfigurable, in terms of data acquisition strategies, and economically sustainable as compared to traditional fixed monitoring stations.

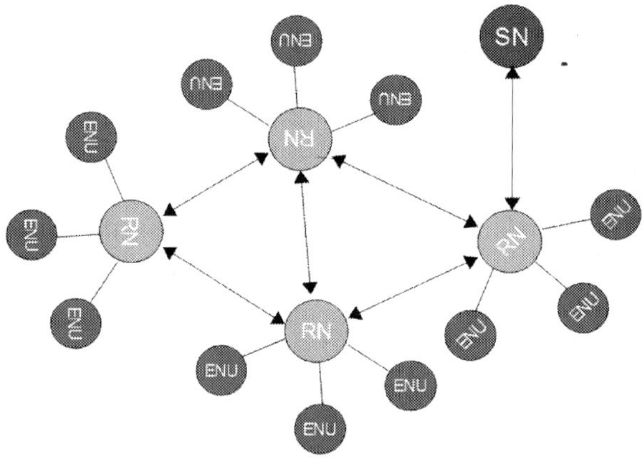

Fig. 3. The hybrid cluster-three network configuration

Critical locations were identified along the perimeter of the industrial sites, and within some specific relevant internal areas potentially involved in emissive processes. Seven SNs and 10 ENs to be described in the following have been deployed so far.

8.1 The SN unit

Each SN unit typically consists of the five components such as sensor unit, analogue digital converter (ADC), central processing unit (CPU), power unit, and communication unit. Communication unit's task is to receive command or query and transmit data from CPU to outside world. CPU is the most complex unit; it interprets the command or query to ADC, monitors and controls power if necessary, processes received data and manages the EN wake-up.

The block diagram of the SN unit is represented in Fig. 4. It consists of a GPRS antenna and GPRS/EDGE quadriband modem, a sensor board, a wireless unit and a micro-controller ARM-9, operating at 96 MHz clock.

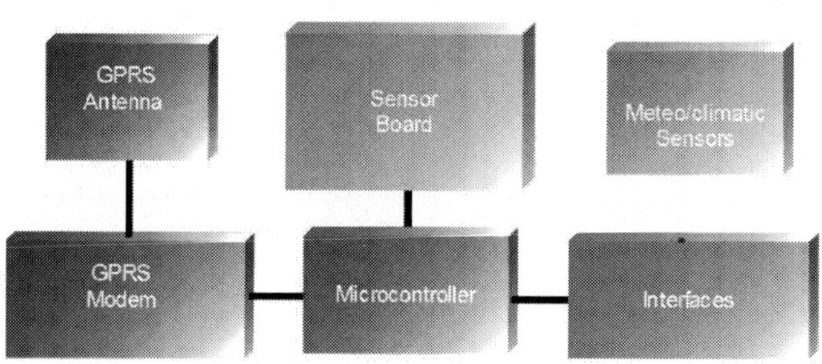

Fig. 4. Block diagram of the Sink Node Unit

The GPRS unit operates on the basis of a proprietary communication protocol over TCP/IP, with DHCP. Dynamic re-connectivity strategies were implemented to provide an efficient and reliable communication with the GSM base station. All the main communication parameters like, IP address, IP port (server and client), APN, PIN code and logic ID can be remotely controlled.

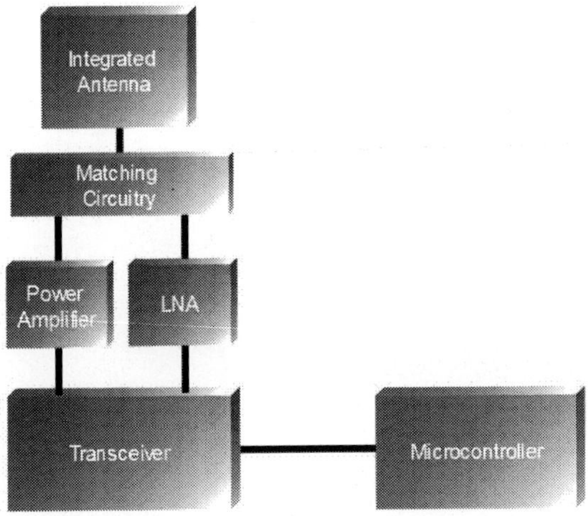

Fig. 5. Block diagram of the wireless interface

The system is based on an embedded architecture with high degree of integration among the different subsystems. The unit is equipped with various interfaces including LAN/Ethernet (IEEE 802.1) with TCP/UDP protocols, USB and RS485/RS422, in addition to a wireless interface, which provides short range connectivity. The sensor acquisition board is equipped with 8 analogue inputs, and 2 digital inputs. The SN unit is also equipped with a Wireless Interface (WI), represented in Fig. 5, providing connectivity with the EN units. The WI operates in the low-power, ISM UHF unlicensed band (868 MHz) with FSK modulation, featuring proprietary hardware and communication protocols. Distinctive features of the unit are the integrated antenna, which is enclosed in the box for improved ruggedness, and a PA and LNA for improved link budget. The PA delivers some 17 dBm to the antenna, while the receiver Noise Figure was reduced to some 3.5 dB, compared with the intrinsic 15 dB NF of the integrated transceiver. As a matter of fact, a connectivity range in line-of-sight in excess of 500 meters was obtained.

This results in a reliable communication with low BER, even in hostile e.m. environments. The energy required for the operation of the unit is provided by a 80 Ah primary source and by a photovoltaic panel equipped with a smart voltage regulator. Owing to a careful low-power design, the unit could be powered with a small (20 W) photovoltaic panel for undiscontinued and unattended operation.

A picture of one of the SN unit installed at the Mantova plant is represented in Fig. 6, left. The battery and photovoltaic panel are clearly visible; the GPRS unit is the grey box close to

the photovoltaic panel, and the WI is the white box on the top. The wind sensor and the RHT sensor with the solar shield are also visible. A concrete plinth serves as base for the unit, thus avoiding the need of excavations, which could be troublesome in the context of the plant due to pollution and contamination issues.
A picture of an EN unit is represented in Fig. 6, right. The photovoltaic panel along with the power supply and sensor board units are visible in the middle, while the VOC detector unit, protected by a metallic enclosure, is visible at the bottom. Also in this case a concrete plinth serves as the base for the unit.

Fig. 6. SN (left) and EN (right) units installed in proximity of the pipeline and of the chemical plant

8.2 The EN unit

The block diagram of the EN is represented in Fig. 7; it consists of a WI, similar to that previously described, and includes a VOC sensor board and a VOC detector. The WI unit is visible on the pole-top. Additionally, that solution allows wired connectivity of multiple VOC unit to the same EN, thus increasing modularity and flexibility of the architecture. The acquisition/communication subsystem of the EN unit is based on an ARM Cortex-M3 32 bit micro-controller, operating at 72 MHz, which provides the required computational capability compatible with the limited power budget available.
To reduce the power requirement of the overall subsystem, two different power supplies have been implemented, one for the micro-controller and one for the peripheral units; accordingly, the microcontroller is able to connect/disconnect the peripheral units, thus preserving the local energy resources. The VOC detector subsystem, in particular, is powered by a dedicated switching voltage regulator; this provides a very stable and spike-free energy source, as required for proper operation of the VOC detector itself.

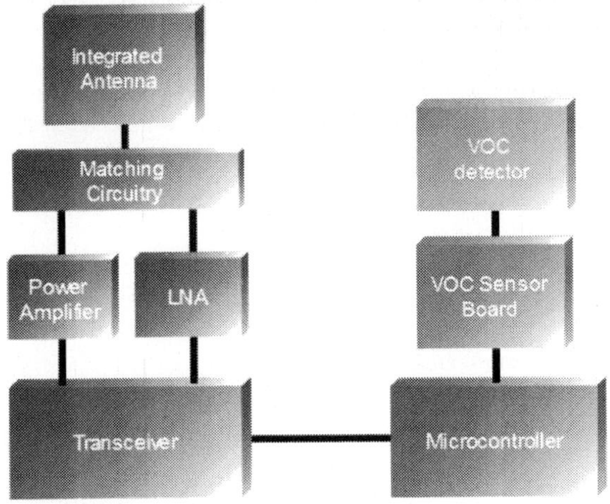

Fig. 7. Block diagram of the End Node Unit

The communication between EN unit and VOC detector board is based on a RS485 serial interface, providing high immunity to interference and bidirectional communication capability, as required for remote configuration/re-configuration of the unit.

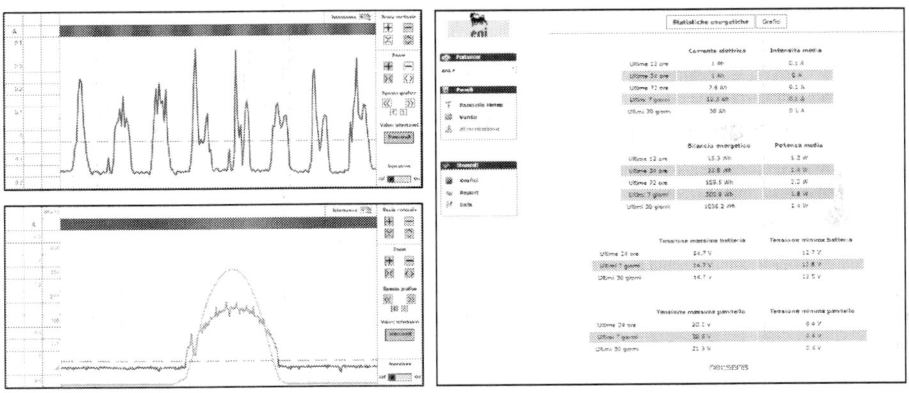

Fig. 8. Energy balance of the photovoltaic subsystem

Thanks to the efficient communication protocols and effective power management strategies, the EN unit has a battery life on some two months of continuous VOC detector operation at 1 minute transmission data-rate, only relaying on primary energy resources. The technologies described above allow for the implementation of monitoring procedures in different ways, namely real-time sampling, continuous or discontinuous measurement, VOC analysis with specific concentration of single compounds, to name a few.The secondary energy source plays a key role in ensuring the stand-alone and unattended operation of the sensor network infrastructure. The photovoltaic power supply unit includes a charge regulator which was specifically designed to provide maximum energy transfer efficiency from the panel to the battery under any operative condition. In Fig. 8 upper left, the weekly graph of the power absorbed/generated by the photovoltaic power supply is represented; the blue line represents the positive balance, i.e. the panel is charging the battery, while the red line represents the negative balance, i.e. the primary source is supplying energy to the subsystem. In Fig. 8, bottom left, a comparison between the current generated by the system and the solar radiation under very clean daylight condition is presented; the right sheet represents the energy budget statistics generated by the system for one of SN unit. In Fig. 8 right, a summary of the daily, weekly and monthly energy balance is represented; more detailed analysis and diagnostics are available.

9. The VOC detector

The VOC detector obviously plays a key role for the real-time monitoring system; the main requirements are listed in Table 1.

Operation mode	Diffusion (no pumped)
Targeted gas	VOCs IP> 10.6 eV
Concentration range (ppb)	2,5 to 5,000
Minimum Detectable Level (ppb)	> 2,5
Sensitivity	> 20 mV/ppm
Accuracy	< 5% in the overall range
Linearity	n.a.
VOC data sampling int. (minutes)	< 15
Power consumption (mW)	< 200
Stabilisation time from power-on T_{90} (s)	< 60
Warm-up time (s)	< 60
Interval between services (days)	> 120
Lifetime (years)	> 5
Specificity to benzene	typically broad band

Table 1. VOC detector requirements

Inspection of Table 1 shows very demanding requirements; an extensive analysis of the state-of-the-art of VOC detectors available on the market was performed to identify the most suitable technology. Different candidate technologies were considered, including Photo Ionisation Detector (PID), Amperometric Sensors, Quartz Crystal Microbalance (QMC) sensors, Fully Asymmetric Ion Mobility Spectrography (FAIMS) based on MEMS, Electrochemical Sensors and Metal Oxide Semiconductor Sensors (MOSS).

It turned-out that PID technology fitted quite well to the requirements of Table I, and thus it was elected as the basic technology to be used for this application. The device chosen for this application was he Alphasense AH, which exhibits 5ppb (isobutylene) minimum detection level.

Both theoretical and experimental investigations of PID operation were carried-out to assess the technology. Two major issues were identified, capable of potentially affecting the use the PID in our application; the first was that in the low ppb range the calibration curve of the PID is non-linear; this would require an individual, accurate and multipoint calibration with inherent cost and complexity; the second was that, when operated in diffusion mode at low ppb and after a certain time of power-off, the detector requires a stabilisation time of several minutes, thus preventing from operating it at minutes duty-cycles.

As for the calibration issue, a linearisation procedure was developed based on a behavioural model of the PID[2]; accordingly, the voltage read-outs received by the detector, V_n, are prior preprocessed by multiplying with a non-linearity compensation factor, $a(C)$, function of the concentration C:

$$V_{cn} = \alpha(C_n)V_n = S_v C_n \tag{1}$$

where V_{cn} is the read-out corrected by the non-linearity compensation factor a, C_n is the concentration in ppm and V_n is the nth read-out in mV, and S_v is the PID sensitivity in mV/ppm. Equation (1) shows that, after compensation, the values V_{cn} can be easily mapped in the corresponding concentration value.

In Fig. 9 and 10 the linearised calibration curves in the range 0-500 ppb are presented for two different PIDs. Fig. 9 represents the experimental calibration curve (read-out vs concentration) of a PID with a relatively high sensitivity, 150 mV/ppm. The non-linearity in the range 0-200 ppb is clearly observed, blue line.

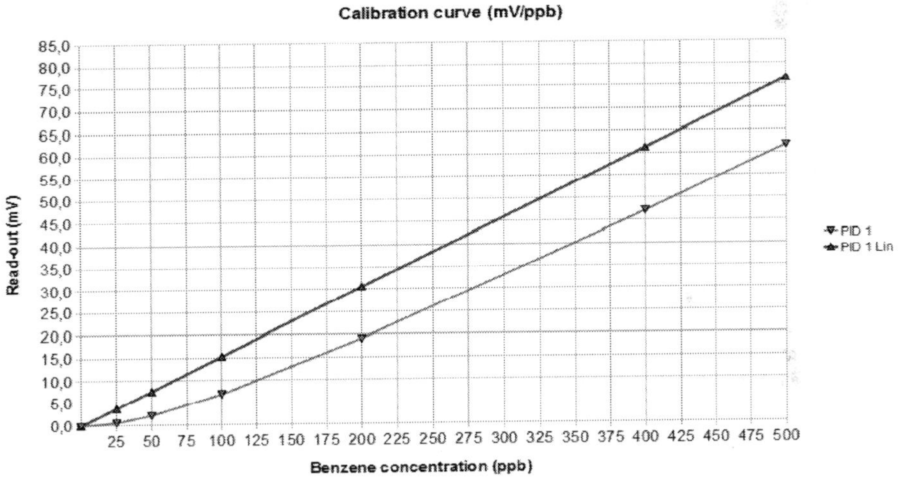

Fig. 9. Calibration curves for a PID with high sensitivity before (blue) and after (red) linearisation

The result of the linearisation process, according to the previously outlined procedure, is represented by the red line. Fig. 10 represents the same as Fig. 9 for a PID with relatively low sensitivity (50mV/ppm). In both cases, the linearisation procedure proved to be effective. The main advantage of the described approach is that for performing the PID calibration, one single parameter is needed, i.e. the value of the PID sensitivity, which is measured at ppm concentrations; this makes much simpler and less costly the calibration process.

---

[2] GF Manes, unpublished results

As for the stabilisation time, several experiments were performed to qualify the PID performance; it was found that at low concentration (tens or hundreds ppb), which represents the area of operation of the VOC detectors in our application and when operated in the diffusion mode, the PID exhibits a stabilisation time of some minutes after a power-off/power-on cycle. A typical PID duty cycled response after storage is represented in Fig. 11. The experimental stabilisation curve is compared with a 80 s decay-time exponential function showing an excellent fitting. After a warm-up of several hours the PID was powered-off for 15 minutes and then powered-on again; thie sequence simulated a 15 minute sampling interval, which was the initial target of our application; in this experiment ambient concentration was around 50 ppb, which represents the average concentration where the PID is supposed to be set up.

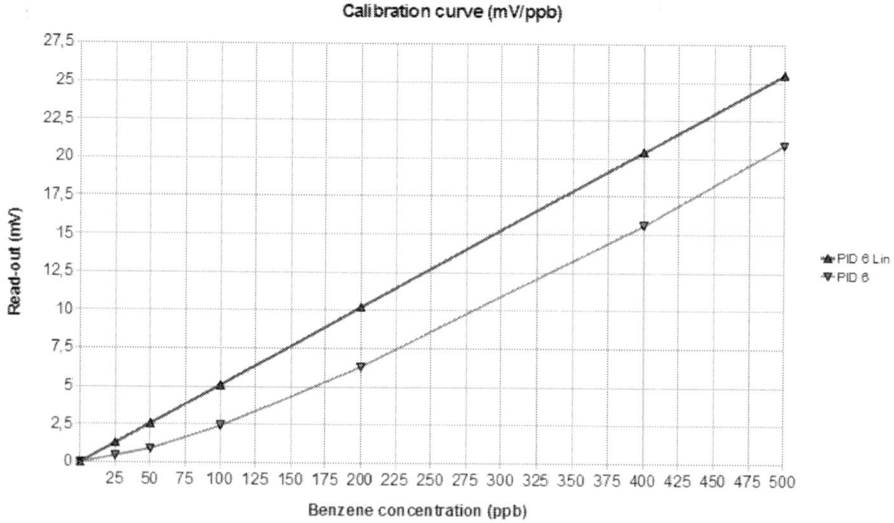

Fig. 10.Calibration curves for a PID with low sensitivity before (blue) and after (red) linearisation

As observed in Fig. 11, a 300 seconds stabilisation time is needed prior the PID can reach a stable read-out value. This experiment shows that a 15 minutes sampling interval calls for a 5 minutes stabilisation time, thus resulting in some 30% duty-cycle. A duty-cycled operation, as compared with a continuous power-on operation, is desirable in principle to prolong both the battery- and lamp-life; however, the benefit of energy saving allowed for by the 30% duty cycle is marginal, when compared with the advantage of achieving a more time-intensive monitoring of VOC concentration, as provided by continuous power-on operation. In terms of energy resources, continuous power-on operation requires some 35 mAh charge, which corresponds to 1 month of full operation with a 30 Ah primary energy source; the corresponding power consumption of 360 mW@12 Vdc can be balanced using a 5 W photovoltaic panel.

The UV lamp expected life is more than 6000 hours of continuous operation; we expect at least a quarterly service for the PIDs, due to environment contamination and related lamp efficiency degradation. For those reasons it was decided to operate the PID in continuous operation mode.

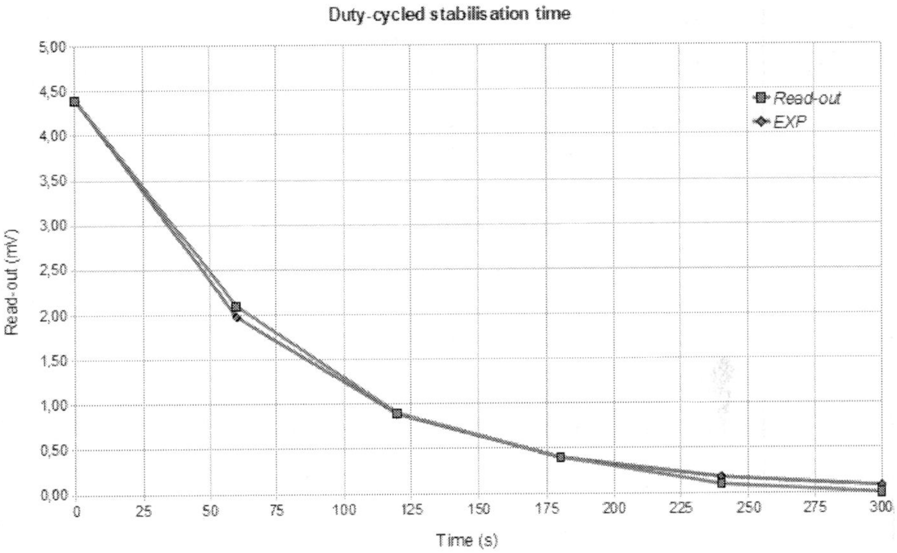

Fig. 11. PID stabilisation curve on duty-cycled power-on

10. Experimental results

Data from the field are forwarded to a central database for data storage and data rendering. A rich and proactive user interface was implemented, in order to provide detailed graphical data analysis and presentation of the relevant parameters, both in graphical and bi-dimensional format. Data from the individual sensors deployed on the field can be directly accessed and presented in various formats by addressing the appropriate sensor(s) displayed on the plant map, see Fig 12 left.

The position of each SN and EN unit is displayed on the map; by positioning the mouse pointer over the corresponding icon, a window opens showing a summary of current parameter values.

A summary of the sensor status for each deployed unit can be obtained by opening the summary panel, Fig. 12, right. The summary panel reports current air temperature/humidity values, along with min/max values of the day (left lower, in Fig. 12), wind speed and direction (left upper, in Fig. 12), and VOC concentration (right, in Fig. 12), in the last six hours. A graphic representation of data gathered by each sensor on-the field can be obtained by opening the graphic panel window, see Fig. 13.

The graphic panel allows anyone to display the stored data in any arbitrary time interval in graphic format; up to six different and arbitrarily selected sensors can be represented in the same graphic window for purpose of analysis and comparison.

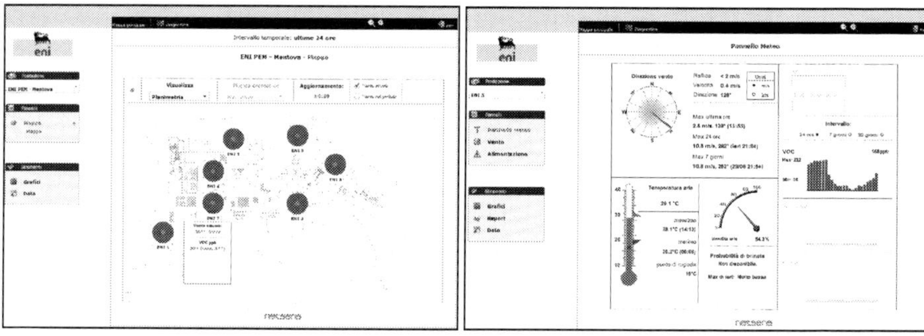

Fig. 12. Plant lay-out and details of the sensors

In Fig. 13 left, the VOC concentration traces of three different detectors are represented in a period of one day; in Fig. 13 right, the same data are displayed in a period of 30 days. By using the pointer, it is possible to select a time sub-interval and to obtain the corresponding graphic representation at high resolution.

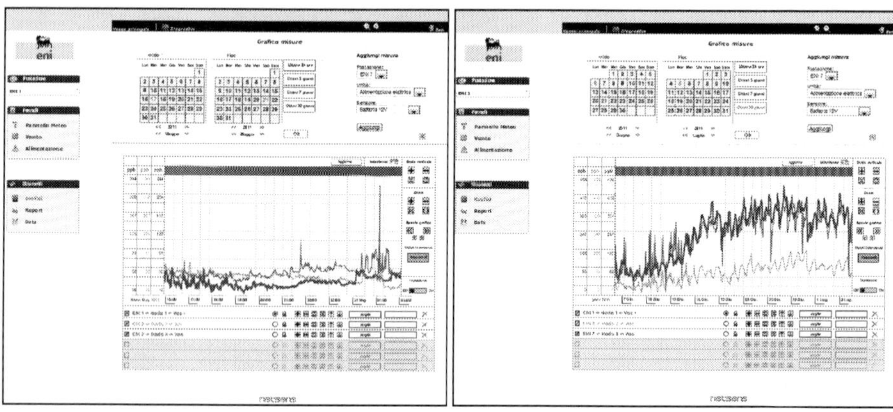

Fig. 13. Representation of sensor data in graphic format

In Fig. 13 left, the VOC concentration background is around 50 ppb; thanks to the very intensive sample-interval, 1 minute, the evolution of the concentration in time, along with other relevant meteo-climatic parameters can be very accurately displayed; it should be noted that the spikes which can be observed in the blue trace, Fig. 13 left, have a duration of some 3 minutes. The multi-trace graphic feature is very useful to perform correlation between different parameters. In Fig. 14 two examples of correlation between WSD and VOC concentration are shown. In Fig. 14 left, the VOC concentration, green line, exhibits a night/day variation; this is compared with the wind speed, rosé line, which increases during the day hours and decreases during the night hours, very likely due to the thermal activity. As it can be observed, in fact, wind speed and VOC concentration are in phase opposition, i.e. the greater the wind speed, the lower the average VOC concentration in the plant, that is in good agreement with what one can expect.

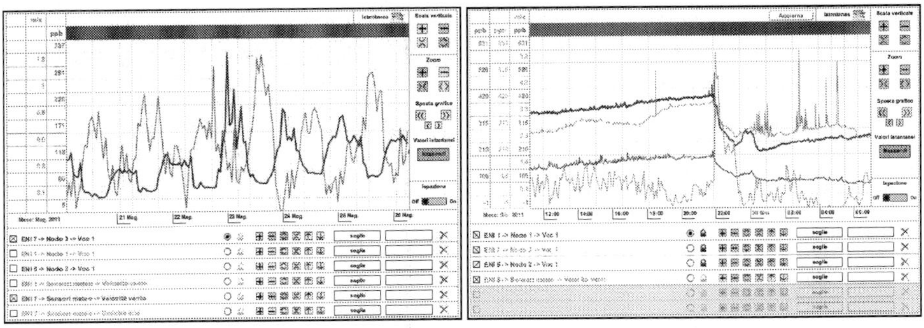

Fig. 14. Correlation between wind speed and VOC concentration

The effect of a sudden wind speed increase, light green line, is shown on the right graph of Fig. 14 right. It can be observed a wind speed increases to some 5m/s and more, green line, around 10 pm; accordingly, the VOC concentration detected by the three PIDs deployed in the plant is suddenly decreased. It should be noted that the three PIDs are located several hundred meters far apart each other.

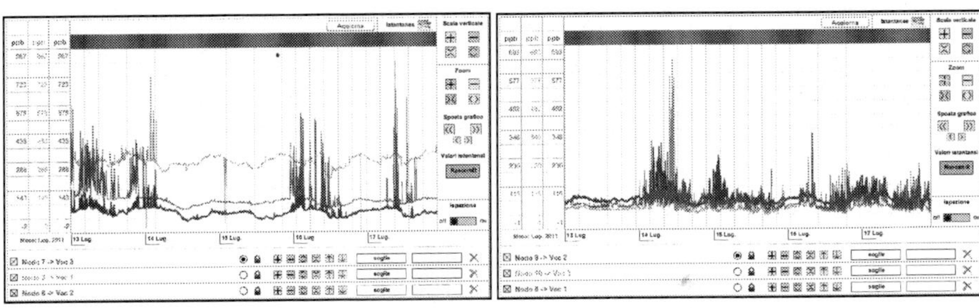

Fig. 15. Multi-trace read-outs of the six VOC sensors deployed around the ST40 plant

In Fig. 15, the read-outs of the 6 VOC sensors deployed around the ST40 plant are represented; it should be noted the very good uniformity among the background concentration levels demonstrating the effectiveness of the calibration procedure.

The user interface can perform various statistics on the data items; in the graphic panel, the user can enter the inspection mode, see the button on the lower right in Fig. 16, and set an user defined inspection window (in white); the window can be set over an arbitrary time interval; parameters like max/min, arithmetic mean and maximum variation can be then obtained for each of the sensor represented in the graphic window, lower right.

The sensitivity of the PID sensor is demonstrated in Fig. 17, where the traces of two different PIDs are shown. The PIDs are located some 500 meters far apart. At the time of data recording, there were some maintenance works going on in the plant's area.

The VOC components due to maintenance works were detected by the PIDs and recorded as small variation of the concentration around the mean value during the working hours (from 8 am to 6 pm, roughly), to be compared with the more smoothed traces recorded during the night. A diagnostic panel is available to evaluate the system Quality of service (QoS) and the gathered data reliability, see Fig. 18; connectivity statistics are displayed along with the current status of connectivity for each of the SN and EN units. The status of the GPRS connectivity and the related statistics are represented in column 3 and 6 from left, respectively.

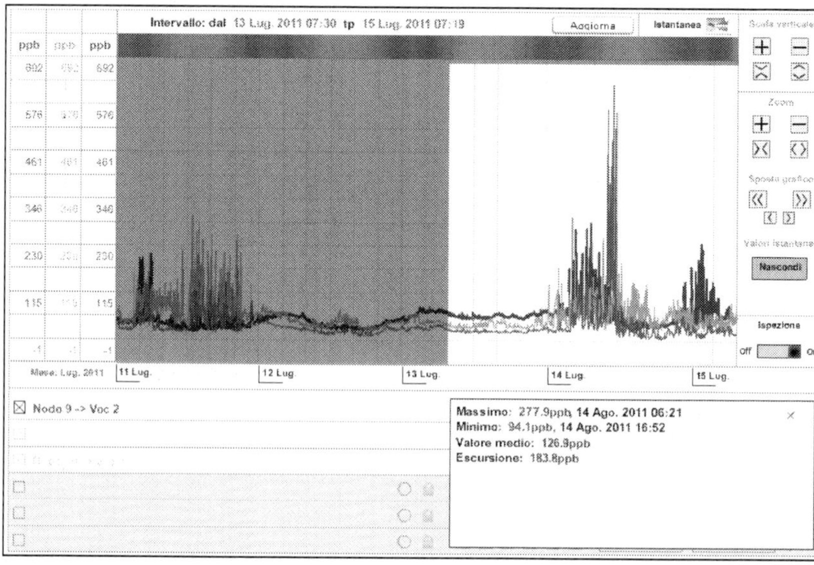

Fig. 16. Statistical parameters analysis

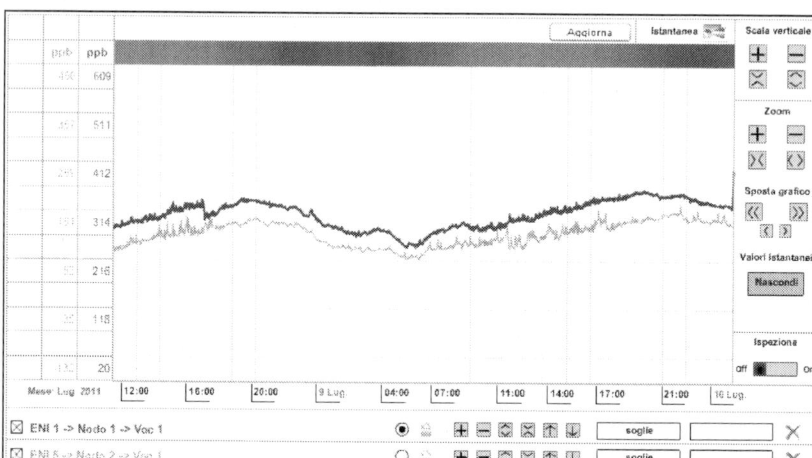

Fig. 17. Day/night VOC read-outs

As it can be observed, GPRS connectivity in excess of 99% is obtained, because of the periodic restart of the SN unites which do not get connected for a short time interval, and thus reducing

the overall GPRS efficiency figure. EN unit status and connectivity are displayed in the columns 4 and 9 from left, while power supply status is showed in column 5 from left.
The diagnostic panel identifies any lack of connectivity and/or reliability of each single SN or EN unit for immediate service action.

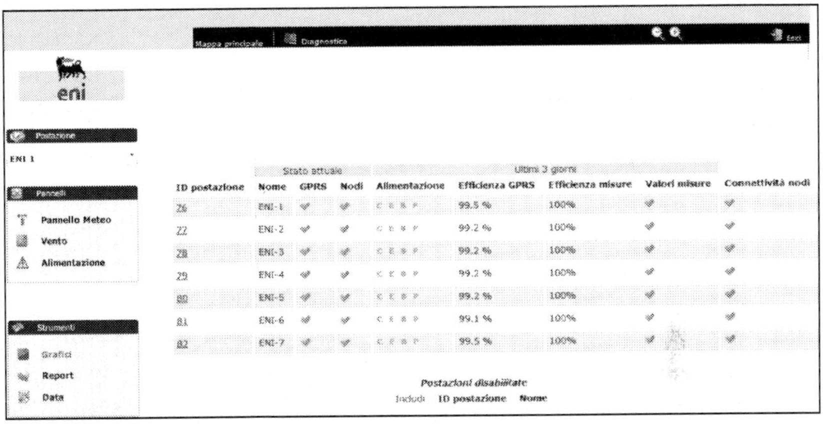

Fig. 18. The diagnostic panel

In addition to the graphic format, data items can be represented in a bi-dimensional format. It is quite difficult to correlate the data in graphic format from different sensors deployed over the plant; a helpful bi-dimensional picture of the area based on an interpolation of algorithms has been implemented, resulting in a very synthetic representation of the parameters of interest over the plant in pseudo-colours. The sensors are basically punctual and, thus, are only representative of the area in their proximity. For that reason the interpolation would be only effective if an adequate number of sensors is deployed on the field, so that the area is subdivided into elementary cells, *quasi- homogeneous* in terms of the parameter values.
This requirement would result in an unnecessarily high number of units to be deployed. A more effective approach is to take into account the morphology and functionality of the different areas of the plant and deploy the sensors accordingly.
As for the VOC, by instance, the potential sources of VOC emissions in the plant are located in well identified areas like, the chemical plant and the benzene tanks; accordingly, the deployment strategy includes a number (6) of VOC sensors surrounding the chemical plant infrastructure, thus resulting in a virtual fence, capable of effectively evaluating VOC emissions on the basis of the concentration pattern around the plant itself.
As for wind speed and direction, which are relevant for correlation with VOC concentration, on the basis of an evaluation of the plant infrastructures, the areas of potential turbulence were identified and the wind sensors were deployed accordingly. Both SN and EN units were equipped with RHT sensors, whose cost is marginal. In Fig. 19 two bidimensional pictures of the temperature (left) and RH (right) in the area of the plant are represented.
Not surprisingly, both temperature and RH are not uniformly distributed; according to the colour scale of air temperature blue means lower temperature and red means higher temperature; in this case the temperature ranges from 28°C (blue) to 31°C (red). Two areas of higher temperature are clearly identified, one on the left around the chemical plant ST40

and the other on the right around the arrival of the pipeline; this is obviously related to the mechanical activity in those areas. The thermal distribution also influences the air RH as demonstrated in Fig. 19, left. In this case the grey colour means lower RH and the blue colour means higher RH.

The RH values range from 26% to 33%, in this case. The temperature gradient among the different areas in the plant, which in some cases grew to up 5°C, is responsible of some thermal activity possibly affecting the VOC concentration distribution.

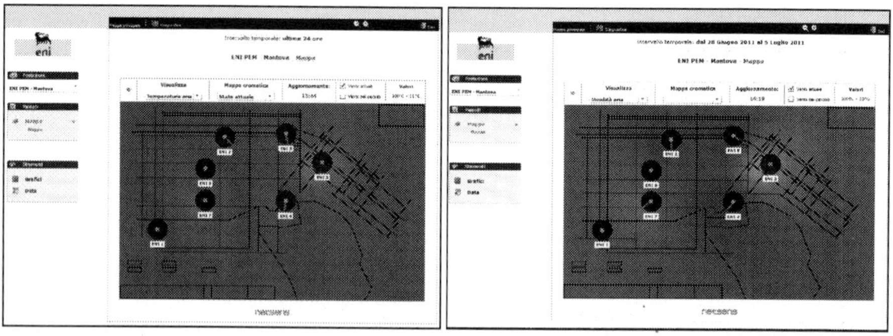

Fig. 19. Bi-dimensional map of air temperature (left) and air RH (right) distribution in the area of the plant

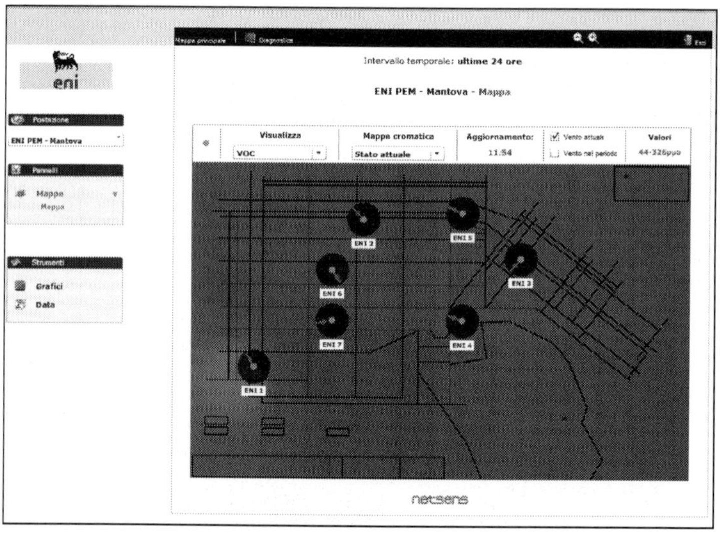

Fig. 20. Bi-dimensional map representing VOC concentration in the plant

VOC concentration is mapped in Fig. 20 in pseudo-colours. In this case blue denotes lower concentration, while red denotes higher concentration; it should be emphasized that the red colour has no reference with any risky or critical condition at all, beings only a chromatic option.

As it can be noted, wind direction represented by blue arrows is far by being uniform over the plant, thus denoting turbulences due to the plant infrastructures and surrounding vegetation.

11. Conclusions

An end-to-end distributed monitoring system integrating VOC detectors, capable of performing real-time analysis of gas concentration in hazardous sites at unprecedented time/space scale, has been implemented and successfully tested in an industrial site

The aim was to provide the industrial site with a flexible and cost-effective monitoring tool, in order to achieve a better management of emergency situations, identify emission sources in real time, and collect continuous VOC concentration data using easily re-deployable and rationally distributed monitoring stations.

The choice of collecting data at minute time interval reflects the need to identify short term critical events, quantify the emission impacts as a function of weather conditions and operational process, and identify critical areas of the plant.

The choice of a WSN communication platform gave excellent results, above all the possibility to re-deploy and re-scale the network configuration according to specific needs, while greatly reducing installation cost. Furthermore, to manage real-time data through a web based interface allowed both adequate level of control and quick data interpretation in order to manage critical situations.

Among the various alternatives available on the market, the choice of PID technology proved to meet all the major requirements. PIDs are effective in terms of energy consumption, measuring range, cost and maintenance, once installed in the field. The installation of weather sensors at the nodes of the main network stations allowed for a better understanding of on-field phenomena and their evolution along with clearer identifcation of potential emission sources.

Future activity will include a number of further developments, primarily the development of a standard application to allow the deployment of WSN in other network industries (e.g. refineries) and an assessment of potential applications for WSN infrastructure monitoring of other environmental indicators.

12. Acknowledgement

This work was supported by eni SpA under contract N.o 3500007596. The authors wish to thank W O Ho and A Burnley, Alphasense Ltd., for many helpful comments and clarifications concerning the PID operation, S Zampoli and G Cardinali, IMM CNR Bologna, for many discussions on PID characterisation and E Benvenuti, Netsens Srl, for his valuable technical support.

Assistance and support by the Management and technical Staff of Polimeri Europa Mantova is also gratefully acknowledged.

13. References

Adler R.; Buonadonna, P. Chhabra, J. Flanigan, M. Krishnamurthy, L. Kushalnagar, N. Nachman, L. & Yarvis M. (2005). *Design and Deployment of Industrial Sensor Networks: Experiences from the North Sea and a Semiconductor Plant* in ACM SenSys, November 2-4, 2005, San Diego, CA.

Alphasense Ltd.; Application Note AAN 301-02

Dargie W.; & Poellabauer, C. (2010). *Fundamentals of wireless sensor networks: theory and practice.* John Wiley and Sons, ISBN 978-0-470-99765-9, 168–183, 191–192

EC Working Group on Guidance for the Demonstration of Equivalence, *Guide to the Demonstration of Equivalence of Ambient Air Monitoring Methods*, January 2010

European Commission, *Integrated Pollution Prevention and Control (IPPC): Reference Document on Best Available Techniques for Mineral Oil and Gas Refineries* , February 2003

European Commission, *Integrated Pollution Prevention and Control (IPPC): Reference Document on Best Available Techniques in the Large Volume Organic Chemical Industry*, February 2003

European Commission, *Integrated Pollution Prevention and Control (IPPC): Reference Document on the General Principles of Monitoring*, July 2003

European Parliament and Council, *DIRECTIVE 2008/50/EC on ambient air quality and cleaner air for Europe*, 21 May 2008

European Parliament and Council, *DIRECTIVE 2010/75/EU on industrial emissions (integrated pollution prevention and control)*, 24 November 2010

ISPRA, *Database of historical emissions of main pollutants in Italy by sectors*

J. Jeong J.; Culler. D.E & Oh. J. H. (2007). *Empirical analysis of transmission power control algorithms for wireless sensor networks* in Proc. 4th Intl. Conf. on Networked Sensing Systems (INSS '07), Piscataway, NJ: IEEE Press, 2007, pp. 27-34.

Karl, H.; & Willig, A. "Protocols and Architectures for Wireless Sensor Networks", Wiley, 1st Edition.

Locke D.C.; & Meloan, C. E. (1965). *Study of the Photoionisation Detector for Gas Chromatography*, in Vol. 37, No. 3, March 1965 pp. 389-397.

Lorincz K.; Malan, D. Fulford-Jones. T.R.F. Nawoj. A. Clavel A. Shnayder, V. Mainland, G. Moulton. S. & Welsh M (2004). *Sensor Network for Emergency Response: Challenges and Opportunities"* In IEEE Pervasive Computing, Special Issue on Pervasive Computing for First Response, Oct-Dec 2004.

Pakzad S. M.; Fenves, G. L. Kim, S. & Culler. D. E. (2008). *Design and Implementation of Scalable Wireless sensor Network for Structural Monitoring.* In ASCE Journal of Infrastructure Engineering, March 2008, Volume 14, Issue 1, pp. 89-101.

Price J. G. W.; Fenimore. D.C. Simmonds, P.G. & Zlatkis A. *(1968). Design and Operation of a Photoionization Detector for Gas Chromatography*, in Analytical Chemistry, Vol. 40, No. 3, March 1968, pp. 541, 547.

R. Szewczyk R.; Mainwaring, A. Polastre, J. & Culler, D. E. (2004). *An Analysis of a Large Scale Habitat Monitoring Application.* ACM Conference on Embedded Networked Sensor Systems (SenSys), November 2004.

Sohraby, K.; Minol, D. & Znati, T. (2007). *Wireless sensor networks: technology, protocols, and applications.* John Wiley and Sons, 2007 ISBN 978-0-471-74300-2, pp. 203–209

Stoianov I.; Nachman, L. & Madden, S. (2007). *PIPENET: A Wireless Sensor Network for Pipeline Monitoring* IPSN'07, April 25-27, 2007, pp. 264-273 Cambridge, Massachusetts, U.S.A.

15

Land Degradation of the Mau Forest Complex in Eastern Africa: A Review for Management and Restoration Planning

Luke Omondi Olang[1] and Peter Musula Kundu[2]
[1]*Department of Water and Environmental Engineering, School of Engineering and Technology, Kenyatta University, Nairobi,*
[2]*Department of Hydrology and Water Resources, University of Venda, Thohoyandou,*
[1]*Kenya*
[2]*South Africa*

1. Introduction

The Mau Forest Complex is the largest closed-canopy montane ecosystem in Eastern Africa. It encompasses seven forest blocks within the Mau Narok, Maasai Mau, Eastern Mau, Western Mau, Southern Mau, South West Mau and Transmara regions. The area is thus the largest water tower in the region, being the main catchment area for 12 rivers draining into Lake Baringo, Lake Nakuru, Lake Turkana, Lake Natron and the Trans-boundary Lake Victoria (Kundu et al., 2008; Olang & Fürst, 2011). However, in the past three decades or so, the Mau Forest Complex (MFC) has undergone significant land use changes due to increased human population demanding land for settlement and subsistence agriculture. The encroachment has led to drastic and considerable land fragmentation, deforestation of the headwater catchments and destruction of wetlands previously existing within the fertile upstream parts. Today, the effects of the anthropogenic activities are slowly taking toll as is evident from the diminishing river discharges during periods of low flows, and deterioration of river water qualities through pollution from point and non-point sources (Kenya Forests Working Group [KFWG], 2001; Baldyga et al., 2007). Augmented by the adverse effects of climate change and variability, the dwindling land and water resources has given rise to insecurity and conflicts associated with competition for the limited resources. It is hence becoming urgently important that renewed efforts are focused on this region to avail better information for appropriate planning and decision support.

Such a process will nonetheless, require an integrated characterization of the changing land and water flow regimes, and their concerned socio-economic effects on resource allocation and distribution (Krhoda, 1988; King, et al., 1999). Assessing the impacts of the environmental changes on water flow regimes generally require provision of time series meteorological, hydrological and land use datasets. However, like in a majority the developing countries, the MFC does not have good data infrastructure for monitoring purposes (Corey et al., 2007; Kundu et al., 2008). A majority of research studies in the area have relied on low resolution land cover datasets, including approximate physically-based procedures to understand the space and time surface alterations. Renewed efforts are thus underway in the MFC at present in order to avail high resolution information to be used for updating the existing databases with a view of improving future forecasts for restoration management as shown in Figure 1. Datasets from relevant research organization such as the World Agro-forestry Centre (ICRAF), Regional Centre for Mapping of Resources for Development (RCMRD), Regional Disaster Management Center of Excellence (RDMCOE) and IGAD - Climate Prediction and Application Centre (ICPAC) are hence being harmonized for use in evaluating the environmental effects of spatial changes, especially within hotspot regions of the complex. Cost effective computer-based techniques, which can efficiently analyze diverse physically-based variables are also under consideration to enhance the application of appropriate distributed-based management interventions (Kundu, 2007; Olang, 2009).

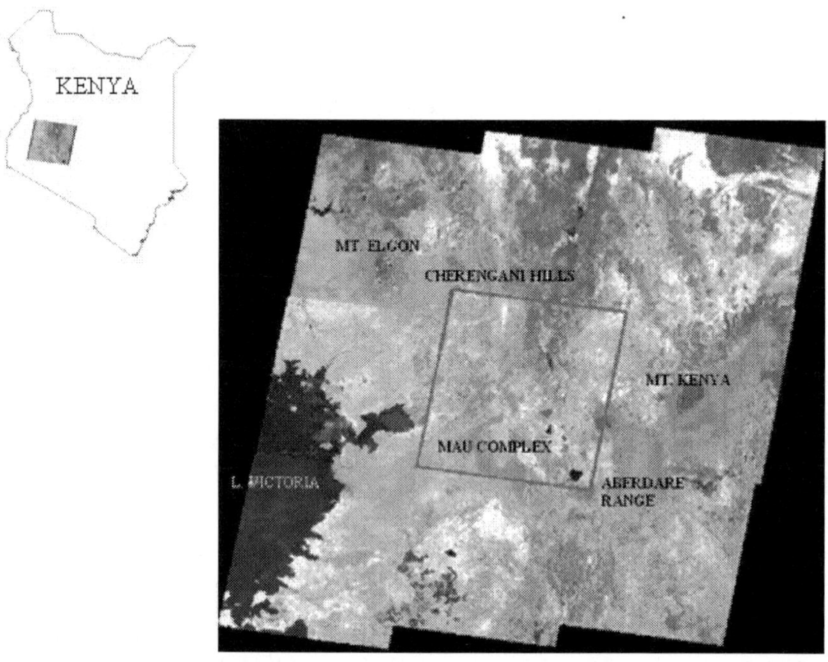

Fig. 1. Location of the five water towers of Kenya, including the MFC region (Mosaiced images of Landsat 2000).

Furthermore, with continued advancements in global remote Sensing (RS) and GIS monitoring techniques, it is increasingly becoming possible to evaluate detailed land cover change trajectories for improved resource management. Relevant contemporary alternatives such as automated extraction of geomorphologic and hydrologic properties from satellite derived Digital Terrain Models (DEM) can thus be undertaken as viable tools for model based simulation of relevant catchment-based properties. Already, there is a general consensus that for such spatial models to be used for successive impact analyses and decision support, the

results should provide detailed information with a good degree of confidence, and where possible, validated through a participatory approach involving ground measurements and indigenous knowledge (Liu et al., 2004; Refsgaard & Henriksen, 2004; Rambaldi et al., 2007).
Generally, most of the existing studies in the MFC were carried-out at catchment-scales with a view to determine the hydrological impacts of the environmental changes. Studies that catalog the land cover alterations to provide time-series trajectories for continued update of the existing water resources master plans are very few. In fact, the existing efforts are often isolated, unpublished and difficult to access to enhance synergistic research geared towards dependable restoration management. In this contribution therefore, the general ecology and deforestation patterns of the MFC are reviewed with the aim of consolidating and documenting the scattered information important for hinging the development of improved tools for sustainable land and water resource management. Emphasis is placed on the findings of previous works employed to monitor surface alterations as a fundamental component of land degradation in the susceptible MFC.

2. Environmental changes and land cover degradation

Environmental changes arise from the fact that most natural and artificial earth surface features are in a state of flux. The rate of these changes is quite often not uniformly distributed, but depends rather on the interactions of the biophysical and human components (Coppin et al., 2004; Jensen, 2005). The need for resource sustainability through proper management has today prompted timely and accurate monitoring of environmental changes to understand their relationships and interactions within a given ecosystem. However, monitoring environmental changes requires a deep understanding of the relevant environmental attributes over time and space to avoid simplistic representations. Common examples of environmental changes largely witnessed today in the developing countries include changes in forest characteristics due to human induced deforestation processes, ecological changes due to the need for agricultural expansion and land use/land cover changes due to factors related to human influences from increased population (Pellikka et al., 2004; Corey et al., 2007). In the last couple of years, significant attention has been given to land use and land cover changes, since they form a major component of global changes with greater impact than that of climate change (Foody, 2001; Olang et al., 2011). Such changes in land cover can be generally differentiated into land cover modification and land cover conversion. Land cover modification generally refers to the full substitution of one cover type by another, as is the case with urbanization.
In a majority of developing countries, land cover conversion which refers to gradual changes affecting the nature of the land cover but not their overall classifications are common. Such conversions may arise from the natural resilience of an ecosystem due to climatic variability and/or from complex land cover changes due to direct or indirect anthropogenic factors. Specifically in the MFC, both land cover modifications and conversions are predominant, and are largely attributed to the increasing human population pressure demanding more land for settlement, pasture and agriculture. This is further aggravated by the dire need for economic sustainance from the within vicinity natural resources without taking into account proper land use management practices. Forest degradation through charcoal burning followed by conversion of the deforested areas into subsistence agriculture is widespread in the headwaters catchments. In addition to this are the uncontrolled cattle grazing, slash and burn farming methods in the midland areas. With

continued diminishing economic alternatives for the rural population, more farms are being put under small scale subsistence agriculture to provide a means of a living for the riparian communities living in the forest complex.

3. The Mau forest complex

3.1 Physiography and geology

The major geomorphological features of the forest complex comprise of the escarpments, hills, rolling land and plains (Figure 2). The topography is predominantly rolling land with slopes ranging from 2% in the plains to more than 30% in the foothills. Geological studies have shown that the area is mainly composed of quaternary and tertiary volcanic deposits (Sombroek et al., 1980). The quaternary deposits include pyroclastics and sediments, and largely cover the Northern part of the complex. Tertiary deposits predominate in the southern parts, and include black ashes and welded tuffs. From field-measurements, the top soils in the plains are of clay loam (CL) to loam (L) in texture, with friable consistence and weak to moderate sub-angular blocky structure. The subsoil texture ranges from silty clay loam (SCL) to clay loam (CL) and clay (C), with pH values ranging from 5.6 to 6.4, making them slightly to moderately acidic in nature (China, 1993). In the upland areas however, the soils are largely of high content of silt and clay consequent of Ferrasols, Nitisols, Cambisols and Acricsols according to the Food and Agricultural Organisation of the United Nations (FAO-UN) soil classification procedure (World Soil Information [ISRIC]/FAO-UN, 1995). In the lowland, Luvisol, Vertisol, Planosol, Cambisol and Solonetz soils from the Holocene sedimentary deposits are primarily prevalent and occur in saline and sodic phases.

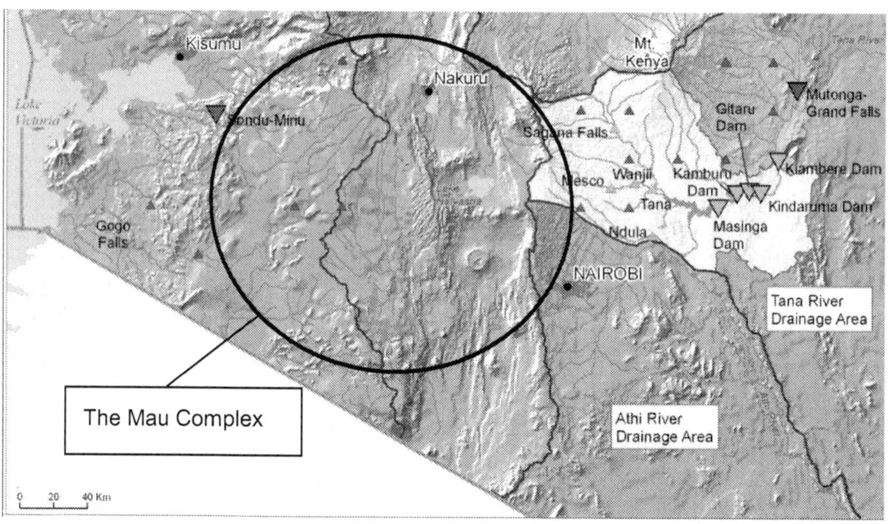

Fig. 2. Physical features, including the drainage network of the Mau Forest Complex (World Resources Institute, 2007).

Similar trends in the soil and geological characteristics of the area were also achieved with processed soils data obtained from the Global Environment Facility Soil Organic Carbon

(GEFSOC) project (FAO-UNESCO, 1998; Batjes & Gicheru, 2004). This dataset is available at a scale of 1:1M for Kenya, and is a modification of the original SOTER soils data of the International Society of Soil Science (ISSS). Other hydrological studies of the headwaters of the MFC have employed remotely sensed datasets to derive the geomorphological characteristics of the region (Kundu, 2007; Baldyga et al., 2007). A 3-Arc second grid based digital elevation model (DEM) acquired from the Shuttle Radar Topographic Mission was used in this context. Through computer aided procedures in a GIS, a raster analysis was performed to generate stream directions and networks, which matched very closely with the actual drainage patterns.

3.2 Climate

3.2.1 Rainfall

The climate of the Mau complex is largely influenced by the North - South movement of the Inter-tropical Convergence Zone (ITCZ) modified by local orographic effects. In terms of seasonality, the complex can be classified as trimodal, with the long rainy season predominant between the months of May and June and the short rainy season prevalent between the months of September and November. Generally, the complex receives an average annual rainfall of about 1300 mm on normal years devoid of climatic extremes such as the El Niño Southern Oscillation (ENSO). Mean monthly rainfall events in the range of 30 mm to over 120 mm are common (Figure 3).

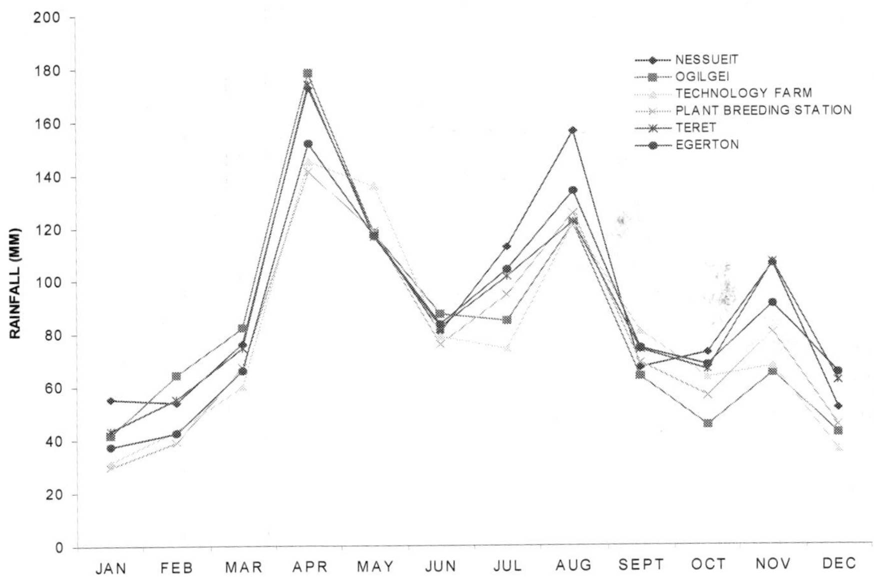

Fig. 3. Monthly rainfall distribution from six selected weather stations within the Mau Complex (Kundu, 2007).

There are a few pluviometric stations in the complex where quality rainfall data can be obtained from. However, due to the diverse topography of the area, the existing gauge density can be considered not sufficient for distributed representation of rainfall induced process. This coupled with uncertanities related to measurement errors and missing data, recent developments by the Kenya Meteorological Department have considered the use of satellite based rainfall estimates (RFE) such as the Tropical Rainfall Measuring Mission (TRMM) for concerned impact studies especially in large areas. In small areas however, RFE require regionalisation through calibration with observed point data to derive region-based adjustment coefficients (Borga, 2000; Krajewski et al., 2002).

3.2.2 Temperature and evapo-transpiration

The Mau Forest Complex generally falls in agro-climatic zones I, II and III when classified according to moisture-indices obtained from average evapo-transpiration rates and annual rainfall amounts. Because of its varied topography, estimation of the actual mean air temperatures for the whole area is often quite complicated. However, based on altitude zones, the monthly air temperature estimates for the basin are as provided in Table 1.

Altitude Zone (m)	Mean monthly air temperatures (°C)			Abs. minimum temp. (°C)
	Maximum	Minimum	Mean	
1 100 - 1 300	27 - 30	15 - 17	21 - 23	10 - 13
1 300 - 1 500	27 - 29	13 - 15	20 - 22	9 - 11
1 500 - 2 000	22 - 28	10 - 13	17 - 20	6 - 9

Table 1. Mean Annual air temperature ranges for different altitude zones

For estimation of localised evapo-transpiration rates, a majority of studies have employed empirical models that incorporate both physical and aerodynamic parameters. The most predominant due to its ability to closely approximate the crop reference evapo-transpiration (ET_o) rates is the FAO Penman-Monteith method (FAO, 1998, 2009). In many cases, the average evapotranspiration of the complex are estimated in relation to the existing land use types. In the entire complex, annual average estimates between 1.3mm/day to 4.2 mm/day, with an average of about 3.85 mm/day, have been recorded. The ETo has also been noted to increase with mean annual rainfall amounts, confirming that the complex is water stressed. The results are also consistent with the reduced infiltration rates owing to the loss of much of the vegetative cover in the area (Owido et al., 2003).

3.3 Hydrology

3.3.1 Drainage and stream network

The Mau Complex is drained mainly by 12 rivers including Rivers Njoro, Molo, Nderit, Makalia, Naishi, Kerio, Mara, Ewaso Nyiro, Sondu, Nyando, Yala and Nzoia. Space and time variations of the stream flows are normally influenced by the morphometry, lithology, land use/cover and rainfall patterns. Normally, the stream flow characteristics are potential indicators of the hydrological status of a region (Calder, 1998). In the last three decades, physical evidence has revealed that the rivers in the MFC have had significant decline in discharges, coupled by dwindling water quality. Other studies have also highlighted the changing hydrological response of the area consequent of the land use/land cover changes (Kundu et al., 2000; Owido et al., 2003). However, the effects associated with climate change cannot be fully ignored in this context as well.

So far, a majority of studies carried out in the MFC have focused more at catchment scales. It will be thus imperative to develop procedures that can be used to asses the hydrological water quality and quantity in the entire forest complex. Considering the weak infrastructural capacities, and hence the poor data quality, the application of remotely sensed datasets provides a great potential in monitoring and management of the area. Conventional Radio Detecting and Ranging (RADAR) based techniques for rainfall and moisture content estimation can be explored in this respect. A reliable and integrated database built in a GIS, can also be used to address various issues related to data management, especially for dependable land and water resource planning. Today, global datasets about surface elevation and land cover characteristics can be freely acquired to enhance simulation studies. Typical synthetic drainage network can equally be derived from Digital Elevation Models (DEM) of the area. Figure 4 illustrates the integration of RS and GIS that could be used for managing the MFC by using freely acquired images. The results generally compared well to the reality despite the medium resolution of the images. With further processing, supported by secondary datasets from high resolution satellite images, it is possible to improve the quality of the derived stream-networks.

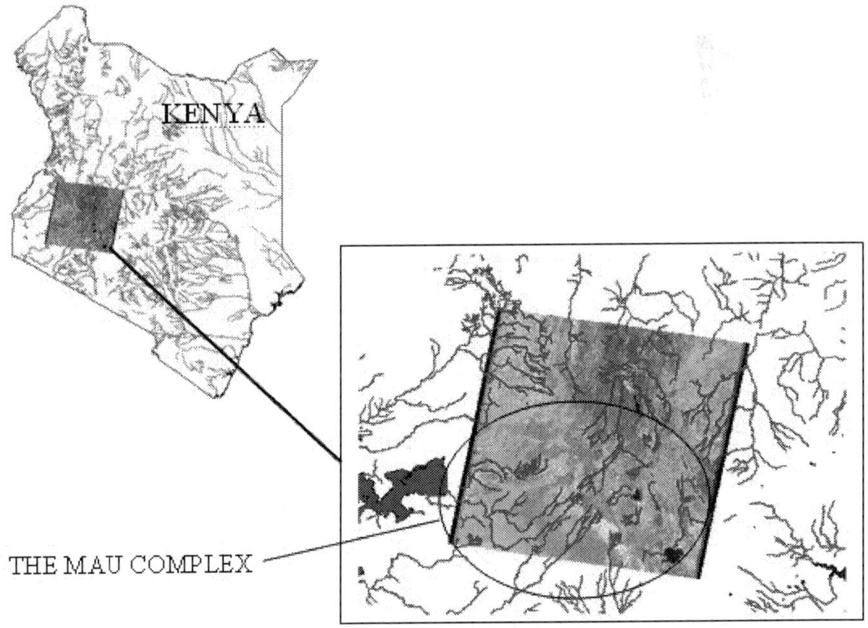

Fig. 4. Drainage network within the Mau Forest Complex.

3.3.2 Velocity and discharge

Normally, to elucidate the mechanisms driving discharge variability, a study period of not less than 30 years is usually recommended. This period is considered long enough to discern effects consequent of either land use changes or climate change and variability. Land use change effects are commonly investigated by understanding the discharge regimes during rainy seasons especially. Depending on the magnitude of the spatial changes however, utmost care must be taken since studies in Kenya have also shown that the effects of land cover change on runoff events tend to diminish with increased magnitude of storm events (Olang & Fürst, 2011). Chemelil (1995) and Kundu et al. (2008) assessed the mean rainfall and discharge characteristics of the Njoro River located in Eastern Mau to understand the influence of environmental changes in the area. The authors used a similar procedure provided by Marcos et al. (2003). The results obtained are illustrated in Figure 5.

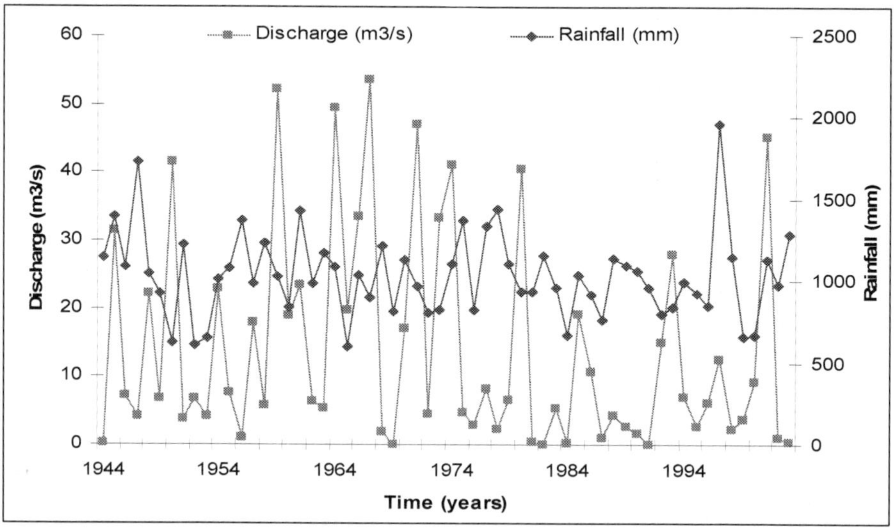

Fig. 5. Long-term annual rainfall and discharge relationship between 1944 and 2001.

From the figure, discharges in the area showed a decreasing trend against a rather consistent rainfall pattern. The frequency of low flow was noted to have increased, especially in the interval between 1980 and 2000. This trend was largely attributed to the changing land use patterns, considering that this period witnessed the highest human encroachment within the MFC. It was hence more likely that the unplanned conversion of forest and woodlands into agriculture and built up area within the headwaters could have influenced the discharge generation processes and hydrological regimes of the area. Other reasons associated with this were increased water abstraction for irrigation and domestic purposes consequent of the rising population.

3.4 Land cover and land use

The Mau complex has the largest indigenous montane forest covering an area of about 2700 km^2 at present. Vegetation in the area varies largely from grasslands with scattered trees in the plains, to shrubland and forests in the hilly uplands. In the higher mountain ranges, bamboo forests are largely predominant. The vegetation around the rivers and lakes mainly comprises of acacia trees and dense bush and shrubs. The escarpments are mainly wooded and bushy with a wide ecological diversity. The higher areas comprise of forests with acacia *xanthophloea, Olea hochstetteri, Croton dichogamus, euphorbia candelabrum* forest and bush land. Previously, the area was largely covered by rich evergreen forests, extending from the highlands of Mau hills, and woodland dominated by acacia trees in the plains. A Landsat image (P169R60) in bands 4-3-2 for February 2000 showing the predominant vegetation and land cover of the Mau complex are provided in Figure 6. The continuous deep red represents highland vegetation on the escarpments and hill ranges. The green-blue patches largely represent open pasture and grassland, bare soil, roads, built-up areas and farms after harvesting.

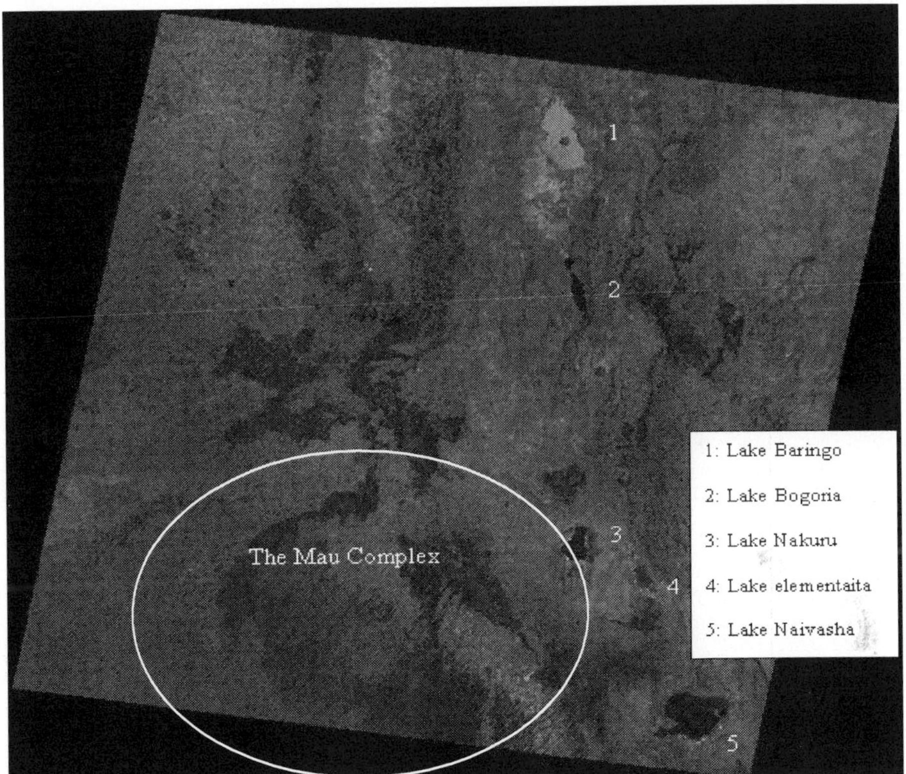

Fig. 6. Landsat image for February 2000 showing Land cover and Rift Valley lakes.

Within the blocks of the MFC, the eastern Mau has witnessed most of the land use changes consequent of the up-surging human population and their associated activities. The former large scale farms have been subdivided and allocated to small scale farmers. For instance, the former Sebiens (Ngondu) and Wright (Njokerio) large scale farms, which produced flowers, wheat and dairy products on commercial basis, have been converted into small arable and grazing plots through the land fragmentation process. In Baruti area near Lake Nakuru, a lunar landscape with the potential for landslide disasters during storm events has been created by the haphazard and unplanned sand mining and quarrying activities carried out in the area (Figures 7 (a) and (b)).

(a)

(b)

Fig. 7. (a): Cleared forested presently used for grazing in Nessueit area (Photo taken in 2005 by Kundu) (b): Sand Harvesting at Baruti area near Lake Nakuru (Photo taken in 2009 by Kundu).

In eastern Mau, more than one-half of the cropland is set aside for subsistence farming. Measures such as contour ploughing and conservation tillage that proved effective in soil and water management have been receiving less and less attention. Consequently, a majority of the land has become susceptible to soil erosion processes with the continued degradation. Coupled with this is deteriorating water quality epitomized by the rising level of dissolved nutrients (NO_3 and PO_4), and eutrophication based processes, especially within Lake Nakuru (Karanja et al., 1986; WWF, 1991; Chemelil, 1995; Shivoga, 2001; Owino et al., 2005).

4. Land cover conversion patterns

Previous studies at the Regional Centre for Mapping of Resources for Development (RCMRD) involving time series analysis of satellite based remote sensing data have revealed significant land cover changes in the MFC (www.rcmrd.org) as shown in Figures 8 and 9. Before 1986, the dominant pre-change land cover types were about 75% of forests, 12% of woodlands and 13% of farms. By 1989, the landscape had changed tremendously giving rise to about 60 % of forest and woodland, and 40 % of agriculture and built-up area. Figure 8 illustrates deforestation patterns between 1986 and 2000. It should be noted that the change statistics were obtained from classification of medium resolution spatial datasets. Depending on the degree of spatial aggregation used during processing therefore, it is likely that the dominant land cover types could have been overestimated at the expense of non-dominant types. However, a ground survey carried out in retrospect revealed that the classification trends generally compared well with the reality on the ground. The dominant land cover types during the period were largely agriculture and built-up area as per the classifications. The indigenous knowledge furthermore divulged that the loss of forests were largely through clear-cut and progressive thinning by the local residents.

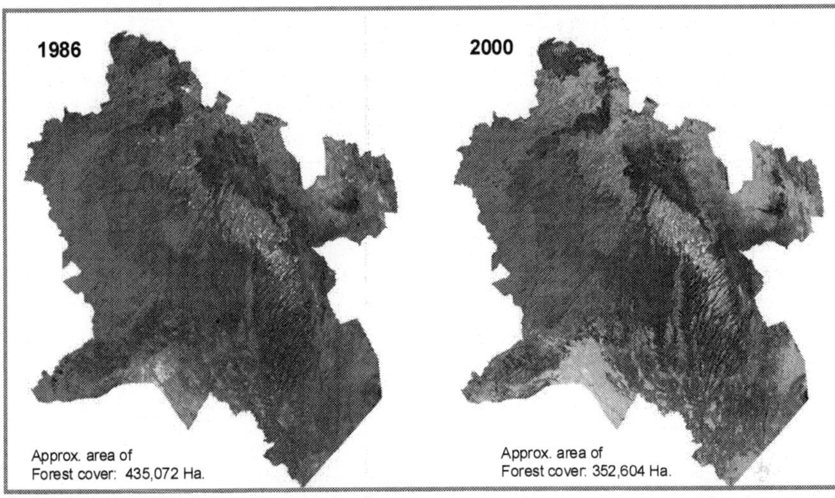

Fig. 8. Deforestation patterns in the Mau complex between 1986 and 2000.

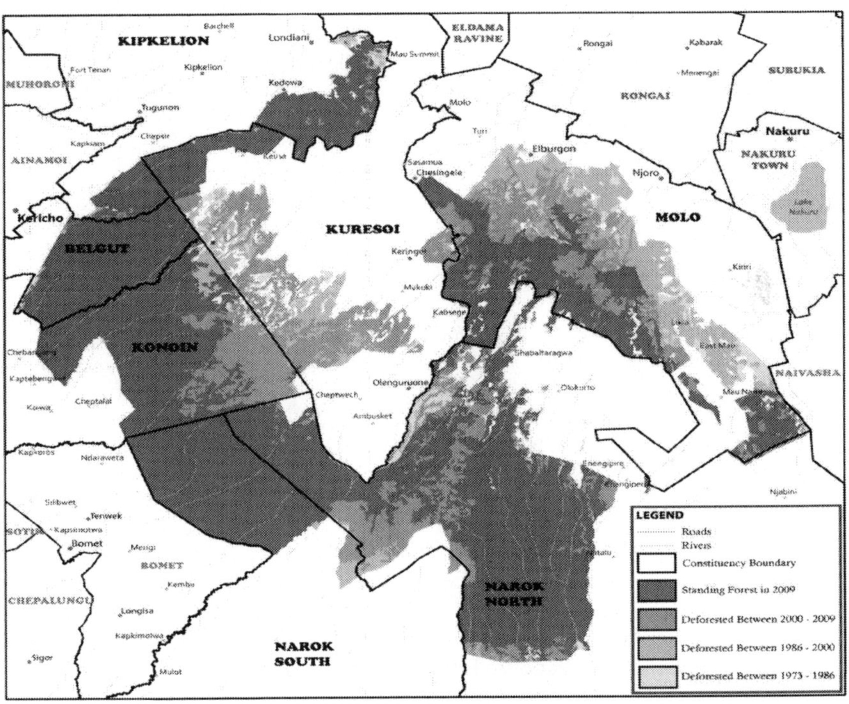

Fig. 9. Deforestation patterns of the MFC located south of Londiani (E. Khamala: 2009)

Further analysis using Landsat satellite images for the period between 1973 and 2009 have revealed that the section of the forest falling south of Kipkelion and Londiani (Figure 9) stood at about 254,100 hectares in 1973, 249,400 hectares in 1986, 226,100 hectares in 2000 and 179,000 hectares in 2009 (http://kenyafromspace.blogspot.com). Relative to 1973, these figures represent percentage deforestation rates of about 2% between 1973and 1986, 11% between 1973 and 2000, and 30% between 1973 and 2009. From the statistics, it could also be deduced that deforestation rates were highest between 2000 and 2009, when about 47,100 hectares of forest were lost. These results generally conform to ground reality since this period witnessed the highest excisions of the forest owing to the human encroachments and settlements followed by irregular and ill planned forest resource exploitation.

5. The environmental impact of land use/cover change

The conversion of forest into agriculture and built-up land in the MFC has led to noteworthy environmental impacts. Generally, increased impervious and hardened surface areas such as roads, parking lots, sidewalks and rooftops diminishes infiltration based processes and, consequently, recharge to the groundwater systems. These processes not only impair the ability of the system to cleanse runoff and protect wetlands, but also amplify the potential for soil erosion and floods, thereby contributing to the degradation of streams and other water bodies. The replacement of forest and woodland by depletive subsistence agriculture has also caused massive inflow of sediments into the nearby Lakes (Ramesh, 1998). The rising nutrient levels from the sediment have affected the growth of blue-green algae (spirulina platensis), which forms the main food for flamingo birds, known to be a major touristic attraction for Lake Nakuru. Apart from reduced revenues associated with ecotourism in the area, the ecological effect of this has been the loss of biodiversity through migration of the birds to other water bodies within the rift valley where complimentary food is available. Conversion from forest to agriculture and grazing land has also disrupted the hydrological cycle of the river drainage basins through increased evaporation and runoff process, especially during rainy seasons.

Generally, low-productivity grass types from natural grassland pastures have lesser leaf area and produce a smaller amount of biomass compared to the forested vegetation. With reduced leaf area and biomass consequent of the land degradation, rainfall interception and surface detention capacity are bound to significantly decrease. This reduces the soil moisture retention capacities, further contributing to the decline in the general evapotranspiration rates (ET_o) of the area. Changes in land use may also affect the groundwater recharge of a system. This however, depends on the groundwater recharge area, which may be different from the surface water catchments. However, studies have also shown that logging or conversion of forest to grassland for grazing can result into rising water table as a result of decreased evapotranspiration. In some cases, the water table may fall as a result of decreased soil infiltration from soil compaction and non-conservation farming techniques. If the infiltration capacity is substantially reduced, the long term effect can be severe cases of drought and desertification (Maidment, 1993; Chemelil, 1995). Removal of forest from a catchment can also cause significant hydrologic consequences such as decreased rainfall interception leading to variations in the stream water quality and quantity (Mutua and Klik, 2007; Olang et al., 2011). Research has shown that tree canopies can intercept 10-40% of incoming precipitation depending on the age, location and density of stand, tree species, rainfall intensity and evaporation rates. Land degradation due to forest logging, forest fires and wind damage can therefore have major and long lasting effects upon the canopy characteristics and consequently, overall hydrological response of an area. Clearing of forests can also cause habitat fragmentation, loss of biodiversity and water related pollution problems.

From a macro-scale perspective however, deforestation as a problem to society is still complex in nature. However, there is no question as to whether deforestation affects the climate system dynamics, atmospheric composition and other ecosystem processes. In the tropical regions, observational studies on the effects of land cover conversions on the climate and hydrology of very large basins greater than 100,000 km^2 are still scarce. The few large-scale catchment-based studies available have found no consistent relation between land cover degradation and climate changes. The studies further suggests that if appropriate land cover, precipitation, and discharge data were available in time and space, then perhaps it would be possible to determine whether the impact of land cover change across very large catchments is similar to that observed in smaller catchments. Figure 10 illustrates the impact of land cover changes of the Mau on the ecology of river Njoro (a) and Lake Nakuru (b).

(a) (b)

Fig. 10 (a): Drying river bed at Baruti area (b): Retreating water levels in Lake Nakuru (photos taken by Kundu in 2009).

6. Environmental conservation strategies

More recently, the Mau Forest Complex has received considerable attention from local and international organizations due to its ecological significance, which is posing a threat to the whole region. In Kenya, most forest areas are now under the management of the Kenya Forest Service (KFS), which has made substantial steps towards addressing the degradation and deforestation threat to all the major water towers. Among the steps is the new forest policy and law, which were promulgated in 2005. The new law lays emphasis on a participative approach to management of forest resources by all stake holders including local communities and the private sector. A further step is the creation of the Task Force for the Mau Forest Complex (TF-MFC) under the office of the Prime Minister of Kenya, with the mandate to recommend strategies for restoring the forest complex in line with Vision 2030. A very important and urgent scheme is the reforestation and restoration through tree planting as shown in Figure 11. Such activities are organized by the concerned governmental chief officers, in collaboration with environmental non-governmental organizations and international organizations.

(a) (b)

Fig. 11. (a): Restoration of the Enderit Block of the Eastern Mau Forest Reserve
(b): Tree planting to restore the forest covers (Photo taken by Kundu in 2010).

A number of international and sub-regional organizations are involved in the conservation and rehabilitation planning. The major international programme is under the UNEP/DEWA, which is also involved in the assessment of the threats to critical montane forests in East Africa including Mt. Kenya, Aberdare range and Mt. Kilimanjaro. Other organizations which have shown interest, directly or indirectly, include the Africa Convention for the Conservation of Nature and Natural Resources (2003), East Africa Community Treaty (1999), Convention on Wetlands of International Importance Especially as Waterfowl Habitat (Ramsar Convention, 1971), Convention on Biological Diversity (1992), International Tropical Timber Agreement (1983, revised 1994) United Nations Forum on Forests, Intergovernmental Authority on Development (IGAD), Johannesburg Plan of Implementation of the World Summit on Sustainable Development (WSSD), Lake Victoria Protocol (2003), Protocol for Environment and Natural Resources, The United Nations Framework Convention on Climate Change (1992), the World Heritage Convention (1972), the United Nations Convention to Combat Desertification (UNCCD) (1994); the Convention on International Trade in Endangered Species (CITES, 1973), The United Nations Convention to Combat Desertification (UNCCD) (1994), The Nile Basin Initiative (NBI) amongst others. So far under the TF-MFC mandated with co-coordinating the rehabilitation planning of the Mau ecosystem, a number of strategic options have been proposed and realized, in part. The major key interventions were categorized into three phases. Phase 1 involves short term options achievable within the first three years. Phases 2 and 3 involve medium and long term interventions aimed at consolidating the management efforts for sustainability reasons. Among the key interventions, the first and second phases include:

- *Development of effective institutional framework and strategic Management Plan*

Under this framework, a Mau Forests Complex Authority (MFCA) was to be established to coordinate and oversee the management of the complex. The authority was to be guided by board of directors comprising representatives of the main stakeholders, including the economic sectors directly dependent on the goods and services of the Mau Forests Complex such as water, energy, tourism and wildlife, agriculture and forestry. Ecological requirements, in conformity with the needs of existing strategic plans, including for Vision 2030 were to be integrated in the development plan. The current status of the Mau ecosystem, including the existing data status for management purposes were to be considered in achieving this. Additionally is the need for assessment studies on the critical catchment areas and biodiversity hotspots, which require immediate and appropriate conservation strategies.

- *Boundary surveys, issuance of title deeds and monitoring and enforcement*

This was to involve the demarcation of the legal boundaries and assessment of the critical water catchment areas, assessment of vegetation cover status and biological diversity hotspots in the MFC. Furthermore was the need for routine monitoring to prevent new encroachment, charcoal burning and tree felling that could further attenuate degradation process. Demarcation and fencing of hydrological and biological hotspots or where significant human-wildlife conflicts could occur was hence imperative in this context.

- *Relocation, resettlement and livelihood support and development*

This activity involved the relocation of all people living in the demarcated protected forests. In the event of resettlement thus, the government was to provide alternative land and funds for the development of the new lands, and livelihoods, while taking into consideration vulnerability of the people within the locations. Immediate livelihood support including water, food, shelter and energy were hence required for the families relocated from the complex to lessen the resentment felt by those aggrieved by their relocation.

- *Public awareness and community sensitization*

The activity was mainly to address the needs of the local communities living around the forest. The restoration process was to be done in consultation with local communities, who were to benefit both through directly employment opportunities and/or indirectly through ecosystem services including water provision through a restored ecosystem. Sustainable livelihood options in the forest, with particular emphasis on employment opportunities and natural resource based income generating activities were to be explored. This was to include, but not limited to, raising most of the required seedlings for rehabilitation, with the balance being produced through institutional nurseries through technical support by private and international organizations.

7. Conclusions

The negative environmental impacts on the MFC, have reached crisis level. Presently, the riparian communities and the Kenya government through key economic sectors that directly depend on goods and services of the region are paying the price of over three decades of negligence and improper land use management. The ongoing restoration efforts, including educating the general public about the need for sustainable environmental conservation in such areas is highly essential and should be sustained. It is imperative that the restoration and rehabilitation efforts are fortified through integration with potential socio-economic activities that can support the survival of the riparian rural communities. Exploring the role of eco-tourism, in relation to natural forested ecosystem, followed by putting in place appropriate and sustainable management framework are hence important in this respect. In order to further support the rural communities it is crucial to initiate long-term agro-forestry based practices such as production of sustainable wood products, and non-timber products such as medicinal plants and honey for commercialization purposes.

Also, worth mentioning as a fundamental aspect of the conservation would be the unavoidable role of continued research in the region. Further studies that go hand in hand with the restoration and rehabilitation process would be a key support tool that enables necessary and appropriate adjustment as need arises. Evaluating the interactions of the rehabilitated forest ecology in relation with the biological and hydrological systems will be important at every stage. With the increasing advancement in RS techniques, future research activities aimed at exploring new and innovative methods for environmental monitoring and management are also imperative in this respect. Other studies related to carbon sequestration to defer the effects of global warming through Reduced Emissions associated with Deforestation and Degradation (REDD) are also necessary. Considering that land degradation due to anthropogenic causes still remains a major threat to ecosystems and natural resource sustainability in Kenya, successful rehabilitation of the Mau Forest Complex will offer a good prototype that can be studied and possibly emulated across other regions experiencing similar environmental challenges.

8. Acknowledgements

The authors would like to acknowledge the support of the relevant authorities at the World Agro-forestry Centre, Regional Centre for Mapping of Resources for Development, Regional Disaster Management Center of Excellence and IGAD - Climate Prediction and Application Centre. Our Sincere gratitude also goes to the editorial team at InTech Publishers for streamlining this work to publication standards.

9. References

Baldyga, T. J., Miller, N. S., Driesse, L. K., & Gichaba, N. C. (2007). Assessing land cover change in Kenya's Mau Forest region using remotely sensed data. *African Journal of Ecology*, 46, 46–54, doi:10.1111/j.1365-2028.2007.00806.x

Batjes, N. H. & Gicheru, P. (2004) Soils data derived from SOTER for studies of carbon stocks and change in Kenya (Version 1). World Soil Information (ISRIC) GEF-SOC/VROM project report no. 2004/01 (February 2004).

Borga, M. (2002). Accuracy of radar rainfall estimates for streamflow simulation. *Journal of Hydrology*, 267, 26-39.

Calder I.R. (1998). Water-resource and land use issues. SWIM Paper 3. Colombo: IIMI

Chemelil, M.C. (1995). The effect of human induced watershed changes on stream flows. PhD Thesis, Loughborough University of Technology, UK.

China, S.S. (1993). Land Use Planning using GIS. Unpublished PhD thesis, University of Southampton

Coppin P., Jonckheere, I., Nackaerts, K., & Muys, B. (2004). Digital change detection methods in ecosystem monitoring: a review. *International Journal of Remote Sensing*, 25, 1565-1596.

Corey, J. A. B., Navjot, S. S., Kelvin, S-HP. & Barry, W. B. (2007). Global evidence that deforestation amplifies flood risk and severity in the developing world. *Global Change Biology*, 13, 2379–2395, doi:10.1111/j.1365-2486.2007.01446.x

FAO. (1998). Food and Agriculture Organization of the United Nations. Crop evapotranspiration: Guidelines for computing crop water requirements - FAO Irrigation and drainage paper 56, Rome, Italy.

FAO. (2009). Food and Agriculture Organization of the United Nations. Crop evapotranspiration: Guidelines for computing crop water requirements - FAO Irrigation and drainage paper 56, Rome, Italy.

FAO-UNESCO. (1988) Soil Map of the World, Revised Legend. FAO World Soil Resources report no. 60. Food and Agricultural Organization of the United Nations, UNESCO, Rome, Italy.

Foody, G. M. (2001). Monitoring the magnitude of land cover change on the southern limits of the Sahara. *Photogrammetric Engineering and Remote Sensing*, 67(7), 841-847. http://www.isric.org/isric/webdocs.Docs/ISRIC_Report_2004_01.pdf.

ISRIC/FAO-UN. (1995). Procedures for Soil Analysis. Technical Paper 9. (5th edition).

Jensen, J. R. (2000). Remote Sensing of the Environment: an earth resource perspective. Prentice Hall series in geographic information science. Upper Saddle River, NJ, USA

Jensen, J. R., 2005. Introductory Digital Image Processing: A Remote Sensing Perspective (3rd edition). Prentice Hall series in geographic information science. Upper Saddle River, NJ, USA.

Karanja, A., China, S. S. & Kundu, P. M.(1986). The influence of land use on the Njoro River Catchment between 1975 and 1985. In: *Soil and Water Conservation in Kenya* - University of Nairobi, Nairobi, Kenya.

KFWG. (2001). Kenya Forests Working Group. Excision and settlement in the Mau Forest. Report of Kenya Forest Working Group, Nairobi, Kenya, pp.15.

King, L.A. & Hood, V. L. (1999) Ecosystems health and sustainable communities: north and south. *Ecosystem Health*, 5, 49–57.

Krajewski, W. F. & Smith, J. A. (2002). Radar hydrology: rainfall estimation. *Advances in Water Resources*, 25(8), 1387-1394.

Krhoda, G. O. (1988). The impact of resource utilization on the hydrology of the Mau Hills forest in Kenya. *Mt. Resources Development*, 8, 193–200.

Kundu P. M., China S. S. & Chemelil, M. C. (2008). Automated extraction of morphologic and hydrologic properties for River Njoro catchment in Eastern Mau, Kenya. *Journal of Science, Technology, Education and Management*, 1 (2), 14-27

Kundu, P. M.(2007). Application of remote sensing and GIS techniques to evaluate the impact of land use and land cover change on stream flows: The case for River Njoro catchment in eastern Mau-Kenya. PhD Thesis. Egerton University, Kenya.

Liu, D., Mausel, P. Brondizio, E. & Moran, E. (2004). Change detection techniques. *International Journal of Remote Sensing*, 25, 2365-2401.

Maidment, D. R. (1993). Handbook of hydrology. McGraw Hill, New York, San Francisco, USA.

Marcos, H. C., Aurelie, B. & Jeffrey, A.C. (2003). Effects of large-scale changes in land cover on the discharge of the Tocantins River, Southeastern Amazonia. *Journal of Hydrology*, 283, 206-217.

Mutua B. M. & Klik, A. (2007). Predicting daily streamflow in ungauged rural catchments: the case of Masinga catchment, Kenya. *Hydrological Sciences*, 52(2), 292-304.

Olang L.O., Kundu P. M, Bauer T. & Fürst, J. (2011). Analysis of spatio-temporal land cover change for hydrological impact analysis within the Nyando River basin of Kenya. *Environmental Monitoring and Assessment (Springer)*, 179, 389–401, doi::10.1007/s10661-010-1743-6.

Olang, L. O. & Fürst, J. (2011). Effects of land cover change on flood peak discharges and runoff volumes: model estimates for the Nyando River Basin, Kenya. *Hydrological Processes*, 25, 80–89, doi:10.1002/hyp.7821.

Olang, L. O. (2009) Analysis of land cover change impact on flood events using remote sensing, GIS and hydrological models. A case study of the Nyando River Basin in Kenya. Dissertation, University of Natural Resources and Applied Life Sciences (BOKU) of Viena, Vienna, Austria.

Owido, S. F. O, Chemelil, C. M., Nyawade, F. O. & Obadha, W. O. (2003). Effects of Induced Soil compaction on Bean (Phaseolus Vagaries) Seedling Emergence from a Haplic phaeozen soil. *Agricultura Tropica. ET subtropica*, 36, 65-69.

Owino, J., Owido, S. F. O. & Chemelil, C. M. (2005). Nutrients in runoff from a clay loam soil protected by narrow grass strips. *Journal of Soil and Tillage Research (Elsevier)*, 88, 116-122.

Pellikka, P., Clark, B., Hurskainen, P., Keskinen, A., Lanne, M., Masalin, K., Nyman-Ghezelbash P. & Sirviö, T. (2004). Land Use change monitoring applying Geographic Information Systems in the Taita Hills, SE-Kenya. In: Proceeding of the 5th African Association of Remote Sensing of Environment Conference, Nairobi, Kenya.

Rambaldi, G., Muchemi, J., Crawhall, N. & Monaci, L. (2007). Through the Eyes of Hunter-Gatherer: Participatory 3D modelling among Ogiek indigenous peoples in Kenya. *Information Development*, 23(2-3), 113-128, doi:10.1177/0266666907078592.

Ramesh, T.(1998). Lake Nakuru Ramsar Project. World Wide Fund for Nature (WWF) (www.aaas.org/international/ehn/biod/thampy.htm)

Refsgaard, J. C. & Henriksen, H. J. (2004). Modelling guidelines--terminology and guiding principles. *Advances in Water Resources*, 27, 71–82, doi:10.1016/j.advwatres.2003.08.006

Shivoga, W. A. (2001). The influence of hydrology on the structure of invertebrate communities in two streams flowing into Lake Nakuru, Kenya. *Hydrobiologia*, 458, 121-130.

Sombroek, W. G.,. Braun, H. M. H. & van der Pouw, B. J. A. (1980). The explanatory soil map and agro-climatic zone map of Kenya. Report No. E.1, Kenya Soil Survey, Nairobi, Kenya.

World Resources Institute. (2007). Nature's Benefits in Kenya: An Atlas of Ecosystems and Human Well-Being. Washington, DC, USA.

WWF (World Wide Fund for Nature). (1991). Conserving Africa's elephants: current issues and priorities for action. (eds. H.T. Dublin, T.O. McShane and J. Newby) , WWF International, 1196 Gland, Switzerland.

16

Concepts for Environmental Radioactive Air Sampling and Monitoring

J. Matthew Barnett
Pacific Northwest National Laboratory, United States of America

1. Introduction

Environmental radioactive air sampling and monitoring is becoming increasingly important as regulatory agencies promulgate requirements for the measurement and quantification of radioactive contaminants. While researchers add to the growing body of knowledge in this area (Byrnes, 2001; Till & Grogan, 2008), events such as earthquakes and tsunamis demonstrate how nuclear systems can be compromised. The result is the need for adequate environmental monitoring to assure the public of their safety and to assist emergency workers in their response. Two forms of radioactive air monitoring include direct effluent measurements and environmental surveillance.

Direct effluent radioactive air sampling is typically conducted at the exhaust point. The considerations for analysis should include particulates and gases in use; one cannot neglect short-lived radioisotopes or hard-to-detect (HTD) radionuclides. An emission point may be in the form of an actively exhausted stack or vent. Emissions may come from several industries, such as medical isotope production, hospital use, research institutes, and industrial processes.

Environmental surveillance is conducted when emissions emanate from a fugitive pathway such as a waste pile, abandoned building, or contaminated land mass or breather tank. Monitoring stations are often located at near the facility boundary or nearby public areas in the affected directions. Often, a combination of direct effluent (point source) sampling and post release environmental monitoring is employed to assure the public, demonstrate low emissions of radioactive material, and comply with regulations.

This chapter presents basic concepts for direct effluent sampling and environmental surveillance of radioactive air emissions, including information on establishing the basis for sampling and/or monitoring, criteria for sampling media and sample analysis, reporting and compliance, and continual improvement.

2. Basis for sampling and/or monitoring

Releases of airborne radionuclides into the environment are typically managed so that they are minimized, utilizing the As Low As Reasonably Achievable (ALARA) concept. These releases encourage the need to demonstrate that the environment is protected, which is usually accomplished through direct effluent sampling at the point of exhaust and/or through environmental surveillance at locations both on and off the site (Fig. 1). A combination of both methods may be employed depending on the facility needs and regulatory requirements.

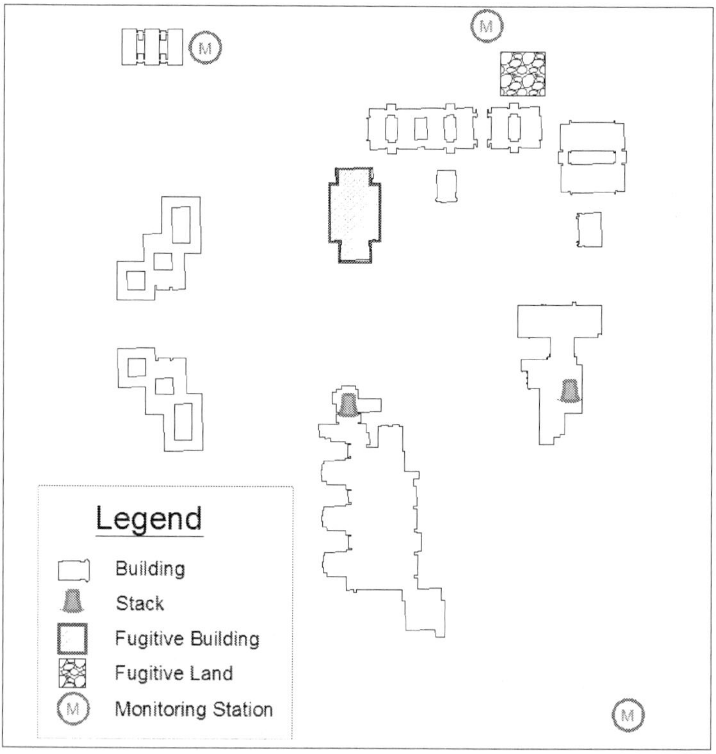

Fig. 1. Example facility showing stacks, fugitive emissions, and on-site monitoring station locations

Exposure to humans from the release of radioactive materials into the atmosphere would generally occur through the inhalation or ingestion pathway; an open wound would be another possible way for internal deposition. Additional exposure comes from immersion, material deposited on the soil and vegetation, and through the resuspension of material when disturbed. Hence, the categories for consideration in establishing radioactive air sampling systems include particulate radionuclides, gases (e.g., tritium and carbon-14), and special categories such as radioiodines and other HTD radionuclides (e.g., those with a short half-life or very weak radiation emission). In-depth implementation methods are available in established standards such as *Sampling Airborne Radioactive Materials From the Stacks and Ducts of Nuclear Facilities* (International Organization for Standardization [ISO], 2010) and *Sampling and Monitoring Releases of Airborne Radioactive Substances From the Stacks and Ducts of Nuclear Facilities* (American National Standards Institute [ANSI], 2011) as well as in *Radioactive Air Sampling Methods* (Maiello & Hoover, 2010). When sampling/monitoring is not conducted, releases may be estimated.

2.1 Sampling point source releases of radioactive substances

Over 40 years ago, proscriptive sampling methods were normative, with an emphasis on the isokinetic sampling of airborne radioactive material from exhaust points (ANSI, 1970). Since then, advances in sampling techniques and improved technology have yielded a new approach to representative sampling (ANSI, 2011; ISO, 2010). Because of these advances, the goal of achieving an unbiased, representative sample now results in a standards-based approach with definitive criteria to establish the sampling at a well-mixed location.

Point sources are discrete, well-defined locations (such as a stack, vent, or other functionally equivalent structure) from which radioactive air emissions originate (Washington Administrative Code [WAC], 2005; U.S. Environmental Protection Agency [EPA], 2002a). Point sources are actively ventilated or exhausted. Emissions from a point source may be captured, treated, monitored, sampled, and/or controlled. At some threshold, direct effluent sampling must be conducted to verify low emissions, and a graded approach based on potential emissions is recommended. Table 1 shows the ANSI N13.1-2011 approach to direct effluent sampling and monitoring requirements based on the U.S. limit of 0.1 mSv/yr (10 mrem/yr) (EPA, 2002a).

Potential Impact Category	Monitoring and Sample Analysis Procedures	Potential Fraction of Allowable Limit
1	Continuous sampling for a record of emissions and in-line, real-time monitoring with alarm capability; consideration of separate accident monitoring system	>0.5
2	Continuous sampling for record of emissions, with retrospective, off-line periodic analysis	>0.01 and ≤0.5
3	Periodic confirmatory sampling and off-line analysis	>0.0001 and ≤0.01
4	Annual administrative review of facility uses to confirm absence of radioactive materials in forms and quantities not conforming to prescribed specifications and limits	≤0.0001

Table 1. Graded approach to sampling and monitoring (ANSI N13.1-2011)

Using a graded approach to determine direct effluent sampling and monitoring needs (Table 1) and to design a robust sampling system (whereby the sample is extracted at a homogeneous location within the point source) requires an evaluation of the sample environment, transport mechanisms, and collection materials. The criteria for the homogeneous sampling location includes a determination of the angular or cyclonic flow, uniformity of the air velocity profile, gas concentration profile, and particle concentration profile (ISO, 2010; Table 2). Scaled tests may be utilized to demonstrate compliance with these criteria; however, as technology improves, modeling techniques such as computational fluid dynamics may be used to validate a well-mixed location without the necessity of field tests conducted in the stack or vent (Recknagle et al., 2009).

Characteristic	Methodology	Recommendations
Measurement to determine if flow in a stack or duct is cyclonic	ISO 10780:1994.	The average resultant flow angle should be less than 20 degrees.
Velocity profile	Selection of points across a section based on the guidance in ISO 10780 for the center 2/3 of the area may be added to adequately cover the region.	The coefficient of variance (COV) should not exceed 20% over the center region of the stack that encompasses at least 2/3 of the stack cross-sectional area.
Tracer gas concentration profiles	Selection of points across a section based on the guidance in ISO 10780 for the center 2/3 of the area of the stack or duct. Additional points or area may be added to cover the region adequately.	The COV should not exceed 20% over the center region of the stack that encompasses at least 2/3 of the stack cross-sectional area.
Maximum tracer gas concentration deviations	Selection of points across a section based on the guidance in ISO 10780 for the entire cross-sectional area.	At no point on the measurement grid should the tracer gas concentration differ from the mean value by more than 30%.
Aerosol particle concentration profile	Selection of points across a section based on the guidance in ISO 10780. Additional points or area may be added to cover the region adequately.	The COV should not exceed 20% over the center region of the stack that encompasses at least 2/3 of the stack cross-sectional area.

Table 2. Summary of recommendations for a stack sampling location (ISO 2889:2010)

2.1.1 Direct effluent sampling

Once the sampling location has been identified and qualified, attention to the sample system design is necessary. A typical stack effluent sampling system includes an in-line sample probe within the stack/vent, a sample transport line to the sample media (e.g., filter paper or cartridge), a rotameter, and vacuum gauge with feedback controls. The sample collection and transport are affected by the nozzle design, performance, and specific sampling use (e.g., particulates, gases/vapors). For reporting purposes, the bulk stream flow through the emission point is also required.

During the sampling process, losses - particle losses in particular - should be minimized. The most effective way to accomplish this is to limit the number of bends and horizontal sections of the sample line and to minimize the total sample line length. For harsh sampling environments such as those containing corrosive gases and vapors, construction material should be resistant to degradation. Because some loss is inevitable, required maintenance activities such as inspection, cleaning, and testing can help maintain effective operations.

The collection media is equally important to obtaining a valid sample. Sampler filters are usually adequate for collecting particulate radioactive air media (Fig. 2). Commercial particulate filters are made of glass fiber, acrylic copolymer, or other robust material and vary in size from 25 mm to 20 cm in diameter. Other potentially necessary specialized collection media include silica gel (Fig. 3) or molecular sieves for tritium collection, activated charcoal or silver zeolite cartridges for radioiodines, and bubblers for other gases. For collection media selection, detection criteria for the measurement of alpha, beta, and gamma radiation must also be established to meet lower limits of detection. The sample volume affects the criteria for detection limits, sample size, sampling frequency, and materials used.

Fig. 2. Fixed head radioactive air stack sampler with 47-mm diameter filter in place

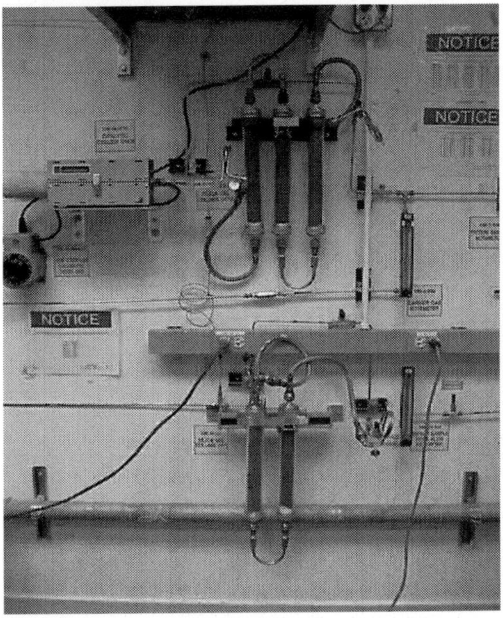

Fig. 3. Silica gel columns in use for tritium sampling system; three columns are used for collecting water vapors, and then the dry gas goes through a catalyst to form the water vapor collected using the two columns (Barnett et al., 2004)

Optimization of the sampling system is the final component of the program development. Balancing the effects and requirements along with a graded approach will generally result in an adequate sample. These considerations are also germane to environmental surveillance sample collection stations and equipment.

2.1.2 Direct effluent monitoring

As identified in the graded approach of Table 1, continuous air monitoring may be required at a point source for real-time analysis and feedback. A continuous air monitor (CAM) provides timeliness in assessing the release of radionuclides to the environment. Fig. 4 shows a combination particulate and gas CAM. While the system specifications require the user to balance the sensitivity, energy response, response time, and accuracy, the CAM should also have alarm capabilities with established thresholds to alert the user to significant releases (DOE, 1991).

Fig. 4. Combination continuous air monitor for particulates and gases

The requirements of sampling at a well-mixed location apply equally to a stack CAM. However, additional maintenance, repair, and calibration are required for a CAM. Maintenance activities can include periodic checks of the system responses to inputs that generate alarms that verify normal operations. Repairs can include replacement of electronics, detectors, or other system components that wear out or become damaged. Finally, an annual calibration is required that covers all aspects of the CAM operations. Calibration activities would include background checks and measurements, source responses to reference standards of given radioisotopes, leak tests, electronics validations, and alarm responses.

A CAM is particularly useful in laboratory work where releases are expected and can be observed and managed, either in normal or upset/accident conditions. For routine work, staff may observe a release to limit the overall activity or bound daily releases. In an upset condition, staff have a near real-time assessment of releases and potentially a second filter for future analysis to confirm releases and potential exposures.

2.2 Airborne radioactive material environmental surveillance

The primary benefits of environmental surveillance for airborne radioactive material are that it identifies emissions from fugitive (and point) sources and provides detailed impacts to the public and the environment. When establishing a site monitoring program, utilization of a data quality objective (DQO) process is recommended, this determines the environmental monitoring needs for routine radiological air emissions to the atmosphere from the emissions/sources of the site in response to regulatory requirements. Assistance with preparing a DQO is available from *Guidance on Systematic Planning Using the Data Quality Objective Process* (EPA, 2006); additionally, the Pacific Northwest National Laboratory (PNNL) used the DQO process to establish three site monitoring locations (Barnett et al., 2010). The development of the DQO includes the following aspects:

1. Stating the problem
2. Establishing goals
3. Assessing inputs
4. Setting boundaries
5. Establishing decision rules
6. Evaluating decision errors
7. Optimizing the results

Besides a DQO, processes such as an implementation plan, sampling and analysis plan, a site environmental monitoring plan, and a data management plan complete a well-managed monitoring program.

Identifying and clearly stating the problem is the first step in the DQO process. This section discusses the background and scope, states the requirements, establishes the problem statement, and identifies the participants and schedule. Once the problem statement is firmly established, the goals of the DQO can be identified, usually a series of supportive questions and actions that specifically address the problem statement.

Assessing inputs and setting boundaries are the next steps in the DQO process. The inputs are used to answer the questions formulated from the goals; include information necessary to meet performance and acceptance criteria; and provide direction for the monitoring, sampling, and analysis methods. Additionally, the boundaries discuss the logistics of implementing the goals and objectives. To provide a viable monitoring program, all seven aspects of a DQO must be considered.

In establishing and evaluating the decision rules and errors, goals and inputs are vital. The decision rules are the answers to questions posed during the goal-setting process, and they utilize the data inputs for the decisions that follow. Decision errors evaluate and discuss potentially incorrect decisions and determine the possible consequences.

The final step in setting up the monitoring program for a facility or site is optimization, which may include requirement compliance, using commercial off-the-shelf equipment, and implementing standard analytical methods. However, when optimized, the goal is to make the operations and systems work efficiently.

Commercial monitoring stations are readily available (Fig. 5), the weather-protected equipment is housed in a small metal portable or stationary cabinet consisting of a pump, flow totalizer, adjustable vacuum gauge, and other equipment and/or electronics as necessary (Fig. 6). The unit's power may be a hard-wired electrical outlet, batteries, or a renewal energy source such as an array of solar panels.

Depending on system and design needs, the filter/sample media may be either inside the monitoring cabinet or external to it. Basic filter papers (Fig. 2) can be fixed to a sample head on the exterior of the cabinet; cartridges (e.g., silver impregnated zeolite and/or activated carbon) can also be fixed to an exterior sample head. Other sample media such as the larger silica gel cartridges may need to be housed inside the cabinet.

Fig. 5. Environmental air monitoring station

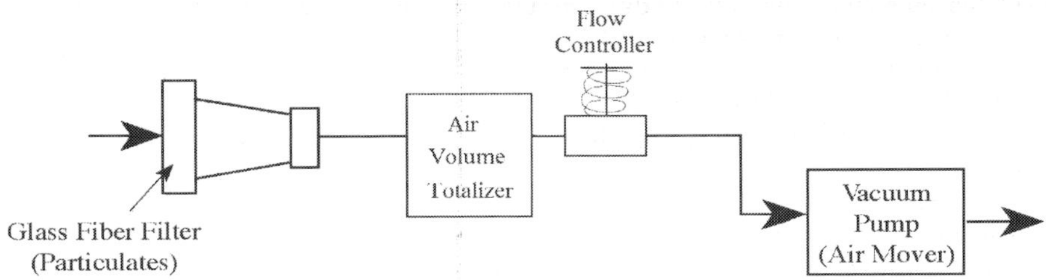

Fig. 6. Air monitoring system schematic for radioactive particulates

2.3 Considerations for assessing hard-to-detect radionuclides

HTD radionuclides have a combination of properties that include a lengthy or very short half-life, low-energy (e.g., weak beta) emissions and detection difficulties, particularly with field instruments but also with laboratory instruments. HTD radionuclides include C-14, Fe-55, I-129, Ni-63, and Tc-99. In the environment, the assessment of HTD radionuclides relies on consideration of alternative approaches such as process knowledge, surrogate/ratio (scaling) measurements, and dose impacts. The overall importance of HTD radionuclides should not be underestimated because they may in fact make a significant contribution to the regulatory dose limit for the public or environment.

The HTD radionuclides are not easily detected because the radiation cannot penetrate outside of its sample matrix or the activity is too low and obscured by background, other radionuclides, or instrument noise. The costs to isolate and analyze for the HTD radionuclides may not be justified when the use of another more readily measurable radionuclide (e.g., Cs-137) can be used to scale the HTD radionuclide measurement. Establishing the scaling factor requires process knowledge about the other radioisotopes available for measurement and the relative quantities of the HTD radioisotopes to the known radioisotopes. Once these are determined, measurement of the HTD radionuclide can proceed as a function of the better known and measured radioisotope.

2.4 Estimating releases in lieu of analytical results

When facility emissions are very low (Potential Impact Category 4, Table 1), an administrative review of the releases to the environment may be used instead of the sampling and monitoring methods described above. This review may employ data logging of actual or estimated releases based on inventory. Emissions may also be conservatively estimated when sampling or monitoring is inadequate.

For gas emissions in particular, the use of data logging is practical and efficient. The facility tracks the known releases of radioactive materials to the environment, and this log becomes the official basis for reporting. In addition to data logging, tracking of a facility's radioactive material inventory can be used to estimate a calculated release (EPA, 1989). In this process, one determines the amount of radioactive material used for the period under consideration. Radioactive materials in sealed packages that remain unopened, and have not leaked during the period are not included. The amount used is multiplied by both a release fraction ([RF]; Table 3) and a decontamination factor ([DF]; Table 4 and Equation 1). If there is more than one abatement control device in series, then multiple DFs are applied. Therefore, it is necessary to know the form of the radioactive material and any abatement controls.

Material Form	Release Fraction (RF)
Gas	1
Liquids	10^{-3}
Particulates	10^{-3}
Solids	10^{-6}

Table 3. Release fractions for estimating radionuclide releases

Abatement Control Device	Type of Radionucides Controlled (i.e., form)	Decontamination Factor (DF)
HEPA filter	Particulates	0.01
Fabric filter	Particulates	0.1
Activated carbon filters	Iodine gas	0.1
Venturi scrubber	Particulates	0.5
Packed bed scrubbers	Gases	0.1
Electrostatic precipitators	Particulates	0.05
Xenon traps	Xenon gas	0.1

Table 4. Typical decontamination factors for estimating radionuclide releases

$$A_{\text{Potentially Released}} = A_{\text{Inventory}} * RF * \pi(DF)_{(i)} \text{ (Bq)} \quad (1)$$

Where:

$A_{\text{Potentially Released}}$ =calculated release in of given isotope in Becquerel

$A_{\text{Inventory}}$ =activity of given isotope in the facility in Becquerel

RF =release fraction for material

$DF_{(i)}$ =decontamination factor for each device used in series

For additional conservatism, one can assign the DF to 1. Also, the EPA (1989) requires that any nuclide heated above 100°C, boils below 100°C, or intentionally dispersed into the environment must have a RF of 1. Other assessment methods include non-destructive assessment, upstream of HEPA filter air concentration measurements, spill release fraction, and back calculation, which may also be used to derive potential radioactive air emissions from a stack (Barnett & Davis, 1996).

3. Criteria for sampling media and correction factors

Sampling media criteria selection must be established. Once the media is selected and evaluated, various correction factors can be applied to the data. For particulate samples, selection of an appropriate filter (paper) is generally acceptable. Sampling for radioactive gases requires special treatment and typically includes the use of activated charcoal, silica gel, or another sampling mechanism based on the characteristics of the gas. Guidance for the selection, optimization, and use of various sampling media are provided (ISO, 2010).

After sampling media selection, subsequent sample collection and analysis are required. In particular, criteria established during the standards based process or DQO process becomes the basis for the analytical laboratory providing results so that meaningful reporting and data trending can be provided to interested stakeholders. Correction factors are often applied to the analytical data to prevent under reported measurements.

3.1 Sample media selection

Several different filter media are available for the collection of aerosol particles: materials include acrylic copolymers, glass fiber, cellulose, and quartz. While most filters are surface collectors and can readily be analyzed, the user should determine the need to dissolve the filter for composite analysis or further specific isotopic analyses. The range of filter flow rates vary, but for environmental applications, a flow rate between 28 and 85 L min^{-1} during the sample collection period is sufficient to collect an adequate sample for analysis. Finally, the overall media efficiency must be considered.

Often, there is a need to monitor tritium, iodines, carbon-14, radon, and krypton, or other gases. Table 5 shows the various elements and types of extraction considerations used. Aspects to consider when monitoring for these special materials include the ability of the media to capture the sample adequately, chemical forms available for sampling, volume necessary to acquire the sample, and the respective efficiencies of the processes employed.

Element	Sampling Method	Analytical Method
C-14	Carbon	Extraction followed by liquid scintillation
	Activated carbon	Extraction followed by liquid scintillation
	Bubblers	Liquid scintillation
Iodines (e.g., I-131)	Carbon	Gamma spectrometry
	Activated carbon	Gamma spectrometry
Radon	Activated carbon	Gamma spectrometry
	Alpha track strips	Alpha track
Tritium	Silica gel	Extraction followed by liquid scintillation
	Molecular sieves	Extraction followed by liquid scintillation
	Bubblers	Liquid scintillation
Argon, Krypton, and Xenon	Activated carbon	Gamma spectrometry
	Cryogenic condensing	Liquid scintillation
	Compressed gas	Gamma spectrometry

Table 5. Gas sampling methods and analytical processes

Useful resources for implementing environmental monitoring of gases include *Radioactive Air Sampling Methods* (Maiello & Hoover, 2010), *Sampling Airborne Radioactive Materials From the Stack and Ducts of Nuclear Facilities* (ISO, 2010), and *Test Methods for Measuring Radionuclide Emissions From Stationary Sources* (EPA, 2002b). These resources provide detailed information on the sampling methods, media, processes, and analytical approaches.

3.2 Applying sample analysis correction factors

Once the quality status of the data is determined (e.g., valid, suspect, invalid, or validated after review), applicable correction factors can be applied to the reported data. Correction factors are applied so that results are not underreported and a conservative approach to emissions estimates is maintained. Depending on the sample method, a variety of correction factors may be applied, including:

1. Radioactive decay factor
2. Self absorption (for filters)
3. Sampler efficiency
4. Transport efficiency
5. Sample collector media efficiency

The radioactive decay factor accounts for the time between the midpoint of the sample collection period and the sample analysis time. In most cases, the radioactive decay factor can be set to 1 because time lapse between collection and analysis is much shorter than the half-lives of the radioisotopes of concern. For short-lived radioisotopes, a correction may be necessary and can vary according to the time and the specific half-life of the isotope.

Self-absorption factor corrects for the bias caused by the absorption of emitted radiation from the collected particles by dust/particulates and the filter media itself. For filters, this factor is dependent on the amount of material collected and is shown in Fig. 7. Other types of self-absorption factors may need to be calculated, for example, those associated with cartridges.

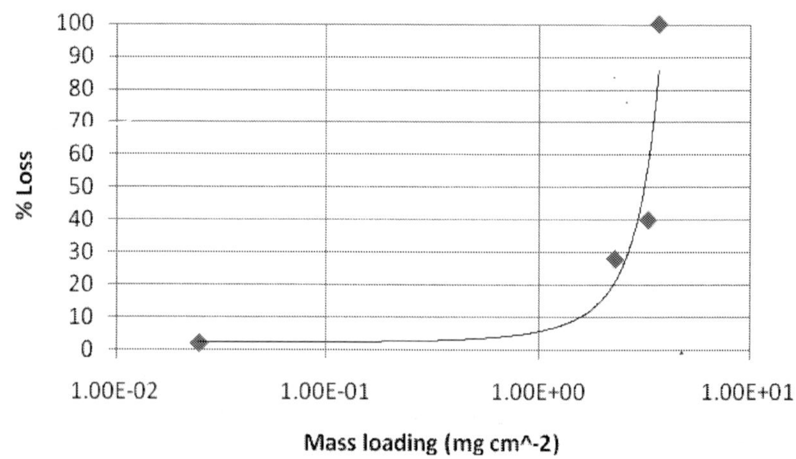

Fig. 7. Percent loss due to self-absorption versus mass loading[1]

The sampler efficiency factor accounts for biases caused by problems with the sampler operation. If the sampler operates without interruption during the sampling period, efficiency is 100% (or 1); however, when operation is incomplete or interrupted, the sampler efficiency factor is determined by the amount of time the sample was collected divided by the entire sample period. If the sampler efficiency factor is too low, an invalid sample may result.

Computer models can be employed to calculate the transportation efficiency correction factor; for example, DEPO has been used in stack monitoring to calculate line losses (McFarland et al., 2000). For environmental monitoring stations that do not have long or complicated transport lines, this factor is often set to 1 and not calculated.

[1] Adapted from Smith et al. (2011) for 47-mm filters when the percent loss is optimized and the exponential function is forced to near zero at very low mass loadings.

The sample collection media efficiency is not to be confused with the total efficiency of the sample media; it is only the part associated with the media itself. Today's filters typically have an efficiency range between 0.8 and 0.9999 for particle sizes in the 0.1 to 10 μm range, depending on the application. Most manufacturers will state rated efficiency for a given range of particle sizes. Silica gel often has 100% retention for tritium sampling until the sample cartridge is fully loaded and breakthrough occurs. In some cases, the unknown media efficiency requires evaluation or estimation.

The calculation and reporting of the final result should include the appropriate correction factors as discussed above. The total activity of a sample is expressed in Equation 2.

$$A_{Total} = A_{Sample} / \pi(E)_{(i)} \text{ (Bq)} \tag{2}$$

Where:

A_{Total} =total activity on sample in Becquerel

A_{Sample} =sample activity in Becquerel

$E_{(i)}$ =efficiency factors, including self-absorption, sampler, transport and media

All data results should be trended against established criteria to evaluate potential changes over time. The repeat measurements at a sampling location can be used to show a normal operating range with the expected statistical deviations. Data trending can also show increasing or decreasing emissions over various cycle times or events. When a data result falls outside of this normal trend, it can then be evaluated. Example causes can be associated with a sampling error (e.g., the wrong sample was reported, or there was a cross contamination of the sample) or a change in the overall emissions characteristics.

4. Reporting and compliance

In many areas, it is mandatory to provide complete and periodic reports to regulatory agencies or customers on the release of airborne radioactive material. The comprehensive report should allow for the discussion of error analysis and provide quantifiable impacts to the public and the environment. Exceeding a regulatory limit, compliance level, or permit condition requires an event notification to the appropriate regulatory agency. Compliance is a cooperative effort between the facility and the local community and regulatory agencies and requires a fully implemented quality assurance (QA) program.

4.1 Annual reporting

An annual report on the emissions of radioactive material has several aspects to consider, and it may be required to include specific information based on applicable regulations or permits and be certified by a responsible individual. Results of reported emissions can then be converted to an off-site dose. Basic elements are identified below:

1. Facility description
2. Emission point description
3. Emissions reporting
4. Input parameters and dose assessment
5. Non-routine releases
6. Supplemental information

A facility description will include historical background on the reporting site, detail the activities conducted resulting in releases of radioactive materials, and offer information on the buildings where operations are conducted. This section provides information on related nearby facilities and their impacts on the results.

The emission point description is used to brief the type of emission unit and the associated characteristics. For example, an emission unit may be a point source that releases radioactive gases (and potentially particulate materials), while a fugitive emission source may be a contaminated waste pile. Careful itemization and clear description of the emission type are essential elements to reveal the impact to overall operations.

Emissions reporting may be in the form of specific sample analyses or theoretical calculations. Specific analyses can be from point sources collected from a sampling system, or they can equally be from environmental surveillance monitoring stations; a combination of both may be necessary to cover all the types of emissions at a particular site. If environmental surveillance monitoring data is not collected, then theoretical calculations can also be used to supplement the reporting of (potentially) released radioactive materials into the environment. The emissions report is a primary factor in the dose assessment.

Input parameters to the dose assessment include the reported emissions. However, other inputs can include meteorological data for joint frequency wind speed distributions, dose conversion models, and exposure pathway parameters (e.g., inhalation, and food stuffs). Dose models such as CAP88-PC also require information on the clearance type, particle size, a scavenging coefficient, and deposition velocity used (EPA, 2007; Simpkins, 2000).

Non-routine releases from upset conditions such as spills or accidents should be reported separately and may be a permit requirement. Stack sampling or environmental surveillance monitoring stations can sample and detect non-routine releases, which would be included in the dose estimates to the public and/or environment.

Supplemental information to an annual report can include collective (population) dose estimates, results from environmental surveillance measurements, and status of methods confirming emissions. The collective dose differs from the dose to the maximally exposed individual, where the latter receives the maximum dose from the reported emissions and the former is the product of the number of persons in a general area (e.g., within 80 km of the facility) and the average dose per person (ENS, 2003). Results of environmental surveillance sampling and other sampling events can be reported in an appendix or as part of the overall results. Finally, the methods of confirming emissions should be discussed in relation to the emission unit; in such cases, a table indicates whether the emissions were measured by a sample or calculation. If sampling was conducted, it should further be noted whether it was continuous or periodic.

4.2 Event reporting

When compliance with permit conditions, emission or concentration limits, or other requirements are not met, the facility must report the information to the appropriate authority. Additionally, non-routine releases or transient abnormal conditions are reported separately and may also be a required by regulation. Often, the stack sampling or environmental surveillance monitoring stations can sample and detect these events, with the results used for dose estimates to the public and/or environment. Specific event reporting may be governed by internal procedures, licenses, and relevant regulations.

It is a good practice to report events to the appropriate regulatory agency within 24 hrs of discovery. It should cite the specific requirement(s) that is out of compliance and the current status of the situation. Immediate actions taken are reported and may include the shutdown of work, additional sampling and monitoring, and estimated impacts to the public and environment.

Regulators may request additional information or formal report and may also assign additional actions. Resuming normal work would be coordinated with the regulators. Depending of the severity of the event, additional actions such as a compliance plan submittal, inspections and assessments, more frequent and additional reporting, and assessment of fines and penalties may be initiated. Work with the regulators and management to identify the appropriate actions and cooperatively agree to the resumption of normal work.

4.3 Compliance aspects

Assessment and conformance to the regulations and permit authorization requirements enable the facility to demonstrate compliance. An organization should evaluate its activities and document its baseline compliance. Additionally, compliance requires the implementation of a robust QA program capable of passing an external audit.

There are two applicable standards for continual improvement and quality: *Environmental Management Systems* (ISO, 2004), and *Quality Management Systems* (ISO, 2008). *Sampling and Monitoring Releases of Airborne Radioactive Substances From the Stacks and Ducts of Nuclear Facilities* (ANSI, 2011) also outlines a basic QA program plan, the standard components of which include:

1. Program Aspects
2. Documentation
3. System and Equipment Characterization
4. Training
5. Maintenance and Inspection Requirements
6. Calibration
7. System Performance Criteria
8. Assessment

Shown in Fig. 8, these QA components form a complete, interdependent program.

The QA program describes administrative and organization roles. The program details data handling and procedures that govern data collection and analysis. It also incorporates the organizations proactive and cooperative relations with the regulators and key stakeholders.

Record keeping is integral to the QA program. A management system for the records is necessary and would include the basis for the collection, identification, storage and retention, and retrieval of the documents related to the program. Documentation related to the program must be available for analyses, audits, and archival purposes.

Characterization of system and equipment components as part of a QA program includes the description of the source term under consideration, the characteristics of the system and equipment, and the design and construction features of the program elements. For example, Fig. 6 could be a QA program drawing showing the basic equipment components of a field-deployed air sampling station.

Individuals involved in the program must be trained to conduct the specific role they have in the program. Training may cover many areas including assessment, data collection or analysis, and reporting. The training records would be managed under the documentation requirements of the QA program.

Fig. 8. Basic components of a QA program

Periodic maintenance and inspection requirements may often be prescribed by regulations. However, the QA program should address the frequency by which maintenance and inspections are conducted. These requirements can easily be adapted into a preventive maintenance program.

In addition to periodic maintenance and inspection, measurement and test equipment are to be calibrated periodically. The specific calibration methods utilize prescribed methods and traceable reference standards. Generally, calibrated equipment is labeled with the calibration and expiration dates.

System performance criteria assures overall satisfactory program operation. Performance criteria can cover the operational requirements, transmission factors, and flow ranges, which are used to identify normal system operations. Tracking and trending of data can supplement and monitor the criteria by enabling the user to see outlier data and observe trends in data over time. The tracking and trending of data can also indicate potential changes to program emissions or in equipment operations.

Self-assessment programs are intended to provide a mechanism for continual improvement in programmatic elements (e.g., procedures, management systems) and operational elements (e.g., monitoring systems, permit compliance) of a program. Periodic review of program elements begins with the planning of an assessment. Once the assessment scope and intent are established, criteria can be evaluated, and strengths and weaknesses identified. Corrective actions can then be assigned and implemented to improve areas of weakness or non-compliance. Once actions are complete, an effectiveness review should be conducted to verify adequate corrective action implementation.

Finally, as a part of the overall QA program, the compliance status should be documented in periodic reports and provided to management and/or appropriate regulatory agencies. These reports should include the status of compliance to the specific permit requirements and regulations. When non-compliance is identified, it must be addressed, and corrective actions should be tracked to completion. Notification to regulatory agencies must also be evaluated and may be required.

5. Case studies for continual improvement

In addition to the periodic use of internal and external assessments, the researcher should prepare to embrace opportunities to improve the sampling and monitoring base of knowledge. The assessment process provides for the necessary feedback to make incremental changes in a program to improve the overall result. The reporting of new or unique operations, special studies, or a resolution to a monitoring question provides information valuable to other programs.

Below are two examples of current, evaluative research areas: air sample volume measurements, and the deposition of material on a sample filter paper. However, there are many areas for improvement, and individuals can make their own contributions.

5.1 Air sample volume measurement evaluation

Determination of the sample volume is critical in collecting ambient air samples for environmental monitoring. Errors in the sample volume measurement are directly proportional to errors in the calculated sample concentration (Fritz, 2009). A variety of instruments are available to measure flow and can include rotameters, electronic mass flow controllers, and venturi meters (Wight, 1994). Fritz (2009) reported on the implementation of a dry-gas meter application to air sample volume measurements in lieu of a more cumbersome and less accurate two-point manual airflow measurement and sample duration. The new method showed improved reliability and measurement resolution, reduced error, and more accurate concentration calculations. The evaluation was conducted over two phases that included a system set-up identical to the field configuration and a testing phase where the new dry-gas meters were installed in the actual sampling network.

With reported results, users can apply the basics of their work into their own evaluations applicable to their particular situation. Consider for example that the air sample volume measurement evaluation is being evaluated for an area without adequate electricity or based on filter flow characteristics. In the first case where electricity is necessary to run a sample pump, solar arrays may be an alternative. One could reasonably create a limited project for the facility to determine the appropriateness of such a system and recommend whether to utilize it in a broader program. For this second case, evaluating sample filters for pressure drop (Barnett & Kane, 1993) may be studied to determine if alternative filter sizes are adequate to meet the air sample volume requirements; however, other considerations may impact the final decision such as the ability to reliably analyze the filter, the overall spectral properties of the radioisotope(s), and the ability to ash the filter easily for additional laboratory analyses.

5.2 Sample filter deposition evaluation

Researchers have probed into the major factors affecting the measurements of radioactivity on air samples collected on filters (Stevens & Toureau, 1963; Higby, 1984). These factors include particles sizes, filter types, filter loading and burial depths, and analysis of energy spectrums. More recently, others have evaluated sample filter deposition characteristics by conducting studies and using computer simulations (Luetzelschwab et al., 2000; Huang et al., 2002; Geryes et al., 2009; Barnett et al., 2009). From recent publications, additional information is now available on the self-absorption that occurs in filters, the measurement losses associated with the filter loading, and the use of Monte Carlo simulations (Fig. 9) to assess the energy spectra in different geometries. Ongoing research in this area is still warranted, given that standards call out a correction factor for self-absorption effects of more than 5% (ISO, 2010; ANSI, 2011).

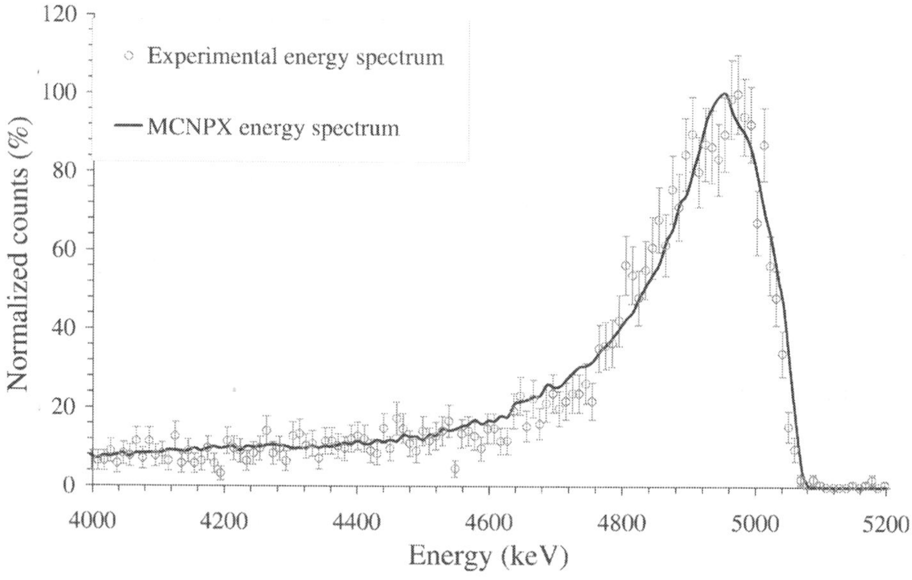

Fig. 9. Comparison of an experimental and simulated energy spectrum in a filter (Geryes et al., 2009)

6. Conclusion

Concepts for environmental radioactive air sampling and monitoring include establishing the basis for sampling/monitoring, criteria for sampling media and analytical requirements, and reporting and compliance. The processes utilized include a standards based and a DQO approach that should be integrated and applied to both direct effluent and environmental surveillance sampling and monitoring. In addition, program improvement can be enhanced through the sharing of knowledge derived from routine operations and the implementation of tested and reviewed ideas. The overall program is used to demonstrated to the stakeholders that the emissions of radioactive materials to the environment is below regulatory limits and that those doses reported from such emissions are reasonably and conservatively accurate.

7. Acknowledgment

This work was conducted at the Pacific Northwest National Laboratory, which is operated for the U.S. Department of Energy by Battelle under Contract DE-AC05-76RL01830.

8. References

American National Standards Institute (ANSI). (1970). *Guide to Sampling Airborne Radioactive Materials in Nuclear Facilities*, ANSI, ANSI N13.1-1969, New York, New York, USA

American National Standards Institute (ANSI). (2011). *Sampling and Monitoring Releases of Airborne Radioactive Substances From the Stacks and Ducts of Nuclear Facilities*, Health Physics Society, ANSI/HPS N13.1-2011, McLean, Virginia, USA

Barnett, J. & Davis, W. (1996). Six Methods to Assess Potential Radioactive Air Emissions From a Stack. *Health Physics*, Vol. 71, No. 5, (November 1996), pp. 773-778

Barnett, J. & Kane, II, J. (1993). Flow Rate Through a Filter with a 25 mm-Diameter Aperture for Hanford Site Alpha Continuous Air Monitors. *Radiation Protection Management*, Vol. 10, No. 4, (July/August 1993), pp. 41-46

Barnett, J.; Cullinan, V.; Barnett, D.; Trang-Le, T.; Bliss, M.; Greenwood, L. & Ballinger, M. (2009). Results of Self-Absorption Study on the Versapor 3000 47-mm Filters for Radioactive Particulate Air Stack Sampling. *Health Physics (Operational Radiation Safety)*, Vol. 97, No. 5, Supplement 3, (November 2009), pp. S161-S168

Barnett, J.; Meier, K.; Snyder, S.; Fritz, B.; Poston, T. & Rhoads, K. (2010). *Data Quality Objectives Supporting Radiological Air Emissions Monitoring for the PNNL Site*, PNNL, PNNL-19427, Richland, Washington, USA

Barnett, J.; True, L.; & Douglas, D. (2004). Review of Tritium Emissions Sampling and Monitoring From the Hanford Site Radiochemical Processing Laboratory, In: *Proceedings of the HPS 2004 Midyear Meeting: Air Monitoring and Internal Dosimetry*, Health Physics Society (Ed.), 199-204, Health Physics Society, McLean, Virginia, USA

Byrnes, M. (2001). *Sampling and Surveying Radiological Environments*, CRC Press, ISBN 1-56670-364-6, Boca Raton, Florida, USA

European Nuclear Society (ENS). (2003). *Collective Dose*. Accessed 02.05.2011, Available from: http://www.euronuclear.org/info/encyclopedia/collectivedose.htm

Fritz, B. (2009). Application of a Dry-Gas Meter for Measuring Air Sample Volumes in an Ambient Air Monitoring Network. *Health Physics (Operational Radiation Safety)*, Vol. 96, No. 2, Supplement 2, (May 2009), pp. S69-S75

Geryes, T.; Monsanglant-Louvet, C.; Berger, L. & Gehin, E. (2009). Application of the Monte Carlo Method to Study the Alpha Particle Energy Spectra for Radioactive Aerosol Sampled by an Air Filter. *Health Physics*, Vol. 97, No. 2, (August 2009), pp. 125-131

Higby, D. (1984). *Effects of Particle Size and Velocity on Burial Depth of Airborne Particulates in Glass Fiber Filters*, Pacific Northwest Laboratory, PNL-5278, Richland, Washington, USA

Huang, S.; Schery, S.; Alcantara, R.; Rodgers, J. & Wasiolek, P. (2002). Influence of Dust Loading on the Alpha-Particle Energy Resolution of Continuous Air Monitors for Thin Deposits of Radioactive Aerosols. *Health Physics*, Vol. 83, No. 6, (December 2002), pp. 884-891

International Organization for Standardization (ISO). (1994). *Stationary Source Emissions – Measurement of Velocity and Volume Flowrate of Gas Streams in Ducts*, ISO, ISO 10780:1994, Geneva, Switzerland

International Organization for Standardization (ISO). (2004). *Environmental Management Systems*, ISO, ISO 14001:2004, Geneva, Switzerland

International Organization for Standardization (ISO). (2008). *Quality Management Systems*, ISO, ISO 9001:2008, Geneva, Switzerland

International Organization for Standardization (ISO). (2010). *Sampling Airborne Radioactive Materials From the Stacks and Ducts of Nuclear Facilities*, ISO, ISO 2889:2010, Geneva, Switzerland

Luetzelschwab, J.; Storey, C.; Zraly, K. & Dussinger, D. (2000). Self Absorption of Alpha and Beta Particles in a Fiberglass Filter. *Health Physics*, Vol. 79, No. 4, (October 2000), pp. 425-430

Maiello, M. & Hoover, M. (2010). *Radioactive Air Sampling Methods*, CRC Press, ISBN 978-0-8493-9717-2, Boca Raton, Florida, USA

McFarland, A.; Mohan, A.; Ramakrishna, N.; Rea, J. & Thompson, J. (2000). *Deposition 2001a, Version 1.0. Deposition: Software to Calculate Particle Penetration Through Aerosol Transportation Systems*, Texas A&M University, NUREG/GR-0006, College Station, Texas, USA

Recknagle, K.; Yokuda, S.; Ballinger, M. & Barnett, J. (2009). Scaled Tests and Modeling of Effluent Stack Sampling Location Mixing. *Health Physics*, Vol. 96, No. 2, (February 2009), pp. 164-174

Simpkins, A. (2000). *Maximally Exposed Offsite Individual Location Determination for NESHAPS Compliance*, Westinghouse Savannah River Company, WSRC-RP-2000-00036, Aiken, South Carolina, USA

Smith, B., Barnett, J. & Ballinger, M. (2011). *Assessment of the Losses Due to Self Absorption by Mass Loading on Radioactive Particulate Air Stack Sample Filters*, Pacific Northwest National Laboratory, PNNL-20098, Richland, Washington, USA

Stevens, D. & Toureau, A. (1963). *The Effect of Particle Size and Dust Loading on the Shape of Alpha Pulse Height Spectra of Air Sample Filters*, Atomic Energy Research Establishment, AERE-R 4249, Harwell, Berkshire, England, United Kingdom

Till, J. & Grogan, H. (Eds.). (2008). *Radiological Risk Assessment and Environmental Analysis*, Oxford University Press, ISBN 978-0-19-512727-0, New York, New York, USA

U.S. Department of Energy (DOE). (1991). *Environmental Regulatory Guide for Radiological Effluent Monitoring and Environmental Surveillance (DOE/EH-0173T)*, Assistant Secretary for Environment, Safety and Health, DE91-013607, Washington, District of Columbia, USA

U.S. Environmental Protection Agency (EPA). (1989). *Methods for Estimating Radionulide Emissions*. U.S. Government Printing Office, 40 Code of Federal Regulations Part 60 (40 CFR 60), Appendix D, Washington, District of Columbia, USA

U.S. Environmental Protection Agency (EPA). (2002a). *National Emissions Standards for Emissions of Radionuclides Other Than Radon From Department of Energy Facilities*. U.S. Government Printing Office, 40 CFR 61, Subpart H, Washington, District of Columbia, USA

U.S. Environmental Protection Agency (EPA). (2002b). *Test Methods for Measuring Radionuclide Emissions From Stationary Sources*. U.S. Government Printing Office, 40 CFR 61, Appendix B, Method 114, Washington, District of Columbia, USA

U.S. Environmental Protection Agency (EPA). (2006). *Guidance on Systematic Planning Using the Data Quality Objective Process (EPA QA/G-4)*, Office of Environmental Information, EPA/240/B-06/001, Washington, District of Columbia, USA

U.S. Environmental Protection Agency (EPA). (2007). *CAP88-PC Version 3.0 User Guide*, Office of Radiation and Indoor Air, Washington, District of Columbia, USA

Washington Administrative Code (WAC). (2005). *Radiation Protection – Air Emissions*, Statute Law Committee, WAC-246-247, Olympia, Washington, USA

Wight, G. (1994). *Fundamentals of Air Sampling*, CRC Press LLC (Lewis Publishers), ISBN 0-87371-826-7, Boca Raton, Florida, USA

17

Multisyringe Flow Injection Analysis for Environmental Monitoring: Applications and Recent Trends

Marcela A. Segundo[1], M. Inês G. S. Almeida[1,2] and Hugo M. Oliveira[1]
[1]*REQUIMTE, Department of Chemistry, Faculty of Pharmacy, University of Porto,*
[2]*School of Chemistry, University of Melbourne,*
[1]*Portugal*
[2]*Australia*

1. Introduction

Multisyringe flow injection analysis (MSFIA) was introduced by Víctor Cerdà and co-workers in 1999 (Cerdà et al., 1999) as a robust alternative to its predecessor flow injection techniques, combining the multi-channel operation of flow injection analysis (Ruzicka & Hansen, 1975) with the possibility of flow reversal and selection of the exact volume of sample and reagent required for analysis as presented in sequential injection analysis (Ruzicka & Marshall, 1990). Generally, flow injection systems are automation tools where, in opposition to batch conventional assays, physico-chemical equilibrium is not attained prior to determination. Hence, flow injection analysis is based in three principles: (1) reproducible sample injection or insertion in a flowing carrier stream; (2) controlled dispersion of the sample zone; and (3) reproducible timing of its movement from the injector point to the detection system.

Since its inception, MSFIA has been the basis for automation of more than 120 different assays, reviewed in several publications (Almeida et al., 2011; Magalhães et al., 2009; Maya et al., 2009; Segundo & Magalhães, 2006). This type of automatic flow injection systems is based on the utilization of a multisyringe burette, depicted schematically in Fig. 1A and 1B. It is a multiple channel piston pump, containing up to four syringes, driven by a single motor of a usual automatic burette and controlled by computer software through a serial port. A two-way commutation valve is connected to the head of each syringe, allowing optional coupling to the manifold lines or to the solution reservoir.

Because the four syringes are driven by the same motor, all pistons move at once in the same direction either delivering (dispense operation) or loading the syringes (pickup operation) with liquids. Considering that the commutation valves can be placed in two positions, there are four possibilities for flow management as depicted in Fig. 1C. Hence, when the pistons are moving upwards, it is possible to dispense liquid into the flow system or send it back to its reservoir. This feature enables that only the necessary amount of reagent solution is introduced into the flow system. Furthermore, when the pistons are moving downwards, it is possible to refill the syringes with solutions present in the respective vessel or to aspirate solutions from the system in order to perform the sampling operation.

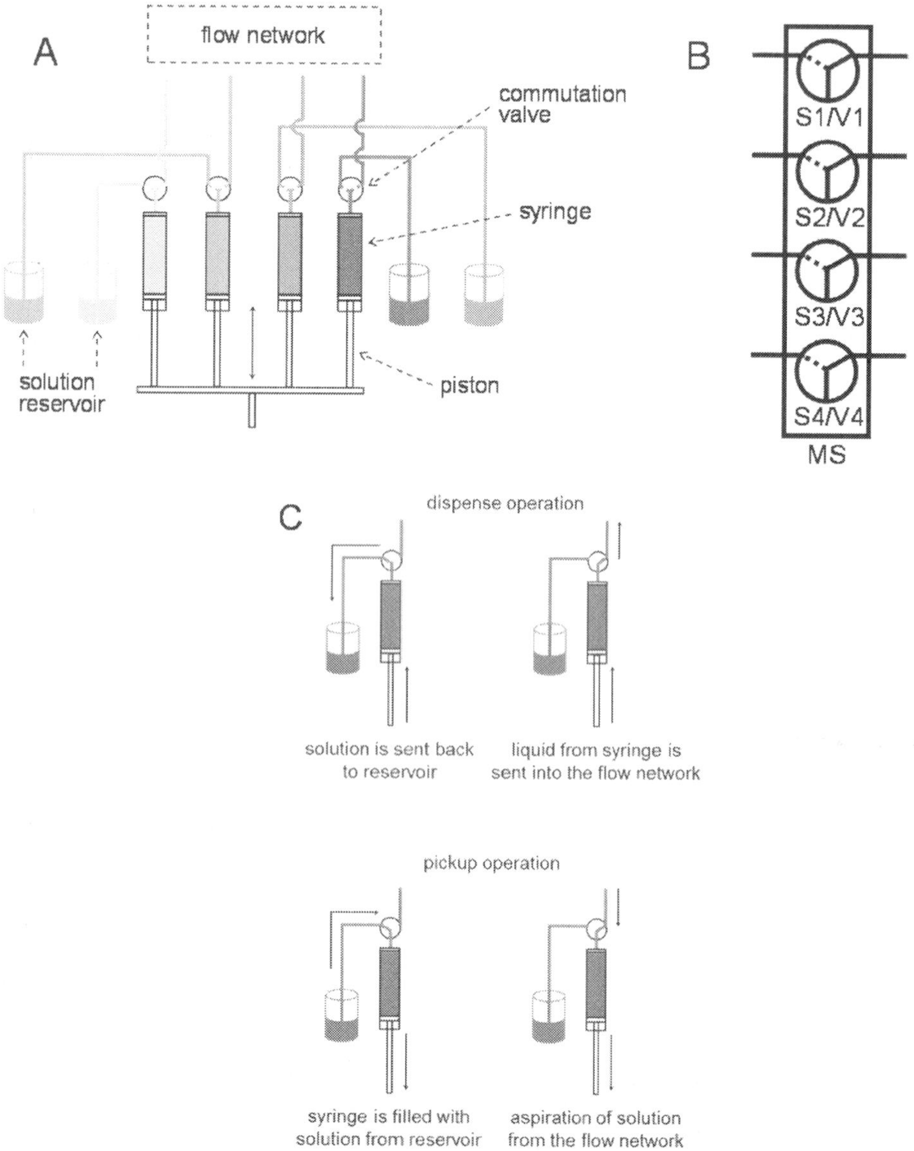

Fig. 1. Schematic representation of multisyringe apparatus, with indication of the different components (A) or simplified (B). Flow management possibilities for one syringe during operation of multisyringe apparatus are also given (C). MS = multisyringe; S = syringe, V = commutation valve.

Syringes with different volumes, ranging from 0.5 to 25 ml are available, enabling the application of a wide range of flow rates. For example, for a 5 ml syringe, flow rates ranging from 0.28 to 15 ml min^{-1} may be attained (Miró et al., 2002). Nevertheless, once the flow rate (and volume) is fixed for one syringe, it is also defined for the other channels, and it will depend on the ratio between syringe capacities as different syringes can be placed in any of the four positions.
Finally, MSFIA manifolds are not restricted to the syringes and the respective commutation valves. The presence of four digital outputs, each capable of providing 12 V/0.5 A, allows the utilization of up to 12 additional commutation valves, also controlled through the multisyringe apparatus. These extra commutation valves are often necessary to assemble a flow network, where analyte determination and sample treatment can be implemented by including confluences for reagent addition, suitable detectors (spectrophotometers, fluorimeters, flame or atomic emission spectrometers) and devices for mass transfer (gas diffusion or dialysis units), for instance.

2. Applications of MSFIA to environmental monitoring

MSFIA systems have been successfully applied to the determination of more than 20 species in environmental samples as illustrated in Tables 1 and 2.
Several applications were targeted to plant macronutrients, such as potassium and phosphorus, and also to micronutrients, including boron, iron and selenium (Table 1). These species were quantified in different types of water (natural and seawater) and also in soil extracts or even soil slurries when applying flame emission spectrometry for determination of potassium (Almeida et al., 2008). The introduction of aqueous samples in flow systems is rather trivial, while manipulation of samples containing suspended solids is not common, requiring a special manifold design employing larger commutation valves and large bore tubing (Almeida et al., 2008).
In fact, solid environmental samples were successfully handled within MSFIA systems. Extraction of potassium contained in 1.8 g of soil was performed in-line, prior to potentiometric determination. The soil was placed in a container where 9 ml of Morgan extractant solution was delivered automatically by one of the syringes. After 6 min, in-line filtration took place, and a small portion of filtrate (100 µl) was sent to the potentiometric detector after in-line addition of an ionic strength adjusting buffer. Different soils were analyzed consecutively without carry-over effects and the filtration unit was reutilized up to 10 times (Almeida et al., 2006).
Besides the determination of total extractable content, MSFIA systems have been employed to dynamic fractionation testing schemes, profiting from its inherent capabilities of controllable flow programming. In fact, Miró and co-workers developed a multiple stirred-flow chamber assembly, containing up to three parallel extractors, to perform sequential extraction of readily mobilizable fractions of trace elements (Cu, Cd, Ni, Pb, Zn) in fly ashes (Boonjob et al., 2008). Though the detection step was performed off-line (not automated) on each 10 ml fraction collected, the MSFIA system still provided information about overall extractable pools in less than 2 hours, a drastic reduction of time when compared to 18 to 24 hours required per fraction in equilibrium leaching tests. Moreover, the implementation of a sequential leaching scheme was easily accommodated in MSFIA, due to its inherent flow features and also by housing different extracting solutions (water, 0.11 M acetic acid, 0.11 M acetic acid/acetate buffer) simultaneously in each syringe of the multisyringe burette.

Analyzed species	Sample type	In-line sample treatment	Determination throughput (h^{-1})	Reference
Boron (available)	Soil extract	---	15	(Gomes et al., 2005)
Iron (available)	Soil extract	---	34	(Gomes et al., 2005)
Iron (total)	Water	Solid phase extraction	12	(Pascoa et al., 2009)
Iron (total) and Fe(III)	Water	---	60	(Pons et al., 2004)
Iron (total) and Fe(III)	Water (natural and seawater)	Solid phase extraction	5 - 10	(Pons et al., 2004, 2005a, 2005b)
Phosphate	Water	Solid phase extraction	11	(Morais et al., 2004)
Phosphate	Soil extract	In-line sequential extraction	Not given	(Buanuam et al., 2007)
Phosphorus (available)	Soil extract	---	15	(Almeida et al., 2005)
Phosphorus	Water	Microwave digestion	12	(Almeida et al., 2004)
Potassium	Soil extracts	Extraction and in-line filtration	13	(Almeida et al., 2006)
Potassium	Soil slurries	---	28	(Almeida et al., 2008)
Selenium	Sea lettuce	---	84	(Semenova et al., 2003)
Selenium	Water (natural and seawater)	Solid phase extraction	8	(Serra et al., 2010)

Table 1. MSFIA methods for nutrient assessment and monitoring

The same group proposed a fully automated strategy for fractionation of orthophosphate in soil samples and in-line spectrophotometric determination resorting to molybdenum blue reaction (Buanuam et al., 2007). The system integrated dynamic sequential extraction using 1 M ammonium chloride solution, 0.1 M sodium hydroxide solution and 0.5 M hydrochloric acid solution as extractants according to the Hieltjes–Lijklema scheme. Solid soil samples were placed in a flow-through customized dual conical chamber (Chomchoei et al., 2004), which could be filled with up to 300 mg of soil. Compared to conventional batch equilibrium procedures, the automatic dynamic fractionation system offered further knowledge on: (i) the extraction kinetics, (ii) the content of phosphorus in available pools, (iii) the efficiency of the leachants and (iv) the actual extractant volume required for quantitative release of orthophosphate. The automatic MSFIA extraction scheme was also validated through application to certified reference material SRM 2704 river sediment and SRM 2711 Montana soil.

Recently, a similar strategy was applied for automated dynamic extraction and determination of readily bioaccessible chromium(VI) in soils. Besides the extraction capabilities, the automatic MSFIA system also fostered in-line quantification of Cr(VI)

after derivatization using 1,5-diphenylcarbazide and in-line adjustment of Cr(VI) concentration prior to determination. This last feature was attained by incorporating a dilution chamber (for extracts from highly contaminated samples) and another flow-through column filled with multi-walled carbon nanotubes for preconcentration of Cr-1,5-diphenylcarbazide (for extracts or fractions with low Cr(VI) concentration). The MSFIA system was successfully applied to SRM 2701 soil, providing the extraction kinetics for sequential extraction using water and an acid rain surrogate. The automatic integration of extraction and Cr(VI) determination also allowed the minimization of interconversion between Cr oxidation states often observed when determination is not carried out immediately after extraction.

In fact, several MSFIA systems were developed for monitoring of environmental pollutants as indicated in Table 2. Both organic and inorganic species were targeted, with a focus on water analysis. In this context, in-line sample treatment is undeniably a requisite when devising monitoring schemes with real environmental samples. There are two main reasons for this. First, analytes, particularly pollutants, are generally present at low concentrations (ppt or ppb range) in these samples, requiring a preconcentration step in order to meet the linear working range offered by the available detection systems. Secondly, the target analyte may be strongly bound or entrapped in the sample matrix or it can present different forms concerning its oxidation state. Hence, solid phase extraction has been frequently implemented in-line, aiming the enrichment and selective uptake of analytes. It has been applied for the determination of trace levels of phosphate (5 - 50 $\mu g\ l^{-1}$ of P) in natural waters combined to chemiluminescence detection (Morais et al., 2004), for determination of selenium (5.7 - 1290 $\mu g\ l^{-1}$) in natural and seawater (Serra et al., 2010) and for determination of trace iron (0.05 - 8 $\mu g\ l^{-1}$; 0.2 - 42 $\mu g\ l^{-1}$) in waters (Pascoa et al., 2009; Pons et al., 2004, 2005a, 2005b) as far as nutrient analysis is concerned.

Solid phase extraction has also been employed in more than half of the applications focusing on pollutants monitoring and it was implemented in several ways. In flow injection systems, sorbents are generally packed in flow-through columns, which are sequentially percolated by conditioning solution, sample, washing solution and eluent, fostering selective retention of target analyte(s), followed by its/their elution after sample matrix removal. This strategy has also been implemented in MSFIA for determination of total phenolics in waters, using Amberlite XAD-4 as sorbent and in-line derivatization with 4-aminoantipyrine (Oliveira et al., 2005). This MSFIA system was further improved and coupled to liquid chromatography, allowing on-line preconcentration and determination of eleven priority phenolic pollutants in water and soil samples (Oliveira et al., 2009).

Extraction membranes, containing different functional groups, have also been employed in MSFIA systems as they are an advantageous alternative to particulate sorbents because they allow higher flow rates (providing high determination throughputs) and low backpressure, avoiding leakages and clogging. Several applications have been reported, namely for preconcentration of nitrophenols and their determination after elution (Manera et al., 2007a) or using optosensing (Manera et al., 2007b) by probing the extraction membrane with a bifurcated optical fiber connected to a CCD spectrometer. In fact, the utilization of optosensors in MSFIA systems is simplified because all solutions required (sample, conditioning and regenerating solutions) can be automatically delivered by the multisyringe burette in a precise and timely way. Hence, optosensing has also been applied in MSFIA systems for trace level determination of 1-naphthylamine in water samples (Guzmán-Mar et al., 2006b) and determination of sulphide (Ferrer et al., 2005b).

Analyzed species	Sample type	In-line sample treatment	Determination throughput (h^{-1})	Reference
Arsenic	Water, fish muscle and liver	---	108	(Semenova et al., 2002)
Arsenic	Water	Solid phase extraction	9	(Long et al., 2006)
Arsenic (total inorganic) and As(III)	Water, sea lettuce	---	10	(Leal et al., 2006b)
Azinphos methyl	Water	Hydrolysis	7	(Ornelas-Soto et al., 2009)
Chlorotriazine herbicides	Water and soil extracts	Solid phase extraction	Not given	(Boonjob et al., 2011; Boonjob et al., 2010)
Chromium(VI)	Soil leachates	In-line extraction	Not given	(Rosende et al., 2011)
Halogenated organic carbons	Water and leachates	Solid phase extraction	9	(Maya et al., 2008)
Mercury	Water, fish muscle	---	44	(Leal et al., 2006a)
Mercury	Water and leachates	Solid phase extraction	30	(Serra et al., 2008)
Mercury (inorganic and organic)	Water, fish muscle	Solid phase extraction	14	(Serra et al., 2009)
1-Naphthylamine	Water	---	90	(Guzmán-Mar et al., 2006a)
1-Naphthylamine	Water	Solid phase extraction	14	(Guzmán-Mar et al., 2006b)
Nitrophenols	Water, seawater and waste leachates	Solid phase extraction	3 - 11	(Horstkotte et al., 2008; Manera et al., 2007a; Manera et al., 2007b)
Nitrophenols	Water	Liquid-liquid extraction	11	(Miró et al., 2001)
Pharmaceutical residues (NSAIDs)	Water	Solid phase extraction	Not given	(Quintana et al., 2006)
Pharmaceutical residues (thiazide diuretics)	Water and solid waste leachates	Solid phase extraction	12	(Maya et al., 2010)
Phenolic compounds (total)	Water	Solid phase extraction	4 – 16	(Oliveira et al., 2005)
Phenolic compounds	Water and soil	Solid phase extraction	4 – 10	(Oliveira et al., 2009)
Polychlorinated biphenyls	Solid waste leachates	Solid phase extraction	Not given	(Quintana et al., 2009)
Sulphide	Water	---	45	(Ferrer et al., 2004)
Sulphide	Wastewater	Analyte separation by gas diffusion	13 - 20	(de Armas et al., 2004; Maya et al., 2007)
Sulphide	Waters (fresh, seawater and wastewater)	Solid phase extraction	5 - 8	(Ferrer et al., 2005a, 2005b; Ferrer et al., 2006)
Warfarin	Water	Solid phase extraction	12	(de Armas et al., 2002)
Sulphonated azo dyes	Water	---	7.5	(Fernandez et al., 2010)
UV filters	Water (seawater and swimming pool)	Solid phase extraction	7	(Oliveira et al., 2010)

Table 2. MSFIA methods for assessment and monitoring of pollutants

MSFIA systems were also devised for speciation of arsenic (Leal et al., 2006b) and mercury (Serra et al., 2009) in environmental samples. For arsenic, As(III) was quantified by atomic fluorescence spectrometry while As(V) was assessed by difference from total As content, determined for the same sample after in-line reduction of As(V) to As(III) by automatic addition of potassium iodide and ascorbic acid. For mercury, atomic fluorescence spectrometry was also employed and speciation between inorganic and organic (methylmercury) forms was performed by selectively retaining mercury tetrachloro complex in an anion exchange membrane, while organic mercury was directed towards a flow-through UV digestor before detection. Inorganic mercury was later eluted from the membrane by in-line reduction with tin chloride, allowing a limit of detection of 16 ng l^{-1}.

Besides the reduction of intervention from laboratory technicians in the analytical operations, automation of environmental assays by MSFIA provided also an acceptable determination throughput, ranging from 7.5 to 108 determinations per hour in automatic systems where no sample treatment was required or where it was performed off-line. For MSFIA systems comprising in-line sample treatment, determination throughputs ranged from 3 to 30 determinations per hour, which are excellent figures. For instance, the determination of total phosphorus in waste water samples was carried out with a determination throughput of 12 determinations per hour by implementing in-line microwave digestion of samples (Almeida et al., 2004). This is a significant reduction of the assay time when compared to the conventional batch digestion that took about 2 hours for quantitative measurements.

3. Recent trends for sample treatment

As mentioned before, environmental samples comprise complex matrices where target analytes are not generally amenable to direct determination by instrumental analysis, requiring sample treatments. Regarding this aspect, solid phase extraction is undoubtedly the most common treatment applied in MSFIA systems as shown in Tables 1 and 2. Besides the examples presented before in the text, MSFIA capabilities have been recently exploited to perform solid phase extraction using bead injection (BI) prior to chromatographic analysis.

The bead injection concept consists of handling solid suspensions in a fully automatic fashion, where the solid-phase sorbent, presented as micrometric beads, is renewed in each individual analytical cycle, rendering a fresh portion of sorbent for each analysis. Moreover, bead injection allows the simultaneous monitoring of both effluent and solid phase itself (optosensing) in real time, which brings complementary and enhanced insight into the solid phase extraction procedure in a single assay (Gutzman et al., 2006).

The bead injection concept is often associated to the lab-on-valve (LOV) platform. The LOV module comprises a monolithic structure with microconduits machined in a polymethylmethacrylate or polyetherimide unit, which is mounted atop a multiposition valve (Fig. 2), representing a step forward towards automation and miniaturization of flow injection systems. The LOV-BI approach offers two main advantages, not matched by any other automatic, flow-based solid phase extraction scheme: (i) the automatic renewal of sorbent, without any intervention of operator or replacement of devices or physical parts of the system, so as to circumvent the progressive deactivation and tighter packing of permanent in-line solid phase extraction cartridges; and (ii) the accurate metering of sorbent and eluate quantities by resorting to bi-directional programmable flow, as precisely controlled by the multisyringe burette (Miró et al., 2011).

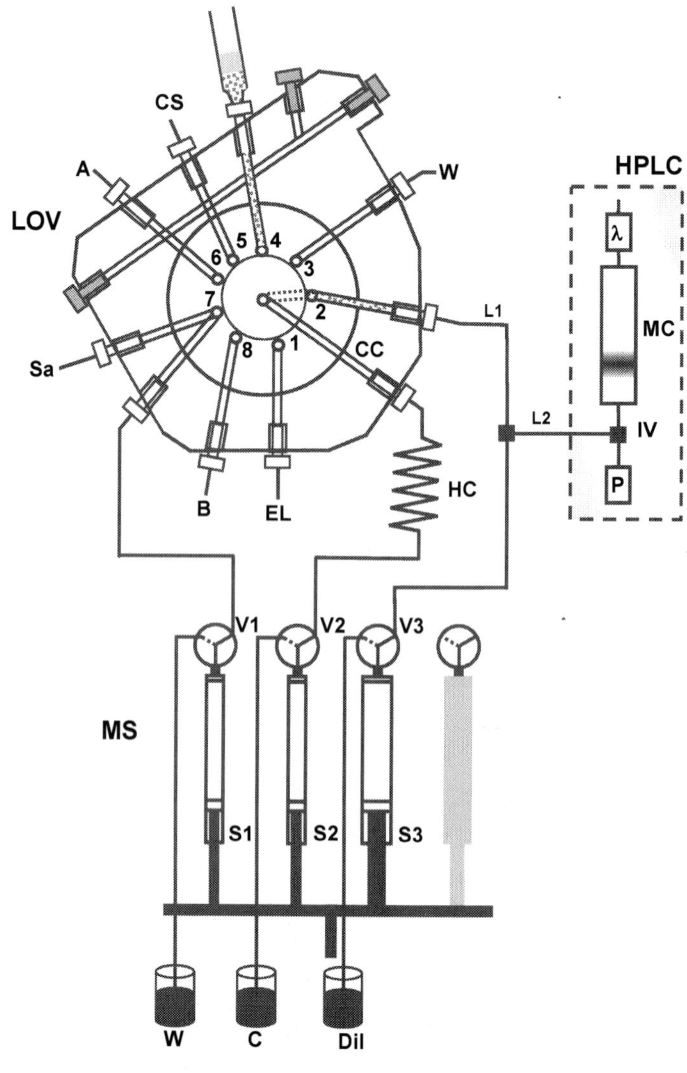

Fig. 2. Schematic representation of multisyringe flow injection system coupled to lab-on-valve for sample treatment, hyphenated to liquid chromatography. LOV: lab-on-valve, MS: multisyringe, HPLC: liquid chromatograph, Si: syringe, Vi: three way commutation valve, A: air, CS: conditioning solvent, C: carrier solution, Dil: diluent, W: waste, CC: central channel, EL: eluent, B: channel for bead discarding, Sa: sample/standard solution, HC: holding coil, P: chromatographic pump, IV: injection valve, MC: chromatographic column, λ: diode array detector. Reproduced with kind permission from Springer Science+Business Media.

However, reliable manipulation of bead suspensions within the flow manifold is the major challenge in mechanized BI protocols for repeatable trapping of beads in microcolumns with subsequent minimization of the uncertainty measurement of the overall analytical method. Initially, spherical shape, uniform size distribution and water-wettability (for reversed-phase materials) were identified as imperative requisites for sorbent selection. Recently, novel strategies for microfluidic handling the sorbent suspensions have been proposed (Oliveira et al., 2011), extending the application of LOV-BI to a larger scope of sorbents, not fitting the previous requirements and opening up new opportunities for preconcentration using molecular imprinted polymers for pharmaceutical residue analysis, for instance.

The hyphenation of LOV-BI-MSFIA to chromatography provided a step further on automation for environmental analysis as sample preparation and analyte separation by chromatography were integrated. Previous automation of sample treatment prior to chromatographic analysis involved robotic analyzers, meaning high equipment costs and expensive operation. By using LOV-BI-MSFIA, while one sample is injected in the chromatographic equipment, the following sample is processed in the MSFIA equipment for matrix removal and analyte enrichment. This is an important, advantageous aspect when dealing with labile analytes that cannot sit on automatic injectors for a long time.

As depicted in Fig. 2, connecting the liquid chromatograph equipment to LOV-BI-MSFIA is rather simple requiring that one of the lateral ports of the LOV platform is directly connected to the injection valve present in the chromatograph, allowing the introduction into the injection loop of all eluate or merely a fraction of it via heart-cut injection protocols. The transfer of the entire volume of eluate into the separation system is essential to reach low limits of detection required by analysis of pollutants in environmental samples, especially when using low-sensitivity detectors, for instance UV spectrophotometers. This approach, along with the handling of a well-defined volume of sample (about 10 ml), fostered the determination of non-steroidal anti-inflammatory drugs (NSAIDs) (Quintana et al., 2006) and chlorotriazine herbicides and some of its metabolites (Boonjob et al., 2010) in environmental samples at the low-μg l^{-1} level (Table 3). The screening of UV filters in swimming pool and seawaters also profited from the combination LOV-BI-MSFIA (Oliveira et al., 2010). In-line dilution was necessary after analyte elution in order to match the eluate composition to the aqueous content of the mobile phase, avoiding band broadening problems.

Analyzed species	Working range	Limit of detection	Precision (RSD%)	Reference
Chlorotriazine herbicides	0.1 - 10 μg l^{-1}	0.02 – 0.04 μg l^{-1}	<5.5	(Boonjob et al., 2010)
Pharmaceutical residues (NSAIDs)	0.4 - 40 μg l^{-1}	0.02 – 0.67 μg l^{-1}	<11	(Quintana et al., 2006)
Polychlorinated biphenyls	2 - 100 ng l^{-1}	0.5 – 6.1 ng l^{-1}	<9	(Quintana et al., 2009)
UV filters	5 - 160 μg l^{-1}	0.45 – 3.2 μg l^{-1}	<13	(Oliveira et al., 2010)

Table 3. Analytical figures of LOV-BI-MSFIA system coupled to chromatographic separation

The hyphenation of LOV-BI-MSFIA to gas chromatography is not as simple as it is for liquid chromatography. First, lower injection volumes are required and the analytes should be eluted in a solvent prone to fast vaporization. In fact, only one application has been described so far, where low values for limit of detection were attained through the automatic, on-line transfer of all eluate to a gas chromatograph equipped with an electron capture detector and a programmable temperature vaporization injector for determination of polychlorinated biphenyls in solid-waste leachates at the 2–100 ng L^{-1} range (Quintana et al, 2009).

4. Conclusions

MSFIA is undoubtedly a suitable automation tool for implementation of environmental analysis. Considering the examples presented here, the proof of concept has been given, shown by application to a large suite of species, comprehending nutrients and pollutants in several environmental matrices. The determination throughputs attained are suitable for most applications sought in environmental monitoring schemes and field deployment is possible for many of the MSFIA systems developed, as long as periodic reagent refilling is guaranteed.

Automation and integration of sample treatment to instrumental quantification of analytes was successfully demonstrated, profiting from the multichannel operation of MSFIA equipment. However, some unique features provided by MSFIA are underexploited for environmental analysis. Recent work regarding LOV-BI-MSFIA coupled to chromatography are still in its infancy and will certainly grow into more reliable, comprehensive analyzers for monitoring of emerging pollutants.

5. Acknowledgement

Financial support from Fundação para a Ciência e Tecnologia (FCT) through grant no. PEst-C/EQB/LA0006/2011 and by European Union through FEDER and QREN 2007-2013 programs (project PTDC/AAC-AMB/104882/2008) is acknowledged. H.M. Oliveira also thanks post-doctoral grant SFRH/BPD/75065/2010.

6. References

Almeida, M.I.G.S.; Estela, J.M. & Cerdà, V. (2011). Multisyringe Flow Injection Potentialities for Hyphenation with Different Types of Separation Techniques. *Analytical Letters*, Vol.44, No.1-3, (Feb 2011), pp. 360-373, ISSN 0003-2719

Almeida, M.I.G.S.; Segundo, M.A.; Lima, J.L.F.C. & Rangel, A.O.S.S. (2004). Multi-Syringe Flow Injection System with in-Line Microwave Digestion for the Determination of Phosphorus. *Talanta*, Vol.64, No.5, (Dec 2004), pp. 1283-1289, ISSN 0039-9140

Almeida, M.I.G.S.; Segundo, M.A.; Lima, J.L.F.C. & Rangel, A.O.S.S. (2005). Multi-Syringe Flow Injection System for the Determination of Available Phosphorus in Soil Samples. *International Journal of Environmental Analytical Chemistry*, Vol.85, No.1, (Jan 2005), pp. 51-62, ISSN 0306-7319

Almeida, M.I.G.S.; Segundo, M.A.; Lima, J.L.F.C. & Rangel, A.O.S.S. (2006). Potentiometric Multi-Syringe Flow Injection System for Determination of Exchangeable Potassium in Soils with in-Line Extraction. *Microchemical Journal*, Vol.83, No.2, (Jul 2006), pp. 75-80, ISSN 0026-265X

Almeida, M.I.G.S.; Segundo, M.A.; Lima, J.L.F.C. & Rangel, A.O.S.S. (2008). Direct Introduction of Slurry Samples in Multi-Syringe Flow Injection Analysis: Determination of Potassium in Plant Samples. *Analytical Sciences*, Vol.24, No.5, (May 2008), pp. 601-606, ISSN 0910-6340

Boonjob, W.; Miró, M. & Cerdà, V. (2008). Multiple Stirred-Flow Chamber Assembly for Simultaneous Automatic Fractionation of Trace Elements in Fly Ash Samples Using a Multisyringe-Based Flow System. *Analytical Chemistry*, Vol.80, No.19, (Oct 2008), pp. 7319-7326, ISSN 0003-2700

Boonjob, W.; Miró, M.; Segundo, M.A. & Cerdà, V. (2011). Flow-through Dispersed Carbon Nanofiber-Based Microsolid-Phase Extraction Coupled to Liquid Chromatography for Automatic Determination of Trace Levels of Priority Environmental Pollutants. *Analytical Chemistry*, Vol.83, No.13, (Jul 2011), pp. 5237-5244, ISSN 0003-2700

Boonjob, W.; Yu, Y.L.; Miró, M.; Segundo, M.A.; Wang, J.H. & Cerdà, V. (2010). Online Hyphenation of Multimodal Microsolid Phase Extraction Involving Renewable Molecularly Imprinted and Reversed-Phase Sorbents to Liquid Chromatography for Automatic Multiresidue Assays. *Analytical Chemistry*, Vol.82, No.7, (Apr 2010), pp. 3052-3060, ISSN 0003-2700

Buanuam, J.; Miró, M.; Hansen, E.H.; Shiowatana, J.; Estela, J.M. & Cerdà, V. (2007). A Multisyringe Flow-through Sequential Extraction System for on-Line Monitoring of Orthophosphate in Soils and Sediments. *Talanta*, Vol.71, No.4, (Mar 2007), pp. 1710-1719, ISSN 0039-9140

Cerdà, V.; Estela, J.M.; Forteza, R.; Cladera, A.; Becerra, E.; Altimira, P. & Sitjar, P. (1999). Flow Techniques in Water Analysis. *Talanta*, Vol.50, No.4, (Nov 1999), pp. 695-705, ISSN 0039-9140

Chomchoei, R.; Hansen, E.H. & Shiowatana, J. (2004). Utilizing a Sequential Injection System Furnished with an Extraction Microcolumn as a Novel Approach for Executing Sequential Extractions of Metal Species in Solid Samples. *Analytica Chimica Acta*, Vol.526, No.2, (Nov 2004), pp. 177-184, ISSN 0003-2670

de Armas, G.; Ferrer, L.; Miró, M.; Estela, J.M. & Cerdà, V. (2004). In-Line Membrane Separation Method for Sulfide Monitoring in Wastewaters Exploiting Multisyringe Flow Injection Analysis. *Analytica Chimica Acta*, Vol.524, No.1-2, (Oct 2004), pp. 89-96, ISSN 0003-2670

de Armas, G.; Miró, M.; Estela, J.M. & Cerdà, V. (2002). Multisyringe Flow Injection Spectrofluorimetric Determination of Warfarin at Trace Levels with on-Line Solid-Phase Preconcentration. *Analytica Chimica Acta*, Vol.467, No.1-2, (Sep 2002), pp. 13-23, ISSN 0003-2670

Fernandez, C.; Larrechi, M.S.; Forteza, R.; Cerdà, V. & Callao, M.P. (2010). Multisyringe Chromatography (MSC) Using a Monolithic Column for the Determination of

Sulphonated Azo Dyes. *Talanta*, Vol.82, No.1, (Jun 2010), pp. 137-142, ISSN 0039-9140

Ferrer, L.; de Armas, G.; Miró, M.; Estela, J.M. & Cerdà, V. (2004). A Multisyringe Flow Injection Method for the Automated Determination of Sulfide in Waters Using a Miniaturised Optical Fiber Spectrophotometer. *Talanta*, Vol.64, No.5, (Dec 2004), pp. 1119-1126, ISSN 0039-9140

Ferrer, L.; de Armas, G.; Miró, M.; Estela, J.M. & Cerdà, V. (2005a). Flow-through Optical Fiber Sensor for Automatic Sulfide Determination in Waters by Multisyringe Flow Injection Analysis Using Solid-Phase Reflectometry. *Analyst*, Vol.130, No.5, (May 2005), pp. 644-651, ISSN 0003-2654

Ferrer, L.; de Armas, G.; Miró, M.; Estela, J.M. & Cerdà, V. (2005b). Interfacing in-Line Gas-Diffusion Separation with Optrode Sorptive Preconcentration Exploiting Multisyringe Flow Injection Analysis. *Talanta*, Vol.68, No.2, (Dec 2005), pp. 343-350, ISSN 0039-9140

Ferrer, L.; Estela, J.M. & Cerdà, V. (2006). A Smart Multisyringe Flow Injection System for Analysis of Sample Batches with High Variability in Sulfide Concentration. *Analytica Chimica Acta*, Vol.573, (Jul 2006), pp. 391-398, ISSN 0003-2670

Gomes, D.M.C.; Segundo, M.A.; Lima, J.L.F.C. & Rangel, A.O.S.S. (2005). Spectrophotometric Determination of Iron and Boron in Soil Extracts Using a Multi-Syringe Flow Injection System. *Talanta*, Vol.66, No.3, (Apr 2005), pp. 703-711, ISSN 0039-9140

Gutzman, Y.; Carroll, A.D. & Ruzicka, J. (2006). Bead Injection for Biomolecular Assays: Affinity Chromatography Enhanced by Bead Injection Spectroscopy. *Analyst*, Vol.131, No.7, (Jul 2006), pp. 809-815, ISSN 0003-2654

Guzmán-Mar, J.L.; Martinez, L.L.; de Alba, P.L.L.; Duran, J.E.C. & Martin, V.C. (2006a). Multisyringe Flow Injection Analysis for Determination of 1-Naphthylamine in Water Samples. *Microchimica Acta*, Vol.153, No.3-4, (Feb 2006), pp. 139-144, ISSN 0026-3672

Guzmán-Mar, J.L.; Martinez, L.L.; de Alba, P.L.L.; Duran, J.E.C. & Martin, V.C. (2006b). Optical Fiber Reflectance Sensor Coupled to a Multisyringe Flow Injection System for Preconcentration and Determination of 1-Naphthylamine in Water Samples. *Analytica Chimica Acta*, Vol.573, (Jul 2006), pp. 406-412, ISSN 0003-2670

Horstkotte, B.; Elsholz, O. & Martin, V.C. (2008). Multisyringe Flow Injection Analysis Coupled to Capillary Electrophoresis (MSFIA-CE) as a Novel Analytical Tool Applied to the Pre-Concentration, Separation and Determination of Nitrophenols. *Talanta*, Vol.76, No.1, (Jun 2008), pp. 72-79, ISSN 0039-9140

Leal, L.O.; Elsholz, O.; Forteza, R. & Cerdà, V. (2006a). Determination of Mercury by Multisyringe Flow Injection System with Cold-Vapor Atomic Absorption Spectrometry. *Analytica Chimica Acta*, Vol.573, (Jul 2006), pp. 399-405, ISSN 0003-2670

Leal, L.O.; Forteza, R. & Cerdà, V. (2006b). Speciation Analysis of Inorganic Arsenic by a Multisyringe Flow Injection System with Hydride Generation-Atomic Fluorescence Spectrometric Detection. *Talanta*, Vol.69, No.2, (Apr 2006), pp. 500-508, ISSN 0039-9140

Long, X.B.; Miró, M.; Hansen, E.H.; Estela, J.M. & Cerdà, V. (2006). Hyphenating Multisyringe Flow Injection Lab-on-Valve Analysis with Atomic Fluorescence Spectrometry for on-Line Bead Injection Preconcentration and Determination of Trace Levels of Hydride-Forming Elements in Environmental Samples. *Analytical Chemistry*, Vol.78, No.24, (Dec 2006), pp. 8290-8298, ISSN 0003-2700

Magalhães, L.M.; Ribeiro, J.P.N.; Segundo, M.A.; Reis, S. & Lima, J.L.F.C. (2009). Multi-Syringe Flow-Injection Systems Improve Antioxidant Assessment. *Trac-Trends in Analytical Chemistry*, Vol.28, No.8, (Sep 2009), pp. 952-960, ISSN 0165-9936

Manera, M.; Miró, M.; Estela, J.M. & Cerdà, V. (2007a). Multi-Syringe Flow Injection Solid-Phase Extraction System for on-Line Simultaneous Spectrophotometric Determination of Nitro-Substituted Phenol Isomers. *Analytica Chimica Acta*, Vol.582, No.1, (Jan 2007), pp. 41-49, ISSN 0003-2670

Manera, M.; Miró, M.; Estela, J.M.; Cerdà, V.; Segundo, M.A. & Lima, J.L.F.C. (2007b). Flow-through Solid-Phase Reflectometric Method for Simultaneous Multiresidue Determination of Nitrophenol Derivatives. *Analytica Chimica Acta*, Vol.600, No.1-2, (Sep 2007), pp. 155-163, ISSN 0003-2670

Maya, F.; Estela, J.M. & Cerdà, V. (2007). Improving the Chemiluminescence-Based Determination of Sulphide in Complex Environmental Samples by Using a New, Automated Multi-Syringe Flow Injection Analysis System Coupled to a Gas Diffusion Unit. *Analytica Chimica Acta*, Vol.601, No.1, (Oct 2007), pp. 87-94, ISSN 0003-2670

Maya, F.; Estela, J.M. & Cerdà, V. (2008). Completely Automated System for Determining Halogenated Organic Compounds by Multisyringe Flow Injection Analysis. *Analytical Chemistry*, Vol.80, No.15, (Aug 2008), pp. 5799-5805, ISSN 0003-2700

Maya, F.; Estela, J.M. & Cerdà, V. (2009). Multisyringe Flow Injection Technique for Development of Green Spectroscopic Analytical Methodologies. *Spectroscopy Letters*, Vol.42, No.6-7, (Dec 2009), pp. 312-319, ISSN 0038-7010

Maya, F.; Estela, J.M. & Cerdà, V. (2010). Interfacing on-Line Solid Phase Extraction with Monolithic Column Multisyringe Chromatography and Chemiluminescence Detection: An Effective Tool for Fast, Sensitive and Selective Determination of Thiazide Diuretics. *Talanta*, Vol.80, No.3, (Jan 2010), pp. 1333-1340, ISSN 0039-9140

Miró, M.; Cerdà, V. & Estela, J.M. (2002). Multisyringe Flow Injection Analysis: Characterization and Applications. *Trac-Trends in Analytical Chemistry*, Vol.21, No.3, (Mar 2002), pp. 199-210, ISSN 0165-9936

Miró, M.; Cladera, A.; Estela, J.M. & Cerdà, V. (2001). Dual Wetting-Film Multi-Syringe Flow Injection Analysis Extraction - Application to the Simultaneous Determination of

Nitrophenols. *Analytica Chimica Acta*, Vol.438, No.1-2, (Jul 2001), pp. 103-116, ISSN 0003-2670

Miró, M.; Oliveira, H.M. & Segundo, M.A. (2011). Analytical Potential of Mesofluidic Lab-on-a-Valve as a Front End to Column-Separation Systems. *Trac-Trends in Analytical Chemistry*, Vol.30, No.1, (Jan 2011), pp. 153-164, ISSN 0165-9936

Morais, I.P.A.; Miró, M.; Manera, M.; Estela, J.M.; Cerdà, V.; Souto, M.R.S. & Rangel, A.O.S.S. (2004). Flow-through Solid-Phase Based Optical Sensor for the Multisyringe Flow Injection Trace Determination of Orthophosphate in Waters with Chemiluminescence Detection. *Analytica Chimica Acta*, Vol.506, No.1, (Mar 2004), pp. 17-24, ISSN 0003-2670

Oliveira, H.M.; Miró, M.; Segundo, M.A. & Lima, J.L.F.C. (2011). Universal Approach for Mesofluidic Handling of Bead Suspensions in Lab-on-Valve Format. *Talanta*, Vol.84, No.3, (May 2011), pp. 846-852, ISSN 0039-9140

Oliveira, H.M.; Segundo, M.A.; Lima, J.L.F.C. & Cerdà, V. (2009). Multisyringe Flow Injection System for Solid-Phase Extraction Coupled to Liquid Chromatography Using Monolithic Column for Screening of Phenolic Pollutants. *Talanta*, Vol.77, No.4, (Feb 2009), pp. 1466-1472, ISSN 0039-9140

Oliveira, H.M.; Segundo, M.A.; Lima, J.L.F.C.; Miró, M. & Cerdà, V. (2010). On-Line Renewable Solid-Phase Extraction Hyphenated to Liquid Chromatography for the Determination of UV Filters Using Bead Injection and Multisyringe-Lab-on-Valve Approach. *Journal of Chromatography A*, Vol.1217, No.22, (May 2010), pp. 3575-3582, ISSN 0021-9673

Oliveira, H.M.; Segundo, M.A.; Reis, S. & Lima, J.L.F.C. (2005). Multi-Syringe Flow Injection System with in-Line Pre-Concentration for the Determination of Total Phenolic Compounds. *Microchimica Acta*, Vol.150, No.2, (Jun 2005), pp. 187-196, ISSN 1436-5073

Ornelas-Soto, N.E.; Guzman-Mar, J.L.; de Alba, P.L.L.; Martinez, L.L.; Barbosa-Garcia, O. & Martin, V.C. (2009). Coupled Multisyringe Flow Injection/Reactor Tank for the Spectrophotometric Detection of Azinphos Methyl in Water Samples. *Microchimica Acta*, Vol.167, No.3-4, (Dec 2009), pp. 273-280, ISSN 0026-3672

Pascoa, R.N.M.J.; Toth, I.V. & Rangel, A.O.S.S. (2009). A Multi-Syringe Flow Injection System for the Spectrophotometric Determination of Trace Levels of Iron in Waters Using a Liquid Waveguide Capillary Cell and Different Chelating Resins and Reaction Chemistries. *Microchemical Journal*, Vol.93, No.2, (Nov 2009), pp. 153-158, ISSN 0026-265X

Pons, C.; Forteza, R. & Cerdà, V. (2004). Expert Multi-Syringe Flow-Injection System for the Determination and Speciation Analysis of Iron Using Chelating Disks in Water Samples. *Analytica Chimica Acta*, Vol.524, No.1-2, (Oct 2004), pp. 79-88, ISSN 0003-2670

Pons, C.; Forteza, R. & Cerdà, V. (2005a). Optical Fibre Reflectance Sensor for the Determination and Speciation Analysis of Iron in Fresh and Seawater Samples

Coupled to a Multisyringe Flow Injection System. *Analytica Chimica Acta*, Vol.528, No.2, (Jan 2005), pp. 197-203, ISSN 0003-2670

Pons, C.; Forteza, R. & Cerdà, V. (2005b). The Use of Anion-Exchange Disks in an Optrode Coupled to a Multi-Syringe Flow-Injection System for the Determination and Speciation Analysis of Iron in Natural Water Samples. *Talanta*, Vol.66, No.1, (Mar 2005), pp. 210-217, ISSN 0039-9140

Quintana, J.B.; Boonjob, W.; Miró, M. & Cerdà, V. (2009). Online Coupling of Bead Injection Lab-on-Valve Analysis to Gas Chromatography: Application to the Determination of Trace Levels of Polychlorinated Biphenyls in Solid Waste Leachates. *Analytical Chemistry*, Vol.81, No.12, (Jun 2009), pp. 4822-4830, ISSN 0003-2700

Quintana, J.B.; Miró, M.; Estela, J.M. & Cerdà, V. (2006). Automated on-Line Renewable Solid-Phase Extraction-Liquid Chromatography Exploiting Multisyringe Flow Injection-Bead Injection Lab-on-Valve Analysis. *Analytical Chemistry*, Vol.78, No.8, (Apr 2006), pp. 2832-2840, ISSN 0003-2700

Rosende, M.; Miró, M.; Segundo, M.A.; Lima, J.L.F.C. & Cerdà, V. (2011). Highly Integrated Flow Assembly for Automated Dynamic Extraction and Determination of Readily Bioaccessible Chromium(VI) in Soils Exploiting Carbon Nanoparticle-Based Solid-Phase Extraction. *Analytical and Bioanalytical Chemistry*, Vol.400, No.7, (Jun 2011), pp. 2217-2227, ISSN 1618-2642

Ruzicka, J. & Hansen, E.H. (1975). Flow Injection Analyses. 1. New Concept of Fast Continuous-Flow Analysis. *Analytica Chimica Acta*, Vol.78, No.1, (Aug 1975), pp. 145-157, ISSN 0003-2670

Ruzicka, J. & Marshall, G.D. (1990). Sequential Injection - a New Concept for Chemical Sensors, Process Analysis and Laboratory Assays. *Analytica Chimica Acta*, Vol.237, No.2, (Oct 1990), pp. 329-343, ISSN 0003-2670

Segundo, M.A. & Magalhães, L.M. (2006). Multisyringe Flow Injection Analysis: State-of-the-Art and Perspectives. *Analytical Sciences*, Vol.22, No.1, (Jan 2006), pp. 3-8, ISSN 0910-6340

Semenova, N.V.; Leal, L.O.; Forteza, R. & Cerdà, V. (2002). Multisyringe Flow-Injection System for Total Inorganic Arsenic Determination by Hydride Generation-Atomic Fluorescence Spectrometry. *Analytica Chimica Acta*, Vol.455, No.2, (Mar 2002), pp. 277-285, ISSN 0003-2670

Semenova, N.V.; Leal, L.O.; Forteza, R. & Cerdà, V. (2003). Multisyringe Flow Injection System for Total Inorganic Selenium Determination by Hydride Generation-Atomic Fluorescence Spectrometry. *Analytica Chimica Acta*, Vol.486, No.2, (Jun 2003), pp. 217-225, ISSN 0003-2670

Serra, A.M.; Estela, J.M. & Cerdà, V. (2008). MSFIA System for Mercury Determination by Cold Vapour Technique with Atomic Fluorescence Detection. *Talanta*, Vol.77, No.2, (Dec 2008), pp. 556-560, ISSN 0039-9140

Serra, A.M.; Estela, J.M. & Cerdà, V. (2009). An MSFIA System for Mercury Speciation Based on an Anion-Exchange Membrane. *Talanta*, Vol.78, No.3, (May 2009), pp. 790-794, ISSN 0039-9140

Serra, A.M.; Estela, J.M.; Coulomb, B.; Boudenne, J.L. & Cerdà, V. (2010). Solid Phase Extraction - Multisyringe Flow Injection System for the Spectrophotometric Determination of Selenium with 2,3-Diaminonaphthalene. *Talanta*, Vol.81, No.1-2, (Apr 2010), pp. 572-577, ISSN 0039-9140

18

Photopolymerizable Materials in Biosensorics

Nickolaj Starodub
National University of Life and Environmental Sciences, Ukraine

1. Introduction

The development of the effective methods for the biological material immobilization is the main problem of biosensorics. This process may be classified as including biological selective components into isolated phase which is separated from free solution but can exchange with its by molecules of substrate, effectors, inhibitors and others (Triven, 1983). The most often biological material is covalently bound with some insoluble polymer, linked together or with some inert protein. In all these cases it is obtained the non-soluble but active complex. It is realization of chemical approaches which have unfortunately a number of disadvantages and main among them is the lost of activity of biological materials (and often very much). Another set of methods is associated with the physical sorption of biological material on the transducer surface at the use of electrostatic or non-covalent mechanisms of binding. In this case, as a rule, the loss of biological material activity does not occur but for the providing a reliable binding there is necessary to complicate immobilization procedure. Application of poly-electrolytes as intermediate layer is the most productive way. From other side, biological material may be not directly bound to the some surface and can be kept inside of a special polymer or double phospholipids (liposomes).

The choice of a particular method of immobilization is a very important moment in the development of biosensors and it must be based on taken into account of the following points: 1) what kind of chemical or physical-chemical reactions will be occurred on the surface; 2) molecules should kept the stability at the process of immobilization and during chip working; 3) chemicals for cross linking should interact with groups of biomolecules which are remote from their active centers; 4) if demands of point 3 can not be fulfilled the bifunctional reagents which are used for the linking should be as large as possible to penetrate to the active centers of biomolecules (for example, activated cellulose is more suitable than glutaraldehyde); 5) active sites must be protected, in particular, by substrates or glutathione, cysteine, papain or others reagents for blocking sulfhydryl groups with their reactivation in advance; 6) the procedure of washing of not-linked biomaterial should not effect negatively for immobilized one, especially, if there are subunit forms to prevent their dissociation; 7) what kind of physical and mechanical abilities may form immobilized material: thin layer, thick film, an amorphous structure, etc. If all these points are correspond to needed conditions the chosen method is well for the creation of biosensor.

2. The application of polymers as immobilization matrix in biosensors

One among of simple and reliable approach for integration of the biomaterial (enzymes, antibodies, antigens, cells) in biosensors is based on the use of polymers (Rehman et al.,

1999; Turner, 1989). The first investigations in this direction were fulfilled shortly after the discovery of this type of instrumental analytical devices by Clark (Turner, 1989). In most cases it was the polymers obtained by chemical way from acrylic acid and used for including enzymes and cells (Freemen, 1986; Starodub et al., 1990; 1995; 1998; 1999; 2001). These investigations shown as perspective application synthetic polymers in biosensor technology and opened all problems in this respect. Up to now a lot of work ware curried out in this aspect.

Recently, miniaturization of the transducers and the application of number of different biological components in the same biosensors are occurred. Unfortunately, the traditional approaches can not fulfill all practice demands in this aspect. Only photochemical formed and cross linked polymers may meet the requirements of technology of biosensors. The application of such approach has the next advantages at the currying out of the immobilization: 1) absence of destructive factors; 2) possibility to work at the room or more low temperatures; 3) accurate given initiation and termination of process; 4) formation biopolymer in precisely defined area, usually in a very small; 5) combination in single technological process of photolithography production of semiconductors. All this could affect significantly on the production costs and expanding its application (Grishchenko et al., 1985; Masljuk & Chranovsky, 1989; Starodub, 1989, 1990; Kuriyamma & Kimura, 1991).

2.1 General characteristics of the formation of photopolymers

For the activation of process of the photochemical polymerization it is used ultraviolet (UV) and not so often fast electrons, roentgen, gamma and plasma radiation. Certainly, these factors damage biological materials so try to use less stringent radiation, for example, UV with wavelength within 300-400 nm. The photochemical transformation of unsaturated compound in the polymer is activated with the help of photo initiators (PhI), which can absorb photons of UV. At that energy of the activation of polymerization process decreases (up to 17-34 kJ/mol) in comparison with the photo initiated solidification. Absorption of light by PhI transforms of its in electronically excited state which causes destruction of molecule with the formation of free radicals initiated polymerization. PhI is usually thermally stable (Masljuk & Chranovsky, 1989). Among of photochemical reactions used for obtaining polymers having practical importance there is necessary to pay attention on the photo destruction (photochemical cross-linking) and photo polymerization. Two types of photocomposition deserve attention. First type of structuring consists in cross-linking of preliminary obtained linear polymer due to taken part of side reaction groups in the presence of PhI (or without it) at the UV irradiation. Second type of photo structuring is photochemical formation of polymers with the participation of bifunctional light sensitive substances. In this case photo cross-linking of linear polymers is accomplished through a special photo sensitive reagent. There is necessary to underline that the obtaining polymers through photochemical initiation is used in industry more often than through photo cross-linking. Oligomers, which are able to polymerization and contended inter chain and terminal double bonds as well as light-sensitive monomer-oligomer compositions based on them, at the exposition to UV radiation solidify, forming a polymer material. The preferential use of these classes of substances at the obtaining of photo polymerisable compositions is connected with the great potential to regular properties of oligomers by changing the nature and structure of the starting compounds for synthesis or as a result of copolymerization with many vinyl monomers, which ultimately determines the possibility of creating complex polymeric materials with different properties. The widespread used compositions of photo polymerizable compositions in biosensors are based on the derivatives of acrylic acid and polyurethanes as well as in the case of photo-cross linkable polymers polyvinyl alcohol and polyvinylpyrrolidone. They are the most dispersed in application. Below we will concentrate our attention on the features of the procedures and preparation photo polymerisable and photo-cross linkable polymers.

2.2 Photo polymerization in biosensorics

We will discuss about two different approaches in the formation of photopolymers, namely, with application of acrylic and urethane derivatives.

2.2.1 Acrylic derivatives

These derivatives as photopolymers were used often for the creation of biosensor elements (Arica & Hasirci, 1987; Kumakura & Kaetsu, 1989; Doretti & Ferrara. 1993; Doretti et al., 1994; Jimenez et al., 1995; Macca et al., 1995; Moser et al., 1995; Gooding & Hall, 1996; Lesho & Sheppard, 1996; Ambrose & Meyerhoff, 1997; Wróblewski et al., 1997; Doretti et al., 1998; Hall et al., 1999; Kolytcheva et al., 1999; Mohy et al., 1999; Rehman et al., 1999). 2-(hydroxyethyl)methacrylate (HEMA) is used the mostet often. The enzyme immobilization in polymeric matrix on the basis of HEMA has some advantages since it has appropriate mechanical abilities and optimal pore size needed for the retention of enzyme molecules, transportation of substrates and products of reaction.

It was reported (Arica & Hasirci, 1987) about β-glucose oxidase (GOD) immobilization in polymeric gel contended HEMA and N,N'-methylenebisacrylamide (gross-linking agent). Azobis-izonitril and ammonia-persulfat served as PhI. Mixture of monomers, PhI and enzyme dissolved in phosphate buffer (0.1 M, pH 7.0) were illuminated by UV lamp (12 W) at the temperature of 25 °C in nitrogen. The immobilized enzyme had maximal activity at pH 7.0 and 35 °C (in opposite 5,5 and 30 °C in free state) with some higher value of K_M (13.33 mM comparatively 6.66 mM) and decreasing maximal reaction rate in 1.6 time. The residual activity of the immobilized enzyme after 60 days of preservation was ~90% of initial level.

The main reason for a significant change in the pH optimum of immobilized GOD (7.0) is, apparently, a local pH decreasing in the membrane during enzyme functioning. We should also mention about some diffusion limitations in the kinetics of reactions catalyzed by immobilized enzymes, leading to an increase in the apparent K_M.

Other researchers (Kumakura & Kaetsu, 1989) used hydroxyethyl, HEMA and tetra-ethylene glycol diacrylate as monomers for immobilization of cellulase. Polymerization was induced by γ-rays of cobalt-60 (exposure time 1 hour, 1 Mrad dose, the temperature of 24° C or -78° C). The polymerization rate increased with the addition of water to HEMA. It was studied the dependence of the enzyme activity on the ratio of water-HEMA and the thickness of the obtained membrane (0.1 - 1 mm). When HEMA content was 20% the cellulase activity increased with the membrane thickness. At the HEMA content of 60% the observed maximum was at 0.6 mm and with pure HEMA - gradual decline of enzyme activity with increasing membrane thickness. When HEMA content was 80% and membrane thickness in 0.1 mm enzyme activity was the same as in case of HEMA content of 20% and 1 mm thick membranes. The reason is that with the raise of water content in the polymer its hydrophilic properties are increased. It promotes to the establishment of native conformation of the enzyme and, moreover, polymer with a water content of 60-90% has a high porosity (pore diameter is 2-5 μm) and enzyme molecules are rapid washed from the membrane. When the water content is about 20-30% the pore diameter reaches 0.2-0.5 mm and although it is larger than the size of the enzyme molecule the polymer keeps an active immobilized molecules. At the use of pure HEMA the obtained polymer has not a porous structure that prevents the penetration of the substrate to the enzyme. Membranes obtained from other monomers (hydroxyethyl acrylate and tetraethylene glycol diacrylate) showed the worst enzyme activity.

A somewhat different method was proposed (Doretti et al., 1998) based on the use of a mixture of HEMA (83%), glycidyl methacrylate (13%) and 4% of trimethyl-propan-trimethacrylate (cross linking agent). This mixture was polymerized by irradiation of γ-rays from cobalt-60 at -78 °C. To the resulting polymer, the enzyme was linked by interaction of amino groups with the methacrylate copolymer (polymer solution contacted with the enzyme). Amperometric transducer (platinum wire) was coated with this polymer to which then butyryl- or acetyl cholinesterase, or cholineoxidase, or peroxidase was linked. It was obtained a linear response to acetylcholine chloride - $5 \cdot 10^{-6}$ – $1.4 \cdot 10^{-4}$ M and for iodide butirilcholine - $2 \cdot 10^{-6}$ - 10^{-4} M, with an optimum at pH 9.5 and 8.0, respectively.

Nylon membrane (with pores of 0.2, 1.2, 3.0 μm) was treated with diethylene glycol dimethacrylate (acetone solution) under the influence of γ-rays (Cesium-137 source). They were then exposed alternately in solutions of glutaraldehyde and β-galactosidase. The best results were achieved using a 15% solution of diethylene glycol dimethacrylate, 2.5% of glutaraldehyde and the enzyme concentration of 10 mg/ml. The optimal response biosensor was achieved at a temperature of 50-60 °C and pH 4-6. It is expected the use of membranes prepared by the above mentioned method to create thermal bioreactor designed for the determination of glucose in the milk (Rehman et al, 1999). There is evidence about linking oligonucleotides to optrodes with help of acrylamide. For this purpose, the surface was treated by 3-methacryl-oxy-propyl-trimethoxy-silane under the ultraviolet light, as well as acrylamide and bisacrylamide in a ratio of 17:1 by weight. Oligonucleotides contended 5'-terminal acrylamide group covalently were linked to above mentioned surfacee. The density of immobilized oligonucleotides was 190 - 200 femtamol/mm^2.

For the immobilization of penicillinase or penicillin amidase it was proposed a approach (Macca et al., 1995) using a mixture of HEMA, N, N'-methylenebisacrylamide and enzyme solution in phosphate buffer. This mixture is sprayed into cooled n-hexane (-78 °C) and irradiated with γ-rays of cobalt-60. As a result of this procedure the spherical polymer beads with a diameter of about 0.5 mm were obtained and used in a flow pH-sensitive bioreactor. It was found that the sensitivity of the biosensor with such membranes was $6.3 \cdot 10^{-3}$ M of benzyl penicillin sodium for both enzymes and a change of its response level reached about 91 mV/decade of concentration for penicillinase and 66 mV/decade of penicillin amidase.

At the creation of an amperometric biosensor for the measurement of glucose the enzyme was mixed in buffer (pH 7.0) with acrylamide, N, N'-methilene diacrylamide, 2,2-dimethoxy-2-phenylacetophenone (PhI) and glycerin. The polymerization was initiated by UV. The biosensor had sensitivity from 45 to 67 nA/mM, the linear response region was located in zone $4 \cdot 10^{-3}$ - 1 mM. About 95% of the maximum response biosensor with a membrane thickness of 10-15 μm was realized within 20-30 sec. It was stable for 3 weeks (Jimenez et al., 1995).

Another amperometric biosensor for glucose (Doretti&Ferrara, 1993) was obtained as follows. HEMA and trimethylol-propan-triacrilate (cross-linking agent) in a ratio of 96:4 was mixed with a solution GOD in 0.1 M phosphate buffer (at a ratio of 2:1). Enzyme concentration in the mixture was 4 mg per g of mixture which was polymerized at -78 °C and with the application of cobalt-60 γ-rays. The optimal level of the response of the obtained biosensor was at pH 6.0 and 40 °C (in compared with pH 5.5 and 45 °C for the free enzyme). Biosensor response time was about 2 min and its linear region was within $5{\cdot}10^{-5}$ – $1.2{\cdot}10^{-3}$ M. The loss of enzyme activity for the month was 25%. It should note that the polymers considered types were used for the creation of the chemical sensors too. Thus, at the development of potentiometric and optical devices for the measurements of polyanions (Ambrose & Meyerhoff, 1997) photosensitive thin films based on decylmetakrilate (DMA) were applied. In accordance with the existing theory, the regulation of the potentiometric response to the polyanion, in particular, to heparin, increases with decreasing amounts of plasticizer and tridodecyl methyl ammonium chloride (exchanger) in the film. By varying the content of cross linker in the DMA film provides an additional mechanism for the regulation of expression of its physical structure and appearance of potentiometric answer on polyanion. Films with low hexanediol-di-methacrylate as a cross linking agent are provided the detection of heparin at 0.04 mM with a low coefficient of diffusion within the polymer due to interactions between adjacent residues of decile groups. Increase of the cross-linking agent violated these interactions and increased the diffusion properties of the polymer. When applying such films to the glass surface of the optical transducer the sensitivity analysis of heparin in non-dissolved human plasma was achieved at 0.5-5.0 units/ml.

It was studied the effects of the immobilization matrix (polyacrylate nature) on the properties of ion-sensitive field effect transistor (IsFET) based sensor at the determination of K+, NO3-, Ca2+ ions (Kolytcheva et al., 1999). The membranes with two matrixes were used. The so-called PA-matrix was prepared on bisphenol-A-diglycidyl-methyl-methacrylate (BIS-GMA) and hexandiol-acrylate (HDDA). PA-matrix II consisted of n-(ethylene oxide)-dimethacrylate ((EO)nDMA). Photo initiators were phenantrenquinon and lutsirin, respectively. To determine the K+, NO3- three types of (EO) nDMA were as most suitable. They contended the three ethylenoxide groups in the monomer ((EO)3DMA). These polymers had a structure polycycles. Homologous derivatives (with the number of ethylenoxide groups) were ineligible due to the short period of existence (15 and 10 min, respectively) as the result of plasticizer leaching. Among of four plasticizers - dibutyl sebacynate (DBS), o-nitrophenyl-n-octyl ether (NPOE), di-octyladipate (DOA) and di-octyl phthalate (DOP) for a polymer matrix based on (EO)3DMA the better chemical sensor characteristics were obtained for DOP. It was found the optimal content of DOP (by weight) for K+ and NO3- membranes (45% and 32%) respectively). For ionophore it was 4% and 3% in the case of valinomycin and tributyl-oktadecyl-phosphonia, respectively. Response of these sensors reached 54.5 and 56.2 mV/decade and the defined minimum was $0.43{\cdot}10^{-4}$ and $0.19{\cdot}10^{-4}$ M for the duration of operation of 3 and 1 month, respectively. Sensors based on polymer PA-matrix II showed better results compared with those which were prepared on the basis of PA-matrix I, and polyvinyl chloride. For Ca2+-sensor it was found the best content (by weight) of plasticizer - 32% for PA-matrix I and 54% for PA-matrix II and ionophore - 3%. The mechanical properties of membranes, their performance and selectivity were the best on the basis of PA-matrix I, while the reproducibility and stability of membranes was improved by the use of PA-matrix II. The best answer had sensors based on PA-matrix I with the inclusion of Ca-ionophores ETH 1001 (Fluka), N', N', N',-N'-3-tetracyclohexyl-3-oxapentandiamide (for 7 studied ionophores) and plasticizer ETH 469 and ETH 2112 (Fluka). When stored sensors contained ETH 469 and ETH 2112 in a solution of 0.125 M KCl and 0.1 M NaCl they remained stable for 7 and 10 days, respectively. Nevertheless, the best results were obtained at the use of PA-matrix I, ETH 469 and

N',N',N',-N'-tetra cyclohexyl-3-oxy-pentane-diamid (in the proportions given above). Linear response of sensors with these membranes was in range of 10-5 - 10-1 M, the slope - 24.5 mV/decade, the response time - 40-100 sec. In the case of PA-matrix II, it was found that the ionophore ETH 1001 and plasticizer DOS were better.

2.2.2 Urethane derivatives

Polyurethane derivatives proved to be very promising in this direction (Munoz et al., 1997; Puig-Lleixaet al., 1999a; 1999b; 1999c). On their basis the sensor for the determination of monochlor acetate was developed (Puig-Lleixaet al., 1999a). As an olygomer it was used aliphatic urethane diacrylate and hexandiol-diacrylate served as gross-linking agent (their ratio was 81:17). 2,2'-dimethoxyphenyl-acetophenone was as PhI. It was chosen the most suitable ion-selective ionophores (tetradecyl ammonium bromide and tetraoctyl ammonium bromide). The plasticizers were selected from bis-(2-ethylhexyl) sebacynate (DOS), dibutyl sebacynate (DBS), di-5-noniladipata (DNA), bis-(2-ethylhexyl) phthalate (DOP) and trioctyl-phosphate (TOP). Ion-selective membrane was formed by applying 100 ml of the membrane cocktail on the surface of the transducer covered with a mixture of epoxy and graphite and irradiated by UV light (365 nm). At the selecting the components of membranes and their relationship it was preferred ammonium bromide-tetradecyl as ionophore. It was found that the best results are obtained by using DOS. Moreover, it was shown that an increase (60%) of plasticizer leads to a significant expansion of the linear response and sensitivity of the biosensor. Optimal content ionophore was at 1%, as it increases to 5-10%, though increases the sensitivity to analyte, but leads to a significant drop of this index during further operation. It was studied the interfering effects of various ions (tris, chloride, nitrate, sulfate, phosphate) in response of the sensor when above mentioned plasticizers were used. The widest range of linear response was obtained when DOS, DNA and TOP were used. Membranes contained DOS and TOP showed the least interference and membranes based on DOS showed a lower limit of the determined substance. The choice was made in favor of DOS using as the plasticizer. Sensitivity biosensor based on the selected components was 54.6±2,3 mV/decade of monochlor acetate, the region of linear response was in the range of $2.1 \cdot 10^{-5}$ - 0.1 M. The response was stable under the pH change from 10 to 4. Response time was less than 18 sec when the 95% of its value was realized (at the concentrations of analyte about 10^{-3}-10^{-2} M). Stability of response was maintained for 90 days.
This same group of authors (Puig-Lleixaet al., 1999c) suggested that the pH-sensitive sensor based on a prepared membrane using a polymer composition, similar to the previous one, but with the inclusion of thri-dodecylamine as ionophore to hydrogen cation and tetrakis-(p-chlorophenyl) borate (KTpClPB) as a cation-exchange site for potassium. As suitable buffer solution it was 0.01 M tris-HCl. Among of four plasticizers (DOS, DNA, dibutyl phthalate - DBP, DOP) the first two were as most suitable for membranes with which the upper limit of sensitivity was more advanced. However, preference is given to DOS because of the greater its resistance to ions of potassium and sodium. Selected photo polymerizable membrane consisted from 42% urethane diacrylate, 55% DOS, 1,5% ionophore, 0,5% KTpClPB and 1% PhI. Sensitivity of the sensor reached 55,4±1,5 mV/pH. When storing the membranes during 9 months their mechanical properties are maintained and reducing the sensor response was only 4%. Notes, that at the operation of the sensor the decrease of its response value was faster than at storage. With the use of the above mentioned acrylurethane oligomers, cross linking agents (tri-propylen-glycol-di-aldehyde - TPGDA and hexandiol-diacrylate - HDDA), plasticizers (DOS, DOP), ionophores and ion exchangers

(valinomycine, nonactine and KTpClPB), urease, and IsFETs it was developed chemosensors for the determination of K^+ and NH_4^+, as well as a biosensor to monitor the urea level (Munoz et al.,1997). PhI was 2,2'-dimetoxyphenil-acetophenon. Membrane for the sensors to determine the K^+ and NH_4^+ ions included 43.1% and 48.2% aliphatic urethane acrylate (molecular weight 1500), 2.2% and 1.9% of ionophores, 15.2% and 13.4% of HDDA, 38.5% and 34.5% of DOS, respectively. The concentration of PI in both cases reached 2%. Sensor for the potassium determination had sensitivity 56 mV/decade (for 3 months). Similar value (about 59 mV/decade) was achieved for ammonia sensor. Region of linear response was 10^{-1}-$3 \cdot 10^{-5}$ M. The membrane sensor for urea was composed from 76.5% urethane acrylate (molecular weight 1700), 3.5% of urease, 16.2% of TPGDA and 4% of the PhI. Biosensor sensitivity was 63 mV/decade of urea, the linear region was within its concentration of $2 \cdot 10^{-4}$-$6 \cdot 10^{-3}$ M. After a week of operation the sensitivity of the biosensor decreased to 55.8 mV/decade. Draws attention to the biocompatibility of polyurethanes used (no annoying process with the implantation of pieces of polymer in the body of rats).

It was reported (Puig-Lleixaet al., 1999b) about the creation of the biosensor for urea determination with the IsFETs and sensitive membrane based on the photo polymerizable urethane derivatives. To do this, monomers and oligomers used in urethane diacrylate aliphatic, cross-linking agent and 2,2-dimetoxyphenyl-acetophenone served as PI. Urease was as biologically sensitive component which hydrolyzed urea to form carbon dioxide and ammonia resulted to local pH increase. Before applying the polymer the gate surface (silicon nitride) of IsFETs was treated with a solution methacryl-oxypropyl-trimethoxy-silane in ethanol. The content of the composition was as follows: 70% of olygomer, 28% of cross linking agent and 2% of PhI. Approximately 0,15 g of this mixture was homogenized in 0.3 ml of ethanol. Membrane thickness of 100 μm was formed applying 2 ml of cooked mixture to the gate. Mixture was irradiated by UV with a wavelength of 365 nm and 22 mW/cm^2 of power in oxygen-free environment and at the temperature of 20 °C. After that the surface was washed with ethanol. Given the fact that the membrane with immobilized biocomponents had insufficient adhesion to the gate surface it was used an original method based on the principle of photolithography. On the edge of the biological membrane and the surface adjacent to it the more hydrophobic polymerizable composition was plotted but without biological components and with the greater adhesion. This area was additional irradiated by UV. As result of this way the central part of the biomembrane remained uncovered by hydrophobic substances and it was polymerized only outdoor areas. It should be noted that the enzyme is not good soluble in the above composition. To solve this problem, tried to pre-dissolve it in water (but at 30% of its content in the composition of the latter is not homogeneous) or glycerol (enzyme dissolved in it worse, but glycerin was better kept in a composition and it was more homogeneous). It was chosen a more appropriate structure of the composition. It was found that at the high content of cross linking agent the membrane is very stiff, quickly exfoliate and at the increasing the water content there is a rapid loss of enzyme activity, membrane is not sufficiently homogeneous and characterized by low adhesion. For further studies was chosen composition which contended 76% of oligomer, 0.5% of cross linker, 22% of glycerol, 1% of the enzyme and 0.5% of PhI. After storage of the prepared mixture for 30 days in the refrigerator the biosensors created on its basis give practically the same answer as that of a fresh composition. The maximum response of the biosensor to urea concentration of $5 \cdot 10^{-2}$-10^{-1} M reaches about 90 mV at pH 5,6. Moreover, at the using of NH_4Cl solution (when the pH depends on the concentration of ammonia and ammonium ion) the biosensor response is linear even at the increasing concentrations of urea (160 mV to 0.1 M urea at pH 5.6). In additional to the response was more pronounced at 1 M than in 0.01 M NH_4Cl. Time to reach 95% response was about 2.5 min for the concentration of 10^{-4} - 10^{-3} M. Sensitivity of the sensor in a solution of 0.01 M NH_4Cl and pH 5.6 was 58.8±1,2 mV/decade, the region of linear response - 0.04 - 36 mM, for 0.01 M Tris-HCl solution - 35 mV/decade in the field of 1-25 mM. The decline of response during the month was 10%.

2.3 Photo linking in biosensorics

Typically photo-linking prepared polymers are used (Jae Ho Shin et al., 1998; Jobst et al., 1993; Nakako at al., 1986; Barie et al., 1998; Dobrikov & Shishkin, 1983a, 1983b; Dontha et al., 1997; Leca et al., 1995; Nakayama & Matsuda, 1992; Nakayama et al., 1995; Navera et al., 1991). Polyvinyl derivatives such as polyvinyl chloride were widely used at the creation of biosensors (Jae Ho Shin et al., 1998). It was communicated about photo-linked polyvinyl alcohol in aqueous solution for the immobilization of cells of *Arthrobacter globiformis*. PhI in this case is not used. Prolonged exposure to UV light (30 min), poor adsorption and mechanical properties of membranes obtained did not allow them to be widely used in the photo immobilization at the manufacture of biosensors.

At the development of amperometric biosensors for the choline determination the immobilization of cholin oxidize was made in the polyvinyl alcohol containing linked styryl-pyridine groups which served as PhI agent (PVA/SbQ) (Leca et al., 1995). The working and measuring electrodes were made from platinum and calomel electrode was as comparative one. The oxidative potential was on the level of 700 mV. The polymer and enzyme solutions were placed on a platinum disk of the working electrode and were irradiated with UV-source with a wavelength of 254 nm during 45 min. Then the polymer was washed in 30 mM of veronal-HCl buffer, pH 8 at 26 °C. It was studied the effect of the polymerization degree and the number of groups on styrylpyrydine on the biosensor response. For this purpose three types of polymer (with a degree of polymerization of 500, 1700 and 2300, and accordingly the number of reactive groups 2.94, 1.31 and 1.06 mol%) were used. The highest sensitivity (21 mA/mol) and the minimum defined limit ($1{,}5 \cdot 10^{-8}$ M) was obtained for a polymer with a longer chain (and less cross-linking groups). This polymer was selected for further studies. The amount of polymer for the electrode in this series of experiments was 0.22–0.39 mg and immobilized cholin oxidase - 0.7 - 1.7 U (at the activity of 17 U/mg). Next it was studied the effect of enzyme content in biosensor response. If the cholin oxidase content was changed from 0.9 to 2.7 U in 0.3 mg of the polymer it was occurred a slight increase of biosensor sensitivity to choline (20 to 22 mA/mol). The response time was about 10-40 sec. When 0.1 M phosphate buffer (contained 0.1 M KCl at pH 8) were used the determined limit reduced to $5 \cdot 10^{-9}$ M, however, narrowed the region and a linear response - $4 \cdot 10^{-8}$ – $4.5 \cdot 10^{-5}$ (vs. $1.5 \cdot 10^{-8}$ - $4.5 \cdot 10^{-5}$).

It was studied the effectiveness of immobilization of butyryl cholin oxidase in the PVA/SbQ-matrix in the comparison with the BSA-matrix cross-linked with glutaraldehyde (Wan et al., 1999). The polymer membrane was manufactured as follows. PVA/SbQ (45 mg) was mixed with the enzyme (5 mg) in phosphate buffer (50 mg, 1 mM, pH 8.0). These mixtures (0.5 ml) were applied to the gate of the IsFET and then irradiated with UV during 25 min. The greatest response of both biosensors to butyryl cholin was found when the phosphate buffer (pH 8.0 at the concentration of 1 mM) was used. Region of linear responses of biosensors measured in dynamic regime was 0.2-1 mM and 0.2 - 5.8 mM and the calculated KM achieved 2 mM and 3.8 mM for BSA- and PVA/SbQ-membranes, respectively. When storing the biosensor with PVA/SbQ-membrane in the dry state and in the dark at 4 °C for 9 months the fall of its response was 20% (similar to the decline in storage in a phosphate buffer at pH 8 in the same conditions was achieved at 1 month). For the biosensor based on BSA-membrane the similar declines of the responses were through 7 and 42 days when it was stored in a dry state and in the buffer, respectively. The field of the determination of such organophosphorus pesticide as trichlorphon was similar for both types of biosensors and amounted to 10^{-3}–10^{-6} M

Navera et al. (1991) reported about the development of the acetylcholine biosensor using carbon fibers. Acetyl cholinesterase and cholin oxidase were co-immobilized in polyvinyl alcohol with a stiryl pyrydine as cross linking agent. Duration of response was 0.8 minutes and the linear region was within 0,2-1,0 mM.

Jobst et al. (1993) created oxygen amperometric biosensor for the application in vivo condition. Selective membrane was made from the poly-N-vinilpirolidon cross linked with 2,6-bis-(4-azidobenziliden)-4-methylcyclohexanone (total 3%) under UV irradiation. For 10 sec 95% of the response is realized and its value in the presence of dissolved oxygen in the water reached about 200 nA.

The biosensor based on the IsFET for the determination of a neutral lipids [34] was developed on the sensitive membrane obtained photo-crosslinking polyvinylpyrrolidone (PVP), 4,4 '-diazidostilben-2,2'-disulfonate sodium (0,1 g of cross-linking reagent in 100 ml of 10% aqueous solution of PVP). To 200 ml of this solution 15 mg lipase and 10 mg BSA were added. This mixture was applied to the IsFET gate, centrifuged at 3000 rev/min for 2 min and irradiated with a mercury lamp during 5 min. Then the mixture was treated during 15 min with a solution of glutaraldehyde at 4 ° C and finally it was kept in 0.1 M solution of glycine (4 °C). The chips were stored in a buffer solution at 4 °C. Linear fields of responses were as follows: for triacetin - 100-400 mM, tributylin - 3-50 mM and triolein – 0,6-3 mM. The minimum detectable concentration of the last was 9 mg/ml. Decline in response for 3 months was 12% only.

At the development of immune biosensors based on surface acoustic waves to detect a specific protein (urease) as photo-crossing agent served bovine serum albumin (BSA) modified aryldiazirine (Barie et al., 1998). Aryldiazirine absorbs light with a wavelength of 350 nm and forms a highly reactive carbenes, which are preferably interact with the C-H, C-C, C=C, N-H, O-H or S-H groups. The surface of the transducer was sialinized by dimethylamino-propyl-ethoxy-silane, then coated with a polyimide film (thermal polymerization mixture *p*-phenylenediamine and 3,3',4,4'-biphenyl-tetracarbocyclic dianhydride) or parilene C (poly (2-chloro-p-xylene). Then, on the surface it was applied the mixture of triftor-methylaryl-diazirine BSA (T-BSA) with dextran and its was irradiated by UV-source (0,7 mW/cm_2, the main emission 365 nm). For the glass surfaces, passivated by parilene the optimum ratio was: 75% T-BSA and 25% dextran at the irradiation time of 45 min. The density of dextran on the surface was 1 ng/mm_2. The special peptides – antibodies to urease were linked to the carboxylated dextran with a mixture of N-hydroxysuccinimide and N-(3-dimethylaminopropyl)-N'-ethylcarbodiimide hydrochloride in 1:4 ratio (passivating layer was polyimide, the operating frequency - 379, .43 MHz, the loss during the passage - 4.89 dB). It was received a response to urea at concentrations of 15-500 µg/ml with a maximum shift of the oscillation frequency transducer 110 kHz.

BSA derivatives were used for cross-linking antibodies to planar optrods (Gao et al., 1995). On the surface of the waveguide (TiO_2/SiO_2) a mixture of 3-(trifluoromethyl)-3-(m-izotiocyanophenil) diazirine derivative of BSA and (Fab')$_2$ fragments of antibodies (4:1) was placed and then it was irradiated with UV-source (0,7 mW/cm2, 20 min) Immobilized antibodies were specific to the prostate antigen. The density of immobilized antibodies was 16.8 fmol permm2 or 1.05 μg per chip. Biosensor sensitivity reached 0.35-3.5 μg of protein per chip. The biosensor had a low non-specific response. Its regeneration was curried out by treatment with glycine buffer (pH 2.3). When storing the biosensor in the presence of 0.5% BSA, and 4 °C during the month there is no significant activity decrease.

2.4 Application of photo polymerisable matrix at the creation of potentiometric enzymatic biosensors

Early (Arenkov et al., 1994a; 1994b; Levkovets et al., 2004; Nabok et al., 2007; Starodub et al., 1999a; 1999b; 2000a; 2002a; Starodub & Starodub, 2002; Starodub, 2006; Starodub et al., 2008) we have developed prototypes of the enzyme- and immune biosensors based on the IsFETs and the electrolyte insulator semiconductors (EIS) structures. Both types of the biosensors are perspective for use in different fields, in particular, medicine, biotechnology and environmental monitoring. Nevertheless, before start of their wide manufacturing there is necessary to optimize the procedure of biological material immobilization on the transducer surface. In generally it is the main problem of biosensorics and for it's solving a lot of different approaches including pure physical, chemical and hybrid physical-chemical methods were proposed (Pyrogova & Starodub, 2008; Starodub et al., 1990; 1995; 1998; 1999c; 2001; 2005). All these methods are directed on more effective fulfillment of the main practice demand which concern the achievement of maximal level of residual activity of biological molecules and exposition of their active centers toward solution, simplification of procedure of immobilization and its combining in unique electronic cycle of transducer manufacturing, preservation of high level of biosensor response during its storage and working, etc.

The application of the liquid polymerisable compositions (LPC) on the basis of monomer-olygomeric substances at the biological membrane creation may be considered as perspective approach directed on providing above mentioned practice demands. These compositions give possibility to form sensitive membranes with adjustable physical-chemical and mechanical abilities without strong temperature and chemical destructive effects on biological molecules. Among the most wide dispersed LPC it is necessary to mention a number of monomeric and olygomeric acrylate compounds (acrylic, metacrylic acids their ethers and derivatives) as well as urethane olygomers and vinyl copolymers (sterol, vinyl acetate, vinylidenchloride, vinylpyrrolidone and others). At the varying of chemical origin and concentration of some components there is possibility to regulate a lot of parameters of biological membranes obtained on the basis of these components (Rebrijev, 2000; 2002; Rebrijev et al., 2001; 2002a; 2002b; Rebrijev & Starodub, 2001; Starodub & Rebrijev, 2002; 2007; Starodub et al., 2002b).

The use of the LPC in biosensors supposes that they should be characterized by number of indexes, namely: they should be non-active concerning biological substances, permeable in respect of determined analytes, as well as have defined hydrophobic-hydrophilic balance and sufficient level of adhesion to the transducer surface. The liquid photopolymerisable composition (LPhPC) causes special interest in biosensorics. Although it's wide application is restricted by the practice demands above-mentioned. As a rule at the biosensor creation the influence of supported substances on the biological materials is not special observed. Usually the excess of biological material is taken and for the estimation of its state the non-direct approaches are used, namely: the determination of biosensor response, the rate of product formation and others. At the same time the change of structure of biological molecules at the creation of biochips or during their preservation reflects disproportionately on the intensity of response and lifetime of biosensor work. Moreover at the multi-layer immobilization of biological material the inner layers may work with the small productivity in comparison with the external ones due to the diffusive restrictions. That is why the main purpose of this work was the elaboration of content of the LPhPC, which is characterized by number of abilities in concordance with the biosensorics demands in respect of above mentioned and some additional ones: simplicity of immobilization procedure and homogeneity of formed membrane. To optimize the conditions of the enzyme including in the LPhPC the absolute level of residual activity of the immobilized molecules was determined and the principal factors affected on this level were characterized.

In experiments it was used: urease from soybean with activity of 200 u/mg (Sigma, USA), GOD from *Penicillium vitale* with activity of 160 u/mg (Kamenskoe distillery, Ukraine), horse radish peroxidase (HRP) of type VI with activity of 275 u/mg (Sigma, USA).

N-vinylpirrolidone (VP) was obtained from "Aldrich" (Germany). 2-hydroxy-2methyl-1-phenylpropan-1-on (Darocure 1173, λ_{max} = 310-350 nm) from "Ciba-Geigy", Switzerland) served as PhI. Monomethacrylate ether ethyleneglycol (MEG) was produced by "BASF" (Germany) and olygocarbonatediethylenglycolmetacrylate (OKM-2) by AOOT "Korund" (Russia). Olygouretane metacrilate (OUM-1000T or OUM-2000T) was synthesised according to (Masljuk & Chranovsky, 1989).

The IsFETs were manufactured in the Institute of Biocybernetics and Biomedical Engineering of PAN (Poland). Each chip contained two IsFETs, which were characterized by 45-48 mV/pH. Construction of the IsFETs, device for registration of their response and the main algorithm of measurement were described early (Starodub et al., 1990). The gate surface of the IsFETs was preliminary cleaned by consecutive washing: sulphuric acid, water and ethanol. On the top of this surface the mixture of the appropriate enzyme and the LPhPC (about 1-5 µl) was dropped. Polymerisation of this mixture was curried out at the effect of the UV radiation in vacuum conditions (0.1-0.2 mm of mercury). As source of the UV it was used lamps: LUF-80-04 (λ_{max} = 300-400 nm, intensity of light on the irradiated surface - about 2.6 Watt/m^2) and DRT-120 (λ_{max} = 320-400 nm, intensity of luminous flux about 12.5 Watt/m^2).

The homogeneity of composition and obtained polymer was determined by visualization, i.e. the absence of visible disseminations at microscopy was taken as maximal level of this index and was marked as (++). Adhesion abilities of the formed polymer were non-direct appreciated on the assumption of time being membrane on the transducer surface without its peeling at the immersion of chip into buffer solution. The extreme positions, i.e. immediate peeling of membrane was marked by (--) and its attaching during two month - by (++). In case of the determination of the residual enzyme activity the LPhPC was presented as two-component mixture containing VP and PhI at 98 and 2 g/100g of concentration, respectively. Then, to 50 µl of this mixture and 20 µl of the enzyme solution was added at the shaking and water was removed in the vacuum conditions (0.1-0.2 mm of mercury). The concentrations of urease, GOD and HRP in the solutions were 0.1, 0.1 and 0.02 mg per 1 ml, respectively. The time of UV irradiation was 11 and 4 min at the application of LUF-80-04 and DRT lamps, respectively. Intensity of luminous flux was measured by the automotive dosimeter (DAU-81). Part of the obtained membrane was dissolved in 2 ml of 10 mM phosphate buffer with pH of 5.5, 7.0 and 6.0 in case of the determination of activity of GOD, urease and HRP, respectively.

It is necessary to mention that at the obtaining of calibration curves the VP, PVP or intermediate products of these substances (depends on duration of irradiation or method of analysis) were added to the analyzed samples. The some details of experiments are given in the text below.
According to the preliminary investigations as main component of the LPhPC it was taken VP as substances with appropriate hydrophilic-hydrophobic balance. The optimal contents of the enzymes and PhI were 3 and 2g per 100g of LPhPC, respectively. Primarily MEG was used as cross-linking polymers. The results of choosing optimal variant of the LPhPC in respect of homogeneity of the obtained polymer, its adhesion to transducer surface and biosensor response are summarised in Table 1.
Applying the above LPhPC and immobilized GOD on the transducer surface it was created biosensor for glucose level control (Fig. 1). It had the following characteristics: linear response region in frame of 0.1 - 10 mM, the slope of the curve 30 mV/pC and response time during 10 -15 min. Km values for GOD immobilized in photopolymer material is 3.1 mM. To calculate Km used graphical method of inverse coordinates. In the literature there is information about the positive experience of the introduction of the LPhPC glycerol, which was injected together with enzyme in a hydrophobic matrix. We also carried out attempts to introduce GOD in the chosen composition of LPhPC using glycerol (in an amount which was 5, 10 and 20 of wt.%). However, it turned out, this led only to a deterioration of the homogeneity of composition and adhesion of the polymer as well as to reducing the latter to the surface of the transducer. So we abandoned the use of glycerol in LPhPC.
Thus, the obtained LPhPC due to its properties for ease of manufacturing and process of biomaterial immobilization may be included in extended technological stages of photolithographic manufacture of semiconductor structures. The created on this basis biosensor may have the characteristics needed for use in laboratory, clinical, food and biotechnology practice.

VP, mas.%	MEG, mas.%	OKM-2, mas.%	OUM-1000T, mas.%	Homogeneity		Adhesion of membrane to ISFET surface	Response on 10 mM GOD**
				mixture with GOD	membranes		
88	10			--	--		-
93	5			-	-		-
88	5	5		-	-	-	12
88		10		++	++	+-	42
78		20		++	+	-	33
78		10	10	+	+	++	46
68		10	20	-	-	++	25
78		5	15	-	+	++	40
78			20	--	--	++	20
78		10	10***	+	+	++	57

Table 1. Some characteristics of the LPhPC based on VP*. *Quantity of PhI in all LPhPC was 2 mas%. ** In 1 mM sodium phosphate buffer, pH 7,0. *** Instead of OUM-1000T it was used OUM-2000T

In literature as a rule, the degree of decrease in activity of biological material in the process of immobilization is not special considered. For the state of biological structures it is using indirect methods such as measurement values the sensor response, speed of formation of different substances, etc. It should be noted that for the immobilization is usually initially taken excess of biological material. However, increased activity of enzymes in the selection of optimal conditions for this process or its decrease in functioning and maintaining biochips disproportionately affects on the efficiency of the measuring device (the intensity of his response, duration of work etc.). Moreover, in most cases the biological material is immobilized often by multilayer and thus the inner layers operate with lower productivity due to diffusion limitations.

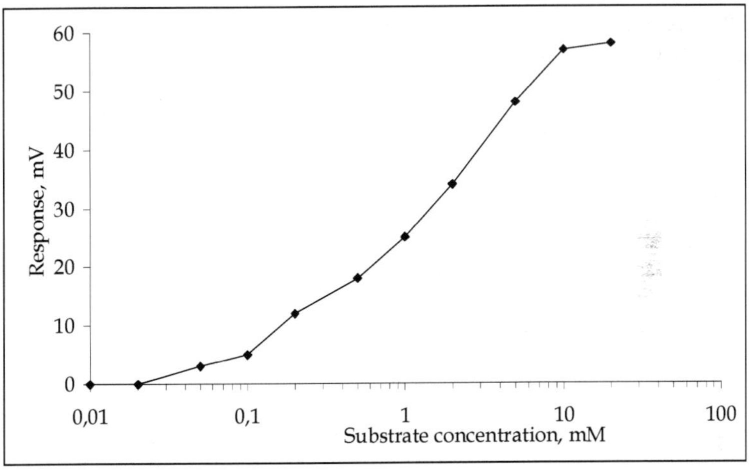

Fig. 1. Response of biosensor with the immobilized GOD (substrate - glucose). Measurements were made in 1 mM of sodium-phospate buffer, pH 7,0.

That is why, the next experiments were fulfilled for the estimation of the absolute level of residual activity of immobilized enzymes, as well as the main factors influencing this level, to determine the optimal conditions for the inclusion of enzymes in photopolymer membrane. For this purpose the enzymes immobilized in LPhPC based on VP. The obtained on this basis polymer was water soluble, so after the dilution of its in buffer solution can there is possible to study the activity of immobilized enzymes.

Fig. 2 presents the results of changes of GOD activity at the including into PVP matrix depending on the source of UV radiation. These data suggest that the decreasing activity of the enzyme occurs to a greater extent when as a source of UV radiation it was used LUF (32.45%) than DRT lamps (37.25%), $p < 0.05$. The presence of VP and PVP in GOD solution made no significant influences on the level of activity, which can serve as an indirect indicator of chemical inactivity of VP and obtained polymer in respect of the enzyme.

It is known that immobilization of biological material is usually preceded by dissolving it in buffer solutions. However, mixing composition, which is able for photo polymerization, with a buffer solution, usually, leads ultimately to a deterioration homogeneity system and mechanical properties of the resulting polymer, due to the presence of salt ions in the system. Therefore, interest was to find out the possibility of eliminating this effect by replacement of buffer solution on distilled water when the preparing compositions contained biological material. First of all, it was necessary to establish the impact of replacing the buffer solution on distilled water for preservation of enzyme activity in the polymer. Consideration of the data is shown in Fig. 2 (UV irradiation LUF for 11 min.) It was shown that the replacement solvent has not affect on the level of residual enzyme activity in the membrane. This was the reason to exclude in these studies the use of buffer solutions with the introduction of the enzyme in the photo polymerizable composition.

The irradiation of the GOD solution (10 mM sodium phosphate buffer, pH 5.5 over time, which corresponds to that given during the course of polymerization, i.e., 11 and 4 min for different powers of UV sources - LUF and DRT) does not significantly affect on the change of activity of the enzyme studied.

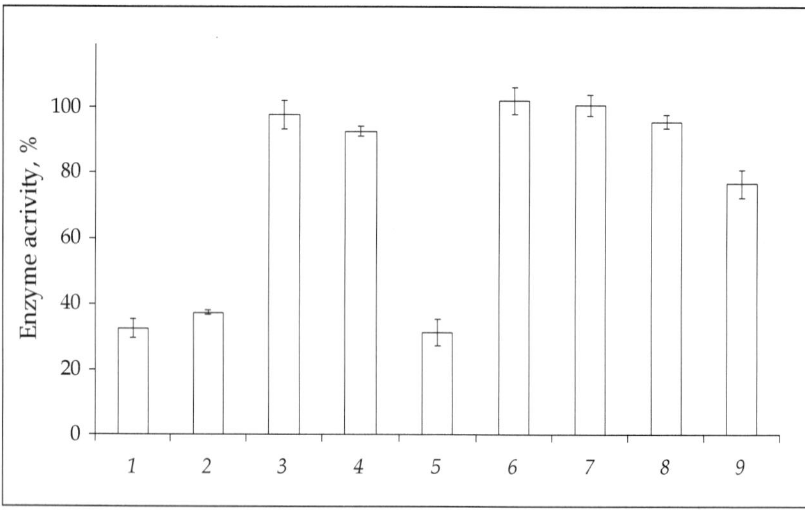

Fig. 2. Residual activity of GOD under different conditions of preparation of membranes. Where: 1, 2, 5 - photo polymerization in VP, 3 - in a mixture of solutions of GOD and VP, 4 - a mixture of solutions of PVP and GOD, 6, 7 - UV-irradiation of buffer solution of GOD, 8, 9 - mixture of solutions of GOD and PhI in glycerin (1, 5, 6, 9 - LUF irradiation; 2, 7 - irradiation of DRT; 1, 2, 3, 4, 8, 9 – GOD was previously dissolved in water, and 5 – GOD was previously dissolved in 10 mM sodium phosphate buffer solution, pH 5.5.

It was interested to study the effect on the GOD activity of another component LPhPC - PhI. For this purpose it was necessary to take into account that the used 2-hydroxy-2-methyl-1-phenylpropan-1-one as PhI is insoluble in water. To this end in LPhPC was used 2% solution (mas.) of PhI in glycerin, which in turn dissolves in water.

As shown in Fig. 2, when entering GOD (water solution) in this composition noticeable change in enzyme activity is not observed. At the same time UV-irradiation of this mixture (source - LUF) leads to a reliable ($p < 0.005$) lower enzyme activity, representing 76.7% of the initial level. However, it is established that at the use of DRT and LUF for photo immobilization the residual activity is according to peroxidase 41.5% and 44% and for urease - 21% and 16.5%, reliable data, $p < 0,05$ (Fig. 3). Conditions of the experiment were the same as in case of GOD immobilization.

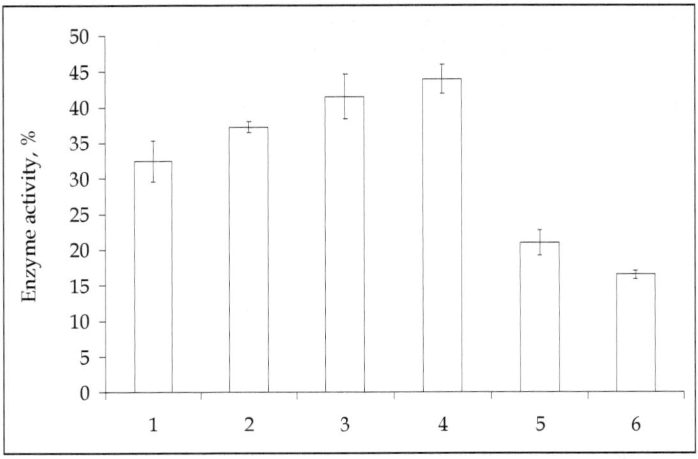

Fig. 3. Residual activity of GOD (1, 2), peroxidase (3, 4) and urease (5, 6) in photo polymerizable matrix. Source of irradiation: LUF - 1, 3, 5 and DRT - 2, 4, 6.

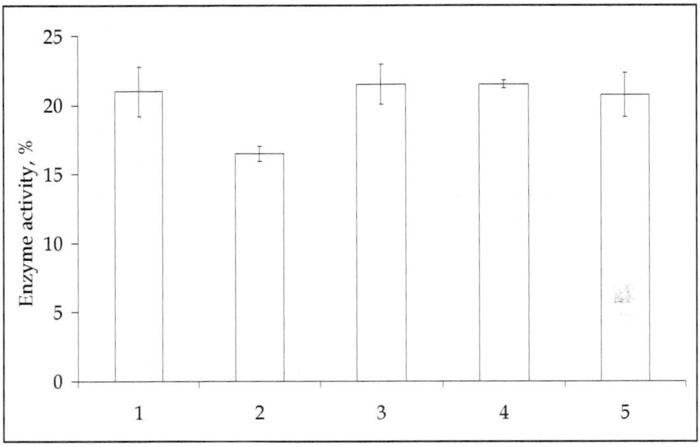

Fig. 4. The level of residual activity of urease after photo immobilization. Where: 1, 2 - without filter for UV; 3, 4 - with application of glass filter; 5 - in condition of low temperature (-8 ^{0}C). Source of irradiation: LUF - 1, 3, 5 and DRT - 2, 4.

Unlike GOD and peroxidase urease reveals itself as the low stable enzyme. The fall of its activity is due, mainly, oxidation sulfhydryl groups present in the active center. This enzyme is subsequently used for working out optimal conditions for immobilization. In addition, interest was to determine the influence of UV radiation of different wavelengths on the amount of residual enzyme activity. For this purpose, the short-wave area up to λ = 300 nm was cut off by a filter (glass). At the using glass (3 mm thick) as the UV-irradiation filter to 300 nm and without it's the enzyme activity in the mixture after irradiation LUF did not change (Fig. 4). However, note that in similar conditions DRT-irradiation the enzyme activity significantly increased ($p < 0.001$), reaching some of the value that was registered using the LUF-irradiation. This experimental fact, most likely due to the fact that short-range (220 - 280 nm) lamp DRT, which has great energy, influences on urease. At the same time, irradiation of LUF with λ_{max} 300 - 400 nm, when the radiation is almost entirely absent in the 220 - 280 nm using a glass filter, did not affect on the activity of the enzyme. Thus the measured power of UV radiation of DRT (220 - 280 nm) was equal to 12 W/m^2, which is 60% of the energy range 300 - 400 nm. Data about the effect of low temperatures (-8 ° C) on urease activity presented in Fig. 4. Given the fact that the freezing point VP is +13 °C, it should be noted that the photo polymerization at -8 °C was carried out in solid phase. Apparently, lowering the temperature of polymerization mixture to -8 °C is not made definite influence on the residual activity of urease.`

To investigate the dependence of the residual activity of urease from time of influence of LUF illumination it was chosen the next time range: 220, 330, 440, 660 and 990 sec. It was found that the enzyme activity decreases after the most exposure for 300 - 420 sec. (Fig. 5). Typically, kinetics process of the polymer solidification had S-shaped character. To measure the degree of polymerization the spectroscopic studies of irradiated RFPK were carried out by infrared spectrophotometer SP-300S Philips with the various time of intervals. The degree of conversion was judged by peak area with a maximum range of 1640 cm^{-1}, which corresponds to the double carbon-carbon bonds in VP that quantitatively reduced in a polymerization composition in the comparison with the relatively quantified not variable carbonyl VP group, which has a maximum peak at 1700 cm^{-1}. The drop in enzyme activity correlates with the polymerization matrix.

It is well known that to preserve the active center of urease during immobilization using blocking its substrate analogs that do not split, for example, thiourea. Thiourea molecule is similar in structure to urea and a urease competitive inhibitor. Introducing thiourea in a mixture and analyzing the activity of the enzyme by the above mentioned method, its impact can not be set because it is constantly present in solution. To avoid this, it was used the following approach. It lies in the fact that the first LPhPC consisting of Oum-2000T - 10 wt. %, VP - 88 wt. % and PhI - 2 wt.% was prepared.

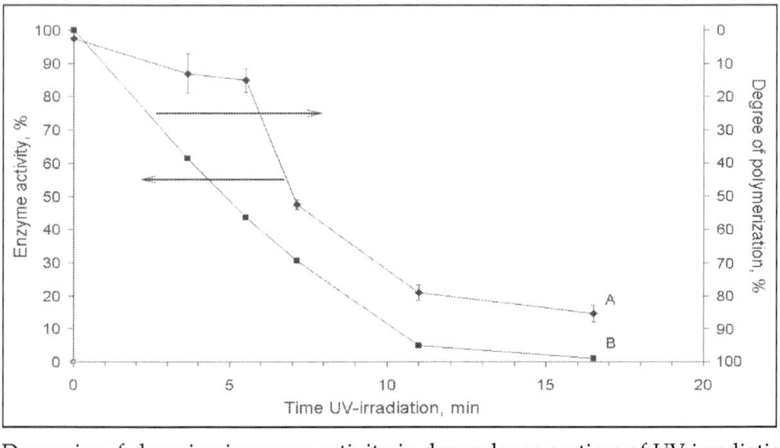

Fig. 5. Dynamics of changing in urease activity in dependence on time of UV irradiation by LUF lamp.

OUM-2000T - is a urethane oligomer with a molecular mass of 2800 with terminal methacrylate groups, i.e. tetra functional compound that performs role of cross linking reagent in this photo polymerizable compositions. Thus, at the photo solidification of this composition the strong three-dimensional polymer is formed, but very flexible. In LPhPC the enzyme solution was injected and this mixture after photo solidification formed the strong elastic film with the thickness of 0.1-0.15 mm. Also, the control film was prepared that does not contain thiourea. Then within two days the films were washed from thiourea. Urease activity was calculated per unit surface of the film. Activity of the enzyme in control films was taken as 100%. The results presented in Fig. 6 shown that at 0.5% (mas.) of the initial contents of thiourea in LPhPC the residual urease activity increases on 11,3% ($p < 0.05$). At the same time increasing the thiourea content in the composition up to 1% stabilized the enzyme in less degree.

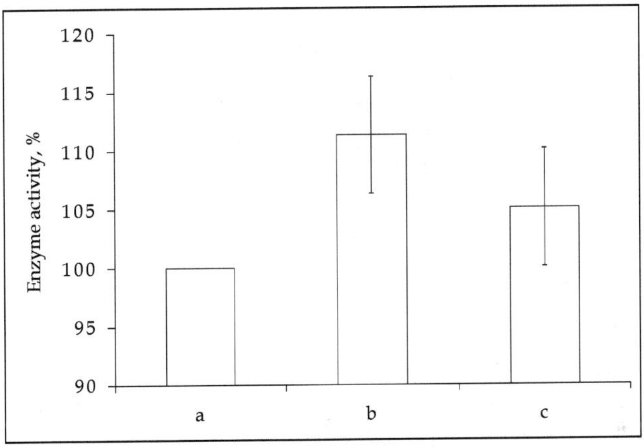

Fig. 6. Influence of thiourea content in phtopolymerizable composition on the activity of the immobilized urease. Content of thiourea according to mass:: *a* – 0 %, *b* – 0,5%, *c* – 1 %.

It was stated that the urease activity decreased in LPhPC at its preservation (at - 4 °C). Trough two months this decreasing reached 15% ($p < 0.05$) (Fig. 7) then this index continued to decline and after six months the reduction was a few less than half (47%) of fresh compositions ($p < 0.005$). At the same time while maintaining the urease in photopolymer matrix (with PVP), a marked decrease in its activity during the two months was not observed. Only after 6 months it was indicated the significant decrease in its activity, which was approximately 30% ($p < 0.01$). Saving GOD over six months in the PVP-matrix leads to a decrease in its activity about 23% ($p < 0.005$).

When the low (-35 - -50 °C) temperature was used for the polymerization the level of residual enzyme activity increased up to 50% at -50 °C in comparioson with the polymerization in ordinary (20 °C) conditions ($p < 0,002$). The required low temperature was achieved using liquid nitrogen (Fig. 8).

Therefore, it was proposed a method of determining absolute enzyme activity during immobilization in a polymer matrix and it was characterized the changes of enzyme activity (GOD, peroxidase, urease) at photo immobilization. The main attention was paid to the dynamics of changes of enzyme activity in the process of photo polymerization when UV irradiation was used. The needed conditions for increasing the activity of enzymes at the immobilization and at the storage prepared membrane were chosen.

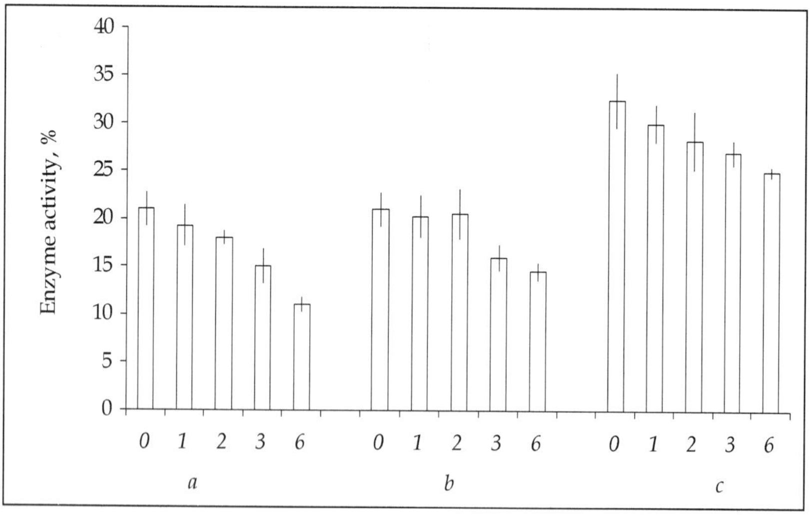

Fig. 7. Dynamics of changes of enzyme activity at the preservation (figures under the columns - quantity of month). Enzyme used: *a, b* – urease, *c* – GOD. Preservation in non polymerised composition (a) and in PVP matrix (b, c). Irradiation – by LUF lamp.

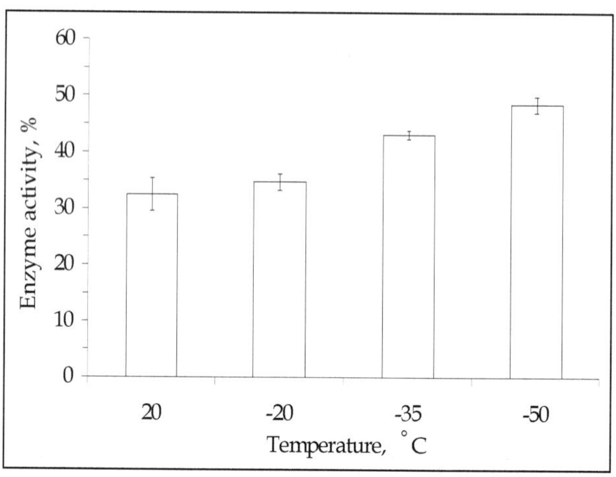

Fig. 8. Effect of temperature during LPhPC polymerization on the residual GOD activity.

2.5 Characterization of work efficiency of urea biosensor with LPhPC

This enzyme was chosen as such which has a much low stability in the comparison with others ones mentioned above. Upon the addition of urea in the test cell the potential at the

IsFET gate decreases as result of pH growth. Noticeable changes are found only during 0.5-3 min after substrate adding. Then, trough a few minutes decreasing voltage signal stops and it goes to the plateau. With increasing concentration of urea the biosensor response time decreases. For example, the duration of the analysis of 0.1 mM of urea solution is 10 min. and at 1 mM of substrate concentration - 4 min. Dependence of the biosensor response on the urease content in the composition is illustrated in Fig. 9, on which is shown that the greatest response observed at the presence in its of 3% of enzyme (mas.). The graph shows that there is a linear relationship between the content of the enzyme in the composition and the biosensor response. In accordance with this relationship it can be concluded that further increase the enzyme content in the composition biosensor response could be larger, and therefore the higher sensitivity of the sensor. However, the attempts to further increase of the enzyme content in the composition led to a sharp deterioration in both its homogeneity and solidity derived from its polymer with immobilized enzyme.

The work of the IsFET based biosensor depends not only on the acidity of the medium and also its ionic strength, but effect of first is much stronger than the second one. It is well known that the work potentiometric biosensors depend on the buffer capacity of solution, which eliminates local changes in pH under the gate region. The developed biosensor showed the largest response in 1 mM sodium phosphate buffer (Fig. 10). However, it should be noted that even at 10 mM buffer, the urea biosensor response was quite significant if the substrate solution was present in concentrations of not lower than 0.5 mM. It is worth noting that the concentration of urea in the blood serum of healthy individuals is 2.50 - 8.33 mM and it increases to 50 - 83 mM in the case of kidney failure as a result of various diseases. So enzymatic biosensor based on the proposed biological membranes can be successfully used for measuring the concentration of urea in the blood without its additional dilution that distinguishes this biosensor from others early proposed (Arenkov et al., 1994a; 1994b; Levkovets et al., 2004; Nabok et al., 2007; Starodub et al., 1999a; 1999b; 2000a; 2002a; Starodub & Starodub, 2002; Starodub, 2006; Starodub et al., 2008).

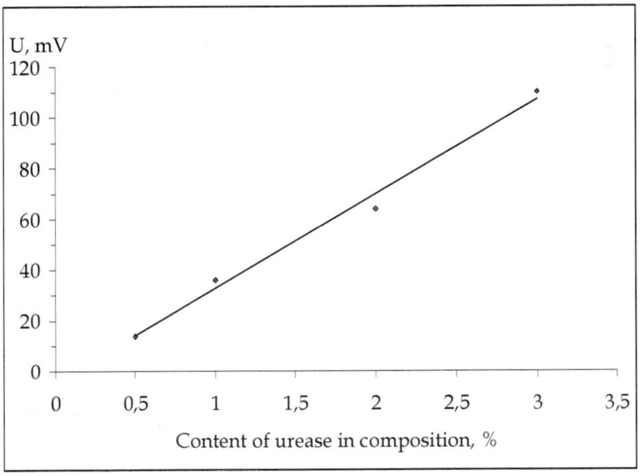

Fig. 9. Dependence of the biosensor response on urease content in the composition. Conditions of measurement: 1 mM of sodium-phosphate buffer, pH 7.3 and 5 mM urea.

Dependence of biosensor response on temperature (Fig. 11) shows that with its increasing from 28 to 41 °C the value of response increases by 15%. Similar data on the dependence of the sensor response on the temperature were obtained by us when the sensitive membrane was cross-linking enzyme with the protein carriers by glutaraldehyde (Soldatkin at al., 1993).

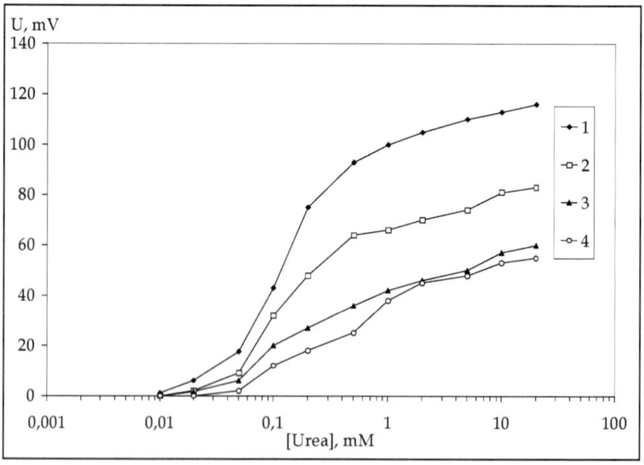

Fig. 10. Dependence of biosensor response on buffer capacity of the analyzed solution. 1-4 - concentration of sodium-phosphate buffer: 1; 2; 5 i 10 mM respectively.

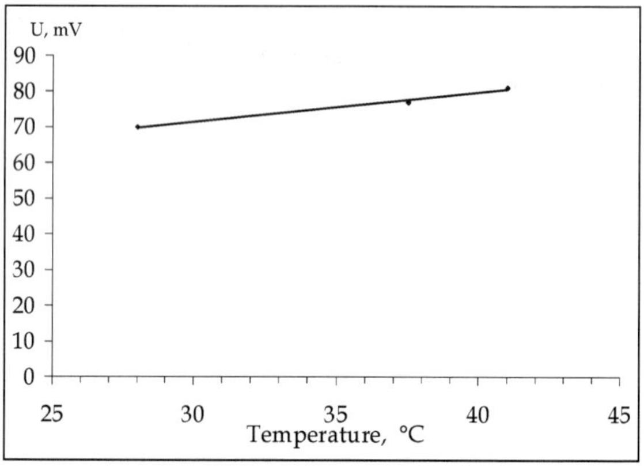

Fig. 11. Dependence of the biosensor response on the temperature. Conditions of measurements: 2 mM sodium phosphate buffer, pH 7.3; 2 mM urea.

It is well known that the optimum pH for urease is at 7.4. Therefore, studying the dependence of sensor response on pH it was conducted in a range from 5.5 to 8.5 at intervals of 0.5. In these experients polimiks-buffer (containing 2.5 mM citric acid, tris hydroxymethyl aminomethane, borax and potassium dihydrophosphate) that supports the buffer capacity in the pH range from 4 to 9. According to the data shown in Fig. 12, the maximum response in this case is achieved when the pH level was in frame of 6 - 6.5. Properties of urease immobilized probably a little different from those which are characteristic for the free enzyme.

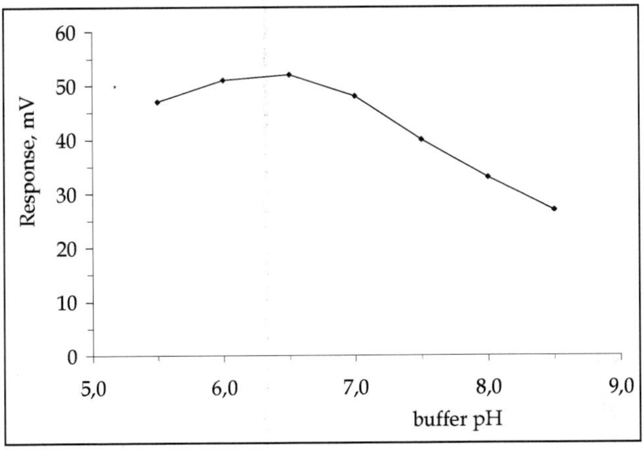

Fig. 12. Dependence of urea biosensor response on buffer pH (1 mM of urea, 10 mM of polymix buffer.

For biological fluids is characterized by the presence of some salts in different concentrations, so it was important to determine the dependence of biosensor response on ionic strength solution of NaCl (basic salt contained in biological fluids). As follows from Fig. 13, increasing concentrations of NaCl in the analyzed solution leads to a decrease in biosensor response for urea (1 mM in 10 mM sodium phosphate buffer, pH 7.0). At NaCl concentration of 300 mM falling response is about 50% but at the next increase of salt concentration up to 500 mM falling response practically does not observe.

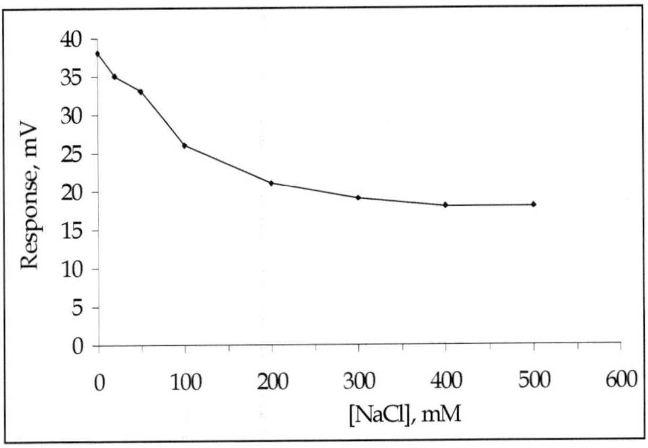

Fig. 13. Dependence of biosensor response on ionic strength of solution to be analyzed (1 mM of urea, 10 mM of sodium phosphate buffer).

In order to verify if the biosensor could be used in real conditions for analysis of human serum the measurements were conducted by both the developed biosensor and a standard colorimetric method using nessler's reagent. The serum blood was preliminary diluted by 10 mM of sodium phosphate buffer (pH 7.3). The data presented in Fig. 14, indicate a high level of coincidence of results obtained by both methods. But for a single measurement differences in test results by these methods were in the range 15-20%.

The special interest at the development of biosensors always the question is aroused about possible time of them operations. It was shown that the intensity of the response of the developed biosensor gradually decreased in course of 40 days. Moreover, during this period reduce of the intensity of response was 20% (Fig. 15). This indicates the possibility of significant extension of time functioning biosensor. As it was mention above urease contains in the active center sulfhydryl groups, which a lot of what determines the loss of enzyme activity over time. The latter are evident in the case of chemical modification or partial denaturation of the enzyme at the formation of biosensor membranes. Under the conditions of experiment the formed enzymatic membrane slowly loses its activity and life can be above or even higher limits.

In the developed photo polymerizable composition enzyme is probably in a stabilized condition. This confirmed by data about the studying responses of the biosensors, biological membranes of which were obtained from the freshly composition and prepared from one preserved in a dark place at 2 °C for 46 days. According to results shown in Fig. 16 the differences in the intensity of responses of biosensors that used these membranes are absent. These data suggest the possibility of long storage of the finished compositions without significant decrease in enzyme activity. In addition, this experimental fact indicates the promising application of compositions in industrial manufacturing sensors with immobilized urease. It seems that pre-prepared photo polymerizable composition can be used for a long time in the process fo the photolithographic formation of biologically active membrane of biosensors. Moreover, this process may be continuing technological production of IsFET using basic approaches of integrated electronic technology

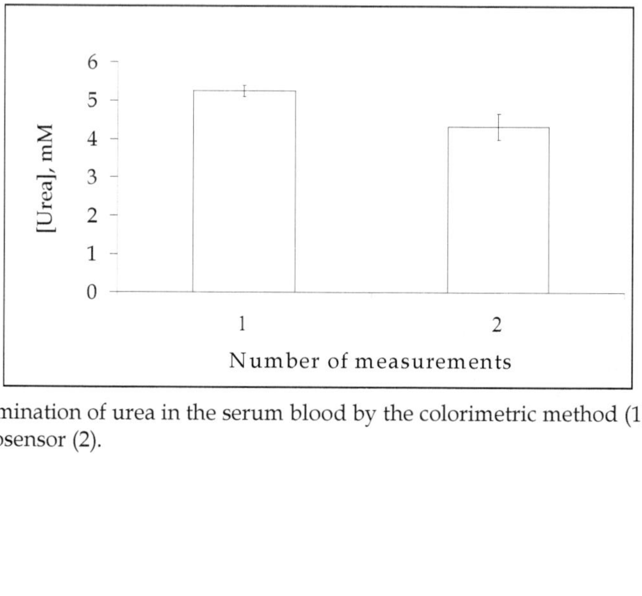

Fig. 14. Determination of urea in the serum blood by the colorimetric method (1) and by the developed biosensor (2).

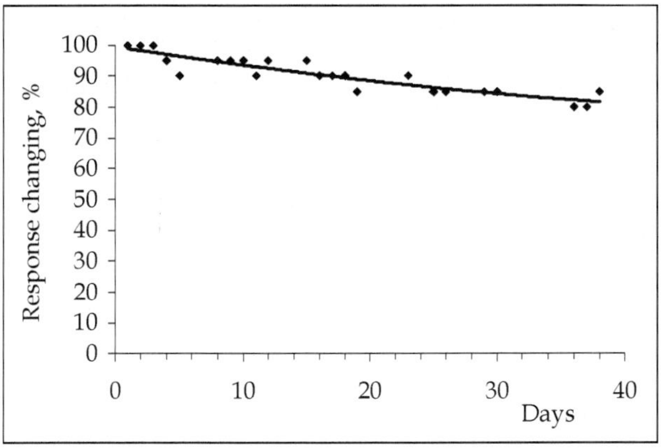

Fig. 15. Changing of response level of urease biosensor during time of its functioning.

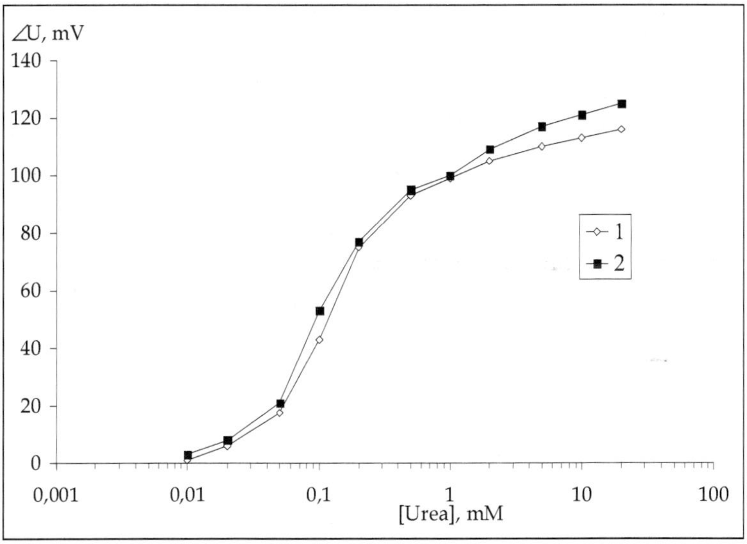

Fig. 16. Level of responses of the biosensors with membranes: fresh prepared (1), preserved (2) composition. Conditions of measurements: 1 mM of sodium phosphate buffer, pH 7,3.

Thus, an easy and fast method for immobilization of the enzyme urease on the surface of the IsFET gate is proposed. Based on the proposed bioactive membrane it was created biosensor for the express determination of urea in solution. Possibility of prolonged operation of the biosensor in real conditions was demonstrated. The conclusion about the possibility of recommendations developed photo polymerizable compositions for combining technologies of bioactive membrane production and manufacture of transducers, in particular, creation of IsFETs.

3. Conclusion

It was demonstrated that the proposed LPhPC is very suitable for the enzymatic biosensor creation. The process of the biological material immobilization on the surface of transducer can be done anyway phasic process and may be served as basis for technology of the biosensor production. Enzymes are long (up to 6 months) remaining active in staying as a part of the developed compositions, capable of photo polymerization and in the polymer membrane obtained from this composition. It was chosen the conditions (temperature, filtration of UV irradiation, the presence of competitive inhibitors) that increase the residual activity of immobilized enzymes. Extensively it was studied the properties of the developed electrochemical biosensors based on the IsFETs for the determination of glucose and urea as well it was show that they have the characteristics needed for use in laboratory, clinical, food and biotech practice.

4. Acknowledgment

This work was supported by State Fond of Fundamental Investigations of Ukraine, grant F28.7/020. Author thanks Dr. V.M. Starodub for assistance in the preparation of this article.

5. References

Ambrose T. M., Meyerhoff M. E. (1997) Photo-cross-linked decyl methacrylate films for electrochemical and optical polyion probes. *Anal. Chem.*, 69. 4092-4098.

Arenkov P.I., Starodub A.N., Beresin V.A. (1994a) Fiber optic immunosensors based on enhanced chemiluminescence and their application to determine different antigens. *Sensors and Actuators (B)*, 18, 1-3, 161-165.

Arenkov P.I., Starodub A.N., Beresin V.A. (1994b) Construction and biomedical application of immunosensors based on fiber optics and enhanced chemiluminescence. *Optical Engeneereng*, 33, 9, 2958-2963.

Arica Y., Hasirci V. N. (1987) Immobilization of glucose oxidase in poly(2-hydroxyethyl methacrylate) membranes. *Biomaterials*. 8, November, 489-495.

Barie N., Rapp M., Sigrist H., Ache H. J. (1998) Covalent photolinker-mediated immobilization of an intermediate dextran layer to polymer-coated surfaces for biosensing applications. *Biosensors & Bioelectronics*. 13. 855-860.

Dobrikov M.I., Shishkin G.V. (1983a) Photo materials on the basis of photo immobilized enzymes. Dependence of light sensitivity on way of photo immobilization. *Avtometry*. 5. 17-23.

Dobrikov M.I., Shishkin G.V. (1983b) Photo materials on the basis photo immobilized enzymes. Dependence of photographic characteristics on conditions of fulfilment of consolidating, strengthening and manifestation of hidden enzymatic image. *Avtometry*. 5. 23-28.

Dontha N., Nowall W. B., Kuhr W.G. (1997) Generation of biotin/avidin/enzyme nanostructures with maskless photolitography. *Anal Chem.*, 69(14), Jul 15. 2619-2625.

Doretti L., Ferrara D. (1993) Enzyme-entrapping membranes for biosensors obtained by radiation-induced polymerization. *Biosensors and Bioelectronics*. 8. 443-450.

Doretti L., Ferrara D., Barison G., Lora S. (1994) Glucose sensors based on enzyme immobilization onto biocompatible membranes obtained by radiation-induced polymerization. *Appl Biochem Biotechnol.*, 49(3), Dec.,191-202.

Doretti L., Gattolin P., Burla A., Ferrara, D.; Lora, S. & Palma, G. (1998) Covalently immobilized choline oxidase and cholinesterases on a methacrylate copolymer for disposable membrane biosensors. *Applied Biochemistry and Biotechnology*, 74, 1-12.

Freemen A. (1986) Gel entrapment of whole cells and enzymes in crosslinked, prepolymerized polyactrylamide hydrazide. *Ann.N.Y.Acad.Sci.*, 434, 418–426.

Gao H., Sanger M., Luginbuhl R., Sigrist H. (1995) Immunosensing with photoimmobilized immunoreagents on planar optical wave guides. *Biosens Bioelectron.*, 10(3-4). 317-328.

Gooding J. J., Hall E.A.H. (1996) Membrane properties of acrylate bulk polymers for biosensor applications. *Biosensors and Bioelectronics*. 11, 10, 1031-1040.

Grishchenko V.K., Masljuk V.K., Gudzera S.S. (1985) *Liquidphoto polimerizable compositions*. Naukova dumka, Kiev.

Hall E. A. H., Gooding J. J., Martens C. E. (1999) Acrylate polymer immobilization of enzymes. *Fresenius J Anal Chem.*, 364, 58-65.

Jae Ho Shin, Sang Yong Yoon, In Jun Yoon, Sung Hynk Choi, Sung Dong Lee, Hakhyun Nam, Ceein Sig Cha. (1998) Potentiometric biosensors using immobilized enzyme laers mixed with hydrophilic polyurethane. *Sensors and Actuators (B)*, 50, 19–26.

Jimenez C., Bartoli J., de Rooij N.F., Koudelka-Hep M. (1995) Glucose sensor on an amperometric microelectrode with a photopolymerizable enzyme membrane. *Sensors and Actiuators (B)*, 26-27, 421-424.

Jobst G., Urban G., Jachimowich A., Kohl F. Tilado O. (1993) Thin-film Clark-type oxygen sensors based on novel polymer membrane systems for *in vivo* and biosensors applications. *Biosensors & Bioelectronics*. 8. 123-128.

Katsev A. M., Abduramanove E. P., Starodub N.F (2009) Immobilization of bioluminescent bacteria on the nonorganic substances and estimation of their ability for the biotesting. *Biotechnology*, 2. 3. 74-78.

Kolytcheva N. V., Műller H., Marstalerz J. (1999) Influence of the organic matrix on the properties of membrane coated ion sensor field-effect transistors. *Sensors and Actuators B.*, 58, 456-463.

Kumakura M., Kaetsu I. (1989) Preparation of immobilized enzyme membrane by radiation-cast-polymerization. *Isotopenpraxis*. 25, 5. 192-195.

Kuriyamma T., Kimura J. (1991) FET-based biosensors. *Bioprocess Technology*. 15. 139–162.

Leca B., Morelis R.M., Coulet P.R. (1995) Design of a choline sensor via direct coating of the transducer by photopolymerization of the sensing layer. *Sensors and Actuators B*. 26-27. 436-439.

Lesho M. J., Sheppard N. F. (1996) Adgesion of polymer films to oxidized silicon and its effects on perfomance of a conductometric pH sensor. *Sensors and Actuators B*. 37. 61-66.

Levkovets I., Adanyi N., Trummer N., Varadi M., Szendro I., Starodub N., Szekacs A. (2004) Development of optical (OWLS) immunosensors for macromolecules and small analytes. *Biokemia*, XXVIII, 7-15.

Macca C., Solda L., Palma G. (1995) Potentiometric biosensing of penicillins using a flow-through reactor with penicillinase or penicillin amidase immobilised by gamma-irradiation. *Analytical Letters*. 28, 10. 1735-1749.

Masljuk A.F., Chranovsky V.A. (1989) *Photochemistry of polymerization able olygomers.* Naukova dumka, Kiev.

Mohy Eldin M. S., De Maio A., Di Martino S. Bencivenga U., Rossi S. (1999) Immobilization of β-galactosidase on nylon membranes grafted with diethylenglycol dimethacrylate (DGDA) by γ-radiation: effect of membrane pore size. *Advances in Polimer Technology*. 18, 2, 109-123.

Moser I., Jobst G., Aschauer E. Svasek P., Varahram M., Urban G. (1995) Miniaturized thin film glutamate and glutamine biosensors. *Biosens Bioelectron.*, 10, 6-7, 527-232.

Munoz J., Jimenez C., Bratov A. (1997) Photosensitive polyurethanes applied to the development of CHEMFET and ENFET devices for biomedical sensing. *Biosensors & Bioelectronics*, 12. 7, 577-585.

Nabok A.V., Tsargorodskaya A., Holloway A., Starodub N.F., Gojster O. (2007) Registration of T2mycotoxin with total internal reflection ellipsometry and QCM impedance methods. *Biosensors and Bioelectronics*, 22, 885-890.

Nakako M., Hanazato Y., Maeda M., Shiono S. (1986) Neutral lipid enzyme electrode based on ion-sensitive field effect transistors. *Analitica Chimica Acta*, 185. 179-185.

Nakayama Y., Matsuda T. (1992) Surface fixation of hydrogels. Heparin and glucose oxidase hydrogelated surfaces. *ASAIO J.* 38(3), jul. 421-424.

Nakayama Y., Zheng Q., Nishimura J., Matsuda T. (1995) Design and properties of photocurable electroconductive polymer for use in biosensors. *ASAAIO J.* 41(3), Jul-Sep. 418-421.

Navera E.N., Sode K., Tamiya E., Karube I. (1991) Development of acetylcholine sensor using carbon fiber (amperometric determination). *Biosens Bioelectron.*. 6(8). 675-680.

Pyrogova L.V., Starodub N.F. (2008) Immobilization of bovine leucosis antigen on the transducer surface of biosensor, *Biotechnology*, 1, 2, 52-58.

Puig-Lleixa C., Ramirez-Garcia S., Bartoli J. (1999a) Development of new photopolymerizable membrane for monochloracetate sensetive potentiometric sensors. *Analytica Chimica Acta*, 386, 13-19.

Puig-Lleixa C., Jimenez C., Alonso J., Bartoli J. (1999b) Polyurethane-acrylate photocurable polymeric membrane for ion-sensetive field-effect transistor based urea biosensors. *Analytica Chimica Acta*, 389, 179-188.

Puig-Lleixa C., Jimenez C., Fabregas E., Bartoli J. (1999c) Potentiometric pH sensors based on urethan-acrylate photocurable polymer membranes. *Sensors and Actuators B*. 49, 211-217.

Rebrijev A.V. (2000) Enzyme sensor on the basis of photopolimeric membranes for the urea determination. *Ukr. Biochem.J.*, 72, 6, 141-142.

Rebrijev A.V., Ivashkevich S.P., Starodub N.F., Kercha S.F., Masljuk A.F. (2001) Electrochemical sensor on the basis of photo polymeric membrans for the determination of urea. *Ukr. Biochem. J.*, 73, 1, 133-142.

Rebrijev A.V., Starodub N.F. (2001) Photopolymers as immobilization matrix in biosensorics. Ukr. Biochem. J., 73, 6. 5-17.

Rebrijev A.V. (2002) Optimization of conditions of the immobilization of enzymes in photo polymerizable membrane. *Ukr. Biochem J.*, 74, 4*b* (addition 2), 194-195.

Rebrijev A.V., Starodub N.F., Masljuk A.F. (2002a) Optimization of the conditions for the immobilization of enzymes in photopolymeric membrane. *Ukr. Biochem. J.*, 74, 3, 82-87.

Rebrijev A.V., Starodub N.F., Masljuk A.F. Liquid photo polimerizable compositions as immobilization matrix in biosensorics. (2002b) *Ukr. Biochem. J.*, 74, 4, 101-106.

Rehman F. N., Auden M., Abrams E. S., Hammond+ Ph. W., Kenney M., Boles T. Ch. (1999) Immobilization of acrylamide-modified oligonucleotides by co-polymerization. *Nucleic Acids Reseach.* 27, 2. 649–655.

Starodub N.F. (1989) Nonelectrode biosensors – a new direction in biochemical diagnostics. *Biopolymers and cells*, 1, 5–15.

Starodub N. F. (1990) Lecture notes of the Int. Sensors Center of Biocybernetics of the Acad. of Sci. of the Socialist. Countries. Jablonna, 3. 173–202.

Starodub N.F., Chustochka L.N., Lazorenko A.V., Bubrjk O.A., Terent'jev A.V., El'skaja A.V. (1990) Integration of biological materials in electrochemical devices. *J. Anal. Chem.*, 45, 1432–1440.

Starodub N.F., Samodumova I.M., Starodub V.N. (1995) Usage of organosilanes for integration of enzymes and immunocomponents with electrochemical and optical transducers. *Sensors and Actuators (B)*, 176, 173-176.

Starodub N.F., Torbicz W., Pijanowska D., Starodub V.M., Kanjuk M.I., Dawgul M. (1998) Optimization of enzyme integration with transducers for analysis of irreversible inhibitors. Eurosensors XII. Proceedigs of the European conference on solid-state transducers and 9 th UK conference on sensors and their applications, Southampton, UK, 13-16 September 1998.

Starodub V.M., Fedorenko L.L., Sisetskiy A.P., Starodub N.F. (1999a) Control of mioglobin level in an immune sensor based on the photoluminescence of porous silicon. *Sensor and Actuators (B)*, 58, 1-3, 409-414.

Starodub N.F., Kanjuk N.I., Kukla A.L., Kanjuk M.I., Shirshov Y.M. (1999b) Multi-enzymatic electrochemical sensor: field measurements and their optimization. *Anal. Chim. Acta*, 385, 461-466.

Starodub N.F., Torbicz W., Pijanovska D., Starodub V.M., Kanjuk M.I., Dawgul M. (1999c) Optimization methods of enzyme integration with transducers for analysis of irreversible inhibitors. *Sensors and Actuators (B)*, 58, 1-3, 420-426.

Starodub N.F., Dzantiev B.B., Starodub V.M., Zherdev A.V. (2000a) Immunosensor for the determination of the herbicide simazine based on an ion-selective field-effect transistor. *Anal. Chim. Acta*, 424, 37-43.

Starodub V.M., Fedorenko L.L., Starodub N.F. (2000b) Optical immune sensor for the monitoring protein substances in the air. *Sensor and Actuators (B)*, 68, 1-3, 40-47.

Starodub N.F., Nabok A.V., Starodub V.M., Ray A.K., Hassan A.K.(2001) Immobilization of biocomponents for immune optical sensors. *Ukr. Biochem. J.*, 73, 3, 16-24.

Starodub N.F., Rebrijev A.V. (2002) *Photopolymerisable membrane for electrochemical enzymaticsensors.* Abstract Book of the Seventh World Congress on Biosensors, Kyoto, Japan, 15-17 May, 2002.

Starodub N. F., Rebriev A. V., Starodub V. M. (2002a) *Biosensors and express biochemical diagnostics of some diseases.* Proc. of the NATO ASI "Biosensors and Express Biochemical Diagnostics of Some Diseases", Eds. N. Marczin and M. Yacoub, IOS Press, 2002.

Starodub N., Rebrijev A., Maslyuk A. (2002b) *Photopolymerisable materils and enzymatic biosensors*. Abstract Book of NATO advanced research workshop "Nanostructured materials and coatings for biomedical and sensor applications". Kiev, Ukraine, 4-8 August 2002.

Starodub N. F., Rebriev A. V. (2002) *Biosensors for diagnostics of some diseases*. Abstract Book of 53rd Annual Meting of the International Society of Electrochemistry "Electrochemistry in Molecular and Microscopic Dimensions". Dusseldorf, Germany, 15-20 September 2002.

Starodub N.F., Starodub V. M. (2002) Biosensor control of water contamination by some chemical organic substances. *Chemistry and Technology of Water*, 6, 112-119.

Starodub N.F., Pyrogova L.V., Demchenko A., Nabok A.V. (2005) Antibody immobilization on the metal and silicon surface. The use of self-assambled layers and specific receptors, *Bioelectrochemistry*, 66, 111-115.

Starodub N.F. (2006) Biosensor for the evaluation of lipase. J. Mol. Catalysis B: Enzymatic, 40, 155-160.

Starodub N.F., Rebriev A.V. (2007) Liquid photopolymerizable compositions as immobilized matrix of biosensors. *Bioelectrochemistry*, 71, 1, 29-32.

Starodub N.F., Kanjuk M.I., Shmyryeva A.N. (2008) Microelectronic multi-parametrical biosensors. *Biotechnology*, 1, 1, 61-73.

Triven M. (1983) *Immobilization enzymes*. Mir, M.

Turner A. P. F. Preface (1989) *Biosensors: Fundamentals and Applications*. Oxford: Oxford Univ. Press. p. 3–12.

Wan K., Chovelon J.M., Jaffrezic-Renault N., Soldatkin A.P. (1999) Sensitive detection of pesticide using Enfet with enzymes immobilized by cross-linking and entrapment method. *Sensors and Actuators (B)*. 58. 399-408.

Wróblewski W., Dawgul M., Torbicz W., Brzózka Z. (1997) Anion-selective CHEMFETs. *Proc. SPIE*. 3054. 197.

Wudvord J. (1988) *Immobilized cells and enzymes*. Methods. Mir, M.

19

Visual Detection of Change Points and Trends Using Animated Bubble Charts

Sackmone Sirisack and Anders Grimvall
Linköping University,
Sweden

1. Introduction

The rapid growth of automatic data collection systems has increased the need for algorithms that can efficiently reveal important features of large or complex datasets. For example, it is often of great interest to examine the occurrence of abrupt changes in long bi- or multivariate time series of data. Several numerical algorithms and statistical tests have been developed to detect abrupt shifts in the mean or other parameters of uni- or multivariate distributions (Caussinus & Mestre, 2004; Hawkins, 1977, 2001; Srivastava & Worsley, 1986; Stephens, 1994). However, there is also a need for visualization techniques that can help the user identify any type of abrupt changes or trends in the collected data. More generally, techniques are needed that can simultaneously highlight important features of the data and filter out irrelevant information (Bederson & Boltman, 1999; Bundesen, 1990; Cleveland & McGill, 1984; Healey, 2000; Ware, 2004). In this chapter, we present flexible and user-friendly animations of bubble charts in which subsets of the collected data are sequentially highlighted on a static background representing all data points.

The basic ideas of interactive visualization of quantitative data were presented before computer technologies were sufficiently developed to enable widespread use of such methods. In 1978, Newton introduced a form of linked brushing that allowed the user to select a subset of observations in one display and simultaneously highlight the same subset in another display. About a decade later, several ground-breaking articles were published. Asimov (1985) introduced the concept of helicopter tours for viewing high-dimensional datasets via a structured progression of 2D projections, and Becker and co-workers (1987a, b) provided a systematic framework for brushing, linking, and other forms of interactive statistical graphics. Moreover, Unwin and colleagues (1988) demonstrated how zooming, rescaling, and overlaying can facilitate visual analysis of multivariate time series data.

More recently, improvements in computing power, display resolution, and numerical algorithms have brought interactive visualization of quantitative data to higher levels and stimulated the development of new applications. The software XGobi and its descendant GGobi set a new standard for interactive modification of linked plotting windows, and an application programming interface made such methods available to the rapidly growing group of R users (Cook & Swayne, 2007; Swayne et al., 2003; the GGobi website, 2011). Zooming and rescaling were established as standard tools in software packages for time series analysis, and visual specification of queries was introduced to facilitate the search for interesting features of time series data (Hochheiser et al., 2003).

Motion charts, or animated bubble charts, represent another breakthrough in data visualization (the Gapminder website, 2011). The basic display is a 2D bubble chart showing observed pairs of two variables x and y that have been recorded annually for a set of objects. By highlighting the positions of the bubbles year by year, changes over time can be visualized. Additional information about the investigated objects can be entered into the graphs by colour-coding the bubbles and letting their size vary with some covariate. A Google gadget (the Google website, 2011) has made motion charts available to any user with a good Internet connection.

The use of animated population pyramids in official statistics (the Australian Bureau of Statistics, 2011) illustrates that almost any static graph in statistics can be animated to visualize changes over time. However, some authors have emphasized that animations are not always superior to static presentations such as a small multiples display (Robertson et al., 2008). Visualization of temporal changes in the size and shape of 2D point clouds represents yet another approach that is particularly suitable for exploring large datasets (Landesberger et al., 2009).

Here, we present a flexible two-stage method for making animated bubble charts in Excel®. In the first stage, a macro written in VBA (Visual Basic for Applications) is utilized to identify data tables in a given worksheet and help the user select and organize the inputs to the animation. This macro also creates a suitable bubble-chart template. Thereafter, a collection of other VBA macros is employed to produce the animation.

The methods and software solutions we propose are designed to handle fairly large datasets with multiple groups of objects and multiple observations per time stamp and group. Furthermore, it can be noted that the order in which different subsets of data are highlighted can be determined by an arbitrary numerical or string variable. In general, bubble charts are used to visualize relationships between interval variables. However, relationships involving categorical or ordinal variables can also be visualized. In such cases, adding a small amount of noise (jitter) to the original data might be helpful, because it will improve the separation of the data points so that each point is made visible. In addition, the visualization can be extended to high-dimensional time series data by using a macro that first performs principal components analysis and then creates 2D animated score charts.

After a brief summary of the general principles of animating bubble charts, and some remarks regarding design issues, we use time series of daily to monthly environmental data to illustrate the power of visual tools to bring out important characteristics of the collected data. Most of our analyses are focused on the occurrence of sudden shifts in the mean or dispersion, and whether or not such shifts can be found in all investigated groups of data. However, the tools presented here are also used to examine temporal trends across seasons and changes along gradients. Moreover, we use a set of multivariate chemical data on olive oils to illustrate how animated score charts can highlight differences between geographical regions.

After presenting a set of useful displays and animation options, we resume our discussion of factors that influence the visual impression of static and animated charts, and we also consider how to achieve a good balance between the information content of a display and perceptual capacity limits. In addition, we address some technical aspects of using spreadsheets with tens of thousands of observations.

2. General principles of animating bubble charts

In Excel® and other spreadsheet programs, graphs added to a worksheet can be updated automatically and almost instantaneously when the content of the worksheet is altered. This enables animations driven by a macro that achieves step-by-step changes in the content of a range of worksheet cells. The speed of an animation can be controlled by making calls to a special function that puts the macro to sleep and wakes it up after a specified amount of time.

Because visual inspection is particularly suitable for detecting motion against a static background, we developed animations in which all data are used to construct a static background, and different subsets of data are sequentially highlighted. In a 2D bubble chart, this type of displays can be constructed by using open markers for the static background and filled markers for the highlighted data. This is illustrated in Figure 1, which shows how the interdependence between reported pH and alkalinity levels in the Baltic Proper has changed over time. In particular, it can be noted that the reported interdependence changed dramatically from 1989–1993 to 1994–1998, most probably due to changes in laboratory practices.

3. Some design issues

A user-friendly implementation of animated bubble charts requires a good balance between flexibility and standardization. The selection of data and the design of the bubble charts should be flexible, whereas efficient updating of spreadsheets and graphs is greatly facilitated if the data tables have a standardized design. This favours two-stage procedures in which a set of user forms first help the user organize the data in a standardized manner and create a suitable graph template; thereafter, the animation can be run and controlled with buttons and scroll bars.

We created a VBA macro that initially determines the position and size of the data tables that are to be visualized, and then utilizes list boxes to select up to five variables for an animated bubble chart. The first variable, which is required and may represent a time stamp, is used to control the highlighting of different subsets of data. Variables two and three, which are also required, represent the x and y variables in a bubble chart. Variable four, which is optional, can be used to partition the set of bubbles into different groups. Finally, another optional variable can be used to size code the bubbles.

The macro that prepares for the animation can also allow the user to select a suitable step length (time step) for the animation and a desired range of animation records (time span). Furthermore, the preparations include automatic scaling of the x- and y-axes of the bubble chart and selection of marker types. The applicability of animated bubble charts can be further increased by performing an optional standardization of the x and y variables to mean zero and variance one, and by calculating the first two principal components of a user-defined set of variables. In the latter case, high-dimensional data can be scrutinized by creating animated 2D score charts.

4. Different types of displays

4.1 Standard bubble charts with groups

The simplest form of bubble charts has a single group of highlighted cases (see Fig. 1). This type of display can easily be generalized to displays in which two or more groups are assigned different coloured markers. Theoretically, the red-green-blue (RGB) system enables colour coding of up to 2^{24} groups. However, static bubble charts with more than eight colours are difficult to perceive (Gilmore et al., 1989), and animated charts are best perceived if no more than four groups of cases are simultaneously highlighted in the same display.

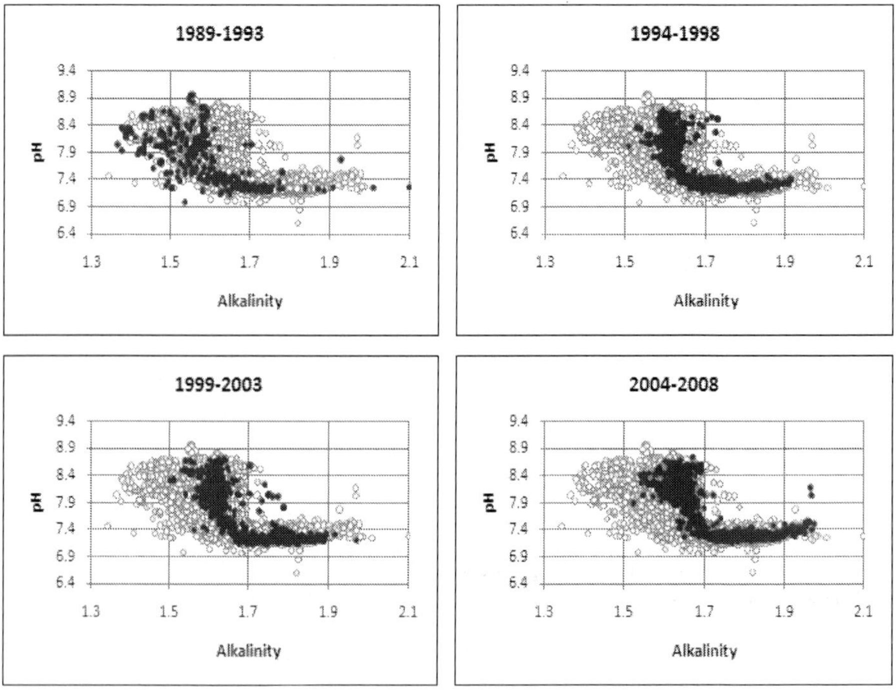

Fig. 1. Four consecutive frames from an animation of pH against alkalinity of seawater samples from the Eastern Gotland Basin in the Baltic Proper (sampling site BY15). Data source: the Swedish Meteorological and Hydrological Institute (SMHI).

Figure 2 shows how the interdependence between pH and salinity of seawater samples varied over time and between laboratories. In particular, it can be seen that in 1989–1993 the variability of pH for a given salinity was unusually large for one of the laboratories involved, which indicates data quality problems. Moreover, there are single outliers in the data that were collected more recently. Further studies are needed to determine whether these outliers represent flawed data or unusual water samples. It cannot be excluded that mixing of seawater due to strong winds can cause rather abrupt changes in pH.

We have already emphasized that multicoloured bubble charts should be used with caution. This advice is further motivated by Figure 3, in which the upper frames with group-specific coloured markers contain more information than the lower frames with black markers only. Nevertheless, the lower frames show more clearly that there was a level shift in the total volume of phytoplankton between the two time periods, although the content of chlorophyll-a changed very little. It should also be kept in mind that if different colours are used in the same panel, they may interfere with each other. Spatial patterns in strong colours may conceal patterns in light colours, if the background is white.

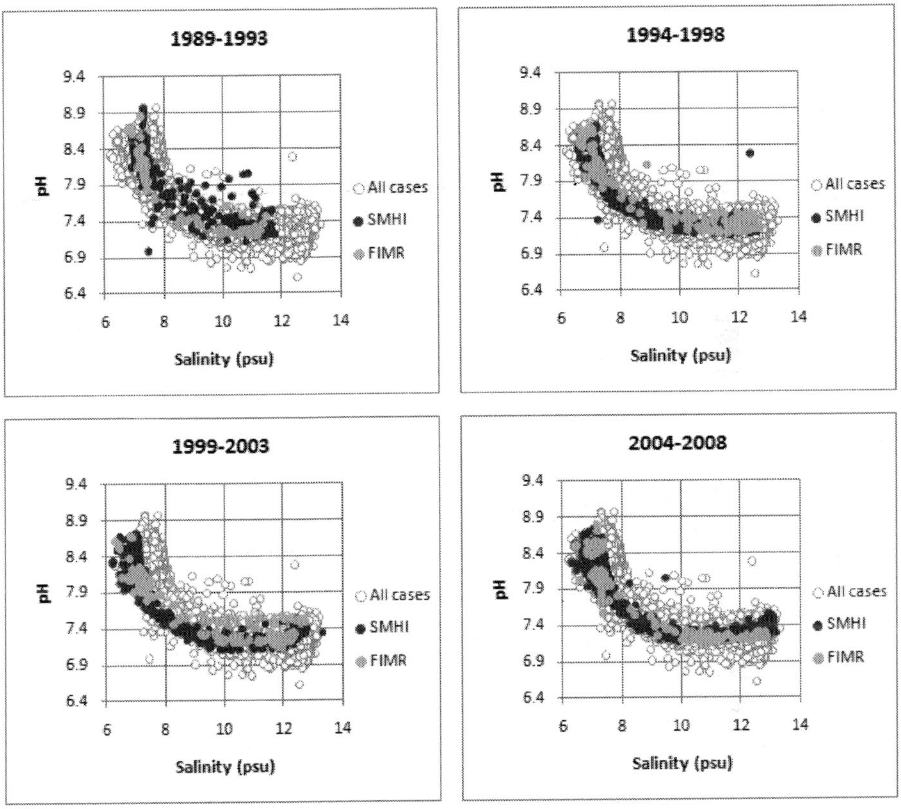

Fig. 2. Four consecutive frames from an animation of salinity and pH data for seawater samples collected in the Eastern Gotland Basin in the Baltic Proper (sampling site BY15) and analysed by the Swedish Meteorological and Hydrological Institute (SMHI) and the Finnish Institute of Marine Research (FIMR).

Size-coding of bubble chart markers is another tool that should be employed with great caution, unless the user actually wants to suppress some data points or the dataset is so small that the markers can be inspected one by one. Furthermore, it is worth noticing that the (average) size of the markers has a strong impact on the perception of a pattern formed by a set of markers . Markers that are too small tend to blur the contours of a cloud of points, and large markers can make it difficult to comprehend the number of points in different subsets of data.

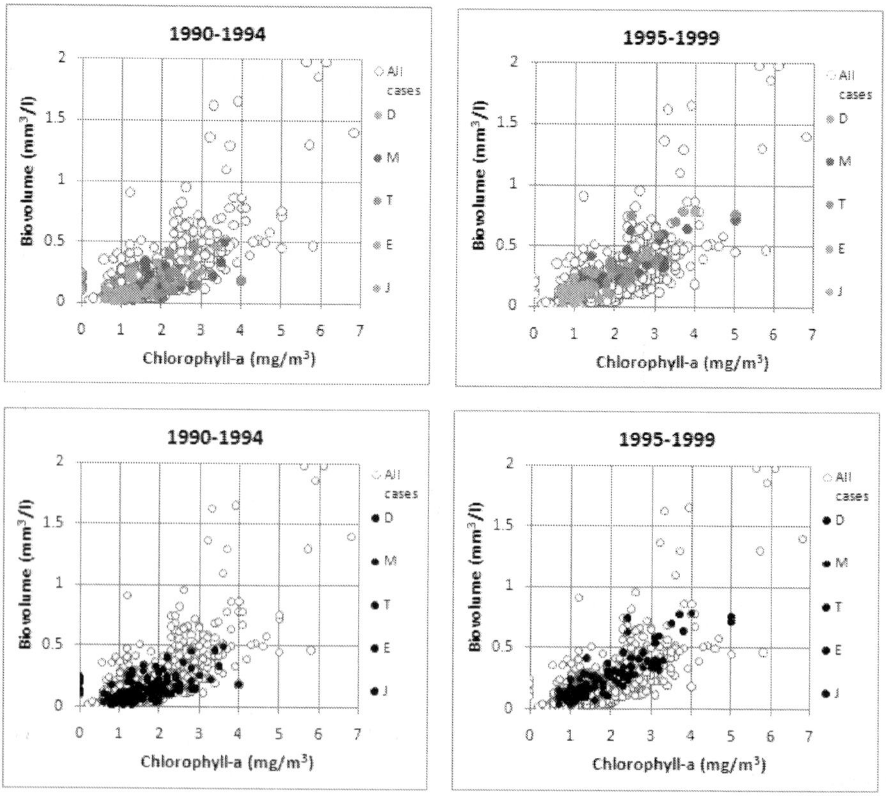

Fig. 3. Bubble charts of phytoplankton data from three sites in Lake Vänern (D, Dagskärsgrund N; M, Megrundet N; T, Tärnan SSO) and two sites in Lake Vättern (E, Edeskvarnaån NV; J, Jungfrun NV) in Sweden. The coloured markers in the upper panels have been changed to black markers in the lower panels. Data source: the Swedish University of Agricultural Sciences (SLU).

4.2 Jittered bubble charts

A jittered plot adds some random noise to the x or the y coordinate, or both. Such plots are particularly useful for categorical and ordinal data, because they can give a realistic visual impression of the number of cases in different parts of the chart. In environmental monitoring, jittered plots are particularly useful when the x coordinate represents a class variable such as month or season, or the y coordinate represents a count variable such as the number of species found in the analysed sample.

Figure 4 illustrates a suspected artificial level shift in temperature data from the Czech Republic. The time series plot indicates that the temperature difference between the two investigated meteorological stations increased in 1998. By using a jittered plot to visualize the differences by month, it can be seen that the level shift was present during all seasons and was particularly pronounced during the warmer months.

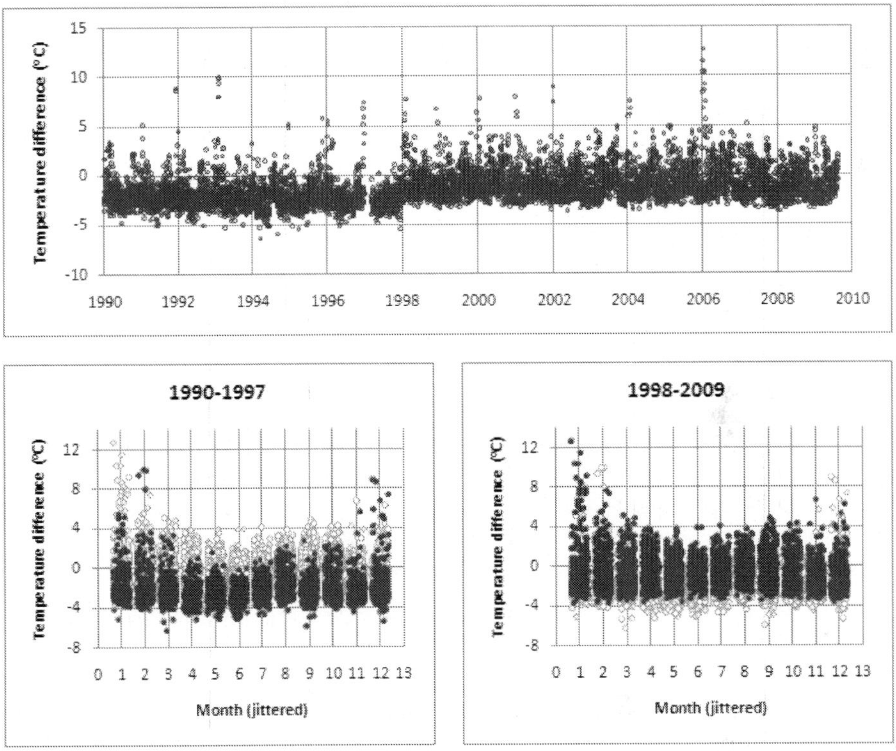

Fig. 4. Ordinary time series plot and jittered bubble charts of the difference in daily mean temperatures between the meteorological stations Protivanov and Jevičko in the Czech Republic. A small amount of noise has been added to the month number. Data source: the Czech Hydrometeorological Institute, Brno.

4.3 Bubble charts with trend lines

When there is pronounced seasonal variation in the collected data, it may be of interest to animate changes in trend slopes by month. This can be achieved by using the month as animation variable and one of the built-in trend line options in Excel®. Figure 5 shows long-term temperature trends in central England, and the four panels draw attention to the fact that the trend slope gradually decreases from March to June. In principle, this pattern could have been revealed by producing a series of static plots. However, this process can be automated by using software for animation. In addition, animation can help to identify between which months of the year that the major changes in trend slopes occur. Such differences in slopes between adjacent months can be further accentuated by standardizing the data so that differences in monthly means are eliminated.

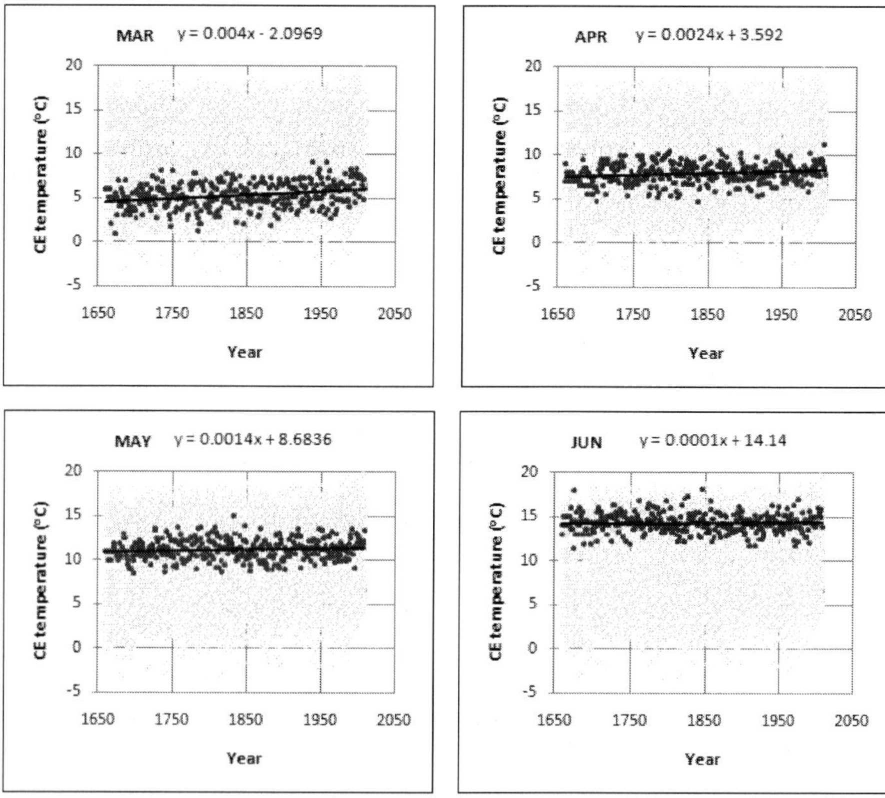

Fig. 5. Four consecutive frames from an animation of trends by month for the Central England Temperature series compiled by the Hadley Centre, UK.

4.4 Gradient charts

In many environmental monitoring programmes, the sampling sites have a natural order. For example, samples from the marine environment are often taken along salinity or depth gradients, air pollutants are measured at different distances from a point source, and river water quality can be measured at different runoff levels. This calls for techniques that can efficiently visualize how relationships between two or more variables change along a gradient.

Figure 6 illustrates in two different manners how the relationship between the concentrations of phosphorus and suspended matter in a small stream varies with the runoff level. It is obvious that, compared to a static chart in which colour- and shape-coded markers are used to indicate runoff levels, an animated display has two advantages. First, there is no perceptual interference between the different subsets of data. Second, the analyst can inspect one highlighted subset while the previous subset is still fresh in memory.

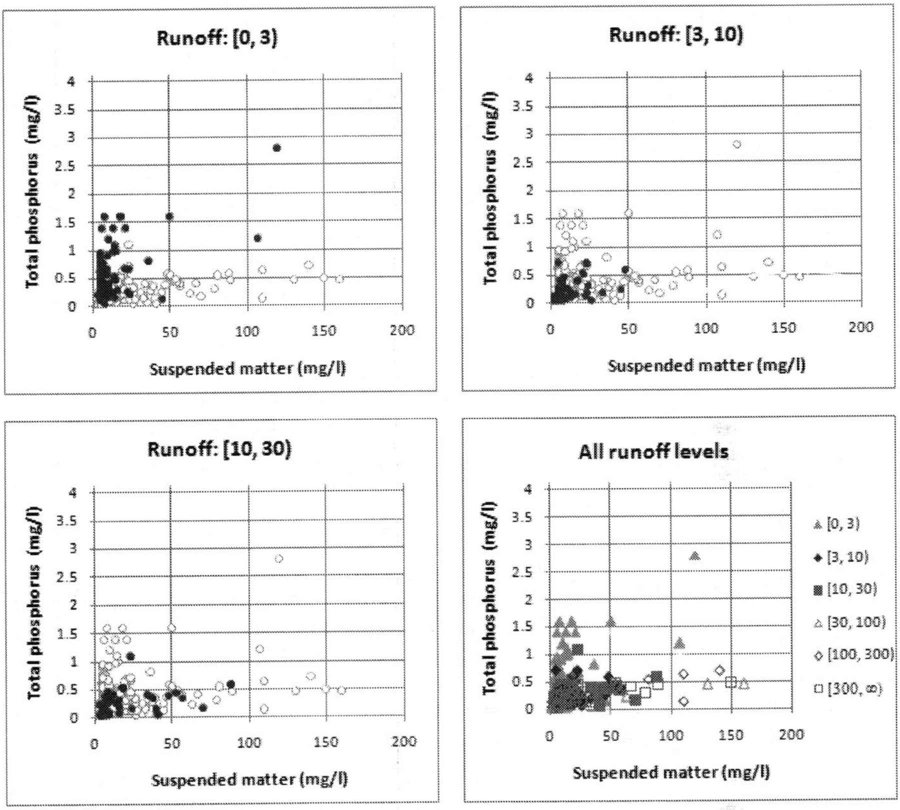

Fig. 6. Relationship between the concentrations of phosphorus and suspended matter in stream water from an agriculture-dominated catchment in southern Sweden. Data source: the Swedish University of Agricultural Sciences (SLU), catchment code N33.

4.5 Score charts for a pair of principal components

When the collected data are multivariate and the coordinates are strongly correlated, important information can be obtained from score charts in the coordinate system determined by the first two principal components. An animation can refine such information by highlighting data points by time or group. As in the gradient plots in the previous section, the advantage of an animated display is that there is no perceptual interference between the different subsets of data.

Figure 7 shows an animation of regional differences in the chemical composition of olive oil from different regions in Italy. The score charts draw attention to the fact that some groups of objects are more heterogeneous than others. By ordering the regions from south to north, or according to some characteristic of the areas, this type of animations can also highlight various gradients in the chemical composition.

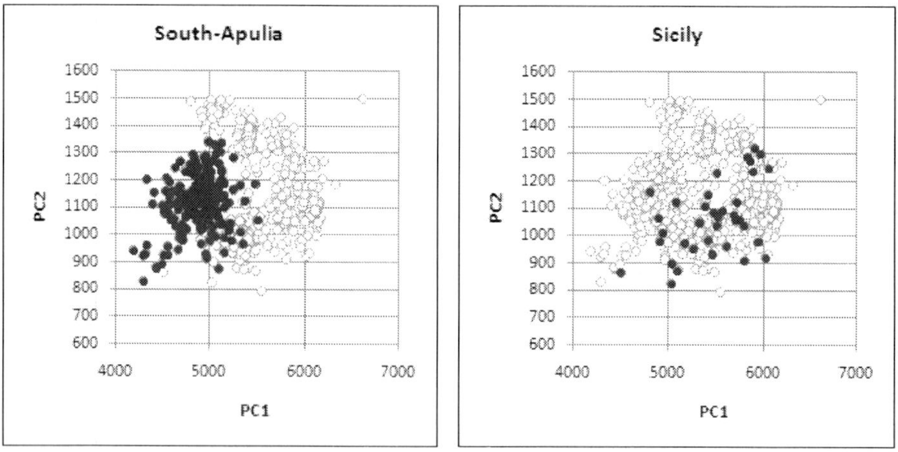

Fig. 7. Two frames from an animation of score charts derived from a dataset containing information about the content of eight different fatty acids in olive oil from nine different regions in Italy. Raw data were obtained from the Ggobi Website.

5. Computational aspects

The technical performance of Excel-based animations is markedly influenced by the technique that is used to update the content of the worksheet cells. In particular, the computational time can be reduced considerably, if large arrays are updated by a single command rather than by creating a loop in which individual cells are updated one by one. The performance can also be improved by turning off the automatic screen updating and the automatic calculation of worksheets during parts of the execution of the animation macro.

The design of the markers in the bubble chart is yet another factor that strongly influences the computational time. It takes longer to update large bubbles than small markers, and more elaborate bubbles that resemble 3D balls can greatly retard the animation.

Test runs using a dataset comprising 10,000 cases showed that a chart with 400 highlighted bubbles could be updated in less than two seconds on a standard PC. If the dataset is substantially larger, it may be preferable to base the animation on a (random) sample of the original data.

6. Discussion

When multiple views or complex graphical coding of multivariate data are used to bring loads of information into a single display, there is a considerable risk that the data representation will be visually impenetrable. Displays with multiple views can suffer from visual fragmentation, and perceptual interference can occur between different graphical codes in the same display (Healey, 2000; Bartram, 2001). The animated bubble charts presented in this article represent an attempt to simultaneously reduce visual fragmentation and perceptual interference.

The static background composed of open markers showing the distribution of the entire dataset enables rapid assessment of the distribution of a highlighted subset of data points. Moreover, the animation facilitates detection of change, because the analyst can inspect the shape and size of a highlighted point cloud while the previous point cloud is still fresh in memory.

Using filled markers of standardized shape makes it easier to discern the colour coding. Further, perception of a scatter plot can be strongly affected by the size of the markers, and hence it is worth noting that the built-in scaling feature in Excel can be used to reduce or increase the size of the bubbles in the charts. However, as emphasized in the introduction, only a few different colours and bubble sizes can be readily distinguished by visual inspection, and there may be perceptual interference between colour and size coding (Healey, 2000; Bartram, 2001). In addition, it should be mentioned that static visualizations, such as a small multiples display, are still viable alternatives to animated graphs (Robertson et al., 2008).

Much of the work presented here was inspired by Rosling and co-workers (Gapminder, 2011), who demonstrated that the animated bubble chart is a powerful tool for visualizing temporal trends in official statistics and other data collected annually for a set of objects. When one variable is plotted against another, and a video is created to simultaneously display changes over the period of data collection, the motion of the bubbles can draw attention to subsets of objects that move simultaneously in the same direction. Similarly, the motion makes it easier to identify deviating objects that move in a completely different direction.

Our work here has demonstrated that animated bubble charts are also very useful for inspecting temporal changes in the shape and size of 2D point clouds. For example, such animations can efficiently reveal changes in the presence of outliers or in the conditional mean and variance of one variable given another. Moreover, detection of change across time or groups can be greatly facilitated if open bubbles representing the entire dataset are allowed to form a static background, while selected subsets of data points are sequentially highlighted at a rate determined by the user.

Also, it should be noted that animated bubble charts can be useful, even if the order of the highlighted subsets lacks meaning. Without writing any computer code, a large number of simple bubble charts can be created and inspected at a pace determined by the analyst. Our animated 2D score charts represent yet another example of a time-saving procedure that can create a good overview of a complex dataset.

This article has focused on construction of animated bubble charts in a spreadsheet program where charts that are added are automatically updated when the contents of some worksheet cells are updated. Other software or programming environments can provide other solutions to animation problems. In R, for instance, a sequence of frames representing different time stamps are combined into a video prior to the animation, whereas the Google gadget *Motion Chart* provides several means of interaction. The main technical advantages offered by the Excel-based animations presented here are flexibility and the capacity to handle fairly large datasets. Test runs showed that, compared to Google *Motion Chart*, our tools can handle larger datasets. Furthermore, they are very flexible in three respects: (i) an arbitrary numerical or string variable can be used to determine the order in which different subsets of data are highlighted; (ii) any Excel tool can be used to modify the design of the bubble chart prior to the animation; (iii) multidimensional data can be scrutinized by first performing a principal components analysis and then animating a score chart in which the observations are plotted in a coordinate system determined by the first two eigenvectors.

7. Conclusions

Our study demonstrated that animated bubble charts can facilitate detection of change points and trends. More specifically, we emphasized that such charts have the following advantages:

i. the analyst can inspect the shape and size of a highlighted point cloud while the previous point cloud is still fresh in memory;
ii. bubble charts in which the entire dataset is allowed to form a static background put the high-lighted subset into a wider perspective;
iii. animations are time-saving procedures that can readily create a good overview of complex datasets.

Furthermore, we showed that our Excel-based software solutions are very flexible in three respects:

i. an arbitrary numerical or string variable can be used to determine the order in which different subsets of data are highlighted;
ii. any Excel tool can be used to modify the design of the bubble chart prior to the animation;
iii. multidimensional data can be scrutinized by first performing a principal components analysis and then animating a score chart in which the observations are plotted in a coordinate system determined by the first two eigenvectors.

In summary, our results demonstrate that animation can simultaneously reduce visual fragmentation and perceptual interference.

8. Acknowledgements

The authors are grateful to the Swedish International Development Cooperation Agency (SIDA) and its Department for Research Cooperation (SAREC), and the Faculty of Science (FOS) at the National University of Laos (NUOL) for providing financial support and facilities for this research. We are also grateful to the Swedish University of Agricultural Sciences, the Swedish Meteorological and Hydrological Institute, and the Czech Hydrometeorological Institute for supplying environmental data.

9. References

Asimov D. The grand tour: A tool for viewing multidimensional data. *SIAM. Journal of Science and Statistical Computing* 1985; 6:128–143.

Bartram LR. *Enhancing information visualization with motion.* PhD Thesis, School of Computer Science, Simon Fraser University, 2001.

Becker RA, Cleveland WS. Brushing scatterplots. *Technometrics* 1987; 29:127-142.

Becker RA, Cleveland WS, Wilks AR. Dynamic graphics for data analysis. *Statistical Science* 1987; 2:355-396.

Bederson B, Boltman A. Does animation help users build mental maps of spatial information? *Proceedings of the 1999 IEEE Symposium on Information Visualization (InfoVis 1999)*, San Francisco, CA, Oct 1999.

Bundesen C. A theory of visual attention. *Phsychological Review* 1990; 97:523 – 547.

Caussinus H, Mestre O. Detection and correction of artificial shifts in climate series. *Applied Statistics* 2004; 53:405-425.

Cleveland WS, McGill R. Graphical perception: Theory, experimentation, and application to the development of graphical methods. *Journal of the American Statistical Association* 1984; 79:531-554.

Cook D, Swayne, DF. 2007. Interactive and Dynamic Graphics for Data Analysis - with R and Ggobi. New York: Springer Verlag.

Gilmore W, Gertman D, Blackman H. User-Computer Interfaces in Process Control: A Human Factors Engineering Handbook. Academic Press: San Diego, 1989.

Hawkins DM. Testing a sequence of observations for a shift in location. Journal of the American Statistical Association 1977; 72:180-186.

Hawkins DM. Fitting multiple change-point models to data. *Computational Statistics and Data Analysis* 2001; 37:323-341.

Healey CG. Building a perceptual visualisation architecture. *Behaviour and Information Technology* 2000, 19:349-366.

Hochheiser H, Baehrecke EH, Mount SM, Shneiderman B. Dynamic querying for pattern identification in microarray and genomic data. *Proceedings of the IEEE Multimedia Conference and Expo,* July 2003, Baltimore, MD, 2003.

Landesberger TV, Bremm S, Rezaei P, Schreck T. Visual analytics of time dependent 2D point clouds. *Computer Graphics International* 2009; pp 97–101.

Newton C. Graphics: From Alpha to Omega in Data Analysis. In *Graphical Representation of Multivariate Data.* Academic Press: New York, 1978; pp 59-92.

Robertson G, Fernandez R, Fisher D, Lee B Stasko J. Effectiveness of animation in trend visualization. *Visualization and Computer Graphics* 2008; 14:1325–1332.

Srivastava MS, Worsley KJ. Likelihood ratio tests for a change in the multivariate normal mean. *American Statistical Association* 1986; 81:199-204.

Stephens DA. Bayesian retrospective multiple-changepoint identification. *Journal of the Royal Statistical Society: Series C* 1994; 43:159-178.

Swayne DF, Lang DT, Buja A, Cook D. GGobi: evolving from XGobi into an extensible framework for interactive data visualization. *Computational Statistics and Data Analysis* 2003; 43:423-444.

The GGobi website. http://www.ggobi.org/ [12 July 2011]

The Gapminder website. http://www.gapminder.org/ [12 July 2011]

The Google website. https://docs.google.com/templates?type=spreadsheets&q=motion+chart&sort=hottest&view=public [12 July 2011]

The Australian Bureau of Statistics website. http://www.abs.gov.au/websitedbs/d3310114.nsf/home/Population%20Pyramid%20-%20Australia [12 July 2011]

Unwin AR, Wills G. Eyeballing time series. *Proceedings of the 1988 ASA Statistical Computing Section* 1988; pp 263–268.

Ware C. *Information Visualization: Perception for Design* (2nd edn). Morgan Kaufmann Publishers, 2004.

20

Environmental Monitoring of Opportunistic Protozoa in Rivers and Lakes: Relevance to Public Health in the Neotropics

Sônia de Fátima Oliveira Santos[1,2], Hugo Delleon da Silva[1,2], Carlos Eduardo Anunciação[2] and Marco Tulio Antonio García-Zapata[1]
[1]*Instituto de Patologia Tropical e Saúde Pública (IPTSP), Núcleo de Pesquisas em Agentes Emergentes e Re-emergentes, Universidade Federal de Goiás*
[2]*Laboratório de Diagnóstico Genético e Molecular, Instituto de Ciências Biológicas II, Universidade Federal de Goiás*
Brazil

1. Introduction

Water is a natural resource of vital importance to living beings, but due to anthropic action several microorganisms are disseminated into aquatic environments. In developing countries, over one billion people do not have access to clean, properly treated water and approximately three billion people do not have access to adequate sanitary facilities (Kraszewski et al., 2001) This scenery is probably a consequence of the increased environmental degradation, depletion of water resources, and constant contamination of bodies of water with wastewater and industrial effluents (Pedro & Germano, 2001), causing microorganisms from soil, faeces, decomposing organic matter, and other pollutant sources to spread into water.

Goiania, the capital of the state of Goiás, located in the Midwestern Region of Brazil, has ca. 1.221.654 inhabitants and is considered a regional metropolis, among the major Brazilian cities that receive a large number of migrants (Alves & Chaveiro 2007). As a result, the city faces problems of disorderly and unsustainable urban growth with a consequent increase in superficial waste, which is a continuous source of contamination of water courses.

The current sources of public water supply for the city of Goiania, the Meia Ponte river basin and its tributary river João Leite, are constantly submitted to degradation processes due to anthropic action, such as agriculture, intensive livestock production, and urbanization. And although all the water supplies of Goiânia come from this basin (52% from the Joa˜o Leite River and 48% from the Meia Ponte River), this municipality is its largest polluter (Silva et al., 2010).

Among the microorganisms that contaminate the aquatic environment, special attention should be given to opportunistic protozoa, such as Coccidea (*Cryptosporidium parvum, Isospora belli, Sarcocystis* sp., and *Cyclospora* sp.) and Microsporidia that infect the

gastrointestinal tract, are considered emergents (Gomes et al., 2002), and also *Giardia* sp., which causes diarrhea episodes (States et al., 1997), can be spread through water.
The magnitude of enteric protozoan to public health should be emphasized because of their high prevalence, cosmopolitan distribution, and deleterious effects on the individuals' nutritional status and immune system. Although children are the most susceptible individuals to these pathogens, they also affect people from other age groups (Geldreich, 1996), mainly in subtropical and tropical areas.
According to Fayer et al. (2000) the *Cryptosporidium* is a protozoan parasite of vertebrates that causes diarrhea in humans in Different Geographical Regions of the world. Through molecular techniques, it is accepted that the *C. parvum* comprises at least two genotypes: 1 or H - only infectious for humans (anthroponotic), 2 or C - infecting cattle, men and various animals, confirming the zoonotic potential initially attributed to protozoa (Kosek et al. 2001).
Among the various water-borne pathogens (viruses, bacteria, fungi and parasites) are noted protozoa *Giardia duodenalis* (synonym *Giardia lamblia* and *Giardia intestinalis*) Thompson (2000) and *Cryptosporidium* sp., which cause gastroenteritis in humans and animals. These infectious agents are derived mainly from infected people and other warm-blooded animals, which undoubtedly pollute water (Gomes et al., 2002), highlighting some that are considered emerging, such as coccidia, *Cryptosporidium parvum, Isospora belli, Sarcocystis* sp., *Cyclospora* sp. and *Microsporidia* sp. (Garcia-Zapata et al., 2003).
For many years, *C. parvum* was considered the only emerging agent of opportunistic human infection. Recently, using molecular techniques was possible to prove that other animals and other genotypes also affect humans, such as *C. felis* (Caccio et al., 2002), *C. Muris* (Katsumata et al., 2001) or *C. meleagridis* (Pedraza-dias et al., 2000), thus showing that other species may also have an impact on public health, especially for people with immune system changes, such as patients infected with the AIDS (Acquired Immunodeficiency Syndrome), transplant recipients or patients undergoing chemotherapy, diabetics, elderly and very young children (Fayer et al., 2000). In developing countries, over one billion people do not have access to clean, properly treated water and approximately three billion people do not have access to adequate sanitary facilities (Kraszewski, 2001). This scenery is probably a consequence of the increased environmental degradation, depletion of water resources, and constant contamination of bodies of water with wastewater and industrial effluents (Pedro & Germano, 2001), causing microorganisms from soil, faeces, decomposing organic matter, and other pollutant sources to spread into water.
The magnitude of enteric protozoan to public health should be emphasized because of their high prevalence, cosmopolitan distribution, and deleterious effects on the individuals' nutritional status and immune system. Although children are the most susceptible individuals to these pathogens, they also affect people from other age groups (Geldreich, 1996), mainly in subtropical and tropical areas.
Criptosporidiosis is an important parasitic disease that can become a public health problem (Cimerman et al., 2000). The main modes of *Cryptosporidium* sp. transmission are frequently associated to contaminated water, which could be either treated or non-treated superficial water, treated water contaminated along the distribution systems, or inappropriate treated water, usually using only a simple chlorination method (Solo-gabriele & Neumeister, 1996).

Human health is likely to be affected either directly by drinking water contaminated with biological agents such as bacteria, viruses, and parasites, indirectly by consuming food or drinks prepared with contaminated water, or accidentally during recreational or professional activities.

A massive waterborne outbreak of cryptosporidiosis occurred in 1993, in Milwaukee, Wisconsin, in the United States. Approximately 403.000 people experienced illness, 4.400 of them were hospitalized, and 100 deaths were registered (Corso et al., 2003). In 1996, the United States American Environmental Protection Agency (U.S. EPA) started a program to identify, standardize, and validate new methods for the detection of *Giardia* sp. cysts and *Cryptosporidium* sp. oocysts in water environments.

From 1984 to 2000, 76 outbreaks of waterborne *Cryptosporidium* sp. have been associated with in countries like USA, England, Northern Ireland, Canada, Japan, Italy, New Zealand and Australia, affecting about 481.026 people, of these 59.2% were related to drinking water and 40.7% to the recreational use of water (Fayer et al., 2000; Fricker et al. 1998; Glaberman et al., 20; Howe et al., 2002). The most frequent causes of contamination are due to operational failures of treatment systems and water contact with sewage or faecal accident in the case of recreational waters In the U.S., factors such as deterioration in raw water quality and decrease the effectiveness of the process of coagulation and filtration of one of the local water supply companies showed an increase in turbidity of treated water and inadequate removal of *Cryptosporidium* sp. (Kramer et al., 1996).

Programs to monitor these pathogens in water have been spontaneously carried out in some countries such as the United States and the United Kingdom (Clancy et al., 1999). Since this, methods 1622 and 1623 (USEPA, 1999) have been used as reference procedures in the United States (Clancy et al., 2003; Franco, 2004).

In Brazil, the concern about water quality prompted the Health Ministry to issue one Decree - Ordinance 518 (Brasil, 2004) - establishing procedures and responsibilities regarding the control and surveillance of water quality for human consumption and pattern of potability, and other measures. Nowadays, in Brazil, routine monitoring of protozoa is not performed in bodies of water used for the abstraction of water intended for human consumption. Nonetheless, the Brazilian Health Ministry recommends the inclusion of methods for the detection of *Giardia* sp. cysts and *Cryptosporidium* sp. oocysts aiming to reach a standard in which the water supplied to the population must be free of these pathogens.

It should be emphasized that the detection of cysts and oocysts in superficial water is a crucial component to control these pathogens. However, the current methods present high variability of recovery efficiency of *Cryptosporidium* sp. oocysts and *Giardia* sp. cysts (Hsu et al., 2001), leading to the need of aggregating other types of methodology to guarantee that water potability achieves a higher degree of reliability. Due to lack of specific techniques for detection of Microsporidia and Coccidea in water and food, the analysis has been carried out by adaptations of methods used for clinical testing (Thurston-enriquez et al., 2002).

The goal of this study was to optimize and use parasitological and molecular techniques in the analysis and seasonal monitoring of opportunistic protozoa in water from fluvial systems for human usage in the municipality of Goiânia, the capital of the state of Goiás, in the Midwestern Region of Brazil, focusing on *Cryptosporidium* sp., *Cyclospora cayetanensis, Isopora belli* and Microsporidia.

2. Materials and methods

This is a descriptive observational study approved by the Human and Animal Research Ethics Committee at Hospital das Clínicas of Universidade Federal de Goiás.

2.1 Spatial and temporal sample delimitation

A total of 72 samples were collected on a monthly basis for one year (February 2006 to January 2007), from one point in the center of each of the following bodies of water: Meia Ponte river, João Leite river, Vaca Brava Park lake, Bosque dos Buritis lake.

Meia Ponte river

In this river two sites were selected for sampling: the first, 1 km after the emission of wastewater treated by the municipal wastewater treatment plant of Goiânia, located at 16°37'40.94"S latitude and 49°16'13.41"W longitude (MP1), and the second, located at 16°38'22.39"S latitude and 49°15'50.68"W longitude (MP2) (Figure 1).

Fig. 1. Photograph of Meia Ponte river at the time of sampling during the rainy season, showing the high volume of water and its coloring (Santos et al., 2008).

João Leite river

In this river two sites were selected for sampling: one located at 16°37'40.18"S latitude and 49°14'26.08"W longitude (JL1) (Figure 2), when this body of water reaches Goiânia, and the other located at 16°19'37.52"S latitude and 49°13'24.53"W longitude (JL2), before Goiânia. Figure 3 shows hydrographic map with the four sampling points in the rivers under study: João Leite (JL1 and JL2) and Meia Ponte (MP1 and MP2).

Fig. 2. João Leite river upstream of Goiania, after interbreeding Jurubatuba stream with the Posse stream, municipality of Goianapolis (Santos et al., 2008).

Vaca Brava Park lake

This park encompasses an area of approximately 72.7 thousand m², distributed among green areas, walking and jogging tracks, sports courts, playground, and exercise facilities. The site selected for sampling is located at 16°42'31.18"S latitude and 49°16'15.67"W longitude (VB) (Figure 4).

Bosque dos Buritis lake

Bosque dos Buritis is an urban park encompassing an area of approximately 125 m² with three artificial lakes supplied by Buriti stream. The site selected for sampling is located at 16°40'58.51"S latitude and 49°15'38.35"W longitude (BB) (Figure 5)

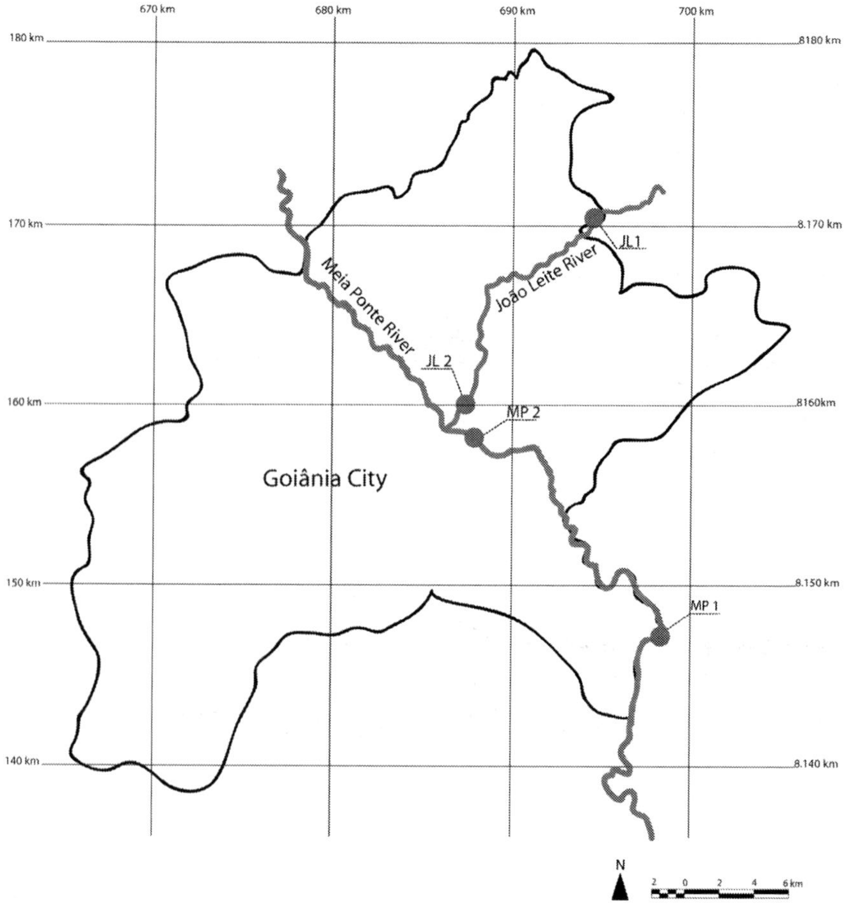

Fig. 3. Hydrographic map showing the four sampling points in the rivers under study: João Leite (JL1 and JL2) and Meia Ponte (MP1 and MP2).

Fig. 4. Photography of Vaca Brava lake, demonstrating the puopulsion system of water (Santos et al., 2008).

Fig. 5. Bosque dos Buritis lake, where we observe the dark Green water (an indicator of eutrophication) (Santos et al., 2008).

2.2 Sample concentration

Each sample was taken in a clean 10-L polyethylene container from one point in the center of the bodies of water approximately 20 cm under the surface and sent within 2 h to the Laboratório de Genética Molecular e Citogenética (Genetics and Molecular Diagnostic Laboratory) of the Universidade Federal de Goiás, and concentrated according to Silva et al. (2010).

Briefly, water samples were pre-filtered in a vacuum filter with qualitative paper filter, a process also called clarification, aiming to remove excessive amounts of organic matter, such as algae, plants, and other organisms, and immediately submitted to microfiltration using a positively nylon membrane with 0.45µm porosity with 47 mm of diameter (Hybond TM-N+, Amersham Pharmacia). The material adsorbed to the membrane was eluted by 5 ml of TE buffer (10 mM Tris-HCl, pH 8.0; 1mM EDTA) and 0.02% Tween-20, aliquoted and stored at -20°C.

2.3 Parasitological analysis

Aliquots of 10 µL of concentrated material were employed to prepare smears in two series of two slides each using the modified Ziehl-Neelsen-stain technique and the Kinyoun hot staining method, fixed in alcohol 70%, and processed for specific detection of Coccidea (*Cryptosporidium* sp., *Isospora belli,* and *Cyclospora caytanensis*).In order to detect enteral Microsporidia, the modified hot-chromotrope technique was used (Kokoskin et al., 1994). All the slides were analyzed in duplicate using a common optical microscope with a 100x oil immersion objective.

2.4 DNA extraction and amplification

The modified method of Boom et al., (1990) was used to extract the genetic material, based on cationic exchange resin processes, simultaneously with the phenol/chloroform method of Sambrook & Russel (2001).

The detection of DNA was performed using Nested-PCR, a variation of the polymerase chain reaction (PCR). The literature was searched to find primers flanking site-specific regions of these opportunistic protozoan genomes (Table 1). The Nested-PCR method was applied only to the positive and/or doubtful samples detected by parasitological methods.

Three primer pairs were used: XIAF/XIAR (*Cryptosporidium* sp. and *C. parvum*), flanking a region of approximately 1325 bp; AWA995f/AWA1206R *(Cryptosporidium* sp.*),* amplifying a region of approximately 211 bp; LAX469F/LAX869R (*C. parvum*), amplifying a chromosomal region of approximately 451 pb.

A conventional PCR was carried out using primers XIAF/XIAR and two aliquots were taken from the resulting product, one for detection of protozoan genera via Nested-PCR, using primers AWA995f/AWA1206R, (Awad-el-Kariem, 1994) and the other for the detection of *C. parvum/C. hominis* using primers LAX469F/LAX869R.

PCR using primers XIAF/XIAR and 28 µL extracted DNA was performed in a final volume of 50 µL with the following reagents: 5.0 µL buffer 10X, 2.0 mM Mg, 200 µM dNTP (dATP, dCTP, dTTP, and dGTP), 0.5 µM of each primer, and 1.25 U Taq DNA polymerase. The reaction conditions were an initial denaturation step for 4 min followed by another denaturation step of 35 cycles of 94°C for 1 min, annealing at 55°C for 45 s, extension at 72°C for 1 min, and final extension at 72°C for 7 min (Xiao, et al., 1999).

Microorganism	Primer	Sequence
Cryptosporidium sp. and	XIAF	5′-TTCTAGAGCTAATACATCCG-3′
C. parvum	XIAR	5′-CCCATTTCCTTGAA ACAGGA-3′
Cryptosporidium sp.	AWA995F	5′-TAGAGATTGGAGGTTGTTCCT-3′
	AWA1206R	5′-CTCCACCACTA AGAACGGCC-3′
C. parvum	LAX469F	5′-CCGAGTTTGATCCAAAAAGTTACGA-3′
C. hominis	LAX869R	5′-TAGCTCCTCATATGCCTTATTGAGTA-3′

Table 1. Primers selected to be used in confirmation/specification of protozoa detected by parasitological methods

PCR using primers AWA995f/AWA1206R and 14 µL DNA amplified by primers XIAF/XIAR was performed in a final volume of 25 µL with the following reagents: 2.5 µL buffer 10X, 1.5 mM Mg, 200 µM dNTP (dATP, dCTP, dTTP, and dGTP), 0.5 µM of each primer, and 1.25 U Taq DNA polymerase. The reaction conditions were an initial denaturation step for 7 min followed by another denaturation step of 40 cycles of 94°C for 1 min, annealing at 54°C for 1 min, extension at 72°C for 3 min, and final extension at 72°C for 7 min.

PCR using primers LAX469F/LAX869R Laxer, (1991) and 14 µL DNA amplified by primers XIAF/XIAR was performed in a final volume of 25 µL with the following reagents: 2.5 µL buffer 10X, 2.0 mM Mg, 200 µM dNTP (dATP, dCTP, dTTP, and dGTP), 0.5 µM of each primer, and 1.25 U Taq DNA polymerase. The reaction conditions were an initial denaturation step for 7 min followed by another denaturation step of 40 cycles of 94°C for 1 min, annealing at 52°C for 1 min, extension at 72°C for 1 min, and final extension at 72°C for 7 min.

The PCR products were separated by electrophoresis on 8% acrylamide gels stained with silver nitrate and on 1.5% agarose gels stained with ethidium bromide. Samples presenting 211-bp and 451-bp bands were considered positive.

2.5 Direct immunofluorescence assay kit

One aliquot of each sample concentrate was tested employing the MERIFLUOR® direct immunofluorescence assay kit using homologous monoclonal antibodies for the detection of *Cryptosporidium* sp. and *Giardia* sp. Each sample was analyzed in duplicate; however, due to a shortage of reagents, this technique was applied to 50% (36/72) of the samples taken at random and the positive samples detected by parasitological methods.

2.6 Statistical analyses

The results obtained in this study were digitalized in spreadsheets using the software Microsoft Office Excel 2007. Statistical analyses were performed using the chi-squared test and the logistic regression analysis. Statistical significance level was set at $p \leq 0.05$ using the Statistical Package for the Social Sciences (SPSS) version 10.0.

3. Results

Among the 72 samples processed, 8.33% (6/72) were positive for the protozoa researched. Using the MERIFLUOR® direct immunofluorescence assay kit, we found six positive

samples: two at JL2 in September and November, one at JL1 in August, two at MP1 in July, and one at VB in September.
Using the modified Ziehl-Neelsen-stain technique, 2.7% (2/72) samples were positive for Coccidea, and the presence of *Cryptosporidium* sp. was detected in two samples and confirmed by the MERIFLUOR® direct immunofluorescence assay kit Figure 6 shows a *Cryptosporidium* sp. oocyst and Figure 7 displays a *Cryptosporidium parvum* oocyst, which is approximately 5 μm in diameter, whereas *Cryptosporidium hominis* oocyst is approximately 4 μm in diameter.

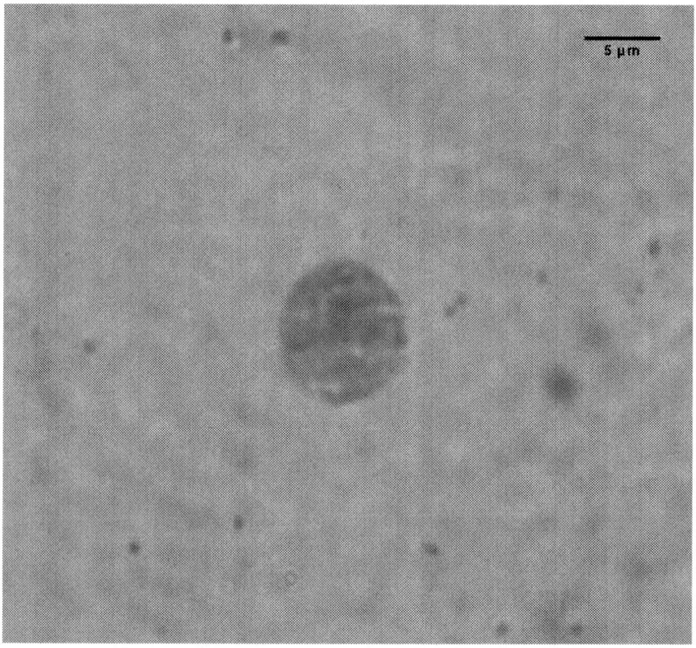

Fig. 6. *Cryptosporidium* sp. oocyst stained by the modified Ziehl-Neelsen (magnitude 100x)technique and confirmed by the MERIFLUOR® direct immunofluorescence assay kit and PCR (Santos et al., 2010).

Using primers AWA995f/AWA1206R we demonstrated that the samples belonged to the genus *Cryptosporidium* sp., and using primers LAX469F/LAX869R, we showed that just the sample collected in July was identified as *Cryptosporidium parvum*. As we detected only two positive samples for *Cryptosporidium* sp., the molecular detection was processed exclusively for them.
Using the Kinyoun hot staining method and the hot-chromotrope method for the detection of protozoa, no samples were found to be positive. Table 2 shows the results of each test carried out for the six sampling sites. Table 3 presents the frequency of protozoa detected in each sampling site.

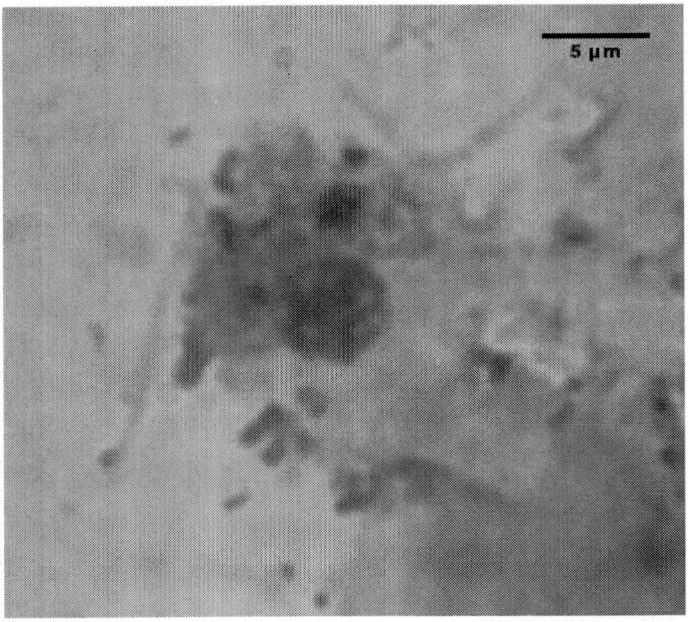

Fig. 7. *Cryptosporidium parvum* oocyst stained by the modified Ziehl-Neelsen technique (magnitude 100x) and confirmed by the MERIFLUOR® direct immunofluorescence assay kit and PCR (Santos et al., 2010).

Sampling site	Method			
	Ziehl-Neelsen	Kinyoun	Hot-chromotrope	MERIFLUOR®
MP1	*C. parvum**	Negative	Negative	*Giardia* sp.
MP2	Negative	Negative	Negative	Negative
JL1	Negative	Negative	Negative	*Giardia* sp.
JL2	Negative	Negative	Negative	*Giardia* sp.**
VB	*Cryptosporidium* sp.*	Negative	Negative	Negative
BB	Negative	Negative	Negative	Negative

MP1: Meia Ponte river, at 16°37'40.94"S latitude and 49°16'13.41"W longitude; MP2: Meia Ponte river at 16°38'22.39"S latitude and 49°15'50.68"W longitude; JL1: João Leite river, at 16°37'40.18"S latitude and 49°14'26.08"W longitude; JL2: João Leite river, at 16°19'37.52"S latitude and 49°13'24.53"W longitude; VB: Vaca Brava Park lake, at 16°42'31.18"S latitude and 49°16'15.67"W longitude; BB: Bosque dos Buritis lake, at 16°40'58.51"S latitude and 49°15'38.35"W longitude. *Confirmation by PCR; ** Two positive samples.

Table 2. Results according to the six sampling sites and the methods used to analyze the 12 samples in each site monitored, in a total of 72 samples (2006/2007)

Protozoa	Sampling site											
	MP1		MP2		JL1		JL2		VB		BB	
	n	%	n	%	n	%	n	%	n	%	n	%
Negative	12	100.0	10	83.4	11	91,7	10	83,3	11	91.7	12	100.0
Cryptosporidium sp.	0	0.0	0	83.4	0	0,0	0	0,0	1	8.3	0	0.0
C. parvum	0	0.0	1	8.3	0	0,0	0	0,0	0	0.0	0	0.0
Giardia lamblia	0	0.0	1	8.3	1	8,3	2	16,7	0	0.0	0	0.0
Total	12	100.0	12	100.0	12	100,0	12	100,0	12	100.0	12	100.0

MP1: Meia Ponte river, at 16°37'40.94"S latitude and 49°16'13.41"W longitude; MP2: Meia Ponte river at 16°38'22.39"S latitude and 49°15'50.68"W longitude; JL1: João Leite river, at 16°37'40.18"S latitude and 49°14'26.08"W longitude; JL2: João Leite river, at 16°19'37.52"S latitude and 49°13'24.53"W longitude; VB: Vaca Brava Park lake, at 16°42'31.18"S latitude and 49°16'15.67"W longitude; BB: Bosque dos Buritis lake, at 16°40'58.51"S latitude and 49°15'38.35"W longitude.

Table 3. General distribution of samples in the six sites according to the presence of protozoa, from February 2006 to January 2007

Average temperature in the period of protozoa occurrence was 26.8°C, while in the period showing no register of this pathogen, it was 25.6°C. The logistic regression analysis for temperature revealed p = 0.262 and OR = 1.227 (Table 4).
Average relative humidity in the period of protozoa occurrence was 42.3%, whereas in the period showing no register of this pathogen, it was 56.3%, a not significant value since the logistic regression analysis for relative humidity revealed p = 0.060 and OR = 0.944 (Table 4).

Protozoa	n	Mean	Standard deviation	p	OR
			Temperature		
Negative	66	25.6	2.5		
Positive	6	26.8	1.5	0.262	1.227
			Relative humidity		
Negative	66	56.3	16.0		
Positive	6	42.3	14.6	0.060	0.944

Table 4. Mean and standard deviation of temperature and relative humidity according to the presence of protozoa in the bodies of water sampled in Goiania during February 2006 to January 2007
(logistic regression analysis)

4. Discussion

This study revealed that the water in all sampling sites monitored during the research is not suitable for human consumption. Despite this evidence, we could observe the presence of people collecting water for human consumption, bathing, washing clothes, and even fishing. This fact is highly worrying because various waterborne diseases, not only related to opportunistic protozoa, but also to several other biological agents, can be transmitted through these contaminated bodies of water. Some sources of pollution observed in the sampling sites were: clandestine sewage discharges, livestock and poultry farms, slaughterhouses, meat processing plants, landfills, among others.

Nonetheless, we detected low recovery efficiency of opportunistic protozoa cysts and/or oocysts, which might be related to environmental influence and physical-chemical factors, such as water pH and turbidity, among others, since the influence of physical-chemical factors on sampling was reported by other researchers (Fricker & Crabb, 1998, McCuin & Clancy, 2003). The influence of physical-chemical factors on sampling was reported by other researchers (Fricker et al., 1998; Clency et al., 2003). Adverse environmental factors have been proven to alter the morphology of cysts and oocysts (Orgerth & Stibbs, 1987) , ; thus justifying the low positivity found in the present study using parasitological methods. Other factors might have had influence as well, such as the concentration of *Cryptosporidium* sp. oocysts, based almost exclusively on particle size (Fricker, 1998). The parasitological techniques employed in our study are not specific and, consequently, concentrate a large amount of several materials that may be present in the water, such as organic and inorganic particles, bacteria, yeast, and algae, which interfere in the detection of the parasites.

However, the methods used in the present study are in accordance with those recommended for concentration and detection of microorganisms by the Standard Methods for the Examination of Water and Wastewater (Clesceri et al., 1998). They are easily applied, do not pose a great risk to the technician, and are low cost techniques, which can be employed by technicians trained to monitor water for human consumption.

Hall and Croll (1997) evaluated the performance of some rapid gravity filters in England using turbidity measurement and particle counts in filtered water as parameters for monitoring and controlling *Cryptosporidium* sp. oocysts as an indicator microorganism, a method similar to the one used in this study.

Some studies have demonstrated that *Cryptosporidium* sp. prevalence is approximately 6% in developed countries (6), around 2-6% in immunodepressed adults (Goldman & Ausiello 2004), and shows a great variation in underdeveloped countries (Casemore, 1990). In industrialized countries, the seroprevalence of oocyst antigens is between 17% and 32% (Goldman & Ausiello 2004). In Canada, a study showed that 21% of the water samples collected were contaminated with *Giardia* sp. cysts and 4.5% with *Cryptosporidium* sp. oocysts (Wallis, 1996). However, in the United States, the contamination of 65% to 97% of superficial water with *Cryptosporidium* sp. oocysts and *Giardia* sp. cysts was reported (Kirkpatrick & Green, 1985), and it was also estimated that 80% of superficial water and 26% of treated water contains oocysts, although their infectivity has not been investigated (Goldman & Ausiello 2004). Nevertheless, we found contamination of 8.33% (6/72) of the samples in the present study, much inferior to the American data, which might be explained by the method applied. Therefore, new methodologies should be tested in order to compare the results in terms of specificity and efficiency to be employed in environmental monitoring of protozoa of public health interest.

Since our sampling points are located before the municipal wastewater treatment plant of Goiânia, the results of this study were considered within the tolerable levels, due to the low protozoan positivity according to the method used, in spite of the clandestine sewage discharges. It is worth mentioning that the water from all sources analyzed in this research is improper for usage *in natura*, because it meets neither the Brazilian standard (Brasil, 2004), which establishes that water for human consumption ought to be free from *Giardia* sp. and *Cryptospridium* sp., nor the American one (McCuin & Clancy, 2003).

The parasitological techniques employed in our study are not specific and, consequently, concentrate a large amount of several materials that may be present in the water, such as organic and inorganic particles, bacteria, yeast, and algae, which interfere in the detection of the parasites. However, the methods used in this study are in accordance with those recommended for concentration and detection of microorganisms by the *Standard Methods for the Examination of Water and Wastewater* (Clesceri, 1998). They are easily applied, do not pose a great risk to the technician, and are low cost techniques, which can be employed by technicians trained to monitor water for human consumption.

The performance of some rapid gravity filters was evaluated in England, using turbidity measurement and particle counts in filtered water as parameters for monitoring and controlling *Cryptosporidium* sp. oocysts as an indicator microorganism (Geldreich, 1996), a method similar to the one used in our study.

The *in vitro* amplification of DNA fragments of *Cryptosporidium* sp. obtained sensibility and specificity. Nevertheless, the amplification was only possible using Nested-PCR primers (AWA995f/AWA1206R and LAX469F/LAX869R). The primer LAX469F/LAX869R amplifies the regions of *C. parvum*/*C. hominis*, but *C. parvum* diagnosis was confirmed by the difference in diameter, since its oocyst is approximately 5 µm in diameter, while *C. hominis* oocyst is approximately 4 µm in diameter.

Nested-PCR presents the advantage of concentrating a smaller quantity of PCR inhibitors (Kirkpatrick & Green, 1985). In environmental samples, there are several Taq DNA polymerase inhibitors, such as fecal hemoglobin and phenolic compounds, and it might have been the case of the samples processed in the present research.

It was possible to obtain satisfactory amplification with the two methods of DNA extraction applied. Furthermore, they are quick and low-cost, although close attention should be paid to the phenol/chloroform method since it is toxic and corrosive.

As adverse environmental factors have been proven to alter the morphology of cysts and oocysts (Hsu BM, 2001), making their detection more difficult, this may justify the low positivity found in the present study using parasitological methods. Other factors might have had influence as well, such as the concentration of *Cryptosporidium* sp. oocysts, based almost exclusively on particle size (Hsu, 2001). Also, the level of protozoa may vary according to the season, and an increase in their resistant forms in rainy periods, winter and beginning of spring has already been reported (Atherholt 1998, Ong et al. 2002)

Temperature has also been considered a factor that influences protozoa and autochthonous microorganism survival in rivers (Howe, 2002). In this study, we observed just small variations of water temperature in the rivers and lakes sampled during the period of study, although within the limits that allow the survival and viability of protozoa. Using univariate logistic regression ($p = 0.066$), we demonstrated that temperature was not a statistically significant variable, whereas humidity ($p = 0.958$) was. In the region of sample collection there are two well-defined seasons, the dry (from April to September) and the rainy (from October to March) seasons, the latter characterized by torrential rain and runoff, which certainly makes the detection of parasites more difficult.

Due to the low number of protozoa found in this work, i.e. two *Cryptosporidium* sp. and four *Giardia* sp., we could not infer if the protozoan levels vary by season, but only observe the qualitative inference of their presence in the bodies of water monitored.

5. Conclusion

- The rivers and lakes of Goiânia are contaminated with opportunistic protozoa;
- Standardization and application of parasitological and molecular techniques in the analysis and seasonal monitoring of opportunistic protozoa were successfully carried out for environmental samples;

- During seasonal monitoring of opportunistic protozoa, with emphasis on Coccidia *Cryptosporidium* sp., *Cyclospora cayetanensis, Isospora belli* and Microsporidia, it was possible to detectic *Cryptosporidium parvum* and *Cryptosporidium* sp. using PCR and Nested-PCR, respectively
- The parasitological and molecular techniques applied are quick, low-cost, and can be employed in laboratories that monitor the microbiological quality of water for human consumption. Considering that the microorganisms studied herein are opportunistic, their persistent contact with humans may generate new parasites able to breach the immune barrier of normal individuals and to produce more aggressive cycles. Our results point to the need for efficient programs to prevent, treat, and monitor the presence of these parasites in rivers and lakes used for abstraction of water intended for human consumption and/or for recreational purposes all over the world. Furthermore, more efficient parasitological techniques, such as PCR, should be adopted in routine analyses in the laboratories of environmental monitoring, water for human consumption should be purified with UV radiation, and the activated sludge generated by wastewater treatment plants and intended for use in agriculture should be monitored.

6. Concluding remarks

Cryptosporidium is considered a coccidia resistant (Carey et al. 2004), because oocysts have characteristics that favor its rapid spread in the environment, such as the ability to withstand the action of commonly used disinfectants (formaldehyde, phenol, ethanol, lysol), able to cross some water filtration systems due to its small size, the ability to float, remain in the environment by a few weeks or months and tolerance in certain temperatures and salinity (Fayer et al. 2004). Given the scope of the aquatic environment coupled with the wide distribution of different species in Brazilian waters, make the control measures of Cryptosporidium limited.

Therefore, to minimize the risks inherent in the spread of cryptosporidiosis in the populations of free-living mammals, it is of fundamental importance to environmental control, through the adoption of agricultural practices to prevent pollution of rivers by the faeces of animals (Graczyk et al. 2000), as well as encouraging the adequacy of sanitation facilities, protection of water sources, education and guidance on waste discharges from vessels during nautical activities. Regarding the control measures of captive aquatic mammals, so as to minimize or eliminate the risks inherent in the spread of coccidian, several studies should be adopted.

Finally, it must be remembered that currently monitoring systems treated water are based on the frequency of fecal coliforms and Escherichia coli as indicators of pollution, and that this methodology is insufficient to predict the presence of other pathogens such as parasites. Thus, it is imperative the use of alternative methods for the diagnosis, investigation and monitoring of large amounts of water of these pathogens. For in this way can be proposed reorganization measures that contribute to reducing the incidence of opportunistic diseases emerging in water of human use, especially for children, elderly, immunocompromised and immunosuppressed patients.

7. References

Alves, T. M., & Chaveiro, E.F. Metamorfose urbana: a conurbação Goiânia-Goianira e suas implicacões sócio-espaciais [Urban metamorphosis: the conurbation of Goiânia and

Goianira cities and its socio-spatial implications]. *Revista Geográfica Acadêmica,* 1 :95–107, 2007.

Atherholt, T.B.; LeChevallier M.W.; Norton W.D.; & Rosen J.S. Effect of rainfall on giardia and crypto. *J Am Water Works Assoc,* 90: 66–80, 1998.

Awad-el-Kariem, F.M.; Warhurst D.C.; & McDonald V. Detection and species identification of Cryptosporidium oocysts using a system based on PCR and endonuclease restriction. *Parasitology,* 109: 19–22, 1994.

Boom, R.; Sol C.J.A.; Salimans M.M.M.; Jansen C.L.; Wertheim-van-Dillen P.M.E.; & van der Noordaa J. Rapid and simple method for purification of nucleic acids. *J Clin Microbiol,* 28: 495–503, 1990.

Brasil. Ministério de Saúde. Portaria 518 de 25 de Março de 2004. Estabelece os procedimentos e responsabilidades relativos ao controle e vigilância da qualidade da água para consumo humano e seu padrão de potabilidade, e dá outras providências. *Diário Oficial da União,* Brasília-DF, 2004.

Carey, C. M.; Lee, H.; & Trevors, J.T. Biology, persistence and detection of *Cryptosporidium parvum* and *Cryptosporidium hominis* oocyst. Water research, 38 : 818-862, 2004.

Caccio, S.; Pinter, E.; Fantini, R.; Mezzarona, I.; & Pozio, E. Human infection with Cryptosporidium felis: case report and literature review. *Emerging Infectious Diseases,* 8: 263-268, 2002

Casemore, D.P. Epidemiological aspects of human cryptosporidiosis. *Epidemiol Infect,* 104: 1–28, 1990.

Cimerman, S.; Castañeda C.G.; Iuliano W.A; & Palacios R. Perfil das enteroparasitoses diagnosticadas em pacientes com infecção pelo vírus HIV na era da terapia antiretroviral potente em um centro de referência em São Paulo, Brasil. *Parasitol Latinoam,* 57: 111–118, 2002.

Clancy, J.L.; Bukhari, Z.; McCuin, R.M.; Matheson, Z.; & Fricker, C.R. USEPA Method 1622. *J Am Water Works Assoc,* 91: 60–68, 1999.

Clancy, J.L.; Connel, K.; McCuin, R.M. Implementing PBMS improvements to USEPA`S Cryptosporidium and Giardia methods. *J Am Water Works Assoc,* 95: 80–93, 2003.

Clesceri, L.S.; Greenberg, A.E.; & Eaton, A.D. Standard methods for the examination of water and wastewater. 20th ed. Washington, D.C., *American Public Health Association,* 1998.

Corso, P.S.; Kramer, M.H.; & Blair, K.A. Addiss DG, Davis JP, Haddix AC. Cost of illness in the 1993 waterborne Cryptosporidium outbreak, Milwaukee, Wisconsin. *Emerg Infect Dis,* 9: 426–431, 2003.

Fayer, R.; Morgan, U.; & Upton, S.J. Epidemiology of Cryptosporidium: transmission, detection, and identification. *International Journal for Parasitology,* 30:1305-1322, 2000.

Franco, R. M.B. (Docente): *Método 1623*: Evolução e Análise Crítica; 2004; Palestra; ; Unicamp/Comitê de Bacias Hidrográficas e Sociedade Paulista de Parasitologia; Hotel Premium Norte; Campinas; BR; Meio digital; www.ib.unicamp.br/parasito/seminario2004.

Fricker, C.R.; & Crabb, J.H. Water-borne cryptosporidiosis: detection methods and treatment options. Adv. *Parasitol,* 40: 242-278, 1998.

Garcia-Zapata, M.; T.A.; Passo, A.; Ruano, A.L.; Souza júnior, E.S.; Cechetto, F. H.; & Manzi, R.S. Ciclosporíase intestinal: relato dos primeiros casos humanos no estado de Goiás, Goiânia, Brasil. *Revista de Patologia Tropical,* 32 : 121-130, 2003.

Geldreich, E.E. Amenaza mundial de los agentes patógenos transmitidos por el agua. In: Craun GF, Castro R. (Ed.) Calidad del agua potable en América Latina: ponderación de los riesgos microbiológicos contra los riesgos de los subproductos de la desinfección química. Washington, DC, *ILSI Press*, p. 21–49, 1996.

Glaberman, S.; Moore, J.E.; Lowery, C.J.; Chalmers, R.M.; Sulaiman, I.; Eiwin, K.; Rooney, P.J.; Millar, BC.; Dooley, J.S.G., Lal, A.A. & Xiao, L. Three drinking water-associated cryptosporidiosis outbreaks, Northern Ireland. *Emerg. Inf. Dis*, 8 (6): 631-633, 2002.

Goldman L, Ausiello D. Cryptosporidiosis. In: Goldman L, Bennett JC (Ed.) *Cecil textbook of medicine*. 22nd ed. Philadelphia, WB Saunders, p. 2092-2095, 2004.

Gomes, A.H.S.; Pacheco, M.A.S.R.; Fonseca, Y.S.K.; Cesar, N.P.A.; Dias, H.G.G.; Silva, R.P. Pesquisa de Cryptosporidium sp em águas de fontes naturais e comparação com análises bacteriológicas. *Rev Inst Adolfo Lutz,* 61: 59–63, 2002.

Graczyk, T.K., Evans, B.M., Zif, C.J., Karreman, H.J. & Patz, J.A. Environmental and geographical factores contributing to watershed contamination with Cryptosporidium parvum oocysts. *Environ. Res,* 82(3):263-271, 2000.

Hall, T. ;Croll, B.Particle counters as tools for managing Cryptosporidium risk in water treatment. *Water Scien Technol*, 36, 143–149, 1997.

Howe, A.D.; Forster, S.; Morton, S.; Marshall, R.; Osborn, K. S.; Wright, P & Hunter, P. R. Cryptosporidium oocysts in a water supply associated with a cryptosporidiosis outbreak. *Emerg. Inf. Dis*, 8 (6): 619-624, 2002.

Hsu, B.M.; Huang, C.; Lai, Y.C.; Tai, H.S.; & Chung, Y.C. Evaluation of immunomagnetic separation method for detection of Giardia for different reaction times and reaction volumes. *Parasitol Res,* 87: 472–474, 2001.

Katsumata, T.; Hosea, D.; Ranuh, I.G.; Uga, S.; Yanagi, & T.; Khono, S. Short report: possible Cryptosporidium muris infection in humans. *American Journal Tropical Medicine Hygiene*, 62: 70-72, 2001.

Kosek, M.; Alcantara, C.; Lima, A.A.M.; & Guerrant, R.L. Cryptosporidiosis: na update. *The Lancet Infectous Diseases,* 1:262-269, 2001.

Kirkpatrick, C.E. ; & Green, G.A. IV. Susceptibility of domestic cats to infections with Giardia lamblia cysts and trophozoites from human sources. *J Clin Microbiol,* 21: 678–680, 1985.

Kramer, ; M.H.; Herwaldt,; B.L.; Craun, G.F.; Calderon, R.L.; & Juranek, D.D. Surveillance for waterborne-disease outbreaks, United States, 1993-1994. MMWR 45(SS-1): 1-33, 1996.

Kokoskin, E.; Gyorkos, T.W.; Camus, A.; Cedilotte, L.; Purtill, T.; Ward, B. Modified technique for efficient detection of microsporidia. *J Clin Microbiol*, 32: 1074–1075, 1994.

Kraszewski J. Water for people supports small systems for impoverished people worldwide. *J Am Water Works Assoc*, 93: 36–37, 2001.

Laxer, M.A.; Timblin, B.K.; & Patel, R.J. DNA sequences for the specific detection of Cryptosporidium parvum by the Polymerase Chain Reaction. *Am J Trop Med Hyg*, 45: 688–694, 1991.

McCuin, R.M.; Clancy, J.L. Modifications to United States Environmental Protection Agency Methods 1622 and 1623 for detection of Cryptosporidium oocysts and Giardia cysts in water. *Appl Environ Microbiol,* 69: 267–274, 2003.

Ong, C.S.L.; Eisler, D.L.; Alikhani, A.; Fung, V.W.K.; Tomblin, J.; BowniE, W.R.; & Issac-Renton, J.L. Novel Cryptosporidium genotypes in sporadic cryptosporidioisis cases: first report of human infection with a corvine genotype. *Emerging Infectious Diseases*, 8:263-268, 2002.

Ongerth, J. E. ; Stibbs, H. H. Identification of Cryptosporidium oocysts in river water. *Appl Environ Microbiology*, 53 : 672–676, 1987.

Pedraza-Dias, S.; AmaR, C.; & Mclauchlin, J. The identification and characterization of an unusual genotype of Cryptosporidium from human faeces as Cryptosporidium meleagridis. *FEMS Microbiology Letters*, 189:189-194, 2000.

Pedro, M.L.G.; & Germano, M.I.S. A água: um problema de segurança nacional. *Rev Hig Alim*, 15: 15–18, 2001.

Sambrook, J.; & Russel, D. Molecular cloning: a laboratory manual. 3rd ed. v. 1, v. 2, v. 3. *New York, Cold Spring Harbor Laboratory Press Section*, 2001.

Santos, Sônia de Fátima Oliveira. Estudo dos parasitos oportunistas em águas fluviais de uso humano no município de Goiânia-Goiás, Brasil, 2006/2007. *Dissertation (Mestrado em Ciências da Saúde - Master in Health Sciences)* – Pós-graduação em Ciências da Saúde, Universidade Federal de Goiás, Goiânia, 2008.

Santos, S.F.O., Silva, H.D., Souza-Junior, E.S., Anunciação, C. E., Silveira-Lacerda, E. P., Vilanova-Costa, C.A.S.T., Garcia-Zapata, M.T.A. Environmental Monitoring of Opportunistic Protozoa in Rivers and Lakes in the Neotropics Based on Yearly Monitoring. *Water Quality, Exposure and Health*, v.2, p.1 - 8, 2010.

Silva, H.D., Wosnjuk, L.A.C., Santos, S.F.O., Vilanova-Costa, C.A.S.T., Pereira, F.C., Silveira-Lacerda, E.P., Garcia-Zapata, M.T.A., Anunciação, C.E. Molecular Detection of Adenoviruses in Lakes and Rivers of Goiânia, Goiás, Brazil. *Food and Environmental Virology*, v.2, p.35 - 40, 2010.

Solo-Gabriele, H.; & Neumeister, S. US outbreaks of cryptosporidiosis. *J Am Water Works Assoc*, 88: 76–86, 1996.

States, S.; Stadterman, K.; Ammon, L.; Vogel, P.; Baldizar, J.; Wright, D.; Conley, L.; & Sykora, J. Protozoa in river water: sources, occurrence, and treatment. *J Am Water Works Assoc*, 89: 74–83, 1997.

Thompson, R.C.A. Giardiasis as a re-emerging infectious disease and its zoonotic potential. *International Journal for Parasitology*, 30:1259-1267, 2000.

Thurston-enriquez, J.A.; Watt, P.; Dowd, S.E.; Enriquez, R.; Pepper, I. L.; Gerba, C.P. Detection of protozoan parasites and microsporidia in irrigation waters used for crop prodution. *J Food Prot.* US Department of Agriculture, Agricultural Research Service, University of Nebraska. United States, Feb. 2002.

Wallis, P.M.; Erlandsen, S.L.; Isaac-Renton, J.L.; Olson, M.E.; Robertson, W.J.; Van Keulen, H. Prevalence of Giardia cysts and Cryptosporidium oocysts and characterization of Giardia spp. isolated from drinking water in Canada. *Appl Environ Microbiol*, 62: 2789–2797, 1996.

Xiao, L.; Escalante, L.; Yang, C.; Sulaiman, I.; Escalante, A.A.; Montali, R.J.; Fayer, R.; & Lal, A.A. Phylogenetic analysis of Cryptosporidium parasites based on the small-subnunit rRNA gene locus. *Appl Environ Microbiol*, 65: 1578–1583, 1999.

Part 3

Environmental Monitoring with Wireless Sensor Network Technology

21

Biosensor Arrays for Environmental Monitoring

Wei Song[1], Si Wei[2], Hong-Xia Yu[2], Maika Vuki[3] and Danke Xu[1]

[1]*State Key Laboratory of Analytical Chemistry for Life Science, School of Chemistry and Chemical Engineering, Nanjing University,*

[2]*State Key Laboratory of Pollution Control and Resource Reuse, School of the Environment, Nanjing University,*

[3]*College of Natural and Applied Sciences, University of Guam, Mangilao, Guam,*

[1,2]*China*

[3]*USA*

1. Introduction

Environmental monitoring involves several steps such as sampling, sample handling and sample transportation to specialized laboratories, sample preparation and analysis. Traditional environmental monitoring approaches are based on discrete sampling methods followed by laboratory analysis. These approaches do not improve our understanding of the natural processes governing chemical species behavior, their transport and bioavailability, or the relationship between anthropogenic releases and their long-term impact on aquatic systems[1]. The challenge of environmental monitoring in situ requires new and improved analytical devices featuring precision, sensitivity, specificity, rapidity, and ease of operation to detect decreasing concentrations of an ever growing array of pollutants. Such devices must be comparable to or better than traditional analytical systems, and must be simple to handle, small, cheap, able to provide reliable information in real-time, and must be sensitive and selective for the analyte of interest, and suitable for in situ monitoring[2]. Biosensors not only fulfill all these requirements but also have applications in many areas such as clinical diagnostics, forensic chemistry, pharmaceutical studies, food quality control and environmental monitoring.

A biosensor is an analytical device for the detection of an analyte that combines a biological component with a physicochemical detector component. It consists of 3 parts: (1) the sensitive biological element (biological material such as tissue, microorganisms, organelles, cell receptors, enzymes, antibodies, nucleic acids, etc.), a biologically derived material or biomimic; (2) the transducer or the detector element (works in a physicochemical way; optical, piezoelectric, electrochemical, etc.) that transforms the signal resulting from the interaction of the analyte with the biological element into another signal that can be more easily measured and quantified; (3) associated electronics or signal processors that are primarily responsible for the display of the results in a user-friendly way[3]. Depending on the type of transduction mechanism applied and the bio-recognition element employed, the potential for these devices for detection can be enormous. The technological development and the success in single analyte detection propelled advances in the miniaturization of sensors along with multi-analyte detection with sensitivities ranging in the nano-mole to atto-mole range. With advances in techniques for biosensor construction, it has been possible to miniaturize the whole biosensor system on a chip to fabricate biosensor arrays. The Biosensor arrays developed at the Naval Research Laboratory (NRL) has successfully been used in the detection of a variety of protein toxins, organic molecules, physiological health markers, a virus and a number of bacteria, initially in buffer but increasingly in food, biological and environmental matrices [4]. These developed biosensors are rapid, simple to perform and require little-to-no sample pretreatment prior to analysis, even for more complex sample matrices. In addition, the two-dimensional nature of the slide sensing surface facilitates simultaneous analysis of multiple samples for multiple analytes. Research on biosensor arrays as multi-analyte bio-systems has generated increased interest in the last decade. The main feature of the micro-array technology is the ability to simultaneously detect multiple analytes in one sample by an affinity-binding event at a surface interface. Fifteen years ago, the gene expression analysis of *c*DNA on micro-arrays was one of the first applications that successfully detected thousands of labeled target DNA molecules in parallel. Also the first immuno-analytical biosensor array was described at the same time. In the meantime, a great variety of target analytes capable of interacting selectively with a bio-molecular receptor has been adapted to arrays[5]. The biosensor arrays have been envisioned as a tool for rapid, on-site screening of pollutants in whatever location they might be found. The goals of automation, weight reduction, minimal size, ease of use, and reliability have remained paramount as the system has been developed[6].

The challenge of continuous in situ monitoring of environmental pollution requires instruments that are robust and with sufficient sensitivity and long lifetime. Commonly used conventional methods are time-consuming, expensive, require skilled operators, and lack the required selectivity. Biosensor arrays have the advantage of being simple, uniform whole structures featuring direct transduction, high bio-selectivity, high sensitivity, miniaturization, electrical/optoelectronic readout, continuous monitoring, ease of use, and cost effectiveness. User advantages include low price, reliability, no sample preparation, disposability, and clean technology. Hence, biosensor arrays show the potential to complement both laboratory-based and field analytical methods for environmental monitoring. Biosensor arrays are based on one general principle—certain bio-molecular recognition elements are defined on a heterogeneous matrix. Each element is dedicated to an analyte and contains quantitative information. The matrix is a patterned surface where the recognition molecules are immobilized by micro-printing such as screen printed technique, micro fluidic or other micro-structuring processes[5]. The type of biosensor arrays involves DNA-based biosensor array, antibody-based biosensor array, aptamer-based biosensor array, enzyme-based biosensor array, and microorganism-based biosensor. Recent progress in the development of analytical detection methods for antibody arrays, enzyme arrays and aptamer arrays as well as microbial arrays are summarized in this review, and their applications in the environment monitoring are also discussed. Detection approach is focused on electrochemical and optical measurements including various electrochemical or florescent probes as well as label-free approach. The numerous fabrication methods of DNA capture probes, antibodies and aptamer for multiplexed biological targets are also discussed.

2. DNA-based biosensor arrays

Deoxyribonucleic acids (DNA) are arguably the most important of all bio-molecules. The unique complementary structure of DNA between the base pairs adenine/thymine and cytosine/guanine has been the basis for genetic analysis over the last few decades. The ability of a single stranded DNA (ssDNA) molecule to 'seek out', or hybridize to, its complementary strand in a sample is the foundation of DNA-based detection systems. There is a great potential market for simple, cheap, rapid, and quantitative detection of specific genes. Areas of application include clinical, veterinary, medico-legal, environmental, and the food industry[7]. Development of DNA biosensors and DNA biosensor arrays has increased tremendously over the past few years as demonstrated by the large number of scientific publications. Numerous DNA detection systems based on the hybridization between a DNA target and its complementary probe, which is present either in solution or on a solid support, have been described[8]. Homogeneous assays allowing the determination of DNA sequences have been developed. These systems can be based on optical[9] or electrochemical[10] detection. However, they do not allow easy continuous monitoring and miniaturization. Heterogeneous DNA biosensors and DNA biosensor arrays offer promising alternatives to these methods. They allow continuous, fast, sensitive, and selective detection of DNA hybridization, and they also can be reused. DNA biosensors arrays (commonly called gene chips, DNA chips, or biochips) exploit the preferential binding of complementary single-stranded nucleic acid sequences. This system usually relies on the immobilization of a single-stranded DNA (ssDNA) probe onto a surface to recognize its complementary DNA target sequence by hybridization. Transduction of hybridization of DNA can be measured optically, electrochemically, or using other devices. The detection process is schematized in Figure 1[8].

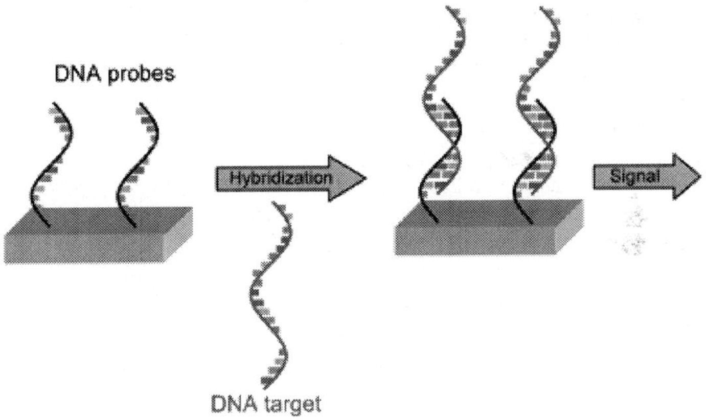

Fig. 1. Steps involved in the detection of a DNA sequence. Reprinted from ref. 8 with permission by the American Chemical Society.

In the case of DNA biosensors arrays, the immobilization of a DNA probe is achieved directly onto a transducer surface. DNA biosensor arrays are made from glass, plastic, or silicon supports and are constituted of tens to thousands of 10 - 100 μm reaction zones onto which individual oligonucleotide sequences have been immobilized. The exact number of DNA probes varies in accordance with the application. DNA biosensor arrays allow multiple parallel detection and analysis of the patterns of expression of thousands of genes in a single experiment.

We have presented an ultra-sensitive and direct electrochemical DNA biosensor array based on Ag aggregate tag and differential pulse voltammtery[11]. The scheme of detection is shown in Figure 2. The silver tags consist of Conjugate 1 (functionalized with capture probes and oligo A and Conjugate 2 (modified with oligo T). Hybridization between complementary oligo (d) A and oligo (d) T anchored on the silver nanoparticles produced aggregate tags. The hybridization-induced tags are successfully applied to bind with the DNA target via sandwich hybridization format and offer direct and amplified readout by differential pulse voltammetric method. We have found that the detection sensitivity by use of the aggregate tags can be improved by 3 orders of magnitude as compared to the single silver nanoparticle labels and a detection limit of 5 amol/L could be obtained.

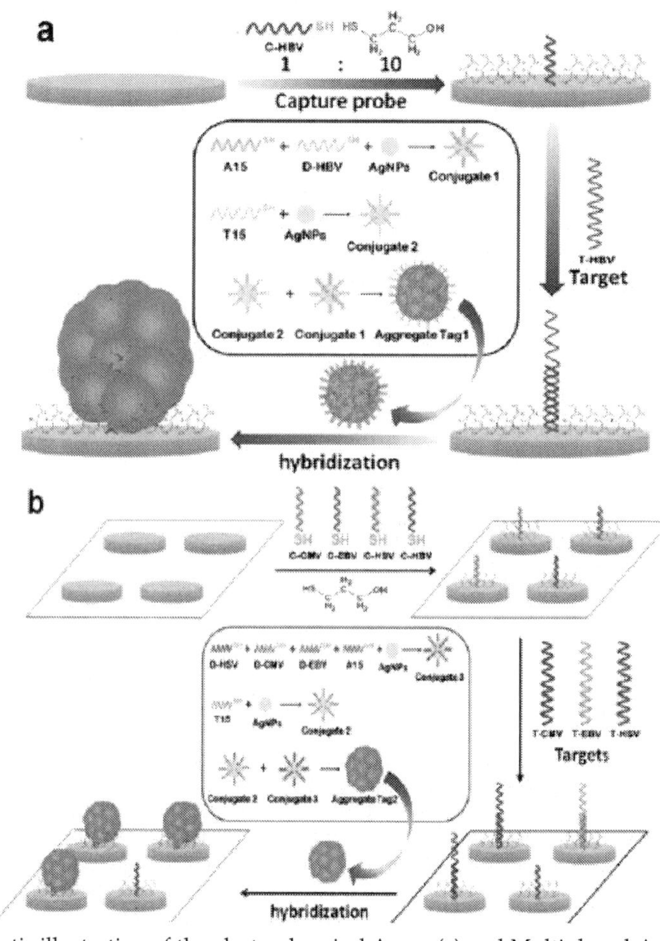

Fig. 2. Schematic illustration of the electrochemical Assay (a) and Multiplexed Assay (b) with silver nanoparticle Conjugates; Preparation of the aggregates is shown as well. Reprinted from ref. 11 with permission by the American Chemical Society.

Environmental applications of DNA biosensor arrays are in the field of species identification. For instance, DNA biosensor arrays are extensively exploited in the detection of pathogenic microorganisms relevant to food, bio-defense and environmental contamination applications. Mainly, DNA biosensor arrays have been coupled to PCR, as a specific detection method of the amplified base sequence. Zhang et al.[12] have developed a label-free electrochemical DNA biosensor array as a model system for simultaneous detection of multiplexed DNAs using micro-liters of sample. A novel multi-electrode array was comprised of six gold working electrodes and a gold auxiliary electrode, which were fabricated by gold sputtering technology, and a printed Ag/AgCl reference electrode was fabricated by screen-printing technology. The DNA biosensor array for simultaneous detection of the human immunodeficiency virus (HIV) oligonucleotide sequences, HIV-1 and HIV-2, was fabricated in sequence by self-assembling each of two kinds of thiolated hairpin-DNA probes onto the surfaces of the corresponding three working electrodes, respectively. The hybridization events were monitored by square wave voltammetry using methylene blue (MB) as a hybridization redox indicator. The oxidation currents of MB accumulated on the array decreased with increasing the concentration of HIVs due to higher affinity of MB for single strand rather than double strands of DNA. Under the optimized conditions, the peak currents were linear over ranges from 20 to 100 nmol/L for HIV-1 and HIV-2, with the same detection limits of 0.1 nmol/L (S/N= 3), respectively. The detection process is illustrated in Figure 3. The biosensor array showed a good specificity without the obvious cross-interference. Furthermore, single-base mutation oligonucleotides and random oligonucleotides can be easily discriminated from complementary target DNAs. Their work demonstrates that different hairpin-DNA probes can be used to design the label-free electrochemical biosensor array for simultaneous detection of multiplexed DNA sequences for various applications.

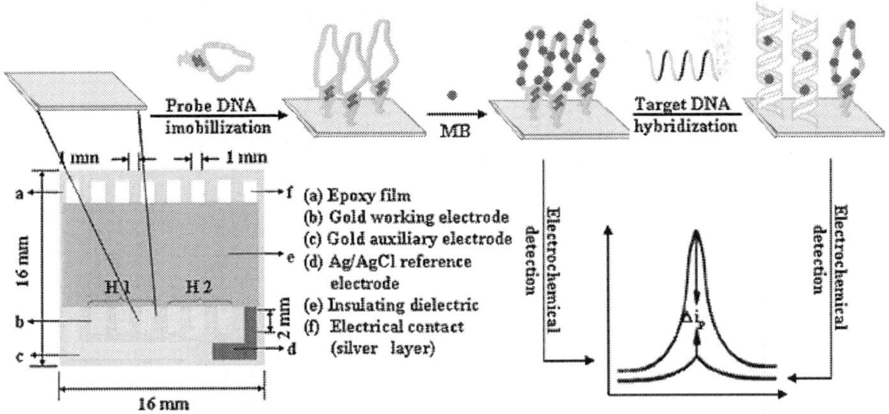

Fig. 3. Schematic diagrams of multi-electrode array and representation of biosensor array with fabrication steps and performance. Reprinted from ref. 12 with permission by the Elsevier.

Using electrochemical impedance spectroscopy (EIS) for biosensing applications typically requires repetitive experiments. To address this need, Bogomolova et al.[13] have designed a multi-specific electrochemical array with eight individually addressable 2 mm-diameter gold working electrodes for rapid biosensing data accumulation by EIS in the presence of redox agent. The array allows to incorporate multiple negative controls in the course of a single binding experiment, as well as to perform parallel identical experiments to improve reliability of detection. The array is fitted with attached electrochemical cell with Ag/AgCl mini reference electrode and can be used to process macro samples of 0.5-1 ml (Figure 4). The reported array is disposable, economical and is easy to use. Examples of array use for label-free genetic sensing of 2.7 kb-long target *Yersinia pestis* DNA and for protein sensing of Ricin Toxin Chain A (RTA) are presented. The authors suggest the reported array design as a tool for researchers in the area of EIS sensing.

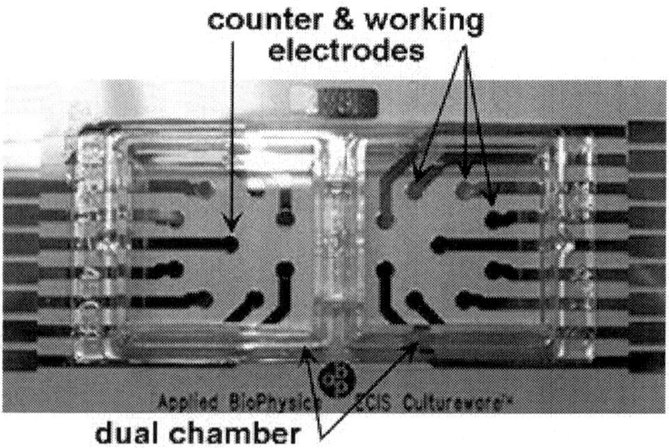

Fig. 4. Custom gold-sputtered dual array electrodes with attached chambers. Reprinted from ref. 13 with permission by the Elsevier.

Fu et al.[14] developed a piezoelectric quartz crystal microbalance (QCM) nucleic acid biosensor array using Au nanoparticle signal amplification to rapidly detect *S. epidermidis* in clinical samples. The synthesized thiolated probes specific targeting *S. epidermidis* 16S *r*RNA gene was immobilized on the surface of QCM nucleic acid biosensor arrays. Hybridization was induced by exposing the immobilized probes to the PCR amplified fragments of *S. epidermidis,* resulting in a mass change and a consequent frequency shift of the QCM biosensor. The results showed that the lowest detection limit of current QCM system was 1.3×10^3 CFU/mL. A linear correlation was found when the concentration of *S. epidermidis* varied from 1.3×10^3 to 1.3×10^7 CFU/mL. In addition, 55 clinical samples were detected with both current QCM biosensor system and conventional clinical microbiological method, and the sensitivity and specificity of current QCM biosensor system were 97.14% and 100%, respectively.

Doong et al.[15] fabricated a sol-gel-derived array DNA biosensors coupled with a fluorescence detection system and a robotic pin-printing platform to detect polycyclic aromatic hydrocarbons (PAHs) in water and serum samples (Figure 5). Parameters including sol percentage, doped-amount of glycerol, dye probes, the surface coating, and DNA concentration were optimized. In their work, two fluorescent dyes, fluorescein isothiocyanate (FITC) and ethidium bromide (EDB) were selected and compared. Results showed that EDB was more sensitive to compete intercalators (PAHs) than FITC, and was selected as fluorescent dye for array-based DNA biosensors. The optimized procedure with *ds*DNA concentration of 23.5 g/mL allowed the fabrication of the DNA biosensor up to 50 spots within 10 min via the developed pin-printing system. For PAH detection, the developed array DNA biosensor effectively detected naphthalene and phenanthrene in the concentration range of 0-10 mg/L in aqueous solution, but was not sensitive to fluoranthene and benzo[a]pyrene. In the serum samples, the apparent water solubility of high-molecular-weight PAHs was greatly enhanced by the dissolved organic compounds in serum, and an obvious DNA toxicity was exhibited in the presence of three-to-five-ring PAHs. Benzo[a]pyrene showed high toxic effect at low concentration in serum samples, clearly showing that the sol-gel-derived array DNA biosensor with EDB as sensing probe can effectively detect PAHs in water and biological samples.

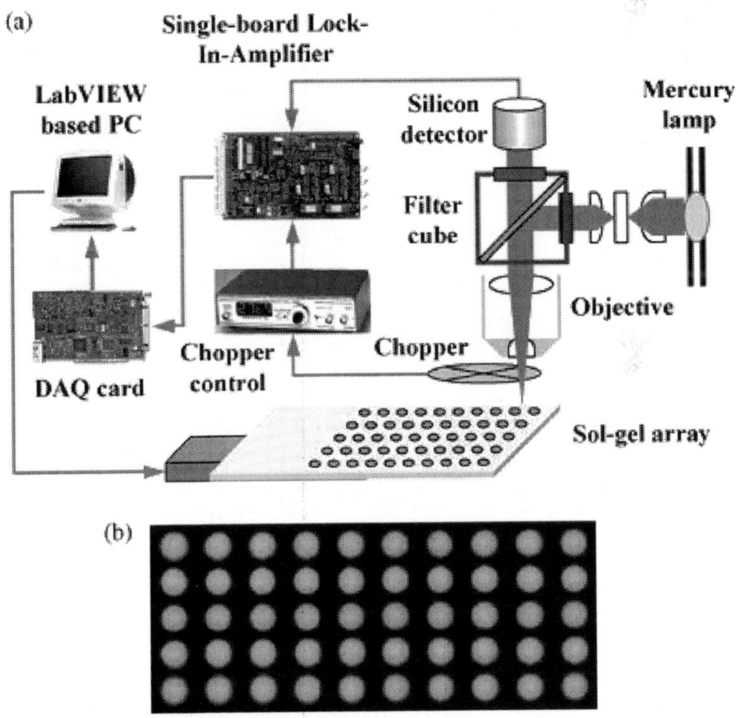

Fig. 5. (a) the schematic diagram of the developed fluorescence detection system and (b) images of the array DNA biosensor using the sol-gel processes. Reprinted from ref. 15 with permission by the Elsevier.

3. Antibody based biosensor array

Immunoassays gained popularity for biomedical applications in the 1970s because of the impressively low detection limits and high selectivity for analyzing complex samples that could be achieved with relatively simple procedures and instrumentation. The availability of highly selective antibodies for an increasingly wide variety of important analytes was also an important factor in the growth of the method over the following decades. The development of more sensitive labels and detection devices also improved the sensitivity of the assays even further. Once immunoassays became more common, the development of more convenient immuno-sensors that are easier and faster to use gained momentum[16].

Antibody-based biosensors are inherently more versatile than enzyme-based biosensors in that antibodies have been generated which specifically bind to individual compounds or groups of structurally related compounds with a wide range of affinities. There are, however, several limitations in the use of antibody-based biosensors for environmental monitoring applications. These limitations include the complexity of assay formats and the number of specialized reagents (e.g., antibodies, antigens, tracers, etc.) that must be developed and characterized for each compound and the limited number of compounds typically determined in an individual assay as compared to the multiple compounds that contaminate environmental samples[17].

Antibody-based biosensor arrays are a powerful tool for analytical purposes. Immuno-analytical micro-arrays are a quantitative analytical technique using antibodies as highly specific biological recognition elements[5]. They can be designed for a variety of analytical applications producing rapid results with low limits of detection (LOD). The detection antibodies are in direct contact with the sample, without prior sample cleanup. Immobilized haptens in combination with an indirect competitive immunoassay (IA) are the common format for the detection of multiple small molecules (e.g. pesticides, pharmaceuticals, small toxin targets). Haptens provoke an immune response if coupled to a protein by use of their functional groups. Hapten micro-arrays use analyte derivatives as immobilized recognition molecules. Antibody micro-arrays quantify proteins, bacteria, or viruses by using a sandwich immunoassay format. Recent advances reported for antibody-based biosensor arrays for environmental applications have primarily been focused toward these. For example, Jin et al.[18] have developed a fluorescent NP-mediated Ab micro-array system for the detection of bioterrorism agents exemplified by ricin, CT, and SEB toxins (results are displayed in Figure 6). High sensitivity, specificity, and reproducibility were achieved by using their antibody biosensor array. They found that substituting monoclonal antibodies (mAb) with highly purified polyclone antibodies (pAb), even though having similar titer by ELISA, could dramatically improve the micro-array performance. A likely explanation is that pAb enhances the Ag capture by binding multiple sites on the analyte. The micro-array format allows a multiplexed, high-throughput, and high-parallel assay. Furthermore, the miniature feature permits low consumption of both sample and reagents, reducing the amounts of biohazard waste as well as the costs to conduct the assay.

Seidel et al.[19] fabricated an automated chemiluminescence (CL) read-out system for analytical flow-through micro-arrays based on multiplexed immunoassays. The micro-array chip reader (MCR 3) is designed as a stand-alone platform, with the goal to quantify multiple analytes in complex matrices of food and liquid samples for field analysis or for routine analytical laboratories. The CL micro-array platform is a self-contained system for the fully automated multiplexed immuno-analysis comprising the micro-array chip, the fluidic system and the software module that enable automated calibration and determination of analyte concentrations during a whole working day. The detection of antibiotics in milk was demonstrated to validate this device. Therefore, an automated multi-analyte detection instrument is needed for the simultaneous and rapid quantification of antibiotics. Also regeneration is required to avoid replacing the assay surface. The European Union, for example, has defined maximum residue levels (MRLs) for a number of antibacterial compounds. However, despite the obvious demand for quantitative multi-residue detection methods that can be carried out on a routine basis, there is currently a lack in the development of such systems. In particular, an automated multi-analyte detection instrument is needed that is capable of quantifying several antibiotics simultaneously within minutes. Seidel's group[20] developed a new hapten based micro-arrays for the parallel analysis of 13 different antibiotics in milk within six minutes by applying an indirect competitive chemiluminescence micro-array immunoassay (CL-MIA). To allow multiple analyses, a regenerable micro-array chip was developed based on epoxy-activated PEG chip surfaces, onto which micro-spotted antibiotic derivatives like sulfonamides, b-lactams, aminoglycosides, fluorquinolones and polyketides are coupled directly without further use of linking agents. Using the chip reader platform MCR 3 (Figure 7), this antigen solid phase is stable for at least 50 consecutive analyses.

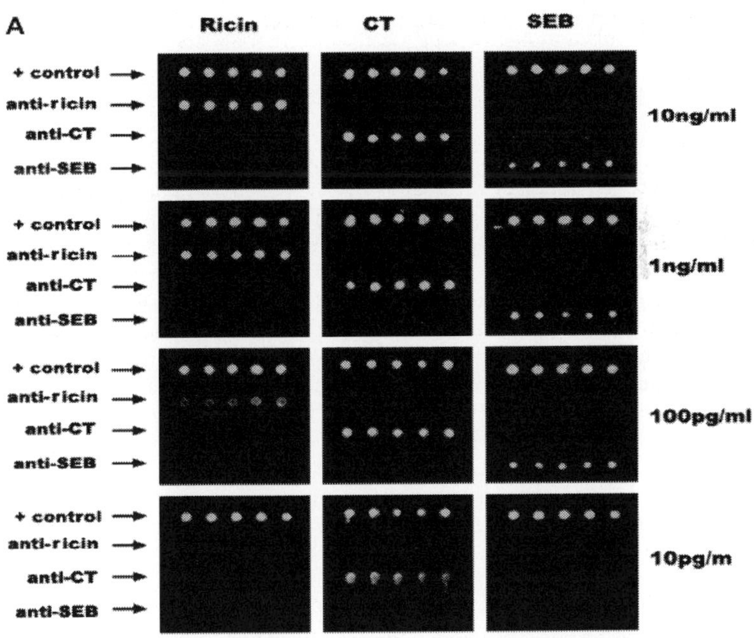

Fig. 6. The detection results of the different toxins by using antibody arrays. Reprinted from ref. 18 with permission by the Elsevier.

An impedance biosensor based on interdigitated array microelectrode (IDAM) coupled with magnetic nanoparticle–antibody conjugates (MNAC) was developed and evaluated for rapid and specific detection of *E. coli* O157:H7 in ground beef samples by Li et al.[21]. MNAC were prepared by immobilizing biotin-labeled polyclonal goat anti-*E. coli* antibodies onto streptavidin-coated magnetic nanoparticles, which were used to separate and concentrate *E. coli* O157:H7 from ground beef samples. Magnitude of impedance and phase angle were measured in a frequency range of 10 Hz to 1 MHz in the presence of 0.1mol/L mannitol solution. The lowest detection limits of this biosensor for detection of *E. coli* O157:H7 in pure culture and ground beef samples were 7.4×10^4 and 8.0×10^5 CFU/mL, respectively. The regression equation for the normalized impedance change (NIC) versus *E. coli* O157:H7 concentration (N) in ground beef samples was NIC=15.55N−71.04 with R^2= 0.95. Sensitivity of the impedance biosensor was improved by 35% by concentrating bacterial cells attached to MNAC in the active layer of IDAM above the surface of electrodes with the help of a magnetic field. Based on equivalent circuit analysis, it was observed that bulk resistance and double layer capacitance were responsible for the impedance change caused by the presence of *E. coli* O157:H7 on the surface of IDAM. Surface immobilization techniques, redox probes, or sample incubation were not used in this impedance biosensor. The total detection time from sampling to measurement was 35 min.

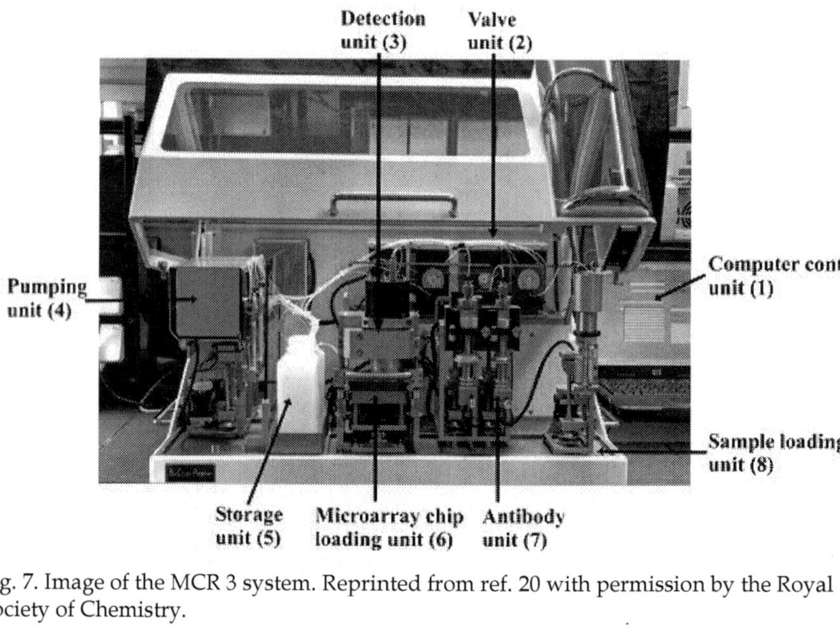

Fig. 7. Image of the MCR 3 system. Reprinted from ref. 20 with permission by the Royal Society of Chemistry.

Because of the potential health risks of aflatoxin B1 (AFB1), it is essential to monitor the level of this mycotoxin in a variety of foods. An indirect competitive immunoassay has been developed using the NRL array biosensor (Figure 8) by Shriver-Lake et al.[22], offering rapid, sensitive detection, and quantification of AFB1 in buffer, corn and nut products. AFB1-

spiked foods were extracted with methanol and Cy5-anti-AFB1 was added to the resulting sample. The extracted sample/antibody mix was passed over a waveguide surface patterned with immobilized AFB1. The resulting fluorescence signal decreased as the concentration of AFB1 in the sample increased. The limit of detection for AFB1 in buffer, 0.3 ng/mL, was found to increase to between 1.5 and 5.1 ng/g and 0.6 and 1.4 ng/g when measured in various corn and nut products, respectively.

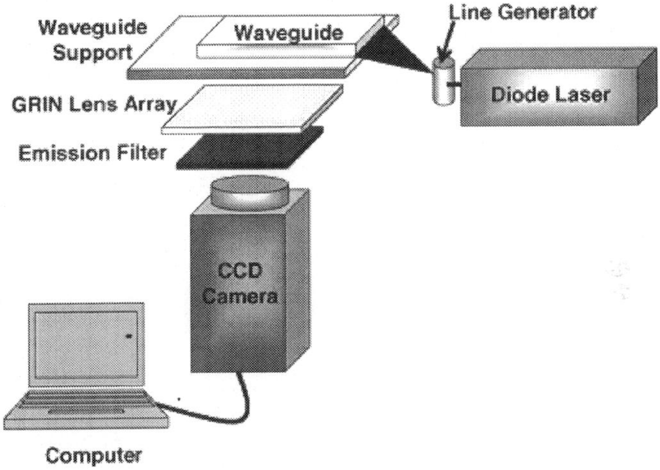

Fig. 8. Schematic of the NRL array biosensor. Reprinted from ref. 22 with permission by the Elsevier.

Ligler et al.[4] has clearly demonstrated the versatility of the biosensor arrays for the detection of both large and small food contaminants either individually or simultaneously. The bacterial pathogen *C. jejuni* was measured using a sandwich immunoassay both in buffer and additional complex food matrices with LODs ranging from 500 to 3780 CFU/ml. The mycotoxins OTA, DON, AFB1 and FB were detected simultaneously on a signal substrate using a competitive-based immunoassay format taking only 15min. The combination of sandwich and competitive immunoassay formats on a single substrate was demonstrated, allowing the simultaneous detection of both large (*C. jejuni*) and small (AFB1) food pathogens with LODs in buffer of 500 CFU/ml and 0.3 ng/mL, respectively.

Deoxynivalenol (DON), a mycotoxin produced by several *Fusaruim* species, is a worldwide contaminant of foods and feeds. Because of the potential dangers due to accidental or intentional contamination of foods with DON, there is a need to develop a rapid and highly sensitive method for easy identification and quantification of DON. In Taitt's study[23], they have developed and utilized a competitive immunoassay technique to detect DON in various food matrixes and indoor air samples using a biosensor array. A DON biotin conjugate, immobilized on a NeutrAvidin-coated optical waveguide, competed with the DON in the sample for binding to fluorescently labeled DON monoclonal antibodies. To demonstrate a simple procedure amenable for on-site use, DON-spiked cornmeal, cornflakes, wheat, barley, and oats were extracted with methanol water (3:1) and assayed without cleanup or pre-concentration. The limits of detection ranged from 0.2 ng/mL in buffer to 50 ng/g in oats. The detection limit of DON spiked into an aqueous effluent from an air sampler was 4 ng/mL.

Parro et al.[24] have developed antibodies and a multi-array competitive immunoassay (MACIA) for the detection of a wide range of molecular size compounds, from single aromatic ring derivatives or polycyclic aromatic hydrocarbons (PAHs), through small peptides, proteins or whole cells (spores). Multiple biosensor arrays containing target molecules are used simultaneously to run several competitive immunoassays. The sensitivity of the MACIA for small organic compounds like naphthalene, 4-phenylphenol or 4-terbutylphenol is in the range of 100–500 ppb, for others like the insecticide terbutryn it is at the ppt level, while for small peptides, as well as for more complex molecules like the protein thioredoxin, the sensitivity is approximately 1-2 ppb, or 104-105 spores of Bacillus subtilis per milliliter. For organic compounds, a water–methanol solution was used in order to achieve a better dissolution of the organics without compromising the antibody–antigen interaction. The above-mentioned compounds were detected by MACIA in water–(10%) methanol extracts from spiked pyrite and hematite-containing rock powder samples, as well as from a spiked-sand sample subjected to organic extraction with dichloromethane–methanol (1/1).

Sandwich immunoassays have also been conducted by running the samples through a multiple channel (one per flow channel) over the array of capture antibodies for a period of time sufficient for the antibody to bind any target agent in the sample. In most of the assays described by Ligler et al.[25], they use 8–10 min for binding in order to balance assay sensitivity with keeping the total assay time short (under 15 min). The sensitivity is clearly greater if longer periods of time are used, particularly if the samples are viscous and the rate of diffusion of the target within the sample is slow. In this array, biotinylated capture antibodies were exposed to the avidin-coated waveguide using flow channels molded out of polydimethylsiloxane as described previously. TNT (trinitrotoluene) antibody spots bound a Cy5-labeled TNB (trinitrobenzene) tracer molecule in the tracer cocktail to provide a positive control. Assays were conducted using staphylococcal enterotoxin B (SEB), ricin toxin, cholera toxin, mouse IgG and Bacillus globigii (*B. globigii* is an anthrax spore simulant) as targets. The Cy5-TNB bound to the appropriate antibodies in all lanes. Samples containing SEB, cholera toxin, mouse IgG and *B. globigii* bound cleanly to the appropriate spots without showing any cross-reactivity against other capture antibodies.

Contamination of food by mycotoxins occurs in minute quantities, and therefore, there is a need for a highly sensitive and selective device that can detect and quantify these organic toxins. Taitt et al.[26] reported the development of a rapid and highly sensitive array biosensor for the detection and quantitation of ochratoxin A (OTA). The array biosensor utilizes a competitive immunoassay format. Immobilized OTA derivatives compete with toxin in solution for binding to fluorescent anti-OTA antibody spiked into the sample. This competition is quantified by measuring the formation of the fluorescent immuno-complex on the waveguide surface. The fluorescent signal is inversely proportional to the concentration of OTA in the sample. Analyses for OTA in buffer and a variety of food and beverage samples were performed. Samples were extracted with methanol, without any sample cleanup or pre-concentration step prior to analysis. The limit of detection for OTA in several cereals ranged from 3.8 to 100 ng/g, while in coffee and wine, detection limits were 7 and 38 ng/g, respectively.

Golden et al.[27] have developed a "do-it-yourself" biosensor array for the simultaneous detection of multiple targets in multiple samples within 15-30 min. The biosensor is based on a planar waveguide, a modiWed microscope slide, with a pattern of small (mm^2) sensing regions. The waveguide is illuminated by launching the emission of a 635 nm diode laser into the proximal end of the slide via a line generator. The evanescent Weld excites Xuorophores bound in the sensing region and the emitted Xuorescence is measured using a Peltier-cooled CCD camera. Assays can be performed on the waveguide in multichannel Xow chambers and then interrogated using the detection system described in their paper. This biosensor can detect many different targets, including proteins, toxins, cells, virus, and explosives with detection limit rivaling those of the ELISA detection system.

Kramer et al.[28] presents a new, versatile, portable miniaturized flow-injection biosensor array which is designed for field analysis. The temperature-controlled field prototype can run for 6 h without external power supply. The bio-recognition element is an analyte-specific antibody immobilized on a gold surface of pyramidal structures inside an exchangeable single-use chip, which hosts also the enzyme-tracer and the sample reservoirs. The competition between the enzyme-tracer and the analyte for the antigen-binding sites of the antibodies yields in the final step a chemiluminescence signal that is inversely proportional to the concentration of analyte in the given range of detection. A proof of principle is shown for nitroaromatics and pesticides. The detection limits reached with the field prototype in the laboratory was below 0.1 g/L for 2,4,6-trinitrotoluene (TNT), and about 0.2 g/L for diuron and atrazine, respectively. Important aspects in this development were the design of the competition between analyte and enzyme-tracer, the unspecific signal due to unspecific binding and/or luminescence background signal, and the flow pattern inside the chip.

The multianalyte array biosensor (MAAB) is a rapid analysis instrument capable of detecting multiple analytes simultaneously. Rapid (15 min), single-analyte sandwich immunoassays were developed for the detection of Salmonella enterica serovar Typhimurium by Taitt et al.[29], with a detection limit of 8×10^4 CFU/mL; the limit of detection was improved 10 fold by lengthening the assay protocol to 1 h. S. enterica serovar Typhimurium was also detected in the following spiked foodstuffs, with minimal sample preparation: sausage, cantaloupe, whole liquid egg, alfalfa sprouts, and chicken carcass rinse. Cross-reactivity tests were performed with *Escherichia coli* and *Campylobacter jejuni*. To determine whether the MAAB has potential as a screening tool for the diagnosis of asymptomatic Salmonella infection of poultry, chicken excretal samples from a private, noncommercial farm and from university poultry facilities were tested. While the private farm excreta gave rise to signals significantly above the buffer blanks, none of the university samples tested positive for S. enterica serovar Typhimurium without spiking; dose-response curves of spiked excretal samples from university-raised poultry gave limits of detection of 8×10^3 CFU/g.

Campylobacter and Shigella bacteria are common causes of food- and water-borne illness worldwide. There is a current need in food, medical, environmental, and military markets for a rapid and user-friendly method of detecting such pathogens. The array biosensor developed at the NRL encompasses these qualities. Ligler et al.[30] reported on a sandwich immunoassay-based biosensor array that was developed for the detection of Campylobacter and Shigella species in both buffer and a variety of food and beverage samples. The limit of detection for Shigella dysenteriae in buffer and chicken carcass wash was 4.9×10^4 CFU/mL, whereas Campylobacter jejuni could be measured at concentrations as low as 9.7×10^2 CFU/mL. The limits of detection and dynamic range were found to vary depending on the sample matrix, but could be improved by running the sample over the waveguide surface for longer periods of time. Samples were analyzed with no pre-concentration or enrichment steps and little-to-no sample pretreatment prior to analysis, and the total analysis run time was 25 min.

Biosensor array that is capable of detecting multiple targets rapidly and simultaneously on the surface of a single waveguide has also been studied. Ligler et al.[31] developed a sandwich and competitive fluoroimmunoassays to detect high and low molecular weight toxins, respectively, in complex samples. Antibodies were first-immobilized in specific locations on the waveguide and the resultant patterned array was used to interrogate up to 12 different samples for the presence of multiple different analytes. Upon binding of a fluorescent analyte or fluorescent immunocomplex, the pattern of fluorescent spots was detected using a CCD camera. Automated image analysis was used to determine a mean fluorescence value for each assay spot and to subtract the local background signal. The location of the spot and its mean fluorescence value were used to determine the toxin identity and concentration. Toxins were measured in clinical fluids, environmental samples and foods, with minimal sample preparation. Results were reported for rapid analyses of staphylococcal enterotoxin B, ricin, cholera toxin, botulinum toxoids, trinitrotoluene, and the mycotoxin fumonisin. Toxins were detected at levels as low as 0.5 ng/mL.

4. Aptamer based biosensor array

Aptamers are single-stranded (ss) DNA or RNA molecules, typically <100 monomer units, which have the ability to bind to other molecules with high affinity and specificity. They are selected from random oligonucleotide pools by a process called Systematic Evolution of Ligands by Exponential enrichment (SELEX). Conceptually, the SELEX process is based on the ability of these small oligonucleotides to fold into unique three-dimensional (3-D) structures which can interact with a specific target with high specificity and affinity through such interactions as van der Waals surface contacts, hydrogen bonding and base stacking interactions. Aptamers can offer a strong and reliable role as biological recognition elements in most analytical applications. The specificity and high affinity of aptamers to a wide variety of targets, coupled with the ease of design and molecular engineering, as outlined earlier, make aptamers highly suitable for development as molecular biosensors. By moving in this direction, investigators have made appreciable efforts in recent years to utilize these unique features of aptamers to devise appropriate strategies with which to effectively and efficiently apply aptamers for biological agent recognition, identification, characterization and quantification[32].
Analogous to immunoassays based on the antigen-antibody interaction, aptamer-based biosensors can adopt different assay configurations to transduce bio-recognition events. Since aptamers have been selected to bind very different targets, ranging from small molecules to macromolecules, such as proteins, various assay configurations have been designed and reported. Nevertheless, the majority of these designs fall into two categories of configuration (Figure 9): single-site binding and dual-site binding[33]. The use of aptamers as new biological receptors in biosensor arrays can accelerate the development of biosensors of practical relevance. Because of their exceptionally high stability, selectivity and sensitivity, aptamer-based biosensor arrays have the potential to overcome the lacking functional and storage stability of most biosensors[34]. For example, an electrochemical impedance spectroscopy method of detection for aptamer-based electrochemical biosensor array (Figure 10) is reported in which the binding of aptamers immobilized on gold electrodes leads to impedance changes associated with target protein binding events by Xu et al.[35]. Human IgE was used as a model target protein and incubated with the aptamer-based array consisting of single-stranded DNA containing a hairpin loop. To increase the binding efficiency for proteins, a hybrid modified layer containing aptamers and cysteamine was fabricated on the photolithographic gold surface through molecular self-assembly. Compared to immunosensing methods using anti-human IgE antibody as the recognition element, impedance spectroscopy detection could

provide higher sensitivity and better selectivity for aptamer-modified electrodes. The results of this method show good correlation for human IgE in the range of 2.5-100 nmol/L. A detection limit of 0.1 nmol/L was obtained, and an average of the relative standard deviation was 10%. The method describes the first label-free detection for arrayed electrodes utilizing electrochemical impedance spectroscopy.

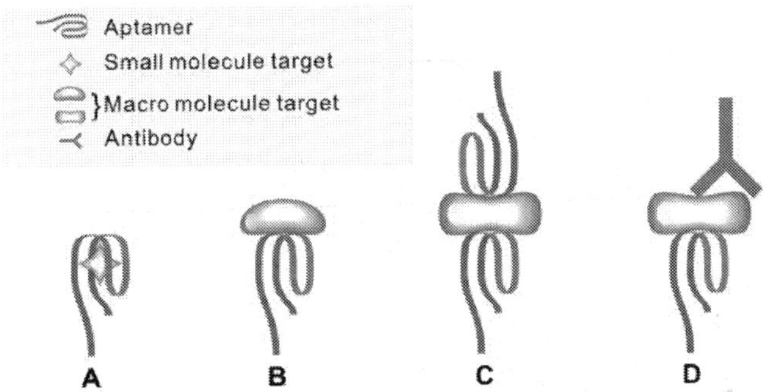

Fig. 9. Aptamer-based assay formats. (A) Small-molecule target buried within the binding pockets of aptamer structures; (B) single-site format;(C) dual-site (sandwich) binding format with two aptamers; and, (D) "sandwich" binding format with an aptamer. Reprinted from ref. 33 with permission by the Elsevier.

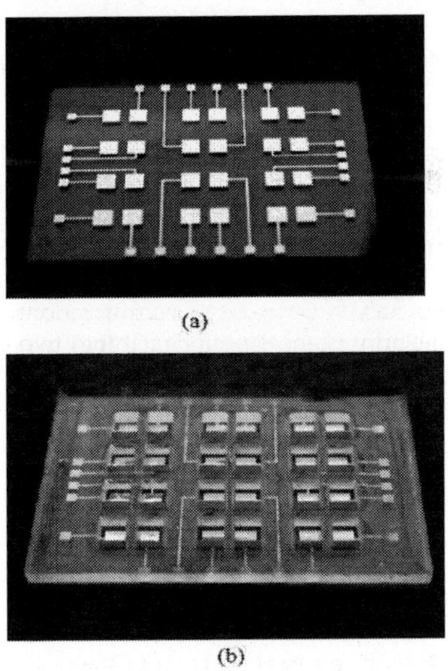

Fig. 10. Construction of the gold array electrode chip: (a) a photolithographic gold film array electrode; (b) a home-constructed PDMS frame containing 24-microwells for the immobilization. Reprinted from ref. 35 with permission by the Royal Society of Chemistry.

Quandt et al.[36] designed a Love-wave biosensor array by coupling aptamers to the surface of a Love-wave sensor chip. The sensor chip consists of five single sensor elements and allows label-free, real-time, and quantitative measurements of protein and nucleic acid binding events in concentration-dependent fashion. The biosensor was calibrated for human-thrombin and HIV-1 Rev peptide by binding fluorescently labeled molecules and correlating the mass of the bound molecules to fluorescence intensity. Detection limits of approximately 75 pg/cm^2 were obtained, and analyte recognition was specific. The sensor can easily be regenerated by simple washing steps. They further demonstrated the versatile applicability of the sensor by immobilizing single-stranded DNA (*ss*DNA) for the detection of the corresponding counter-strand.

The large quantity of aptamers which have been selected to bind complex molecules of low molecular weight leads to the possible use of these aptamers not only in diagnostic assays, but also in a wider range of applications, such as environmental analytical chemistry[37]. Selection of DNA ligands to the chloroaromatics, 4-chloroaniline (4-CA), 2,4,6-trichloroaniline (TCA) and pentachlorophenol (PCP), was performed by a novel method utilizing magnetic beads (MBs) having a linker arm for immobilization[38]. Moreover, Labuda et al.[39] reported for the first time the selection of RNA aptamers for the recognition of hydrophobic aromatic carcinogens. In particular, RNA aptamers with a K_d in the low micro-molar range have been selected for aromatic amines residues using as a model methylendianiline, which is a common industrial chemical employed to manufacture plastics, glues and foams.

A toxin-related work based on aptamers arrays have been published by Ellington et al.[40] . The authors reported the adaptation of a chip-based micro-sphere array (the "electronic taste chip") to aptamer receptors. Their detection system is illustrated in Figure 11. Unlike most protein-based arrays, the aptamer chips could be stripped and reused multiple times. The aptamer chips proved to be useful for screening aptamers from in vitro selection experiments and for sensitively quantitating the bio-threat agent ricin. The system composed of a flow cell connected to a fast performance liquid chromatography pump and a fluorescence microscope for observation. The flow cells contained silicon chips with multiple wells in which beads modified with the sensor elements were deposited. Commercially available streptavidin agarose beads were modified with biotinylated aptamers; RNA anti-ricin aptamers were used to demonstrate the possibility of quantifying the labeled protein. A sandwich assay format was also optimized using anti-ricin antibodies, to directly detect the unlabelled protein. In the first type of assay, the aptamer was bio-tinylated, immobilized and put in contact with the solution containing fluorescently labeled ricin, once introduced into the chip wells. The fluorescence intensities of the captures proteins were used to construct a calibration plot for ricin and a detection limit of 8 mg/ml was obtained. In the sandwich assay, the anti-ricin aptamer acted as a capture reagent and unlabelled ricin bound to the aptamer could interact with fluorophore-labeled fabricated an aptamer-based biosensor array for protein detection.

Environmental allergenic disease is a major cause of illness and disability, and there is broad consensus that the prevalence of type I allergy is increasing worldwide. Recent advances in biotechnology have yielded potentially useful functional binding aptamers that can enable low cost, high affinity allergen measurement. Aptamers are selected in vitro from combinatorial oligonucleotide libraries and therefore have several advantages over the traditionally used antibodies for detection of allergens. Aptamer-based methods could be used for measuring environmental allergens. Integrating the resulting aptamer-based allergen measurements to enhance quantization in an ongoing and complementary environmental childhood asthma epidemiological study forms the basis for the third and final aim. Successful use of aptamers for measuring environmental allergens should lead to a more cost effective, flexible, and health relevant method and thereby provides the potential for a more fundamental understanding of the role of environmental allergens in respiratory health.

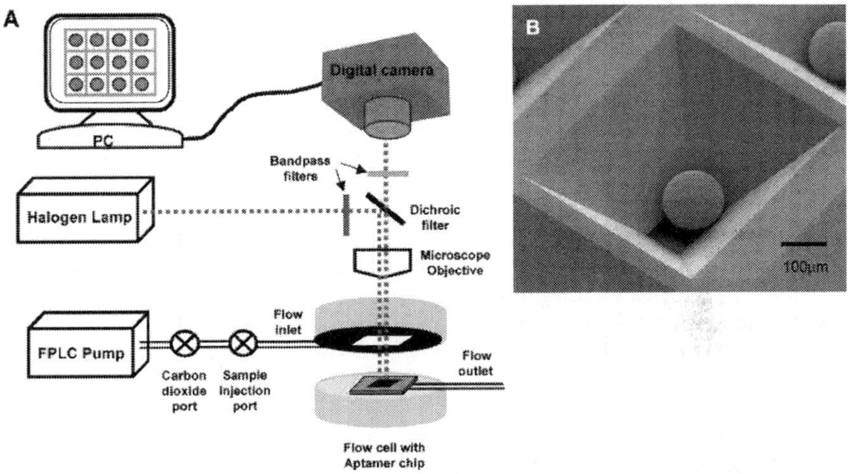

Fig. 11. Detection systems. (A) The electronic tongue setup contains a fluid delivery system, fluorescence microscope, digital camera, flow cell in which the aptamer chip will be loaded, and computer for data analysis. (B) Close-up look at a bead in a rectangular-shaped micro-machined well of the aptamer chip. Reprinted from ref. 40 with permission by the Royal Society of Chemistry.

5. Enzyme based biosensor array

Enzyme-based technology relies upon the natural specificity of given enzymatic protein to react biochemically with a target substrate or substrates. Like ion channels, there are many enzymes that participate in cellular signaling and, in some cases, are targeted by compounds associated with environmental toxicity. In general, enzyme-based biosensors employ semi-permeable membranes through which target analytes diffuse toward a solid-phase immobilized enzyme compartment. Ion selective, amperometric, or pH electrodes measure reaction components such as hydrogen peroxide (from oxidation of glucose by glucose oxidase) or ammonium ions (from urease metabolism of urea)[41]. Enzymes were historically the first molecular recognition elements included in biosensors and continue to be the basis for a significant number of publications reported for biosensors in general as well as biosensors for environmental applications. There are several advantages for enzyme biosensors. These include a stable source of material (primarily through bio-renewable sources), the ability to modify the catalytic properties or substrate specificity by means of genetic engineering, and catalytic amplification of the biosensor response by modulation of the enzyme activity with respect to the target analyte[17].

Recent progress with respect to enzyme biosensors for environmental applications has been reported in several areas[42]. These areas include the following: genetic modification of enzymes to increase assay sensitivity, stability and shelf life; improved electrochemical interfaces and mediators for more efficient operation; and introduction of sampling schemes consistent with potential environmental applications. More recently, enzyme-based biosensor arrays also have been used in the application of environmental monitoring. For example, Kukla et al.[43] developed a multi-enzyme electrochemical biosensor array. Their sensor array is based on capacitance pH-sensitive electrolyte–insulator–semiconductor (EIS) sensors with silicon nitride ion-sensitive layers and different forms of cholinesterase, urease and glucose oxidase as sensitive elements. With this sensor array, the authors used a multi-enzyme analysis to recognize the heavy metal ions in solutions containing a mixture of different metal ions, as well as for determination of the metal ion content in the analyzed samples. The content of toxic elements was determined by estimation of the residual activity of enzymatic membranes after the injection of analyzed samples. The conditions for enzyme sensors operation, such as buffer capacity, substrate concentration, time of incubation and time of response signal measurement, were optimized to reach the maximal sensitivity of multi-sensor for analysis of heavy metal ions in the investigated solutions. The results show that multi-enzyme analysis followed by mathematical processing is an efficient approach to develop biosensor arrays for toxic substrates detection.

Organophosphate pesticides (OPs) used to be widely used in agriculture due to their high efficiency as insecticides. OPs have been shown to result in high levels of acute neuron-toxicity and carcinogenicity, with the majority being hazardous to both human health and to the wider environment. A rapid, reliable, economical and portable analytical system will be of great benefit in the detection and prevention of OPs contamination. A biosensor array based on six acetylcholinesterase enzymes coupled with a novel automated instrument incorporating a neural network program has been reported by Hart et al.[44]. The biosensor array and the instrument is illustrated in Figure 12. Electrochemical analysis was carried out using chronoamperometry and the measurement was taken 10 s after applying a potential of 0 V vs. Ag/AgCl. The total analysis time for the complete assay was less than 6 min. The array was used to produce calibration data with six organophosphate pesticides (OPs) in the concentration range of 10^{-5} mol/L to 10^{-9} mol/L to train a neural network. The output of the neural network was subsequently evaluated using different sample matrices. There was no detrimental matrix effect observed from water, phosphate buffer, food or vegetable extracts. Furthermore, the sensor system was not detrimentally affected by the contents of water samples taken from each stage of the water treatment process. Their biosensor array system successfully identified and quantified all samples where an OP was present in water, food and vegetable extracts containing different OPs. There were no false positives or false negatives observed during the evaluation of the analytical system. Their biosensor arrays and automated instrument were evaluated in situ in field experiments where the instrument was successfully applied to the analysis of a range of environmental samples.

Recently, many studies have focused on the development of biochemical sensors, which are well suited for the rapid, simple and selective analysis of pesticides. Specially, they combine the selectivity of the enzymatic reactions with operational simplicity and simple detection schemes. Valle et al.[45] developed an electronic tongue, employing an array of inhibition biosensors and Artificial Neural Networks (ANNs). The array of biosensors was made up of three amperometric pesticide biosensors that used different acetylcholinesterase (AChE) enzymes: a wild type from electric eel (EE) and two different genetically modified enzymes

(B1 and B394). In order to model the response to dichlorvos and carbofuran mixtures, a total amount of 22 solutions were prepared, with random concentrations. Chronoamperometric responses of the biosensor array were used in order to obtain the inhibition bioelectronic tongue. Mean values of concentration of pesticides evaluated were 0.79 nmol/L for dichlorvos and 4.1 nmol/L for carbofuran. Good prediction ability was obtained with correlation coefficients better than 0.918 when the obtained values were compared with those expected for a set of 6 external test samples not used for training.

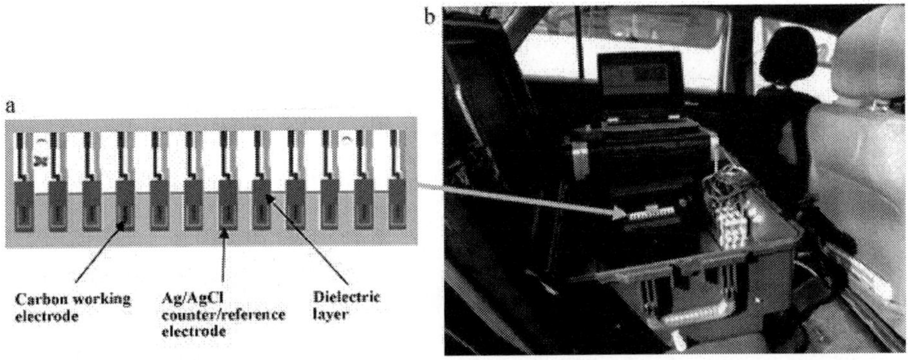

Fig. 12. (a) electrode array comprising 12 screen-printed carbon electrodes and an Ag/AgCl counter/reference electrode printed on an alumina substrate; (b) array in the prototype biosensor system operating in the field powered from a car battery via the lighter socket. Reprinted from ref. 44 with permission by the Elsevier.

Another approach is by using Organophosphorus hydrolase (OPH). OPH is a 72 kDa homodimeric, metalloenzyme, containing two zinc ions in the active site involved in catalytic and/or structural functions. OPH catalyzes the hydrolysis of Organophosphates (OPs) resulting in its detoxification. Some of the biosensors that were developed exploiting OPH as the bio-recognition element on different detection platforms have been reported. Though highly sensitive and selective towards different OPs, their inability to provide simultaneous measurements of different analytes was a major shortcoming. Simonian et al.[46] developed a biosensor array (Figure 13) with the potential for direct detection of organophosphates using OPH, conjugated with a pH-sensitive fluorophore, carboxynaphthofluorescein (CNF). The presence of reference spots allows the discrimination of the enzymatic and non-enzymatic based pH changes; bovine serum albumin (BSA) was used as a non-enzymatic scaffold protein for CNF attachment at the reference spots. An array biosensor unit developed at the Naval Research Laboratories (NRL) was adopted as the detection platform and appropriately modified for enzyme-based measurements. A planar multi-mode waveguide was covered with an optically transparent TiO_2 layer to increase the surface area available for immobilization. The biosensor enabled the detection of 2.5 μmol/L paraoxon, and 10 μmol/L parathion respectively. Very short response time of 30 s can be achieved with a total analysis time of less than 2 min. When operated at room temperature and stored at 4 ℃, the waveguide retained reasonable activity for greater than 45 days.

An array-based optical biosensor for the simultaneous analysis of multiple samples in the presence of unrelated multi-analytes was fabricated by Doong et al.[47]. The authors used

Urease and acetylcholinesterase (AChE) as model enzymes, which were co-entrapped with the sensing probe, FITC-dextran, in the sol-gel matrix to measure pH, urea, acetylcholine (ACh) and heavy metals (enzyme inhibitors). Environmental and biological samples spiked with metal ions were also used to evaluate the application of the array biosensor to real samples. The biosensor exhibited high specificity in identifying multiple analytes. No obvious cross-interference was observed when a 50-spot array biosensor was used for simultaneous analysis of multiple samples in the presence of multiple analytes. The sensing system can determine pH over a dynamic range from 4 to 8.5. The limits of detection of 2.5-50 μmol/L with a dynamic range of 2-3 orders of magnitude for urea and ACh measurements were obtained. Moreover, the urease-encapsulated array biosensor was used to detect heavy metals. The analytical ranges of Cd(II), Cu(II), and Hg(II) were between 10 nmol/L and 100 mmol/L. When real samples were spiked with heavy metals, the array biosensor also exhibited potential effectiveness in screening enzyme inhibitors.`

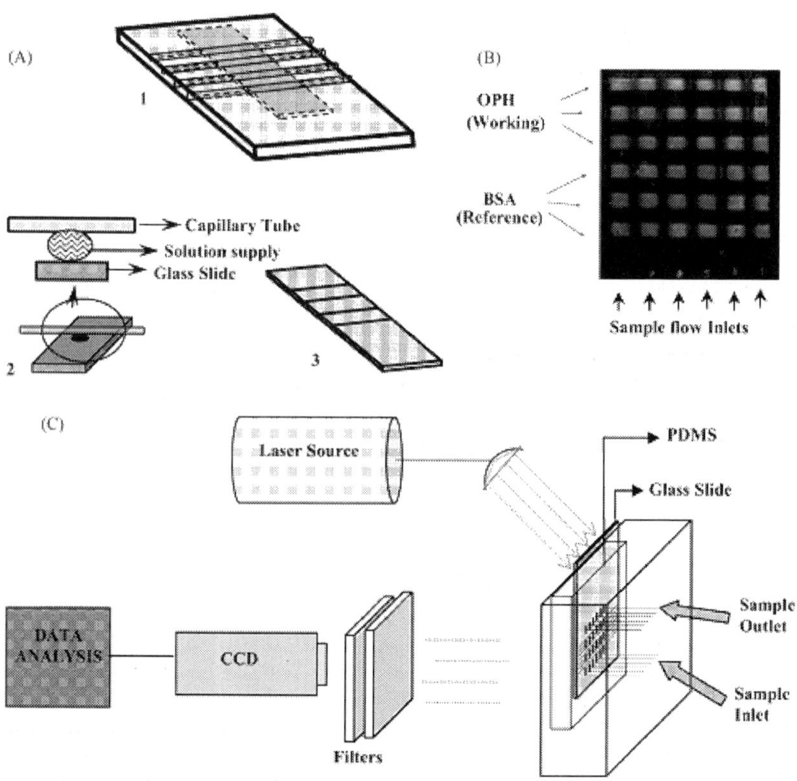

Fig. 13. (A) Schematic of modified process for incubation using thin glass tubes. (B) Schematic of the glass slide with immobilized proteins and fluorophores. (C) Schematic of the array biosensor. Reprinted from ref. 46 with permission by the Elsevier.

Solna et al.[48] use screen-printed four-electrode system as the amperometric transducer for determination of phenols and pesticides using immobilized tyrosinase, peroxidase, acetylcholinesterase and butyrylcholinesterase. Acetylthiocholine chloride was chosen as substrate for cholinesterases to measure inhibition by pesticides, hydrogen peroxide served as co-substrate for peroxidase to measure phenols. In their work, the compatibility of hydrolases and oxidoreductases working in the same array was studied. The detection of p-cresol, catechol and phenol as well as of pesticides including carbaryl, heptenophos and fenitrothion was carried out in flow-through and steady state arrangements. It was demonstrated that electrodes modified with hydrolases and oxidoreductases can function in the same array. The limit of detection for catechol using tyrosinase was equal to 0.35 and 1.7 µmol/L in the flow and steady systems. Lower limits of detection for pesticides were achieved in the steady state system: carbaryl 26 nmol/L, heptenophos 14 nmol/L and fenitrothion 0.58 nmol/L. Similar multi-enzyme-based electrochemical biosensor arrays for the determination of pesticides[49-52] and phenols[53] have been reported by other workers.

6. Microorganism-based biosensor array

A microbial biosensor is an analytical device which integrates microorganism(s) with a physical transducer to generate a measurable signal proportional to the concentration of analytes. In recent years, a large number of microbial biosensors have been developed for environmental, food, and biomedical applications[54].

Enzymes are the most widely used biological sensing element in the fabrication of biosensors. Although purified enzymes have very high specificity for their substrates or inhibitors, their application in biosensors construction may be limited by the tedious, time-consuming and costly enzyme purification, requirement of multiple enzymes to generate the measurable product or need of cofactor/coenzyme. Microorganisms provide an ideal alternative to these bottle-necks. The many enzymes and co-factors that co-exist in the cells give the cells the ability to consume and hence detect large number of chemicals; however, this can compromise the selectivity. They can be easily manipulated and adapted to consume and degrade new substrate under certain cultivating condition. Additionally, the progress in molecular biology/recombinant DNA technologies has opened endless possibilities of tailoring the microorganisms to improve the activity of an existing enzyme or express foreign enzyme/protein in host cell. All of these make microbes excellent biosensing elements[55].

Microorganism-based biosensor arrays classically used for environmental biosensing are mainly bacteria and yeasts, and to a lesser extent algae. Various strains have been exploited, from commercial and well-characterized cells harboring a broad range of substrates to genetically engineered organisms specially constructed to detect specific molecules or groups of molecules, passing through environmental cells isolated from polluted sites offering greater robustness and more specific enzymatic properties[56].

Rapid identification of *Escherichia coli* strains is an important diagnostic goal in applied medicine as well as the environmental and food sciences. Mikkelsen et al.[57] reported an electrochemical, screen-printed biosensor array, where selective recognition is accomplished using lectins that recognize and bind to cell-surface lipopolysaccharides and coulometric transduction exploits non-native external oxidants to monitor respiratory cycle activity in lectin-bound cells. Ten different lectins were separately immobilized onto porous membranes that feature activated surfaces. Modified membranes were exposed to untreated *E. coli* cultures for 30 min, rinsed, and layered over the individual screen-printed carbon electrodes of the sensor array. The membranes were incubated 5 min in a reagent solution that contained the oxidants menadione and ferricyanide as well as the respiratory substrates succinate and formate. Electrochemical oxidation of ferrocyanide for 2 min provided chronocoulometric data related to the quantities of bound cells. These screen-printed sensor arrays were used in conjunction with factor analysis for the rapid identification of four *E. coli* subspecies (*E. coli* B, *E. coli* Neotype, *E. coli* JM105 and *E. coli* HB101). Systematic examination of lectin-binding patterns showed that these four *E. coli* subspecies are readily distinguished using only five essential lectins.

The last decade has witnessed a significant increase in interest in whole-cell biosensors for diverse applications, as well as a rapid and continuous expansion of array technologies. The combination of these two disciplines has yielded the notion of whole-cell array biosensors. Belkin et al.[58] presented a potential manifestation of this idea by describing the printing of a whole-cell bacterial bioreporters array (Figure 14). Exploiting natural bacterial tendency to adhere to positively charged abiotic surfaces, they describe immobilization and patterning of bacterial "spots" in the nanoliter volume range by a non-contact robotic printer. They show that the printed Escherichia coli-based sensor bacteria are immobilized on the surface, and retain their viability and biosensing activity for at least 2 months when kept at 4°C. Immobilization efficiency was improved by manipulating the bacterial genetics, the growth and the printing media and by a chemical modification of the inanimate surface. The result suggests that the methodology presented by them may be applicable to the manufacturing of whole-cell sensor arrays for diverse high throughput applications. In the course of the study, they have also described a novel specific reporter for the detection of respiratory inhibitors. Sodium azide, a chemical with a constantly increasing world distribution, served as the model toxicant. The sensor's response was rapid (20 minutes after exposure) and dose-dependent, and could be maintained for at least 2 months at 4 °C.

Li et al.[59] developed a double interdigitated array microelectrodes (IAM)-based flow cell for an impedance biosensor to detect viable *Escherichia coli* O157:H7 cells after enrichment in a growth medium. Their study was aimed at the design of a simple flow cell with embedded IAM which does not require complex microfabrication techniques and can be used repeatedly with a simple assembly/disassembly step. The flow cell was also unique in having two IAM chips on both top and bottom surfaces of the flow cell, which enhances the sensitivity of the impedance measurement. *E. coli* O157:H7 cells were grown in a low conductivity yeast–peptone–lactose–TMAO (YPLT) medium outside the flow cell. After bacterial growth, impedance was measured inside the flow cell. Equivalent circuit analysis indicated that the impedance change caused by bacterial growth was due to double layer capacitance and bulk medium resistance. Both parameters were a function of ionic concentration in the medium, which increased during bacterial growth due to the conversion of weakly charged substances present in the medium into highly charged ions. The impedance biosensor successfully detected *E. coli* O157:H7 in a range from 8.0 to 8.2×10^8 CFU/mL after an enrichment growth of 14.7 and 0.8 h, respectively. A logarithmic linear relationship between detection time (T_D) in h and initial cell concentration (N_0) in CFU/mL was $T_D = -1.73 \log N_0 + 14.62$, with $R^2 = 0.93$. Double IAM-based flow cell was more sensitive than single IAM-based flow cell in the detection of *E. coli* O157:H7 with 37–61% more impedance change for the frequency range from 10 Hz to 1 MHz. The double IAM-based flow cell could be used to design a simple impedance biosensor for the sensitive detection of bacterial growth and their metabolites.

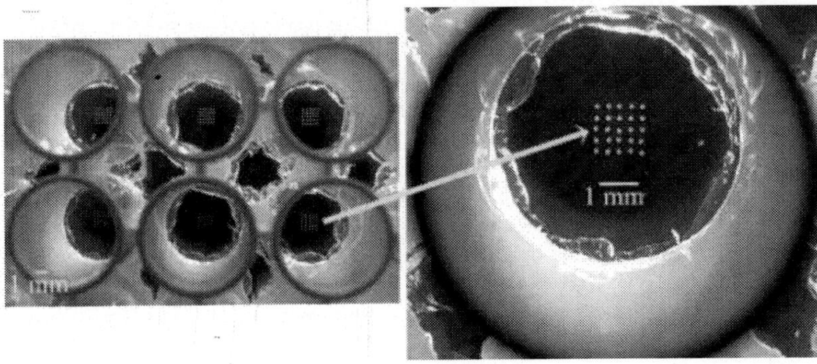

Fig. 14. Twenty five spots, 1 nl each, of strain SM118 in ectoine, printed onto the wells of 96-well plate with an APTES coated glass bottom. Reprinted from ref. 58 with permission by the Royal Society of Chemistry.

Worldwide herbicide discharge into the aquatic environment is also a growing concern. Adverse effects induced by herbicide contamination are impacting a great variety of organisms and ecosystems, ranging from the primary producers to animals and humans. Biosensors for the rapid detection of herbicides in the environment have also been explored. A multiple-strain algal biosensor was constructed for the detection of herbicides inhibiting photosynthesis by Podola et al.[60]. Nine different microalgal strains were immobilized on an array biochip using permeable membranes. The biosensor allowed on-line measurements of aqueous solutions passing through a flow cell using chlorophyll fluorescence as the biosensor response signal. The herbicides atrazine, simazine, diuron, isoproturon and paraquat were detectable within minutes at minimal LOEC (Lowest Observed Effect Concentration) ranging from 0.5 to 100 μg/L, depending on the herbicide and algal strain. The most sensitive strains in terms of EC50 values were Tetraselmis cordiformis and Scherffelia dubia. Less sensitive species were Chlorella vulgaris, Chlamydomonas sp. and Pseudokirchneriella subcapitata, but for most of the strains no general sensitivity or resistance was found. The different responses of algal strains to the five herbicides constituted a complex response pattern (RP), which was analyzed for herbicide specificity within the linear dose-response relationship.

Recombinant bioluminescent bacterial strains are increasingly receiving attention as environmental biosensors due to their advantages, such as high sensitivity and selectivity, low costs, ease of use and short measurement times. Gu et al.[61] use a cell-based array technology that uses recombinant bioluminescent bacteria to detect and classify environmental toxicity followed by developing two biosensor arrays, i.e., a chip and a plate array. Twenty recombinant bioluminescent bacteria, having different promoters fused with the bacterial lux genes, were immobilized within LB-agar. About 2 μl of the cell-agar mixture was deposited into the wells of either a cell chip or a 384-well plate. The bioluminescence (BL) from the cell arrays was measured with the use of highly sensitive cooled CCD camera that measured the bioluminescent signal from the immobilized cells and then quantified the pixel density using image analysis software. The responses from the cell arrays were characterized using three chemicals that cause either superoxide damage (paraquat), DNA damage (mitomycin C) or protein/membrane damage (salicylic acid). The responses were found to be dependent upon the promoter fused upstream of the lux operon within each strain. Therefore, a sample's toxicity can be analyzed and classified through the changes in the BL expression from each well. Moreover, a time of only 2 h was needed for analysis, making either of these arrays a fast, portable and economical high-throughput biosensor system for detecting environmental toxicities.

Because of their ability to perform functional sensing, living cell-based biosensors are drawing increased attention. The work reported by Walt et al.[62] demonstrates the ability to fabricate an optical imaging fiber-based living bacterial cell array for genotoxin detection. A biosensor composed of a high-density living bacterial cell array was fabricated by inserting bacteria into a micro-well array formed on one end of an imaging fiber bundle. The size of each micro-well allows only one cell to occupy each well. In this biosensor, *E. coli* cells carrying a *recA::gfp* fusion were used as sensing components for genotoxin detection. Each fiber in the array has its own light pathway, enabling thousands of individual cell responses to be monitored simultaneously with both spatial and temporal resolution. The biosensor was capable of performing cell-based functional sensing of a genotoxin with high sensitivity and short incubation times (1 ng/mL mitomycin C after 90 min). The biosensors demonstrated an active sensing lifetime of more than 6 h and a shelf lifetime of two weeks. Their group reported another live cell biosensor array[63], which was fabricated by immobilizing bacterial cells on the face of an optical imaging fiber containing a high density array of micro-wells. Each microwell accommodates a single bacterium that was genetically engineered to respond to a specific analyte. A genetically modified *Escherichia coli* strain, containing the *lacZ* reporter gene fused to the heavy metal-responsive gene promoter *zntA*, was used to fabricate a mercury biosensor. A plasmid carrying the gene coding for the enhanced cyan fluorescent protein (ECFP) was also introduced into this sensing strain to identify the cell locations in the array. Single cell *lacZ* expression was measured when the array was exposed to mercury and a response to 100 nmol/L Hg^{2+} could be detected after a 1-h incubation time. The optical imaging fiber-based single bacterial cell array is a flexible and sensitive biosensor platform that can be used to monitor the expression of different reporter genes and accommodate a variety of sensing strains.

7. Conclusion and future direction

In recent years, there have been dramatic advances in a new analytical format, the biosensor array, a tool that has revolutionized our ability to characterize and quantify biologically and envitonmetally relevant molecules. The biosensor arrays address the need for rapid, sensitive, and specific screening for multiple pollutants at the site of sample collection. The biosensor arrays have several very significant advantages for such applications: (1) The number of analyte which can be detected simultaneously can be expanded as need dictates and specific analyte become available. (2) The biosensor arrays and tracer reagents are reusable if no target agent binds to the array surface. This feature significantly decreases the cost and operational burden for the user and simplifies automation for extended monitoring applications. (3) The biosensor array is simple to use. It is easily portable for first responder applications. The insertion of the sensor array, tracer reagents and samples is very simple with no requirement for alignment operations by the user. (4) The biosensor array is a low-cost system which can be made even more cost effective with mass production. (5) The biosensor array can be easily adapted for continuous monitoring operations by integration with a computer-controlled sampler to format automatic analytical system. Because of these advantages, more and more biosensor arrays are applied in varied areas including environmental monitoring. An overview of the applications for environment by using biosensor arrays, which are not mentioned in this review, are listed in Table 1.

Target	Biosensor array type	LOD	Reference
Herbicide Subclasses	Array of photosystem II mutants	3×10^{-9} mol/L	[64]
Metal ions	All-solid-state potentiometric biosensor array	10^{-6} mol/L	[65]
Microbial species	Electrochemical biosensor array	Not given	[66]
Escherichia coli	Quantum dot-based array	10 CFU/mL	[67]
Bio-hazardous agents	Planar waveguide biosensor array	5×10^5 CFU/mL	[68]
aflatoxin B_1	NRL biosensor array	0.6 ng/g	[69]
Ochratoxin A	Antibody-based biosensor array	3.8 ng/g	[70]
Odour	Colorimetric biosensor array	Not given	[71]
Escherichia coli	Antimicrobial Peptides based biosensor array	10^7 CFU/mL	[72]
Yersinia pestis F1	Antibody-based biosensor array	25 ng/mL	[73]
Bacillus globigii	Antibody-based biosensor array	10^5 CFU/mL	[74]
Shigella dysenteriae	Antibody-based biosensor array	5×10^4 CFU/mL	[75]

Table 1. Applications of biosensor arrays for environmental monitoring

Despite the high number of biosensor arrays under development and the amount of research literature on this area, few practical systems are currently enjoying market acceptance for environmental applications. The Naval Research Laboratory (NRL) biosensor arrays are the most successful type of biosensor arrays that have found commercial application not only in environmental monitoring but also in the monitoring of bio-molecular interaction events in general. Biosensor arrays still need more research and development in order to achieve the stability, sensitivity, specificity, and versatility that will attract confidence of potential users, especially for biotechnology and environmental applications.

8. References

[1] G. Hanrahan, D.G. Patil, J. Wang, J Environ Monitor, 6 (2004) 657-664.
[2] M. Badihi-Mossberg, V. Buchner, J. Rishpon, Electroanal, 19 (2007) 2015-2028.
[3] A. Cavalcanti, B. Shirinzadeh, M. Zhang, L. Kretly, Sensors, 8 (2008) 2932-2958.
[4] K.E. Sapsford, M.M. Ngundi, M.H. Moore, M.E. Lassman, L.C. Shriver-Lake, C.R. Taitt, F.S. Ligler, Sensor Actuat B-Chem, 113 (2006) 599-607.
[5] M. Seidel, R. Niessner, Analytical and Bioanalytical Chemistry, 391 (2008) 1521-1544.
[6] F.S. Ligler, K.E. Sapsford, J.P. Golden, L.C. Shriver-Lake, C.R. Taitt, M.A. Dyer, S. Barone, C.J. Myatt, Anal Sci, 23 (2007) 5-10.

[7] J. Zhai, H. Cui, R. Yang, Biotechnology Advances, 15 (1997) 43-58.
[8] A. Sassolas, B.D. Leca-Bouvier, L.J. Blum, Chemical Reviews, 108 (2007) 109-139.
[9] Y.-C. Cao, Z.-L. Huang, T.-C. Liu, H.-Q. Wang, X.-X. Zhu, Z. Wang, Y.-D. Zhao, M.-X. Liu, Q.-M. Luo, Anal Biochem, 351 (2006) 193-200.
[10] J. Wang, G. Liu, A. Merkoçi, J Am Chem Soc, 125 (2003) 3214-3215.
[11] H. Li, Z. Sun, W. Zhong, N. Hao, D. Xu, H.-Y. Chen, Analytical Chemistry, 82 (2010) 5477-5483.
[12] D.D. Zhang, Y.G. Peng, H.L. Qi, Q. Gao, C.X. Zhang, Biosens Bioelectron, 25 (2010) 1088-1094.
[13] E. Komarova, K. Reber, M. Aldissi, A. Bogomolova, Biosensors and Bioelectronics, 25 (2010) 1389-1394.
[14] H. Xia, F. Wang, Q. Huang, J.F. Huang, M. Chen, J. Wang, C.Y. Yao, Q.H. Chen, G.R. Cai, W.L. Fu, Sensors, 8 (2008) 6453-6470.
[15] R.A. Doong, H.M. Shih, S.H. Lee, Sensor Actuat B-Chem, 111 (2005) 323-330.
[16] N.J. Ronkainen, H.B. Halsall, W.R. Heineman, Chem Soc Rev, 39 (2010) 1747-1763.
[17] K.R. Rogers, Analytica Chimica Acta, 568 (2006) 222-231.
[18] W. Lian, D. Wu, D.V. Lim, S. Jin, Anal Biochem, 401 (2010) 271-279.
[19] K. Kloth, R. Niessner, M. Seidel, Biosensors and Bioelectronics, 24 (2009) 2106-2112.
[20] K. Kloth, M. Rye-Johnsen, A. Didier, R. Dietrich, E. Martlbauer, R. Niessner, M. Seidel, Analyst, 134 (2009) 1433-1439.
[21] M. Varshney, Y.B. Li, Biosens Bioelectron, 22 (2007) 2408-2414.
[22] K.E. Sapsford, C.R. Taitt, S. Fertig, M.H. Moore, M.E. Lassman, C.A. Maragos, L.C. Shriver-Lake, Biosens Bioelectron, 21 (2006) 2298-2305.
[23] M.M. Ngundi, S.A. Qadri, E.V. Wallace, M.H. Moore, M.E. Lassman, L.C. Shriver-Lake, F.S. Ligler, C.R. Taitt, Environmental Science & Technology, 40 (2006) 2352-2356.
[24] P. Fernández-Calvo, C. Näke, L.A. Rivas, M. García-Villadangos, J. Gómez-Elvira, V. Parro, Planet Space Sci, 54 (2006) 1612-1621.
[25] L.C. Shriver-Lake, F.S. Ligler, Ieee Sens J, 5 (2005) 751-756.
[26] M.M. Ngundi, L.C. Shriver-Lake, M.H. Moore, M.E. Lassman, F.S. Ligler, C.R. Taitt, Analytical Chemistry, 77 (2005) 148-154.
[27] J. Golden, L. Shriver-Lake, K. Sapsford, F. Ligler, Methods, 37 (2005) 65-72.
[28] I.M. Ciumasu, P.M. Krämer, C.M. Weber, G. Kolb, D. Tiemann, S. Windisch, I. Frese, A.A. Kettrup, Biosensors and Bioelectronics, 21 (2005) 354-364.
[29] C.R. Taitt, Y.S. Shubin, R. Angel, F.S. Ligler, Appl. Environ. Microbiol., 70 (2004) 152-158.
[30] K.E. Sapsford, A. Rasooly, C.R. Taitt, F.S. Ligler, Analytical Chemistry, 76 (2004) 433-440.
[31] F. Ligler, C. Taitt, L. Shriver-Lake, K. Sapsford, Y. Shubin, J. Golden, Analytical and Bioanalytical Chemistry, 377 (2003) 469-477.
[32] K. Sefah, J.A. Phillips, X. Xiong, L. Meng, D. Van Simaeys, H. Chen, J. Martin, W. Tan, Analyst, 134 (2009) 1765-1775.
[33] S. Song, L. Wang, J. Li, C. Fan, J. Zhao, TrAC Trends in Analytical Chemistry, 27 (2008) 108-117.
[34] B. Strehlitz, N. Nikolaus, R. Stoltenburg, Sensors, 8 (2008) 4296-4307.
[35] D. Xu, D. Xu, X. Yu, Z. Liu, W. He, Z. Ma, Analytical Chemistry, 77 (2005) 5107-5113.

[36] M.D. Schlensog, T.M.A. Gronewold, M. Tewes, M. Famulok, E. Quandt, Sensors and Actuators B: Chemical, 101 (2004) 308-315.
[37] S. Tombelli, M. Minunni, M. Mascini, Biomol Eng, 24 (2007) 191-200.
[38] J.G. Bruno, Biochemical and Biophysical Research Communications, 234 (1997) 117-120.
[39] U. Brockstedt, A. Uzarowska, A. Montpetit, W. Pfau, D. Labuda, Biochemical and Biophysical Research Communications, 313 (2004) 1004-1008.
[40] R. Kirby, E.J. Cho, B. Gehrke, T. Bayer, Y.S. Park, D.P. Neikirk, J.T. McDevitt, A.D. Ellington, Analytical Chemistry, 76 (2004) 4066-4075.
[41] J.J. Pancrazio, J.P. Whelan, D.A. Borkholder, W. Ma, D.A. Stenger, Ann Biomed Eng, 27 (1999) 697-711.
[42] S. Rodriguez-Mozaz, M. Lopez de Alda, D. Barceló, Analytical and Bioanalytical Chemistry, 386 (2006) 1025-1041.
[43] A.L. Kukla, N.I. Kanjuk, N.F. Starodub, Y.M. Shirshov, Sensors and Actuators B: Chemical, 57 (1999) 213-218.
[44] A. Crew, D. Lonsdale, N. Byrd, R. Pittson, J.P. Hart, Biosens Bioelectron, 26 (2011) 2847-2851.
[45] M. Cortina, M. del Valle, J.-L. Marty, Electroanal, 20 (2008) 54-60.
[46] M. Ramanathan, A.L. Simonian, Biosens Bioelectron, 22 (2007) 3001-3007.
[47] H.-c. Tsai, R.-a. Doong, Biosensors and Bioelectronics, 20 (2005) 1796-1804.
[48] R. Solná, S. Sapelnikova, P. Skládal, M. Winther-Nielsen, C. Carlsson, J. Emnéus, T. Ruzgas, Talanta, 65 (2005) 349-357.
[49] R. Solná, E. Dock, A. Christenson, M. Winther-Nielsen, C. Carlsson, J. Emnéus, T. Ruzgas, P. Skládal, Analytica Chimica Acta, 528 (2005) 9-19.
[50] S. Sapelnikova, E. Dock, R. Solná, P. Skládal, T. Ruzgas, J. Emnéus, Analytical & Bioanalytical Chemistry, 376 (2003) 1098-1103.
[51] J.J. Pancrazio, S.A. Gray, Y.S. Shubin, N. Kulagina, D.S. Cuttino, K.M. Shaffer, K. Eisemann, A. Curran, B. Zim, G.W. Gross, T.J. O'Shaughnessy, Biosensors and Bioelectronics, 18 (2003) 1339-1347.
[52] S.J. Young, J.P. Hart, A.A. Dowman, D.C. Cowell, Biosensors and Bioelectronics, 16 (2001) 887-894.
[53] S. Sapelnikova, E. Dock, T. Ruzgas, J. Emnéus, Talanta, 61 (2003) 473-483.
[54] L. Su, W. Jia, C. Hou, Y. Lei, Biosensors and Bioelectronics, 26 (2011) 1788-1799.
[55] Y. Lei, W. Chen, A. Mulchandani, Anal Chim Acta, 568 (2006) 200-210.
[56] F. Lagarde, N. Jaffrezic-Renault, Anal Bioanal Chem, 400 (2011) 947-964.
[57] P. Ertl, M. Wagner, E. Corton, S.R. Mikkelsen, Biosens Bioelectron, 18 (2003) 907-916.
[58] S. Melamed, L. Ceriotti, W. Weigel, F. Rossi, P. Colpo, S. Belkin, Lab Chip, 11 (2011) 139-146.
[59] M. Varshney, Y.B. Li, Talanta, 74 (2008) 518-525.
[60] B. Podola, M. Melkonian, J Appl Phycol, 17 (2005) 261-271.
[61] J.H. Lee, R.J. Mitchell, B.C. Kim, D.C. Cullen, M.B. Gu, Biosens Bioelectron, 21 (2005) 500-507.
[62] Y. Kuang, I. Biran, D.R. Walt, Analytical Chemistry, 76 (2004) 2902-2909.
[63] I. Biran, D.M. Rissin, E.Z. Ron, D.R. Walt, Anal Biochem, 315 (2003) 106-113.
[64] M.T. Giardi, L. Guzzella, P. Euzet, R. Rouillon, D. Esposito, Environmental Science & Technology, 39 (2005) 5378-5384.
[65] W.Y. Liao, C.H. Weng, G.B. Lee, T.C. Chou, Lab Chip, 6 (2006) 1362-1368.

[66] P. Ertl, S.R. Mikkelsen, Analytical Chemistry, 73 (2001) 4241-4248.
[67] N. Sanvicens, N. Pascual, M. Fernández-Argüelles, J. Adrián, J. Costa-Fernández, F. Sánchez-Baeza, A. Sanz-Medel, M.P. Marco, Analytical and Bioanalytical Chemistry, 399 (2011) 2755-2762.
[68] C.A. Rowe-Taitt, J.W. Hazzard, K.E. Hoffman, J.J. Cras, J.P. Golden, F.S. Ligler, Biosensors and Bioelectronics, 15 (2000) 579-589.
[69] K.E. Sapsford, C.R. Taitt, S. Fertig, M.H. Moore, M.E. Lassman, C.M. Maragos, L.C. Shriver-Lake, Biosensors and Bioelectronics, 21 (2006) 2298-2305.
[70] M.M. Ngundi, L.C. Shriver-Lake, M.H. Moore, M.E. Lassman, F.S. Ligler, C.R. Taitt, Analytical Chemistry, 77 (2004) 148-154.
[71] N.A. Rakow, K.S. Suslick, Nature, 406 (2006) 710-713.
[72] N.V. Kulagina, M.E. Lassman, F.S. Ligler, C.R. Taitt, Analytical Chemistry, 77 (2005) 6504-6508.
[73] C.A. Rowe, S.B. Scruggs, M.J. Feldstein, J.P. Golden, F.S. Ligler, Analytical Chemistry, 71 (1998) 433-439.
[74] C.A. Rowe, L.M. Tender, M.J. Feldstein, J.P. Golden, S.B. Scruggs, B.D. MacCraith, J.J. Cras, F.S. Ligler, Analytical Chemistry, 71 (1999) 3846-3852.
[75] K.E. Sapsford, A. Rasooly, C.R. Taitt, F.S. Ligler, Analytical Chemistry, 76 (2003) 433-440.

22

Environmental Monitoring Supported by the Regional Network Infrastructures

Elisa Benetti, Chiara Taddia and Gianluca Mazzini
Lepida SpA, Viale A. Moro 64, 40127 Bologna
Italy

1. Introduction

The aim of this chapter is the presentation of studies and research results concerning environmental monitoring techniques promoted by Lepida SpA across a wide area, the Italian Emilia-Romagna Region.

Lepida SpA *Lepida SpA* (2011) is an in house providing company established by a Regional Law (11/2004, "Regional Development of the Information Society") of Emilia-Romagna region, which consolidates a common vision and a collaborative approach with the local Public Administrations.

Lepida SpA was created in the end of 2007 by the Emilia-Romagna Regional Government, as unique shareholder and founder. Currently has 395 Public Administrations and Public Entities as shareholders. Lepida SpA is involved in the governance of the Regional ICT Plan which defines the regional ICT strategies and policies within the regional territory, acting as innovation facilitator among its partners.

The core business of Lepida SpA is represented by the regional ICT infrastructure but its operations range between telecommunication networks, digital divide and broadband networks strategies and ICT applications and services. Among the main activities and experiences pursued by Lepida SpA we can mention: the planning, development, management and monitoring of the telecommunications networks (fixed and mobile) of the P.A., including the deployment of new broadband networks (wired and wireless) within the region; the definition and implementation of suitable solutions for the Digital Divide topics and for the Next Generation Access Networks in order to ensure high speed internet for the citizens and businesses; the realization of ICT platforms and services for the Public Adminitrations (federation of authentication, payments, ..) that enable a large number of on-line services in favor of citizens and Enterprises; the realization of on-line services for e-Governement purposes and interaction between the P.A. and the Enterprises and citizens.

The infrastructure provided by Lepida and owned by the Public Administrations partners of Lepida spA, is an heterogeneous interconnected network covering the whole regional territory (more than twentytwo thousand square kilometers of area). It includes a regional area network (Optical Fiber) called *Lepida*, wireless networks (Hyperlan) that are extensions of *Lepida* which allow to solve Digital Divide in some mountain territories, and a regional emergency digital radio network (TETRA) called *ERretre*. A map of the Optical Fiber and Hiperlan link is illustrated in Figure 1.

The availability of this powerful infrastructure offers many opportunities for the P.A. to deploy and provide useful and interesting services to the citizens. Furthermore it represents a unique great regional test bed for the development and testing of new applications and services exploiting the potential of the ICT infrastructures.

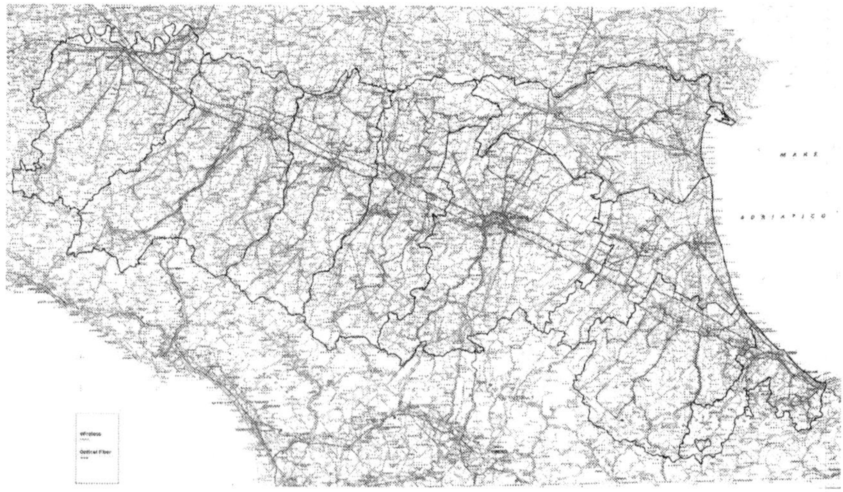

Fig. 1. Optical Fiber and Hiperlan link

In particular, this chapter will present efficient sensor network applications promoted by Lepida SpA and based on the regional hybrid access network, with the aim to realize environmental monitoring through an efficient usage of the territorial assets, by reaching therefore the important goal of public resources savings. The effort of Lepida SpA has been directed towards two primary directions: the first one is the exploitation of the Lepida SpA networks as a communication infrastructure that enables the messages exchanged by the softwares of data management that the Public Administrations already owns and uses for their environmental monitoring activities; the second one, besides the exploitation of the Lepida SpA networks like described in the first model, also proposes the usage by the Public Administrations partners of a proper software and/or hardware platform of data management, planned, tested and promoted by Lepida SpA.
In order to achieve this aim Lepida SpA has adopted a research method based on the following steps: 1) census of the sensor networks and communication networks used for environmental monitoring purpose, existent and operating across the whole regional territory 2) proposal of architectural, infrastructural and application service solutions 3) realization of experimental test-beds 4) adaptation and tuning of the solutions proposed during the second step in view of the results obtained during the third step 5) realization of a full service.
The census activity has been performed all over the Emilia-Romagna territory, by taking into consideration all the Public Organizations. This investigation has highlighted the presence of a huge amount of small sensor networks deployed all over the regional territory, consisting of spatially distributed devices for the monitoring of environmental conditions, such as temperature, sounds, pollutant, traffic, river and basin and also a lot of cameras for the video surveillance and video environmental monitoring. Typically they have been realized in the

past as independent and autonomous systems, each one by using its own communication network to transport the collected data, each one by using its own sink to elaborate the data and each one belonging to a specific local Public Administration.

This scenario often brings the local Public Administrations to inefficient and expensive managements and maintenance of the data transmission, collection and elaboration. In such a scenario, the two working directions followed by Lepida SpA and mentioned above, can represent an effective way for the Public Administrations to pursue environmental monitoring activities while saving as much as possible resources and while following economies of scale. In particular Lepida SpA has defined a centralized architecture Taddia et al. (2009) based on a centre of collection, elaboration, management and diffusion of the sensor data that, by exploiting the hybrid access regional network, beside solving the inefficiencies can also provide further benefits that would be impossible to realize with independent and separate management systems. Let us mention just a few of the possible benefits enabled by the architecture promoted by Lepida SpA: data sharing among different Public Administration by saving the data property thanks to authentication and profilation solutions; correlation of data belonging to different Administrations. Lepida SpA has tested this architecture with some Public Administrations Taddia et al. (2010).

This chapter starts with a description of the adopted research method, by giving a comprehensive description of the first step of this research, the census of the resources available inside the Emilia-Romagna region. The rest of the chapter will describe more in detail how the aforementioned research method has been applied to three scenarios, by presenting three test bed actived by Lepida SpA in collaboration with three Public corporations: River Basin Consortium of the River Po affluents, Drainage Consortium of the western Romagna, River Monitoring for the Civil Protection of the Emilia Romagna Region. The three cases all exploit different network technologies among the ones offered by the the hybrid regional infrastructure, depending on factors such as the geographical position of the monitoring systems and the amount of data exchanged during the monitoring process.

2. Research methods

The method adopted by Lepida has performed, as a first step, an exhaustive census of all the automatic sensor networks deployed in the regional territory, not already integrated with regional sensor networks (sensor networks owned and managed by a regional Entity called ARPA *ARPA* (2011), Regional Agency Prevention and Environment for the Emilia-Romagna region). The Public Administrations in fact, may acquire and use their own networks in order to meet local needs that are within their competence. In order to carry out the census, all municipalities, provinces, the River Basin Consortium of all the provinces and the civil protection have been contacted. For each network, the following items have been surveyed: type of measured data, number of sensors used, number of data loggers used, transmission media and the Administration involved. Offices for environment and mobility, farming, civil protection and provincial police, have been consulted in main cities of each province. Received responses have been inserted in a database containing the following information: the owner Administration, the service manager, the operator, type of monitoring, number of stations installed, number and type of sensor used and the transmission media. Subsequently an analysis of these responses has highlighted different trends and consolidated needs, depending on the responsible Administration and its skills and jurisdiction. Various types of networks, used by different Administrations, that have been found thanks to the census, are shown in Figure 2.

Types of Monitoring Systems	Entity					
Type of Monitoring	Municipality	Province	River Basin Consortium	Mountain Association	Regional Civil Protection	Total
Videosurveillance	14			1		15
Accesses control	8					8
Traffic Monitoring	7	1				8
Landslides Monitoring		1	6			7
River Monitoring	1	4			1	6
Metereological station	1	4				5
Traffic light infraction	4					4
Messages	3					3
Traffic light optimization	3					3
Speed control	1	1				2
Acoustic	1					1
Bycicles count	1					1
Parking control	1					1
water monitoring		1				1
Building stability	1					1
Total	46	12	6	1	1	66

Fig. 2. Types of monitoring systems related to different entities

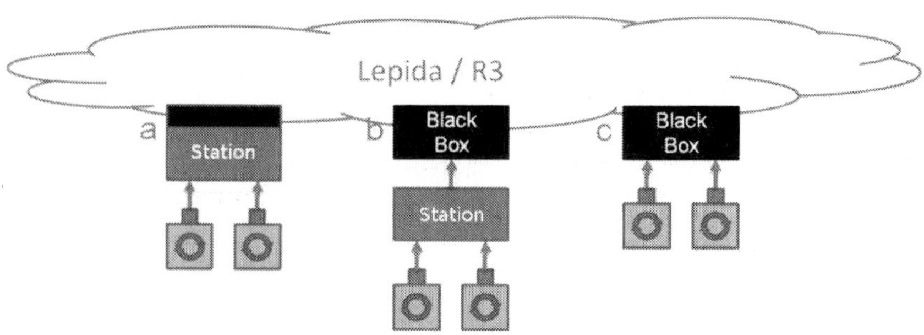

Fig. 3. Models of integration.

Afterwards, for each type of monitoring system, the type and number of sensors used have been mapped, so that their spread could be better understood. As a result was noted that the most common sensors are: the inductive coil (its low cost and its simplicity of use have made it the leader in sensor networks for traffic monitoring); the camera (used by local Public Administrations in response to a need of an improved security for citizens, furthermore the wealth of information intrinsic in its data detected, that is a stream of images, makes this sensor suitable also for other applications such as traffic monitoring or rivers flow control); the inclinometer (its purpose is related to applications for landslides monitoring).

A further analysis about possible efficient architectures that could be proposed to shareholders Administrations, pointed out that is desirable to integrate all existing networks, both for surveillance systems, which are increasingly spreading throughout the territory, and for landslides monitoring, currently managed in a summary way. The presence of a unique wide regional network on the territory, composed by Lepida and *ERretre*, makes this integration possible and it represents also the opportunity to have a uniform and guaranteed transmission of data gathered by all sensor networks. Three different models of integration with Lepida network have been proposed, as shown in Figure 3. Two of them exploit a small hardware and software module programmed by Lepida SpA and called BlackBox, which is mainly devoted to the integration between the communication infrastructure and the sensors.

(a) IP and TETRA driver: a monitoring station, provided by third-parties, on one hand interfaces to sensors and on the other hand to the most suitable telematic infrastructure, chosen between *Lepida* and *ERretre*, through suitable management drivers;

(b) Gateway: a control board interfaces to the monitoring station provided by third-parties through a proprietary protocol or through the standard protocol Modbus. The BlackBox, on the transport network side, provides the most suitable driver depending on the transmission media that will be chosen;

(c) Direct interface: the BlackBox could directly interface to sensors and at the network side performs the gateway functionalities as described in step (b).

The results obtained by the census activities have given the room of defining a suitable architecture able to face the problematic arisen, both in terms of data management system and in terms of communication technologies and infrastructures. Starting from this architectural solutions, some test-beds have been activated nad they will be described in detail on the following Sections.

3. River basin consortium

The subject involved in this testing is the River Basin Consortium of the "Po" River, an agency that deals with the emergency activities related to the water channels and seismic events of "Piacenza", "Parma", "Reggio-Emilia" and "Modena" territories.
The current sensor network that the River Basin Consortium owns and uses presents a lot of problematic aspects: these are particularly correlated to the communication networks currently used, and to the management and storage of data. The data management and storage are fully delegated to private companies that do not offer a system able to ensure the necessary levels of availability and persistence of data. Furthermore, data are distributed on different servers that differ in technology and data representation: there is not a single centralized system that could gather all available information in a standardized format.
Lepida SpA in this case has proposed to the River Basin Consortium of the "Po" River a test-bed activity based both on an interface to the communication infrastructure provided by the *ERretre* network, and on a prototype of a data management center that could satisfy all the needs requested by a full monitoring system.

3.1 BlackBox

The BlackBox prototype has been realized through a control board based on ARM Linux. As shown in the second model of Figure 2 it could be connected transparently to all proprietary tracking stations which export the Modbus interface. This is an open serial communication protocol, master-slave or master-multislave, developed to transmit

information between several PLCs (Programmable Logic Controllers) through a network connection and has become, over the years, a de facto standard communication protocol for the industry. Otherwise, in the third model schematized in Figure 2, the BlackBox provides the management of three different types of sensors: digital sensors, that could also be connected in a multiple modality through a multi-master and multi-slave communication bus; a single generic alarm button; a single serial sensor.
In order to properly handle these three types of sensors, for each one of them a dedicated parallel task has been implemented in the BlackBox: this ensures the management of any kind of warning, even asynchronous, from sensors. Furthermore, the BlackBox interfaces to the network both to transmit data and receive commands, through two different ways: either using the Ethernet connection for communication via IP or the serial connection for communication via Tetra terminal, in this case by SDS. The software is based on a task that periodically requests a measure to all the sensors connected and sends them to the data collection center, also managing the reception of any command configuration parameter, such as changing the sampling rate or actuating connected devices, for example an acoustic or light signal. A software unit receives as input the messages sent by the BlackBox, interpreting and storing them properly. The server where this unit resides, is interfaced both to the IP network and ERretre through a modem connected to a ttyUSB port. In particular, when a message is received the unit, according to the opcode message and to the sender sensor typology, properly extracts the information and stores them in a table or in another textual file available in the system and used by the entity, considering them as a single sensor in a unique instant of sampling. A single message, in fact, could also contain several measures of a unique sensor but related to subsequent sampling instants, or measures sent by different sensors but related to the same sampling instant.
The experimentation with the River Basin Consortium is based on the second model of integration and, due to the isolated location of the test-bed site which does not allow an ethernet connection to the Lepida Network, the communication is done via SDS.

3.2 Landslides monitoring

The test-bed organized by Lepida SpA was installed on the 16th of July, 2010, at the landslide by Fosso Moranda, in the Polinago municipality, province of Modena. It consists of a proprietary survey station (Datalogger) with two biaxial inclinometers at different depths, which perform accurate measures related to millimetric movements of the ground, and a piezometer, which measures the hydrostatic pressure, attached to it. The BlackBox is connected to a Tetra modem for the transmission of data, according to the configuration where the detection station acts as a slave and the BlackBox is both the master and the gateway towards the Tetra transmission network, as shown in Figure 4. The system is powered by a photovoltaic panel and is normally turned off. At a scheduled sampling rate, typically every hour, the monitoring station will "wake up" and control the power supply of the entire system: both Tetra modem and Blackbox. The BlackBox requests to the station data from sensors, then sends the response message to the data management center and commands the proprietary station, that supervises the power control, to shut down the system. The communications between the proprietary station and the BlackBox physically occur through a serial connection and logically exporting at both sides the standard interface Modbus, as previously explained. In addiction to specific parameters the system also includes the monitoring of the backup battery level, which is useful in checking the functioning of the whole automated measurement system. All processed data have a low weight, that is about

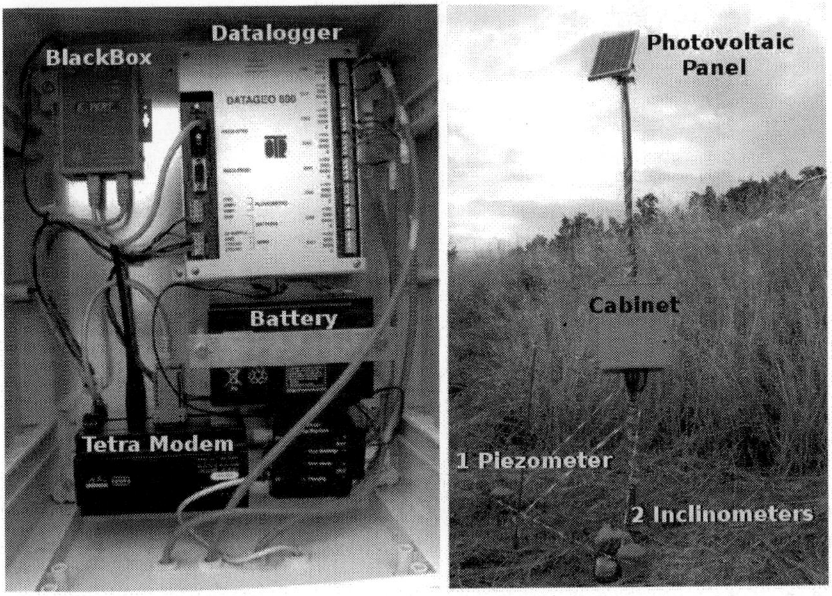

Fig. 4. Cabinet and installation site

20 bytes for each transmission. However the test-bed is highly significant because it is related to a real installation site characterized by particularly hostile conditions, located in an isolated area without any continuous electrical power available. The activation of the whole system has been made possible thanks to a survey about Tetra modems on the market and the identification of which one of them are compatible with the regional network. These could be, unlike ordinary terminals, turned on and off through a simple contact, providing less current absorption and having a lower price.
As a consequence of the good results achieved, the River Basin Consortium and Lepida Spa has arranged a second experimentation phase that should include three new installations connected to multiple sensors and an extension of the BlackBox features, such as remote log retrieval, remote change of the frequency sampling.

3.3 LabICT and Data Management Center

In a previous research phase, a prototype of a unified Data Management Center (DMC) was internally carried out at Lepida SpA R&D Laboratory, in order to receive data, normalize and validate them depending on operation thresholds according to their type and brand. A further analysis of data also allowed a cross-checking of different sensors to trigger alarms for values exceeding from defined thresholds, or for failures. An initial authentication foresaw a base profiling that determined primarily two types of users: basic and operator. for the basic one, thanks to a web interface, a real-time graph with the last samples gathered could be visualized, an historic archive including all measurement done could be consulted and these values could be sent, in a graphic format or through a pdf table, to an e-mail address. Moreover a map showed the location of the stations and the BlackBoxes installed all over the regional territory; for the operator one, in addition to the basic features, this type of user could

insert new units and sensors pertaining to his entity or his partners. Finally he could define new alarm thresholds.
Although this system was quite complete, it had been implemented with the aim to show its potential in environmental monitoring and some features were in an embryonic stage of development. As a result of an increasing interest and a great satisfaction showed by the entities, at the end of the experimental phase, starting from this previous experience the prototype is evolved into a more complex and efficient solution taking advantage of the LabICT-PA (Laboratory for Information and Communication Technology for Public Administration). The LabICT-PA, created in 2007 by the Emilia Romagna region, is part of the Regional High Technology Network and aims to accelerate innovation in public administration. Since 2011 LabICT-PA is also a member of Europeean Network of Living Labs *ENoLL* (2011). The organizational model of LabICT-PA is based on the living labs, where the functional requirements and specifications are defined by and with the users, that is Public Administrations. Design and testing phases will be also carried out through a continuous dialogue with end users. The main partners and their roles in this living lab are: the Emilia Romagna Regional Government that determines the police through the ICT plan; Lepida SpA that, as in house providing company established by a regional law, coordinates activities and provides technical competences and effort; almost 400 public shareholders of Lepida SpA that represent end users; almost 100 business partners, called the club of stakeholders Lepida, that are the think tanks that create added value for PA and for the market; finally universities and research institutes serve as research partners for the laboratory. In this sensor networks context, LabICt-PA has created a fully working prototype, non-engineered, of data management center for sensor networks..

3.3.1 Architecture

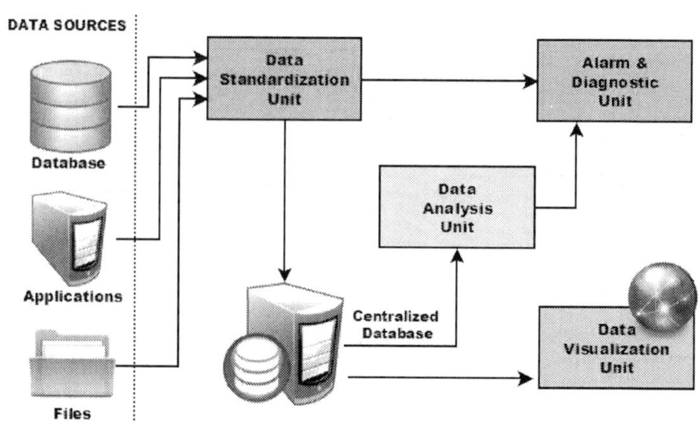

Fig. 5. Data Management Center Architecture

This project aims to integrate all sensor networks deployed in the region through the implementation of a shared platform that could uniformly handle all kinds of environmental data. Firstly, the database of the previous prototype was completely revised to improve the management of the data, intended as a single measure detected by the sensor, making it the most generalized as possible. In fact, the main architectural features are:

- Modularity: each block is independent and communicates through the exchange of XML files;
- Scalability: each module can be implemented on different physical machines;
- Configurability: main operating parameters could be defined in a database, including the definition of new types of sensors, thresholds, alarms, and so on.

All sensors have been schematized in a hierarchical way so that multiple sensors may depend, whether or not, on a BlackBox, which can be connected to another unit, too, for example proprietary stations. Each one of these elements is categorized as a sensor, this is because they are all able to send and receive signals, moreover each sensor can perform different types of measurement with different timing for the acquisition. Finally, measures may be punctual, aggregated or their avarage is calculated, depending on various time intervals. In addition to the tables dedicated to sensors management, the database also includes additional tables necessary to provide addresses, ticketing, alarms, profiling, logging. The Middleware, the Control Center and the Monitoring Center consist of opensource units (Figure 5): each one has its own characteristics, in order to satisfy all the features proposed and also maintain a huge flexibility, in fact each unit inside t he project is independent from the others. The whole managing of data within the Data Management Centre can be divided into three main phases, acquisition, processing, viewing, and this allows to describe each single functional unit. Heterogeneous data sources will be homogenized by the first standardizing unit and then the measures will be evaluated by the analysis unit that will validate them and will check all alarm thresholds. The alarm and diagnostic unit will be contacted by both units and manages and logs the events. Finally, the validated data will be displayed by the visualization unit through a web interface. Communications between two different units are done by using Web Services.

3.3.2 Data standardization unit

This module is the interconnection and standardization middleware between the data and the central unit, therefore plays the role of collector and uniforms data sent from different sources storing them in the database of the DMC. It is based on the following elements:

- Atomic modules for data retrieval: are used to retrieve the data, both automatically at a preset timeslot and on-demand, gathering data from various sources or databases. Inside each atomic unit the access procedure and the detailed commands used to retrieve data from a specific source are specified.
- Atomic units manager: is always active and coordinates the required units. It also serves as a collector for messages sent by the individual atomic modules and redirects them through the units of communication, alarming, diagnostics and data analysis.
- Communication unit: it allows the manager to communicate with other modules inside the platform, on one hand by collecting the total number of messages and errors from the manager, on the other hand receiving as input all requests sent by the DMC and directed to the manager.

Output messages produced by this unit are: the standardized data subsequently stored on a centralized database, the notification messages that new data has been inserted in the database so that the proper unit could start to analyze them, errors and log messages that are transmitted to the diagnostic unit.

3.3.3 Data analysis unit

Its purpose is to control the last data processed by the unit of standardization and to do periodic monitoring on the centralized database in order to trigger the following types of alarms:

- Failures: they occur in two cases, when the unit detects that a certain sensor does not send values in a timeslot that is longer than the sampling rate specified for that sensor, or when the measure is not performed correctly in respect to the working range of the sensor.
- Alerts: several alert situations can be assigned to a unique measure and they may depend on the overcoming of a minimum or maximum threshold, or on an excessive increase or decrease of the measure compared to the previous value stored. The amount of subsequent occurrences of the same state of alert, that must be verified before triggering the proper signaling to the unit of alarms and diagnostics, could be also specified.
- Simplex: this event is triggered as a result of the simultaneous testing of multiple alarm conditions. In a unique simplex both alerts and failures could be associated, linked together by logical operators (and, or, not) so that an event could be characterized by critical conditions based on multiple sensors in very complex relationships.

When one of these alarms occurs, it is communicated to the alarm and diagnostic unit specifying which sensor has triggered the alarm event, the type of event and which alert message has been associated to the event, so that all information needed are forwarded to the dedicated unit, due to simplify and speed up its alarm procedures.

3.3.4 Alarm and diagnostic unit

In addition to alarms generated by the analysis unit, all units part of the system architecture could send error messages in case there is a generic malfunctioning in the DMC such as database connection errors, query failed, units that are not working and so on. The diagnostic unit is implemented using a web service SOAP and handles all the incoming XML requests storing and logging them properly. If they are associated with one or more alarm procedures the unit sends the warning message to one or more users by an email, an SMS or an SDS on a Tetra terminal. Finally, the unit manages generic events that could be scheduled at certain timeslots and which may be linked to the linear chart of a sensor so that when a value exceeds from its alarm, an e-mail should be sent not only including a warning message but also with the graph related to the sensor involved as attachment, due to have a visual feedback of the current situation.

3.3.5 Data visualization unit

This unit is based on a web site consisting of several forms that allow the user to query and monitor the various data structures included into the DMC. All the forms have been integrated into a single portal and are made up of different tabs, available on the main screen of the site. A tree view in the left side of the web site represents all the system control stations and sensors connected to them, then each sensor will match one or more type of measures. This tree is generated according to the initial login: in fact an association is possible between a profile and a user, that specifies which sensors he could visualize. The icons of the tree have different colours to provide visual indications about the status of each sensor: green if the sensor works correctly, red in case of alert, yellow in case of failure and gray if is disabled. The tree view allows the selection of multiple components. A geo-referenced map of the region is also provided in the homepage and the markers shown on it indicate the stations installed

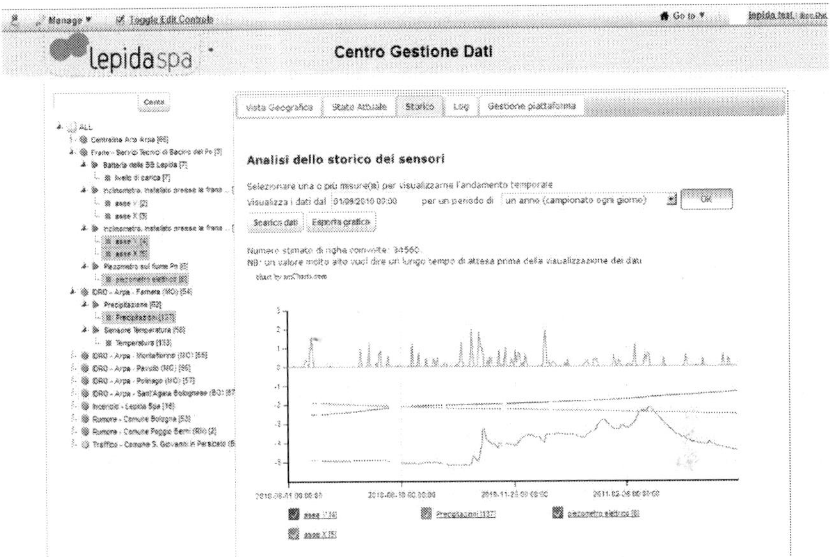

Fig. 6. Data Management Center homepage

using colors in agreement with those defined for the tree view icons. Clicking on a marker a description of the unit and a description of the sensors connected to it are shown. The additional tabs are:

- Real-time monitoring: it provides a graphical and tabular representation of the last data sent by the sensors. The measures to be displayed can be selected through the tree view. The chart adapts its time scale according to a selection done in a drop down menu and then automatically updates itself every 5 seconds. In Figure 6, for example, a multiple real-time chart related to one inclinometer, the piezometer and one ARPA pluviometer is shown.
- Analysis of historical data: in this tab, data could be analyzed with an historical depth that is greater than the one on the real time tab, selecting a start date and a period to display. It 'can be downloaded locally both in a graphic and a tabular format.
- Logs viewing: provides a list in chronological order of all the significant events detected in relation to sensors failures (started or stopped), alerts (started or stopped), invalid values, and so on.
- Platform management: supplies some statistics about the current state of the system, for example the status of the various units involved and an overview of all detected events.

4. Drainage consortium of western Romagna

A Drainage Consortium is a public corporation that coordinates both public actions and private activites concerning the drainage of its territory of scope. For example, hydraulic

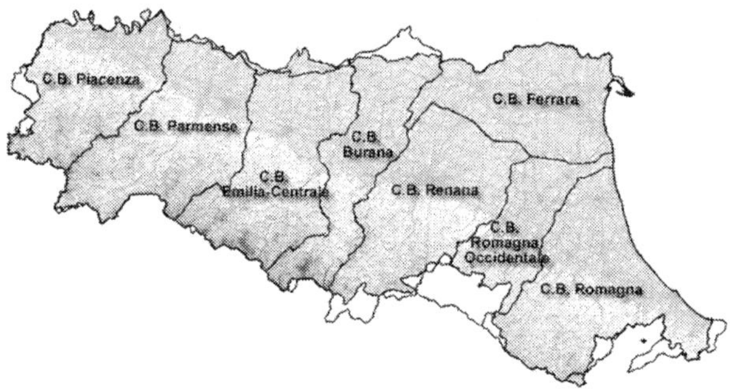

Fig. 7. Drainage Consortiums in Emilia-Romagna region

security, management of the waters intended to the irrigation, involvement into urban planning, environmental and agricultural heritage protection can be considered typical activites and actions covered by a Drainage Consortium.

In Emilia-Romagna region eight Drainage Consortiums exist, subdivided depending on their area of scope, as illustrated in Figure 7. All of them are partners of Lepida SpA, therefore Lepida SpA is legitimized to be involved for the support of their activities, by favouring economies of scale.

Currently each Consortium manages a suitable small sensor network, consisting of a set of data logger, devoted to hydrographical detection and remote control functions, thanks to the use of Programmable Logic Controllers (PLCs) and sensors connected to the data loggers. Furthermore each Consortium has got a suitable monitoring system (typically a server hosting a software system of data management) devoted to the collection of all the gathered data. Data are exchanged between data logger and server and among the data loggers (often there is the need to spread some specific control command from a data logger to other data loggers, by following as a sort of tree communication path) by using analog or GSM technologies (generally GSM is used to send alarm messages to people that need to be activated in case of danger or alarm situations while analog communication channels are used for the data collected by the sensors). Economies of scale could be found in such a scenario, by exploiting the network infrastructures owned by Lepida SpA.

For this purpose Lepida SpA will support the Consortiums, by starting from the Drainage Consortium of the Western Romagna *Lugo* (2011), which has been involved in a test-bed stage. The condition of the equipement managed by the Drainage Consortium of the Western Romagna, before the mentioned test bed stage, can be summerized as follows. It is composed by fifteen data loggers, each one including a PLC with some sensors for the hydraulic data collection and an analog communication module. Each module communicates the monitored data trough UHF channel while the alarm signals are sent through GSM network, by means of Short Message Service (SMS). The monitoring activity is mainly performed by following a polling communication protocol: a central server, devoted to the data collection and elaboration, polls each data logger every thirty minutes, by receiving the data that the sensors connected to the data logger have recorded at a one minute frequency during the last thirty minutes; every poll requires about 300 bytes for each data logger. Besides the polling scheduled activity, the central monitoring system has also the possibility of directly poll a specific data logger at any instant, for example in case of emergency; in this case the polled data logger will send the data collected since the last polling.

The effort of Lepida SpA has been addressed to the communication technology used by the described system. In particular by offering the opportunity of exploiting the regional tetra radio infrastructure of the *ERretre* network as a data transportation driver. This represents for the Consortium an opportunity of economic and resource saving, by replacing the old analog radio with the modern digital tetra radio. Furthermore the *ERretre* network can offer other fundamental advantages: it can guarantee a full coverage inside the whole regional territory, allowing also intercommunications among the devices of different Drainage Consortiums in Emilia-Romagna region; it can offer guaranteed communications, avoiding the congestion events that instead may occur in a GSM network. Another role of Lepida will be the support in defining the technical specifics for a future furnishing of digital tetra radio to be used in a long-term solution when the test-bed stage will be definitely ended.

The test bed activated in collaboration with the Drainage Consortium of the Western Romagna exploits the *ERretre* network, with single slot packet data communication policy.

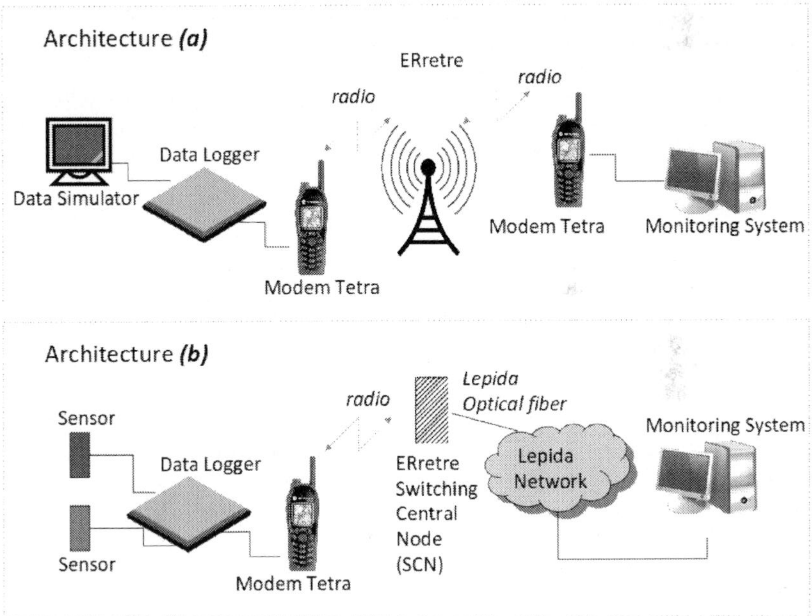

Fig. 8. System architectures tested in collaboration with the Drainage Consortium of the Western Romagna

The test bed has been conducted through two phases. The first one has been realized at Lepida Spa R&D Laboratories, by implementing two different architectures, illustrated in Figures 8, which leads to two different network performances. In architecture *(a)* both the data logger and the monitoring system are connected to two tetra radio modem: in this case any data tramission requires the activation of two radio links (the first one between the data logger and the *ERretre* base station covering the area in which the data logger is located and the second one between the monitoring system and the *ERretre* base station covering the area in which the monitoring system is located). A more efficient usage of the network resources is presented by architecture *(b)*, that exploits the existent cabled connection (optical fiber) between the *ERretre* Switching Central Node and the Lepida network infrastructure. The main advantage of this solution is represented by the reduction of the traffic routed inside the *ERretre* network, which institutional role is mainly represented by emergency communications, and the exploitation of the bandwith offered by the optical fiber link. concerning the data logger element, this has been emulated during the tests realized at the Lepida Laboratories, with a specific data simulator connected to the PLC of a real data logger. Both the two architectures have shown very good results when tested at the laboratories. These are both fundamental: one optimizing the network performances, the other enabling the implementation of sequential commands to more data loggers.

The second experimental phase has been activated with a real data logger, collecting real data from the sensors connected to it. The remote data logger was located at "Mordano" city and the central monitoring system was located at"Lugo" city, each one equipped with a tetra radio. The network architecture implemented is the *(a)* type (by referring to Figure 8 *(a)*, the only difference in terms of the real sensors that were used instead of the data simulator appliance), mainly because of two reasons: the head office of the Drainage Consortium of Western Romagna, located at "Lugo", was not yet connected to the Lepida optical fiber; the Drainage Consortium needs to activate also communications among each single data logger, by following a sort of tree path, in order to allow a data logger to send particular commands directly to other data loggers (for example a data logger being able to sequentially command a more draining installations).

Basically the communication is realized by following a polling communication protocol: the central server polls the data logger every thirty minutes, by receiving the data that the sensors connected to the data logger have recorded at a one minute frequency during the last thirty minutes; every poll requires about 300 bytes. Nevertheless the communication is bidirectional: the PLC is devoted both to the data gathering from sensor like temperature, water levels and water flow and to the reception of commands from the central server, like bulkheads closing or pumps activations; similarly the server receives the data gathered by the sensors and it can send remote commands to the PLC.

Figure 9 shows the place where the draining pump of "Mordano" is located, the control systems with PLC and data loggers and the tetra modem used for the test bed. Figure 10 shows the panel observable at "Lugo" with the monitoring system. The data management software and the data logger have been implemented and supplied by private enterprises that directly collaborate with the Drainage Consortium of Western Romagna.

This experimental phase has shown optimal results in terms of network performance and communications reliability. The test bed will be soon extended to fifteen other PLCs. The long-term installation will see the architecture *(b)* for the connection of the monitoring system and the architecture *(a)* among the single data loggers.

5. Civil protection

A further test bed has been created with the aim of realizing integrated systems for data and video communication on the Emilia-Romagna Regional territory, particularly for hydro geological and hydraulic risk, in cooperation with the Civil Protection of the Emilia-Romagna

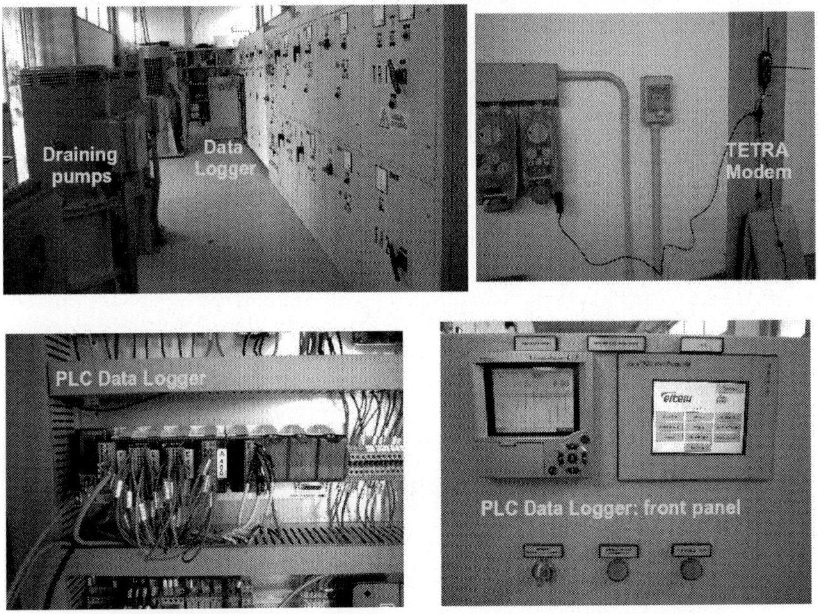

Fig. 9. Draining pump located at "Mordano" city.

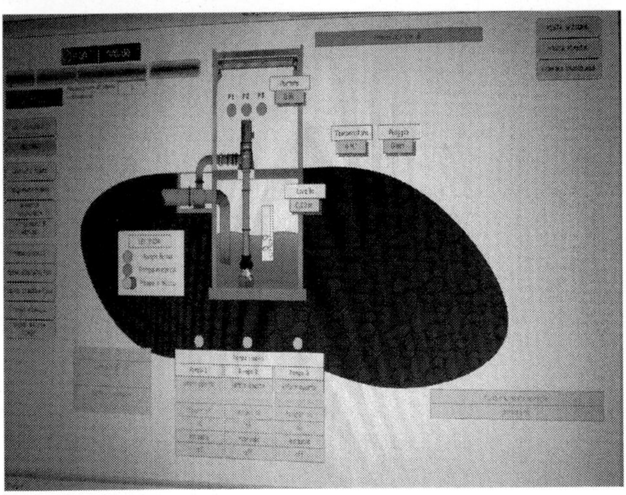

Fig. 10. Monitoring systema located at "Lugo" city.

region. Generally speaking, Civil Protection works around the whole national territory and it is divided in regional agencies, which are subdivided in smaller agencies, devoted to the

execution of pregressively more local aspects. Its main working acivities are forecasting and preventions, rescue activities, emergency trainings.
The project, promoted by the Civil Protection of the Emilia-Romagna region, aims to the realization, with the technologies at disposal nowadays, of a set of integrated systems devoted essentially to the monitoring and to the communication of data/video over the Emilia-Romagna Regional territory, to fulfill the requirements of the Civil Protection: prevention and management of hydrogeological-hydraulic risk, emergencies, disasters, trainings. More precisely, the project includes the following items:

- system for the visualization of large area and for the visual integration of images and data, both newly generated than preeexistant, trough a rear projection videowall located at the Operating Center of the Civil Protection of the Emilia-Romagna Region, in Bologna City;
- capturing systems of territorial images for the permanent monitoring of critical area such as bridges or other infrastructure located near rivers or torrents or dry detention basins;
- capturing systems of territorial images for emergency or temporary situations, devoted to the monitoring of rivers or landslides or disasters area or even training activities, by exploiting a transportable device and a motor vehicle.

The project has started with a first stage implementation that has been developed and employed as an experimental test bed, usefull to understand the limits and strength of the architecture of the whole system. More in detail, this first stage has realized the videowall and a sort of "mini network" of video monitoring, which specfications are explained as follows:

- One video camera and related data/video transmission systems (in Hiperlan technologies at 5.4GHz) devoted to the permanent video monitoring of the river "Savio" near "Cesena" city and located at the railways bridge over this river 11.
- One kit of video monitoring and recording and related data/video multistandard transmission systems (Hiperlan, TETRA, UMTS, Ethernet, etc...). This kit has been installed over a transoprtable device (a trailer truck) devoted to temporary video monitoring activities in area such as rivers, torrents, landslides or other situations of emergency or temporary training exercises.
- One kit of video monitoring and data/video transmission installaed over a motor vehicle (in this case a vehicle owned by the Group of Amateur Radio Operators Volunteers of the city of "Imola" was used) devoted to monitoring activities during training exercises or disasters or emergency situations.
- One hardware and software centralization system (server) for the centralized recording and management of the data/video collected by the remote video systems listed in the previous items; the mentioned server was located at the Data Elaboration Center of the Emilia-Romagna Region, in "Bologna" city;
- Four workstations related to the four systems listed above: two located at the Operating Center of the Civil Protection of the Emilia-Romagna Region, together with the videowall; one located at the of Operating Center of the River Basin Consortium of the city of "Forlí"; the last one located at the Operating Center of the River Basin Consortium of the city of "Cesena".

Lepida SpA has been involved in this first stage of the project for the definition of network architectural specifics and for the configuration of the network active devices involved (switches, routers...), besides offering to the Civil Protection the usage of its network

Fig. 11. Design of camera installation at the railways bridge over the "Savio" river, near "Cesena" city.

infrastructures. A full description of the architecture realized for this first stage test-bed is illustrated in Figure 12. More precisely, Lepida SpA fiber network has been used as a backbone between the camera and the centralized server located at Operating Center of the Civil Protection of the Emilia-Romagna Region, while the connection between each single camera and the closest point of presence of Lepida network has been designed and implemented as an ad hoc wireless link, with the proper and better technology (Hiperlan, TETRA...), depending on the specific use case (temporary or permanent monitoring station).

For this reason, this stage has represented for Lepida SpA mainly a test bed in which testing the network performances serving as collector of many data/video distributed all over the regional territory. The cameras used for the test bed stage are professional Megapixel IP cameras supporting H.264 video streaming. As far as the centralized system of data recording and management is concerned, the first stage used a solution based on a commercial product (Genetec Omnicast).

The item monitored by the camera is the "Savio" river. The amount of traffic over the network is about one image of thirty kBytes every fifteen minutes. When the TETRA channel is used, each image is sent by exploiting the single slot packet data type of communication, that offers a 3kbits/sec of bandwith. The amount of traffic exchanged is compliant with the performances offered by the TETRA channel.

The first stage test bed has shown positive and promising results, leading to the definition of guideline for a second stage of this project. This second stage will involve the realization of at least thirty other video camera located at different river area already defined and spread all over the Emilia-Romagna regional territory. The main requirement of the future new installation is the full compatibility with the infrastructures, the hardware and software equipments already used during the first stage of the project.

5.1 Video management center

During the test bed realized in collaboration with the Civil Protection, the role of Lepida SpA has been mainly focused over the definition of the proper architecture of the communication

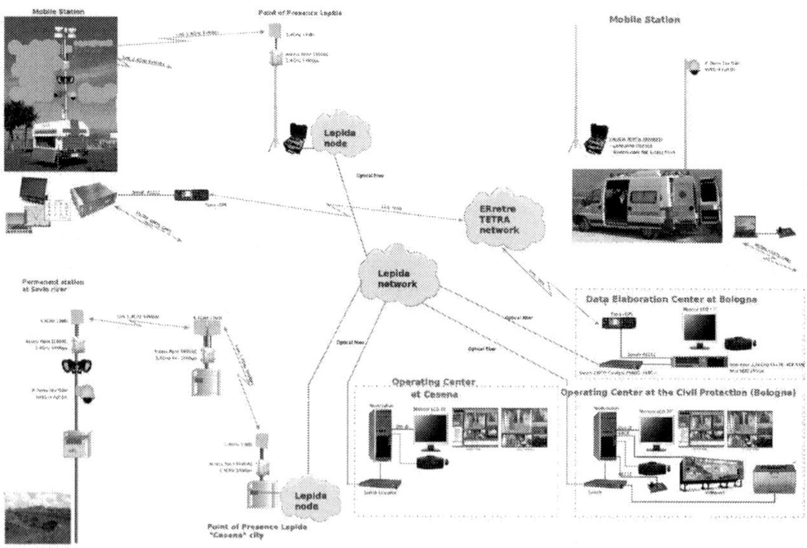

Fig. 12. Architecture of the first stage of the test bed realized in collaboration with the Civil Protection.

network to be used, in terms of the selection of the proper communication driver technology (hiperlan, TETRA, fiber...) and in terms of the proper addressing policy to be adopted among the different Entities (Regional Administrations and Regional Data Elaboration Center, Civil Protections...) involved. During the first stage test bed, differently than during the activity with the River Basin Consortium, no central data management system has been proposed and tested by Lepida SpA, since a commercial one was currently at disposal of the Civil Protection, at least for the purpose of the first stage activity.

As the census results has shown, a lot of Public Administrations in the Emilia-Romagna region are equipped with videosurveillance cameras for the purpose of public security maintenace. As the census activity has highlighted, these video show the same behavior of the environmental sensors installations (temperature, landslides, pollutions...): they often have been installed in former times by each single Administration, as separate and independent systems, resulting in a lot of camera managed by autonomous, independent and very expensive videosurveillance appliances. Furthermore, during the more recent years, expecially the smaller Public Administrations located in the Emilia-Romagna region, have expressed the necessity of providing themeselves with videocamera systems, asking support to Lepida SpA, the natural and institutional reference for the resource efficient development of their activities. More precisely these Administrations needs support concerning the definition of the technical specifics to include in public announcements devoted to the cameras, ad hoc communication infrastructures and video management systems. Such a scenario, like the scenario explored with the River Basin Consortium, points out the necessity of a central management system that could help resource saving policies in the management activities correlated to the video data. Lepida SpA, in order to readily face and solve the necessities that arise from the regional territory has designed a global architecture for an efficient management of all the video data that potentially could be generated by the cameras and videosurveillance systems installed by all the Public Administrations in Emilia-Romagna region.

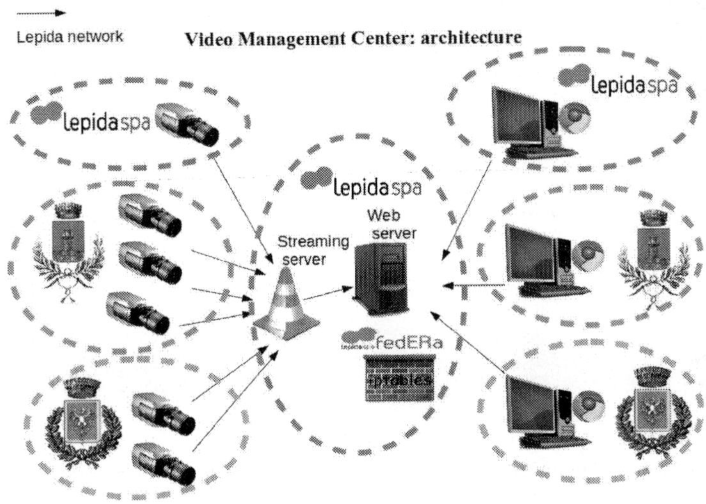

Fig. 13. Architecture for the management of camera strems produced by the Public Administrations.

The architecture proposed by Lepida SpA is illustrated in the Figure 13: Lepida SpA will provide a central server (or a proper cluster of servers) devoted to the storage of all the video streams recorded by the cameras, that will be collected by exploiting the Lepida network infrastructures; this centralized system will host also a full video management service, completely designed and implemented by Lepida SpA with the exploitation of open source technologies, that thanks to the Lepida network infrastructures, will offer to the remote Public Administrations client workstations features like: live streaming view, recorded video of at least one week old (in accordance to the italian law), downloading of recorded video.

Lepida Spa has implemented these features in a Video Management Center prototype. Figures 14 and 15 show some snapshot. More precisely this prototype is composed by a streaming server that collects all the streming video produced by the cameras and streams them to a management server. The management server functions as a web broadcaster of the data streamed by the streaming server, a choice that allows a big scalability in terms of network traffic: each client can watch a live video directly from this streaming server, avoiding the multiplication of network traffic that would be necessary if each client would refere directly to each camera stream. Management and streming server can be hosted in the same hardware server (specifically, the prototype realized follows this policy). The management server is implemented with web services technologies (php, apache, html, javascript, ajax) and it provides a web interface with tabs for the live view, for the recorded video view and for the downloading of the recorded video. In order to be compliant with all the requirements contemplated by the italian law in terms of privacy related to the management of video

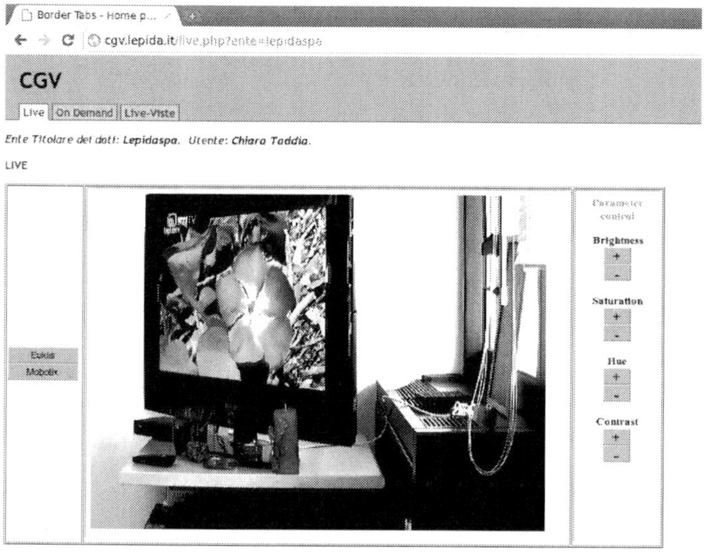

Fig. 14. Video Management Center prototype (CGV): streaming live view feature available on the web interface. The camera shown and listed are used in Lepida SpA R&D laboratories as a test-bed.

data, the prototype also implements other features such as: strong authentication of all the users that can access the web interface; profilation of the users that can access the web interface, in order to limit and specifically select the features each users is enabled to access; logs collection related to all the operations that each user makes on the web interface. The authentication is performed by exploiting the "fedERa" system. "FedERa" *fedERa* (2011) is the regional authentication system designed and promoted by Lepida SpA; it is a federated authentication system that allows the access to all the online services offered by the federated Public Administrations, with the usage of a single username-password that is valid for all the federated services. The profilation system has been implemented ad hoc for this prototype and it allows to differentiate the features that are available to each user: live view, view on demand of the recorded video, download of the recorded video. The profilation system, strictly correlated to the authentication system, allows the creation of logs to trace the activities of video data managements, as required by the italian privacy laws: the log systems collects information about the timestamp, identity of the user and specific operation made on the Video Management Centre (live view, on demand view, download). Actually the Video Management Center prototype has been tested by Lepida SpA internally. In a couple of months it is scheduled the start of a test bed usage of such a system in collaboration with a small Public Administration of the Emilia-Romagna region, for public security purpose. The architecture designed for security videosurveillance purpose and the developed prototype could nevertheless also be used in scenarios such as the one described by the project of the Civil Protection, so for environmental monitoring purposes that are based on video images.

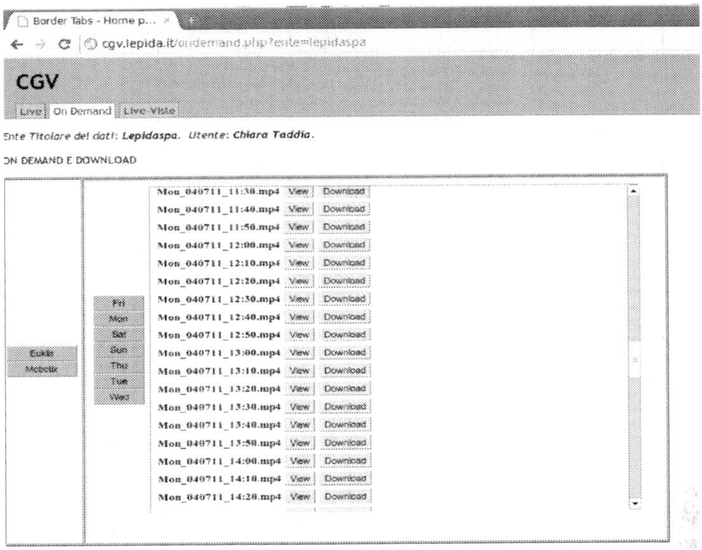

Fig. 15. Video Management Center prototype (CGV): video recorded on demand feature available on the web interface. The camera shown and listed are used in Lepida SpA R&D laboratories as a test-bed.

6. Conclusions

This publication has highlighted some interesting aspects of environmental monitoring mainly related to the Public Administrations context. Let us stress that, expecially in the Public context, an effective usage of the available resources and a development based on economies of scale are fundamental. The research method proposed and adopted by Lepida SpA has represented, as the described test-beds have shown, a valid instrument for the efficient planning, without any resource wasting, both phisical and economic, of all the environmental monitoring activities managed at a regional level. In particular the method proposed has given the opportunity of increasing the knowledge of all the infrastrucutures and system concerning the environmental monitoring field, active and used inside the whole regional territories, resulting in a more aware development of new hardware and software solutions able to face with the practical problems and necessity raised by the Public Administrations of the Emilia-Romagna region. The test-beds activated and described in this chapter will offer in the next months the chance to properly set and define all the systems variables and parameters in order to become useful models for the definition of future full services devoted to the Public Administrations.

7. Acknowledgment

This chapter was prepared in collaboration with Eng. Stefania Nanni, Laboratories ICT Manager at Lepida SpA, Eng. Anna Lisa Minghetti, Lepida network Manager at Lepida SpA, Eng. Federico Marcheselli, ERretre network Manager at Lepida SpA, Eng. Fabio Brunelli, Services Development area at Lepida SpA.

Authors would like to thank the following people and organizations, for their competence and their data: River Basin Consortium of the "Po" River, Civil Protection of Emilia-Romagna Region, Drainage Consortium of the Western Romagna, Marconi Labs, Dab, Misa, Eltel4, IEIIT-CNR Bologna, IConsulting, MEEO.

8. References

ARPA (2011).
URL: *http://www.arpa.emr.it*

ENoLL (2011).
URL: *http://www.openlivinglabs.eu/*

fedERa (2011).
URL: *http://www.lepida.it/lepida-per-attivita/servizi/autenticazione-federata-federa*

Lepida SpA (2011).
URL: *www.lepida.it*

Lugo (2011).
URL: *http://www.bonificalugo.it*

Taddia, C., Nanni, S. & Mazzini, G. (2009). Technology integration for the services offered by the public administrations, *IARIA Neutral, Cannes, France* .

Taddia, C., Salbaroli, E., Benetti, E. & Mazzini, G. (2010). Centralized management of data collection over hybrid networks, *IEEE ACCESS, Valencia, Spain* .

23

ICT for Water Efficiency

Philippe Gourbesville
Nice Sophia Antipolis University / Polytech Nice Sophia, France

1. Introduction

Global change poses unprecedented threats to society through impacts on both the natural environment and engineered infrastructure. Specifically, growing global population requires urban and infrastructure development at the same time as global warming demands massive investment in measures for both adaptation to future climate and mitigation through reduced emissions. The water sector is at the heart of this 21st century challenge, and the need of the hour is to have a major revision of our approaches and implementation of technology for the management of water resources, flood risk and pollution.

As mentioned recently by the Water Supply and Sanitation Technology Platform (2005) - WSSTP - representing all the European water sector actors, "water supply, storm-water drainage, wastewater collection and treatment, as well as quality and quantity management of natural water resources need to be efficiently secured or, where necessary, improved. Only through a paradigm shift from fragmented towards integrated urban water management economic development, social balance and ecological integrity can be secured. [...] During the last three decades the European water industry has built up a great competitive strength based on innovative supply and sanitation concepts, technology, knowledge and skills; availability of financial resources; wide experience in many industrial sectors; close cooperation with European R&D organisations and universities, including active involvement in R&D projects in the various European Union R&D Framework Programmes; expanding markets in the European Union and outside; European Union policy on sustainability, environment and energy; a broad spectrum of efficient governmental structures, tailored to specific local needs. The three largest companies providing water supply and sanitation services in the world are European. In addition, a large number of European Small & Medium Enterprise's (SME's) export their expertise and equipment across the world. Several European firms and institutes have prominent positions in the open market for major water and sanitation studies and implementations. The European water sector is a major economic player - 1% of GDP - with a turnover in the European Union of about 80 billion Euro and an average growth rate of 5% per year, compared to 2.5% per year average growth rate for the European Union economy."

The diagnostic provided by the profession at the European level and with the support of the WSSTP mentions that sustainable approaches for the development of water projects are needed to deliver social, economic and environmental benefits. These demands are pressing issues in the new European Member States, and in developed and developing countries outside Europe. Technologies need to be properly integrated with social, economic and organisational measures. Until now a sectoral approach in water resources management has been dominating and is still prevailing. Many actors are not fully integrated, and many stakeholders remain uninvolved. This has led to fragmented and un-coordinated implementation of policies and technologies, and often leads to inefficient or even unsustainable solutions. To achieve sustainability, Europe, as all countries, has to apply an integrated and participatory approach for water resource management. The water industry is too slow in studying and eventually adopting new technologies. The World Water Council (2009) states: "Without major technological innovations there is little hope of bringing the water equation into balance. There is no doubt that many technological changes can help improve services for millions and reduce the stress on water systems around the world."

To remain in the forefront of this competitive business, innovative skills are essential. The knowledge and experience in water supply and sanitation that is available for example in Europe is dispersed across a large number of small utilities and enterprises. Although not directly visible to the outside world, a considerable body of knowledge has been developed in designing and optimising water infrastructure and management systems over the past 150 years. This diversity of solutions adapted to local conditions in Europe is quite valuable assets in the world market. The energies of all actors in the sector must be combined to merge the dispersed knowledge and expertise and use it to enhance the competitiveness of the water sector.

The challenges faced by the water sector in Europe and worldwide are serious and well-documented. Future water shortages require immediate action on development of resources, reduction of demand and higher efficiency in treatment and transmission. Future flood risk management requires immediate action in risk assessment, defence and alleviation systems, forecasting and warning systems and institutional and governance measures. Such development requires considerable investment in research from governments and large corporations and this is now becoming apparent in many countries. The challenge is made even more difficult, however, by the requirement for solutions to be sustainable and moving towards a "low carbon economy" which are also increasingly being stipulated by government and European Union Directives. For example, the drive for higher reliability in water resource is therefore accompanied by a drive for reductions in cost, emissions, ecological and environmental impacts.

Technology has been revolutionised over recent years and now, matured with mass production allowing wider uptake of methods and devices (Gourbesville, 2009). After the development phase, technology is now entering an application and implementation phase which is targeting several fields including environment. A relevant example is given by the European Union who has defined a major priority for the next 20 years on "ICT for sustainable growth" with the ambition to lead innovation at the worldwide scale. In such context, ICT refers to technologies that provide access to information through telecommunications. It is similar to Information Technology (IT), but focuses primarily on communication technologies. This includes the Internet, wireless networks, cell phones, and other communication mediums. As defined by the European Commission, improving the quality of life should not damage the environment for future generations. Achieving sustainable growth requires better management of all natural resources, from energy to water and ICT - Information and Communication Technologies - can enable this far more efficiently (Holz, 2004), so improving environmental protection without holding back economic development.

Key concerns are the impact of climate change and the inefficient use (or over-use) of natural resources, such as drinking water and energy supplies. However, in order to achieve these objectives, the European Commission focuses its efforts on several specific areas such as Energy Efficient Buildings, Smart Electricity Grids and Smart Metering, Freight, Logistic and Transport, Greener ICT, Water Management. In this last domain, the European Union wishes to recognize the added value of ICT solutions and to support their implementation in the water domain by elaborating, validating and disseminating recommendations, guidelines and specifications on specific technologies and uses. This strategy is duplicated at the international level with the priorities of the National Science Foundation (NSF) in USA and the Green Growth project developed in South Korea.

If the diagnostic is now shared globally, it request coordinated efforts in order to implement the various ICT solution into the water sector. This sector is complex and requires a careful analysis able to underline needs and to identify the added value provided by ICT solutions according to a realistic roadmap for implementation.

2. Methodology for assessing priorities

Obviously, in the coming years the new technologies from the IT sector will affect the full water cycle and the management of the water related services. This process represents a major challenge for the 21st century. However, the impact of these new technologies - from sensors to Decision Support Systems - could be stronger and really significant if priorities are properly defined and implemented within the R&D strategies. The main driver of the strategy has to be to achieve a comprehensive architecture of an Information System (IS) dedicated to water uses and connected to others systems involved in human activities.

By definition, Information systems are implemented within an organization for the purpose of improving the effectiveness and efficiency of that organization (Silver, 1995). Capabilities of the IS and characteristics of the organization, its work systems, its people, and its development and implementation methodologies together determine the extent to which that purpose is achieved. The IS is associated to an architecture which provides a formal definition of the business processes and rules, systems structure, technical framework, and product technologies for a business or organizational information system.

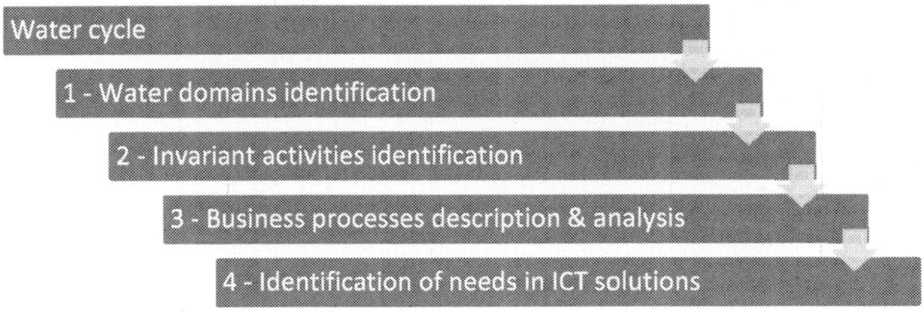

Fig. 1. General methodology for development of ICT solutions in the water sector.

In order to elaborate a specific IS for the management of the water cycle, a methodology is needed for identifying priorities and strategic investments to do in the ICT domain. The

requested approach has to investigate all domains and provide a map of the various process taking places in the different domains of the water uses cycle. This formalization exercise, using mainly concepts and processes, is now requested in order to ensure the coherence of technical choices in a holistic approach.
The methodology has to start from the water cycle, to identify the various water domains and the associated activities. The activities can be then defined with business processes which can be analysed regarding the need of ICT solutions. The proposed methodology is summarized on the Figure 1.

2.1 Domains of the water cycle

The water cycle is frequently defined as the hydrologic cycle which describes the continuous movement of water on, above and below the surface of the Earth. The hydrologic cycle involves the exchange of heat energy, which leads to temperature changes and drives states of water. The water cycle figures significantly in the maintenance of life and ecosystems.

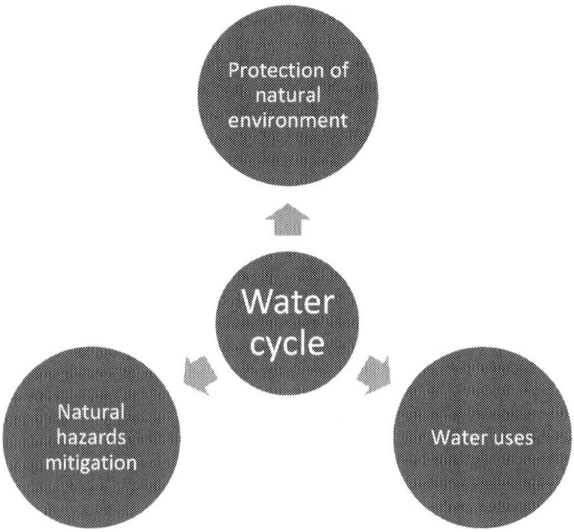

Fig. 2. Domains of water cycle.

In order to preserve this essential resource, the concept of Integrated Water Resources Management (IWRM) has been developed (Jønch-Clausen T. & Global Water Partnership (GWP), 2004). The purpose of the approach is to "promotes the coordinated development and management of water, land and related resources, in order to maximize the resultant economic and social welfare in an equitable manner without compromising the sustainability of vital ecosystems." Operationally, IWRM approaches involve applying knowledge from various disciplines as well as the insights from diverse stakeholders to devise and implement efficient, equitable and sustainable solutions to water and development problems. As such, IWRM is a comprehensive, participatory planning and implementation tool for managing and developing water resources in a way that balances social and economic needs, and that ensures the protection of ecosystems for future generations. In such approach, ICT solutions can play a key role but focus has to be given to the most demanding and relevant domains of the water cycle.

In order to identify which and how ICT solutions can be implemented, it is necessary to look at the water cycle through an approach based on functional domains and business processes. This methodology allows considering each action involved into the resource management and identifying the potential needs of ICT.

The water cycle can be divided in three domains which are associated to specific activities and business processes:

- Protection of natural environment and ecosystems;
- Natural hazards mitigation and disaster prevention;
- Water uses.

The first domain considers all actions needed to assess and advice on the environmental impacts of development proposals and projects related to specific water uses. Results are used by regulatory services. The domain covers also all conservation actions of water related ecosystems.

The second domain is focused on water related natural hazards mitigation actions. Floods, water-borne and vector disease outbreaks, droughts, landslide and avalanche events and famine are the processes covered by this domain. Every year, disasters related to meteorological, hydrological and climate hazards cause significant loss of life, and set back economic and social development by years. The disaster is defined as a serious disruption of the functioning of a community or a society causing widespread human, material, economic and/or environmental losses.

The last domain covers the added influence of human activity on the water cycle. Generally, the water uses refer to use of water by agriculture, industry, energy production and households, including in–stream uses such as fishing, recreation, transportation and waste disposal. All of those uses are directly linked to specific activities and processes which are potential targets for deployment of ICT solutions. In order to stick to the reality oft he water management operated by entities in charge of water services, the traditional classification can be reviewed. The main water uses appear then as: agriculture, aquaculture, industry, recreation, transport/navigation, and urban.

2.2 Water uses, activities and business processes

According to the defined water domains, the water uses represent the largest field where ICT solutions can be developed and implemented. The various uses may be classified and defined as follow.

- Agriculture: Irrigation water use is water artificially applied to farm, orchard, pasture, and horticultural crops, as well as water used to irrigate pastures, for frost and freeze protection, chemical application, crop cooling, harvesting, and for the leaching of salts from the crop root zone. In fact, irrigation is the largest category of water use worldwide.
- Aquaculture: Aquaculture is the farming of aquatic organisms including fish, molluscs, crustaceans and aquatic plants. Farming implies some sort of intervention in the rearing process to enhance production, such as regular stocking, feeding, protection from predators and so forth. It also implies individual or corporate ownership of the stock being cultivated. This activity uses part of the water bodies in order to develop activities.

- Industry: This water use is a valuable resource for such purposes as processing, cleaning, transportation, dilution, and cooling in manufacturing facilities. Major water-using industries include steel, chemical, paper, and petroleum refining. Industries often reuse the same water over and over for more than one purpose.
- Recreation: It often involves some degree of exercise as well as visiting areas that contain bodies of water such as parks, wildlife refuges, wilderness areas, public fishing areas, and water parks. Some of the activities that imply the uses of water for this purpose are: fishing, boating, sailing, canoeing, rafting, and swimming, as well as many other recreational activities that depend on water. Recreational usage is usually non-consumptive; however recreational irrigation such as gardening or irrigation of golf courses belongs to this category of water use. Besides, recreation and tourism represent a growing sector for industry at the worldwide scale.
- Energy: Derived from the force or energy of moving water, which may be harnessed for useful purposes, such as Energy production. There are several forms of water power currently in use or development. Some are purely mechanical but many primarily generate electricity. Broad categories include: conventional hydroelectric (hydroelectric dams), run-of-the-river hydroelectricity, pumped-storage hydro- electricity and tidal power.
- Transport/navigation: It refers to the transport of goods or people using water as a means of transportation. This water use refers only to commercial transport, since recreational transports such as sailing is considered above in Recreation water use.
- Urban: Urban water use is generally determined by population, its geographic location, and the percentage of water used in a community by residences, government, and commercial enterprises. It also includes water that cannot be accounted for because of distribution system losses, fire protection, or unauthorized uses. For the past two decades, urban per capita water use has levelled off, or has been increasing. The implementation of local water conservation programs and current housing development trends, have actually lowered per capita water use. However, gross urban water demands continue to grow because of significant population increases and the establishment of urban centres. Even with the implementation of aggressive water conservation programs, urban water demand is expected to grow in conjunction with increases in population. The urban environment is associated to a high dynamic which implies a growing complexity related to number of inhabitants and management of water resources in order to fulfil the needs of population.

The water uses are associated to business processes and are linked to economical and social values. In most of the cases, five major activities are taking place within each water use and appear as invariants. These key activities are: Investigating /surveying, observing / monitoring, designing, building and decommissioning, operating. Each activity could be defined.

- Investigating/surveying: Consists in the gathering of information of the previous and actual state and/or working of the domain in study. This assembly of information can be done either by a systematic collection of field data (survey) or a collection of information or data from a methodical research of available documents and/or the production of new ones in order to understand or to improve the actual state of the domain.

- Observing/monitoring: From a general point of view, this activity refers to be aware of the state of a system. It describes the processes and activities that need to take place to characterise and monitor the quality and/or state of the domain in study. All monitoring strategies and programmes have reasons and justifications which are often designed to establish the current status of the domain or to establish trends in its parameters. In all cases the results of monitoring will be reviewed and analysed. The design of a monitoring programme must therefore have regard to the final use of the data before monitoring starts.
- Designing (including risk assessment): Refers to the process of devising a system, component, or process to meet desired needs. It is a decision making process (often iterative) in which the basic sciences, risk assessment and engineering sciences are applied to convert resources optimally to meet a stated objective. Among the fundamental elements of the design process are the establishment of objectives and criteria, synthesis, analysis, construction, testing and evaluation. In order to obtain a design that achieves the desired needs for the domain in study, the two previous steps should have been accomplished and taken into account.
- Building & decommissioning: Consists in carrying out the proposed solution (design) for the domain. In order to execute this design, construction and/or decommission activities may be executed. It is essential a minimal environmental impact when accomplishing these activities. The tolerable environmental impact will be obtained from the risk assessment of the designing step.
- Operating: It refers to the action of manoeuvring a system. It may include the combination of all technical and corresponding administrative, managerial, and supervision actions. Operation may also include performing routine actions which keep the system in working order. This latest actions might turn out as response of problems detected during monitoring.

Fig. 3. Invariant activities taking place in the various domains and water uses.

The final step of the approach is dedicated to the identification of the various business processes which are taking place in each activity. A business process is defined as a collection of related, structured activities or tasks that produce a specific service or product (serve a particular goal) for a particular customer or customers. It implies a strong emphasis on how the work is done within an organization, in contrast to a product's focus on what. A process is thus a specific ordering of work activities across time and place, with a beginning, an end, and clearly defined inputs and outputs: a structure for action. Some processes result in a product or service that is received by an organization's external customer. These are called primary processes. Other processes produce products that are invisible to the external customer but essential to the effective management of the business. These ones are called support processes. In keywords, a business process has a goal, has specific inputs and specific outputs, uses resources, has a number of activities that are performed in some order, may affect more than one organizational unit - horizontal organizational impact - and creates value of some kind for the customer. An example of a business process for a water utility can be meter reading. It has to be done in concordance of the billing period. The goal of this process is to give inputs to the billing department, and see the progress of the customer's consumption. Depending on the technology used for the metering (smart or manual metering), different resources (technology, personnel) are used.
The uses in urban environment, carried out by water utilities, can be defined with a limited number of business processes – 29 in total - summarized into the table 1 and which are covering drinking water, waste water and storm water management. The final step of the approach is then to identify for each business process how ICT solutions can be implemented and provide added value. This diagnostic has to be shared by professionals and operators in order to ensure a coherent deployment. This validation process can be made through an associative body gathering representatives from all involved sectors.

1 - Asset management	16 - Water primary network management and water balance
2 - Crisis management	17 - Water secondary network management
3 - Field intervention management	18 - Leak detection
4 - Field works	19 - Meter reading (AMR & MMR)
5 - Use of GIS	20 - AMR & MMR management
6 - Maintenance of GIS	21 - Public service contract management
7 - Management of plant maintenance	22 - Waste water network management
8 - Electro mechanical maintenance	23 - Storm water network management
9 - Laboratory activity and quality control	24 - Waste water treatment plant management
10 - Automation & sensors	25 - Sewer inspection and sewer cleaning
11 - Real time network management	26 - Billing
12 - Planning and design of new assets and plants	27 - Customer care & communication
13 - Water resources management	28 - Innovation & pilots
14 - Environment management	29 - Supports
15 - Drinking water treatment plant management	

Table 1. Business processes for urban uses.

3. The @qua approach

The European Union has defined a key objective for his industrial development on interoperability of systems. This approach is dedicated to various domain including environment and water. In order to support this vision, the European Commission has launched a Thematic Network called @qua under the CIP-ICT PSP Programme. The ICT Policy Support Programme (ICT PSP) under the Competitiveness and Innovation Programme (CIP) aims at stimulating innovation and competitiveness through the wider uptake and best use of ICT by citizens, governments and businesses, particularly Small and Medium-sized Enterprises (SMEs). The approach is based on leveraging innovation in response to growing societal demands.

In his programme frame of ICT Policy Support Programme (ICT PSP) 2011, the General Direction Information Society (DG INFSO) of the European Commission has launched a new theme network dedicated to Innovation Communication Technologies for water management. This domain represents a sector which the European Union wishes to develop during the next 10 years and it's contemplated in different initiatives of the Digital Agenda for Europe 2020 which will allow at the same time improving the user's services quality and developing a sustainable management of resources. These objectives will be achieved with the improvement of already available technologies, adaptation of the existing solutions and the identification of R&D axes to work on the next years.

@qua Innovation Network (http://www.a-qua.eu), founded by 17 partners and managed by Nice Sophia Antipolis University gathers thus ICT and water services leading actors from SME to majors, research entities developing competences in both sectors, local and regional authorities directly responsible for water policy and water management. Partners have developed significant expertise about the interface of ICT and water and at the same time, covering the full spectrum of the water related domain. @qua provides a forum to exchange and to share expertise in deploying innovative ICT solutions for water management, studies feasibility of standardized ICT solutions and interoperability in the field of water management across the EU and develops specifications and guidelines according to a jointly defined "level of sharing" among representatives of professional sectors. Focus of @qua is on gathering and sharing experiences on how to overcome barriers to the introduction of ICT solutions for innovative water management and on how to ensure their wider uptake and best use. Partners have the ambition to develop and to promote the interoperability principle and the use of common standards in the water industry. In a holistic and consistent approach, @qua addresses all the issues of the water management from resources to societal changes, using a wide range of ICT solutions: data acquisition, numerical modelling, real-time monitoring and field operation management.

3.1 The @qua methodology

The @qua thematic network members have developed a general methodology based around few steps which can be summarized as follow:

Step 1. Water business processes and ICT solutions: identification of gaps and expectations of the water domain professionals on ICT solutions;

Step 2. Identification and validation of innovative ICT solutions by the ICT professionals with the objective to bridge the identified gaps during the Step 1;

Step 3. Develop the "level of sharing" of each ICT solution in order to address interoperability, standards, architecture and roadmap for implementation issues;

Step 4. Produce guidelines, standards and specifications on specific ICT solutions needed by the water domain in order to achieve a more efficient water management.

The two main characteristics of the defined approach are:

- the global analysis based on "business processes" and associated added value;
- the definition and the use of concept of "level of sharing" to decide which ICT innovations could be widely disseminated throughout the water profession.

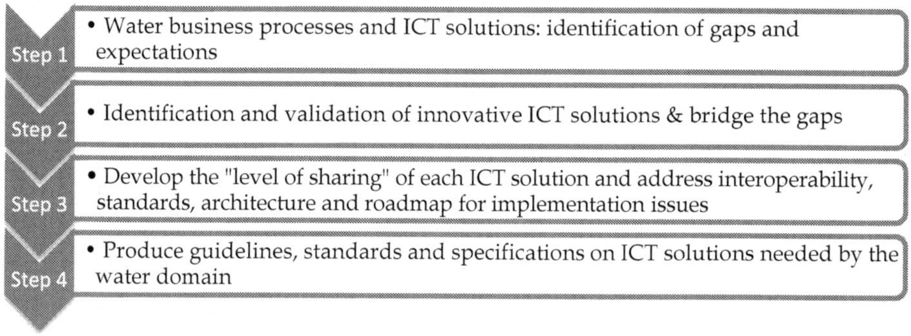

Fig. 4. The @qua methodology.

The initial step, led by water utilities and water engineering companies, is dedicated to the analysis of the business processes, both for the artificial cycle and the natural cycle of water, and both for design and for operations. The business processes are described at a macro scale, where the tiny differences between entities are not seen and where just the common "backbone" is visible. These business models are used as "base maps" in order to show the unequipped - or poorly equipped - steps in terms of ICT. A special attention is turned to the analysis of added value of these unequipped steps. The diagnostic characterizes the added value not only on the economic point of view, but also on sociological and ecological dimensions. In addition to the common map of the water business processes itself, the result of this step is the list of the steps / processes that "deserve" to be equipped with new ICT tools. This effort of analysis according to the business processes vision represents an essential input in the water domain. Until now this diagnostic was not established for several reasons and especially due to the low maturity of water industrial domain regarding ICT solutions and uses.

The second step is led by the ICT sector representatives and consists in a technologic analysis of the needs and requests written by the water companies' representatives. The step includes not only the assessment of the feasibility, the potential availability and the cost of the requests, but it will also propose other tracks, unimagined or not foreseen during the previous step. The water companies have a partial vision of ICT solutions and they need a better knowledge of the current trends of the ICT industry / market. Alternating the leadership of the steps between the "water people" - water companies and other stakeholders - and the "ICT people" brings an efficient synergy.

The third step is focused on the determination of the "level of sharing". This concept is a central element which is developed and used by the @qua network. For the time being, the use and the implementation of existing ICT solutions in the water domain is made case by case, with a quite variable customization which is covering a simple technical adaptation like wavelength, to in depth R&D development like the use of alternative energy sources for power supply in waste water monitoring actions. The partners of the @qua network have significant experience of implementation and development actions. The spectrum of their expertise is covering most the business processes involved in the water domain. From this experience and according to their identified needs in innovative ICT solutions, they define, for each technology identified as a priority, the requested level for developing an efficient interface between the different components involved into the business process. Such work represents a major output for the @qua network and constitutes clearly an added value provision by the network to various professional communities. It is clear that in a wide community as the European water profession, the status of the various Information Systems has a very high variety. This step will analyse the "IS/IT context" parameters in the profession: maturity of the IS, level of integration (integration of the IS itself as well as integration in the business processes), level of alignment with the strategy, and the local parameters (ERP/ software already installed, other relevant IT projects, trends of the local IS/IT market, etc.). This step proposes the ideal "level of sharing", i.e. the level which will maximize the effectiveness and efficiency of the new ICT tools by taking into account the actual current IT/IS situation. This output defines the outcomes of the @qua network, which could go from the very theoretical - methodologies, data models, architectures, principles of standardization, etc. - to the very concrete elements such as list of devices compliant with the selected telecom standards, deployment of a common software and instructions of customization, etc.

In a final step, the production of the guidelines and specifications whose needs are identified in the previous steps. According to the results of the previous step, these results can go from very generic guidelines to more precise technical specifications such as hardware requirement for sensors, software architecture, strategy for implementation and deployment in water services, metadata architecture, business process description and standards. A similar approach has been partly applied with HarmonIT project (http://www.harmonit.org) on the specific field of the hydroinformatic systems interoperability and the development of the OpenMI standards (http://www.openmi.org). In the case of the @qua approach, the spectrum is much more wider because it's addressing most of the business processes involved in all water uses and domains.

3.2 The expected results and impacts

The water domain - and water stakeholders - is very wide and covers a huge number of business processes especially if all domains and activities are considered. This situation legitimates the mapping process and the prioritization of gaps that need to be bridged. Clearly the efforts have to be focused on five major areas directly linked to the urban water use which where both expectations and possibilities are the highest:

a. Real time monitoring
   - Specially real time networks monitoring including Automated Meter Reading(AMR);
   - Installation of leak detectors in the network;
   - Real time quality management (disinfectant, turbidity, pH, temperature, conductivity, RedOx, etc.);
   - Sensors at all Points Of Use (POU);
   - Real time information of customers and stakeholders;
   - Related technologies such as Supervisory Control And Data Acquisition (SCADA), GIS, telecommunications, sensors (especially low cost sensors), inverse models, decision support systems.

b. Cities of Tomorrow
  - In the current vision , there is an absolute need of generalized ICT in the operation of the cities of the future, or sustainable cities, or water-sensitive cities;
  - Cascading usages of water (incl. re-use and recycling), rainwater harvesting, storm water management, desalination, managed aquifer recharge, micro treatment plants, etc. are the core techniques of the cities of the future These techniques need a very high level of monitoring and thus, a sophisticated density of ICT;
  - Leakage reduction in distribution networks;
  - Improving water efficiency in cities.

c. Asset Management and Field Work Management
  - In-pipe and "through road" condition assessment sensing technologies;
  - Continuous performance, condition and risk assessment sensors and prediction models;
  - Optimised network operation and "just in time" repairs and investment programmes;
  - GIS/GPS information;
  - Buried asset electronic identification and tagging devices, wireless communication through road materials;
  - "Wearable computers" for field workers, giving access in real time to all data bases of the company, with interfaces consistent with field conditions.

d. Energy Efficiency
  - Smart grid in water distribution systems (real time management of pumping strategy, refined demand forecast, optimization of network management and of operating costs);
  - Tools for energy saving in treatment plants;
  - Real time status monitoring (open/closed) of manual valves (cf. above : equipment of field operators);
  - Monitoring and control of heat recovery in wastewater;
  - Tools for Smart Metering / Smart Pricing (e.g. condition-based tariffs).

e. Water efficiency
  - Improving water efficiency in cities;
  - Improving water efficiency in agriculture, including detection of illegal abstraction;
  - Ecosystems and land-use management in perspective of project scope and available resources.

4. Some ICT solutions for water efficiency

The analysis of the domains and the business processes demonstrates the relevance and the key position of the data acquisition process through sensors located in the various sectors of the water cycle. This need is recurrent and could be seen in the three domains and takes a central position in surveying, monitoring and operating activities.

4.1 The sensor revolution

The analysis of the domains and the business processes demonstrates the relevance and the Following the PC revolution in the 1980s and the Internet revolution in the 1990s, the on-going revolution is connecting the Internet back to the physical world, creating that world its first electronic nervous system or Information System. The sensor revolution is based on devices that monitor environment - natural & built - in ways that could barely imagine a few years ago.

A sensor is any device that can take a stimulus, such as heat, light, magnetism, or exposure to a particular chemical, and convert it to a signal. Sensors have certainly been around for a very long time with scales (weight sensors), thermometers (temperature sensors) and barometers(pressure sensors). More recently, scientists and engineers have come up with devices to sense light (photocells), sound (microphones), ground vibrations (seismometers), and force (accelerometers), as well as sensors for magnetic and electric fields, radiation, strain, acidity, and many other phenomena.

While the concept of sensors is nothing new, the technology of sensors is undergoing a rapid transformation. Indeed, the forces that have already revolutionized the computer, electronics, and biotech industries are converging on the world of sensors from at least three different directions:

Smaller. Rapid advances in fields such as nanotechnology and (micro electro-mechanical systems (MEMS)) have not only led to ultra-compact versions of traditional sensors, but have inspired the creation of sensors based on entirely new principles. The reduced size fits perfectly with the constraints of the water supply and open possibilities into the monitoring and operating activities.

Smarter. The exponentially increasing power of microelectronics has made it possible to create sensors with built-in "intelligence." In principle, at least, sensors today can store and process data on the spot, selecting only the most relevant and critical items to report. One of the emerging concepts in this domain is the ubiquitous computing paradigm. This approach is highly relevant for the water domain especially for all warning and monitoring systems which may avoid the centralized design.

More Mobile. The rapid proliferation of wireless networking technologies has cut the tether. Today, many sensors send back their data from remote locations, or even while they're in motion.

In the urban water domain, the new sensors are already deeply impacting several business processes with Automated Meter Readers (AMR), water quality control devices and operating supervision. Such trend is following the recent evolution observed in energy distribution sector. An emblematic evolution is the one taking place with the introduction of the smart metering concept for water consumption monitoring.

4.2 From mechanical meters to smart metering

Water meters reading remains one of the core business process of water utilities or public services in charge of drinking water supply. This activity requests a good level of organization and a good management of the devices. To date, water meters have been accumulation meters, pulse meters or interval meters which are all mechanical devices. The data are collected directly regularly on the field. This process can report about consumption and can detect some leakages into the network. However, reactivity is low due to the limited visits on the field. The past decade has seen an evolution of conceptual design of advanced or smart metering and its terminology. Driven by electricity investment, metering has evolved from accumulation meters to interval meters with simple communications, to advanced or smart metering with an increased range of metering functionality. This increase in electricity meter functionality and complexity has started to be mirrored in the water industry.

Interval metering is comparatively more expensive than pulse metering, as the interval meter is required to constantly monitor the water flows through the meter and record this volume at the expiration of the metering interval. By using a fine pulse quantum and analysing the time stamps of these pulses, pulse metering data can be used to approximate interval water metering data and hence deliver similar benefits. Use of pulse metering where a time stamp is made when a certain quantum of water is consumed, is more common in the water industry and these pulse meters are available at reasonable cost.

Smart water metering for the water industry will extend beyond the capability of Automated Meter Reading (AMR). Smart water metering is expected to, as a minimum, establish more granular - within a day - water usage data, two-way communications between the water utility and the water meter, and potentially include communications to the customer. With respect to a customer's household, smart water metering could enable:

- Recording of water consumption within a day;
- Remote meter reading on a scheduled and on-demand basis;
- Notification of abnormal usage to the customer and/or the water utility;
- Control of water consumption devices within a customer's premise;
- Messaging to the customer;
- Customised targeting of segments.

The options to be considered for smart water metering are:

- Choice of communication to the water authority/water utility and the home;
- Choice of consumption data measurement (pulse or interval metering).

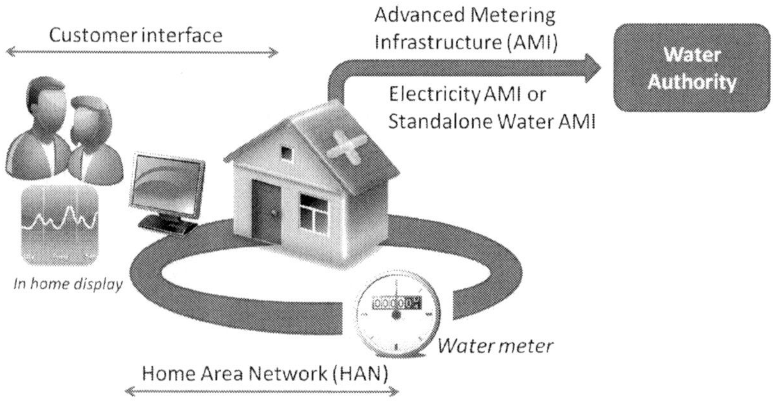

Fig. 5. Smart water metering logical architecture.

Options for the implementation of smart water metering communications arise through choices on:

- Water authority/water utility communications: The method and frequency of data collection through either drive-by collection, leveraging electricity Advanced Metering Infrastructure (AMI) communication networks or standalone water AMI communications networks;
- Customer communications: The method of communicating consumption information to customers: either in real-time across a Home-Area-Network (HAN), or in a historical manner through bills.

Since 2006, various pilot projects - from 100 to 500 smart meters – have been implement worlwide and espacially in Europe within France, Italy, Spain and Malta. The projects are carried out by the water utilities who are supporting development and implementation in various municipalities and for different situations (type of building, type of cities, ...). Most of the projects are based on wireless devices and very few are deployed on the wire networks. Following the first experiments, the main water utilities have already initiated the implementation of smart meters at a large scale with for example more than 350 000 units for France.

The pilot studies and experiments carried out since several years by the water utilities have demonstrated the savings in water consumption due to the use of the smart metering. The savings are taking place at various levels such as:

- Reduction of individual consumption. The details of the consumption are accessible through various media such as a specific website or a small electronic terminal. The information provided to the consumer immediately generates a reduction up to 15%;
- Reduction of water consumption at the macro scale (city to block). The smart metering allows to identified non conform water consumption and consequently help to reduce leakages after and before the smart meter itself. Text messages could be sent to consumers when the consumption is initiating a non coherent pattern with the previous consumption. The water utilities can also detect major leakages on the networks.
- The knowledge in real time of the water consumption allows to identify seasonal needs of the population and to anticipate the volumes of resources to mobilize. This approach allows a more functional use of resources and contributes globally to reduce the consumption.
- The knowledge in real time of the water consumption opens the doors to a new approach about pricing, based on seasonal and even hourly values.

Today, according to various publications and sources (Oracle, 2011), about a third of water utility managers in USA say they are in the early stages of adopting smart meters, despite the fact that 71 percent of water users say that having more detailed information on their water consumption would promote better water conservation. This figure is representative of the worldwide situation. From the water utilities point of view, the following benefits to adopting smart meters could be identified:

- enabling early leak detection ;
- supplying customers with tools to monitor/reduce water use;
- providing more accurate water rates;
- curbing overall water demand;
- improving the ability to conduct preventative maintenance.

The financial efficiency of the smart metering has been already demonstrated through various study cases and pilots (Marshment Hill Consulting, 2010) In developing countries where development of infrastructures and management of water resources represent a great challenge, the opportunity to invest in the smart metering concept is clearly a key issue which request an integrated effort in the global urban management.

5. Conclusion

The water sector represents a major challenge for the 21st century. The climate evolution combined with the growing of pressure of populations will generate new stresses on a limited resource which has to be carefully managed and protected. The fast development of ICT solutions allows today to enter a new area which may be characterized by the idea to move from a scarcity of data to a continuous flow of data - "data rich world" - about natural and built environment. This new situation will become a reality in the coming two decades

and will allow potentially improving, globally, the water management. However, if this perspective represents a clear benefit both for natural and manmade environments, it request the development of a coherent vision based on a process allowing to integrate the fragmented activities developed until now in the water sector. The ICT solutions will allow this integration process but they have to be coordinated under guidelines and standards which have to be jointly defined by the various actors of the water sectors. Regulating bodies, public services, water utilities and IT producers are invited through organisations like @qua, to engage an active dialog in order to develop a coherent strategy. The suggested approach, based on business processes, represents a solution which has to be extended to all activities and domains of the water sector. It implies a real mobilization of all actors from who have to formalize their processes. Of course this effort requests a maturity in the process itself in order to be able to characterize the tasks and their dynamic.
The water sector represents a vast area where ICT solutions can be implemented and provide a real improvement. In order to benefit of these solutions, the water sector has to be pro active and structured in order to express needs. This challenging and exciting task will mobilize many professionals from both sectors and will request debates within the society on choices regarding water and its management.

6. Acknowledgment

The @qua thematic network and this work is funded under the ICT Policy Support Programme of the 7th Framework Program (FP7) of the European Commission.

7. References

Gourbesville, P. (2009) *Data & hydroinformatics: new possibilities and new challenges.* Journal of Hydroinformatics, Vol 11 No 3-4 pp 330-343, ISSN: 1464-7141

Jønch-Clausen T. & Global Water Partnership (GWP) (2004) *IWRM and Water Efficiency Plans by 2005: Why, What and How?*, GWP, 45p, Sweden, ISSN: 1403-5324

Holz, K.P., Hildebrandt G., Weber L. (2006) *Concept for a Web,-based Information System for Flood Management,* Natural Hazards, 38, pp 121-140, ISSN: 0921-030X

Marshment Hill Consulting (2010) *Smart Water Metering Cost Benefit Study,* Marshment Hill Consulting, Melbourne, Available from: http://www.water.vic.gov.au/__data/assets/pdf_file/0003/61545/smart-water-metering-cost-benefit-study.pdf

Oracle (2011) Smart Grid Challenges & Choices, Part 2: North American Utility Executives' Vision and Priorities, Oracle, USA, Available from: http://www.oracle.com/us/dm/h2fy11/utilities-survey-report-400044.pdf

Silver, M.S.; Markus M. L.; Mathis Beath C. (1995) The Information Technology Interaction Model: A Foundation for the MBA Core Course, *MIS Quarterly,* Vol. 19, No. 3, Special Issue on IS Curricula and Pedagogy (Sep., 1995), pp. 361-390, ISSN 1937-4771

Water Supply and Sanitation Technology Platform - WSSTP (2005) *Water Safe, strong and sustainable. European vision on water supply and sanitation in 2030,* WSSTP, Brussels, ISSN-1725-390X

World Water Council (2009) *Politics gets into water. Triennal report 2006-2009,* World Water Council, Marseille. *Available from:* http://www.worldwatercouncil.org/fileadmin/wwc/Library/Publications_and_reports/Activity_reports/TriennalReport_2006-2009.pdf

24

Monitoring Information Systems to Support Adaptive Water Management

Raffaele Giordano, Giuseppe Passarella and Emanuele Barca
Water Research Institute - National Research Council, Bari, Italy

1. Introduction

Decision making in water resources management is widely acknowledged in literature to be a rational process, based on appropriate information and modeling results. Information plays a fundamental role in improving our understanding of the consequences of, and trade-off among, the alternatives in water resources management.
Environmental monitoring networks have the potential to provide a great deal of information for environmental decision processes. Monitoring is widely used to increase our knowledge both of the state of the environment and of socio-economic conditions. Environmental monitoring has demonstrated its capacity within resource management to support decision processes providing knowledge of baseline conditions, to detect change, to establish historical status and trends, to promote long-term understanding or prediction, and to establish the need for, or success of, interventions.
Our knowledge of the complexity of water system processes is increasing, together with our awareness of the uncertainty and unpredictability of the effects of water management on system dynamics. Consequently, the demand for environmental information is growing posing new challenges to monitoring system design. This chapter discusses these new challenges and proposes an innovative monitoring design approach to deal with complexity. The conceptual architecture of an Adaptive Monitoring Information System (AMIS) is proposed. The AMIS properties are used in this work to define a framework to assess the capabilities of current monitoring systems to support water managers to cope with complexity and uncertainty. The framework is used to identify the main limitations and to define the potential improvements of TIZIANO monitoring system, developed to monitor the state of groundwater monitoring in the Apulia Region (South Italy).

2. New challenges for monitoring systems and information management in Adaptive Management (AM)

Incorporating uncertainties about future pressures on river basins into water resources management sets new challenges for environmental resources management. One learning process being developed to address this challenge is Adaptive Management (AM) (Holling 1978). Learning more about the resources or system to be managed and its responses to management actions, in order to develop a shift in understanding, is an inherent objective of AM (Walters, 1997; Fazey et al., 2005). Learning in AM leads to a focus on the role of feedback from the implemented actions. Such feedback-base learning models stress the need for monitoring the discrepancies between intentions and actual outcomes (Fazey et al., 2005). Monitoring becomes the primary tool for learning about the system and its performance under different management alternatives (Campbell et al., 2001).

To this aim, we assume that learning can be defined as a change in a person-system relationship, that is, the understanding of a person's place in the system and how they perceive it (Fazey et al., 2005). This definition implies that, because understanding is the goal which is achieved by the learner, each person may understand the environmental system differently and, therefore, act differently (Fazey et al., 2005). From the information production and management point of view, this implies that mental models influence an actor's perception of a problematic situation by influencing not only what data the actor perceives in the real world and what knowledge the actor derives from it (Timmerman and Langaas, 2004; Pahl-Wostl, 2007; Kolkman et al., 2005), but also what is noticed and what is taken to be significant (Checkland, 2001). It is important in information production and management that there should be a clear understanding and sharing of information users' mental models.

Therefore, contrarily to the traditional approach, in which information needs elicitation was intended in a top-down perspective, the design of a monitoring system for AM should begin by bringing together the interested parties to discuss their understanding of the system, the management problem, the information needed and how this information should be used. This implies involving a wide variety of stakeholders (i.e. scientists, managers, policy makers and members of the public at large) in a debate in which assumptions about the world are teased out, challenged, tested and discussed (Checkland, 2001), leading to the establishment of a common understanding about the system to be managed (Pahl-Wostl, 2007). This shared understanding can be structured in a system cognitive model, which allows the emergent properties of the system (i.e. variables to be monitored, thresholds, etc.) to be identified.

Among the different methods for Cognitive Modelling, an integration between Cognitive Maps (CM) and Causal Loop Diagrams (CLD) would seem particularly interesting to support monitoring system design. Given the peculiarities of the two modelling devices, CM can be used to disclose individual understanding of the system and to support the debate among participants, whereas CLD has great potentialities to simulate system dynamics.

When defining the cognitive model to be used as basis for a monitoring system, it is essential to address certain issues related to complex system dynamics. Firstly, the issue of scale must be tackled, since complex systems have structures and functions that cover a wide range of spatial and temporal scales. The impact of a given management action may vary at different scales (Campbell et al., 2001). Moreover, structures and processes are also linked across scales. Thus, the dynamics of a system at one particular scale cannot be analysed without taking into account the dynamics and cross-scale influences from the scales above and below it (Walker et al., 2006).

To deal with interaction between scales, we assume that the complex web of interacting systems can be broken down recursively into a network of individual systems, each of which determines its own fate and affects that of one or more other systems. The hierarchical structure of relationships between systems and subsystems (Campbell et al., 2001) implies that working on a particular scale often requires insights from at least two other scales, i.e. the level below, to understand the important processes that lead to the emerging characteristics of the level considered, and the level above it. Two sets of variables have to be considered for every system-subsystem pair. One set is required to describe the properties of the subsystem, whereas the second set is needed to describe the contribution of the subsystem to the performance of the whole system. This duality should be repeated at every level of the system hierarchy (Bossel, 2001).

Therefore, during the participatory process aimed at developing the cognitive model, participants should be required to think about their understanding of the total system, its essential component systems and the relationships that exist between them. The variables forming the cognitive model have to be able to describe the performance of the individual system and its contribution to the performance of the other systems. Using this inter-scale cognitive model as a basis for the design phase allows us to define a monitoring system capable of dealing with complex relationships between different scales, thus overcoming one of the main drawbacks of traditional monitoring practices.

However, adopting this inter-scale approach usually results in a demand to monitor a broader set of monitoring variables than traditional monitoring approaches. Some of these variables are fairly cheap to measure, but others, such as trends in very rare and important species, can be very expensive to monitor (Walkers, 1997). Thus, the development of an affordable monitoring program to support Adaptive Management involves substantial, scientific innovation in both method and approach, aimed at simplifying the set of monitoring variables by identifying the key components of the system.

The key components of the system, or key variables, are those that influence the system dynamics and bring about the most important changes (Walker et al., 2006; Campbell et al., 2001). Since these variables influence the overall dynamics of the system, they are of direct interest to managers, who are frequently focused on fast variables. These variables operate at different scales and with different speeds of change. The slowly changing variables determine the dynamics of the ecological system, whereas the social systems can be influenced by slow and/or fast variables (Walker et al., 2006). The conceptual models developed integrating the stakeholders' understanding of the system can be used as a basis for identifying the key variables (Campbell et al., 2001). To this aim, the analysis of CM can provided information about the relative importance of the different variables, by analysing the complexity of the causal chain. Those nodes whose immediate domain is most complex are taken to be those most central and, thus, the most important.

The identification of the key variables can also be supported by a strict integration between system monitoring and system modelling. This, in turn, is essential to any analysis of the implications of water policies. It allows the difficulties in understanding the dynamic feedback of the systems to be overcome, a particularly difficult task in an environmental context because of the number of factors involved. Moreover, humans have a limited capacity to understand the complexity of feedback in ecological systems (Fazey et al., 2005). This leads to erroneous connections between cause and effect and, thus, to erroneous conclusions about the impact of management actions. Conversely, models suggest which variables may be critical to monitor the impact of management actions, by posing elaborate hypotheses of which variables and relationships are critical to understanding the problem in question. The models then consider the dynamic implications of these hypotheses through the simulation of different scenarios. This allows monitoring networks to be designed (and re-designed) according to the model results. The potential of models to simulate future scenarios can be exploited to support the categorisation of the variables according to speed of change, i.e. slow changing variables and fast changing variables. Scenario simulation can draw attention to the role of the slow-changing variables in influencing system dynamics (Walker et al., 2006). The categorisation of variables according to speed of change can be used to program the frequency of data collection, making it easier to identify each variable's trend.

The integration between monitoring and modelling has to be considered as an iterative process. In fact, while models can simulate system dynamics, allowing the identification of key variables, the availability of new data allows the revision and updating of models. Moreover, the speed of change of the variables can also be considered iterative. Indeed, variables classified as slow changing in the model may be identified as fast changing by the monitoring system. In this case, the monitoring sample interval has to be changed. Thus, clearly a re-assessment process is needed both in models and in monitoring.

Simulation of system dynamics facilitates the identification of thresholds, which can be broadly defined as a breakpoint between two states of a system. When a threshold is exceeded, a change in system function and structure results. Such changes regard the nature and extent of feedback, resulting in changes of directions of the system itself. The changes can be reversible, irreversible or effectively irreversible (Walker et al., 2006). Two different types of thresholds can be defined, i.e. positive and negative. A positive threshold represents a desirable change in the state of the system. Such a change can be due to implemented management actions. A negative threshold can be considered as the starting point of a non-acceptable system trajectory. The recognition of these thresholds is particularly important in the case of irreversible changes. In this situation, actions are needed in order to avoid exceeding the threshold. The integration between monitoring and modelling provides information about the current state and the future trajectory of the system.

The position of the threshold is strictly linked to past experience. There are no examples where a new kind of threshold has been predicted before it has been experienced. Typically, the identification of thresholds is based on an analysis of systems similar to the one under investigation (Walker and Meyers, 2004). To this aim, a database is going to be implemented to collect empirical data on possible regime shifts in socio-ecological systems (Walker and Meyers, 2004). Some authors suggest using variances in variable trends to detect an impending system change (Brock and Carpenter, 2006). Integrating these two different approaches can be very useful. In other words, the existing experience regarding regime shifts, coming both from other systems and from the tacit knowledge of experienced and highly skilled people, can be structured and included in the system model. The variance can be calculated using monitoring data and the position of the threshold can be changed.

Integrating system modelling and monitoring iteratively highlights the importance of collecting information on trends. In fact, the availability of time series of data on the different variables allows the behaviour of the system variables and the trajectory of the system to be defined. The detection of trends can support the revision of the hypothesis concerning system dynamics, which is at the basis of the models. For these reasons it is fundamental to develop a monitoring system which is sustainable over time. To this aim, two important issues needs to be addressed, i.e. the need firstly to increase the adaptability of the monitoring system to policy and learning processes, and secondly to reduce monitoring costs through the adoption of scientific and technical innovation in information collection.

3. Adaptive monitoring and information system

Considering the issues described in the previous section, the conceptual architecture of a monitoring system for AM was defined (figure 1). From now onward, we refer to this system as Adaptive Monitoring Information System (AMIS).

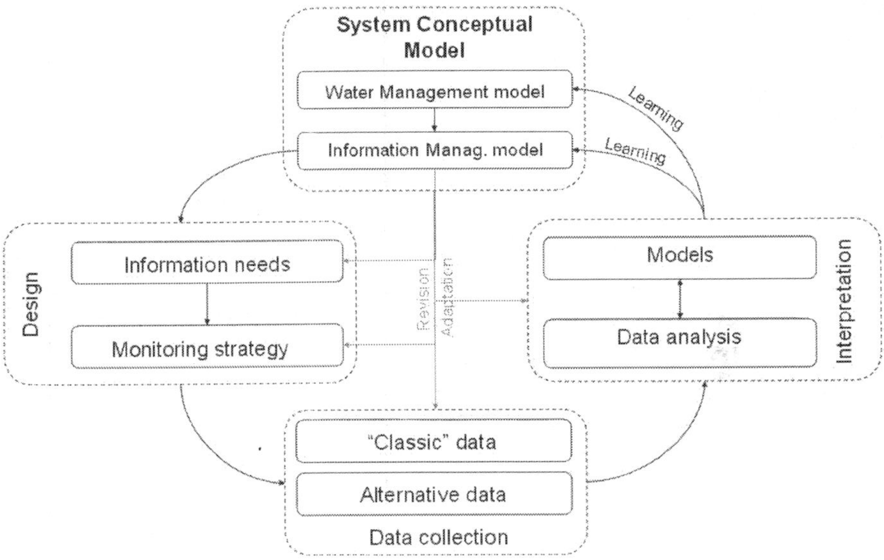

Fig. 1. AMIS conceptual architecture. The figure has been adapted from the Information cycle elaborated by Timmerman and others (2000), to emphasise the two learning processes.

As described previously, the basis for AMIS design is the conceptual model of the system, which simplifies the system and makes the key components and interactions explicit. The definition of this model is based on the integration between a participatory process, allowing experienced stakeholders to provide their understanding of the system, and models able to simulate future scenarios. The conceptual model is structured using the integration between Cognitive Maps and Causal Loop Diagrams.

Two different conceptual models, i.e. the "water management conceptual model" and the "information management conceptual model" are defined as the basis of AMIS. The former concerns the interpretation of the problem considered, while the latter concerns the information needed to solve the problem considered, and the "frames" used to interpret the information (Pahl-Wostl, 2007; Kolkman et al., 2005).

The AMIS architecture consists of four main boxes, i.e. Conceptual model elicitation, Design, Data collection and Interpretation. The links between them represent the iterative process of monitoring design, which is at the basis of AMIS. The figure was elaborated starting from the information cycle developed by Timmerman et al. (2000). This cycle depicts a framework where information users and producers communicate information needs that link the monitoring and decision processes. The monitoring program needs to be adapted to the different stages of the policy definition process, because each stage requires different types of information (Cofino, 1995; Ward, 1995) to make water management and governance adaptive.

Two possible learning processes can be identified. The first one concerns the water management conceptual model. Once information has been examined, a perspective is developed, and an insight is gained and integrated into the conceptual model itself (Kolkman et al., 2005). Information may prove initial models to be wrong and support the debate between actors, which may lead to a revision of models, through reflection and negotiation, in a social learning process. This learning may, in turn, support changes in the water management conceptual model. Moreover, feedback on management actions may generate new questions or new insights. This may make the originally agreed upon information appear inadequate, resulting in new information needs. Thus, the information needed to support a decision process evolves according to the actors' learning process, leading to revision/adaptation in monitoring strategies and data interpretation.

The second learning process relies on feedback from applied monitoring practices. As a result of experience in implementing the monitoring program and assessing its results, adaptation to monitoring may be needed (Cofino, 1995; Smit, 2003). The causes for adaptation can be found within monitoring practices: too little attention may have been spent on specifying the information needs; the information needs may have been specified in such a way that no adequate information can be produced from it, or so that it does not reflect the actual information users' needs; the selected indicators may not adequately measure what they are purported to measure; or the strategy to collect information may not have produced the right information. Furthermore, the available budgets may restrict the number of indicators that can be measured or the intensity of the network in terms of locations and frequency. New information sources may become available (e.g. progress in remote sensing technologies, etc.).

To this aim, an important innovation in AMIS concerns data collection methods. AM often results in a demand to monitor a broad set of variables, with prohibitive costs if the monitoring is done using only traditional methods of measurement. This is particularly true in developing countries, where financial and human resources are limited. In these areas, the monitoring network may cover only small part of the territory or the grid may be too sparse, making the monitoring data unsuitable for the decision process. Furthermore, traditional monitoring is costly, reducing its sustainability over time. The resulting works may be still valuable as one-off assessments, but they do not provide information about the trends of environmental resources and the evolution of environmental phenomena. Thus, the outcomes of environmental policies are often difficult to assess.

To deal with these issues, AMIS is based on the integration of alternative sources of knowledge. Thus, AMIS can be considered as the shared platform through which traditional monitoring information and innovative information sources (e.g. remote sensing monitoring, community monitoring, etc.) are integrated. Therefore, AMIS is able to adapt to data and information availability, supporting adaptive management even in data poor regions.

In Table 1, a comparison between the conventional approach and monitoring to support IWRM and AM is proposed.

Current monitoring practices	Needs for IWRM
- Based on monitoring objectives and disciplinary needs - Information users have unrealistic expectations of the information that will be produced - Data accessibility is limited - Abundant and detailed information is provided - The information provided is highly specialised - The available information is divided over various organisations - Information is transferred to the information users	- Based on policy objectives and information users' needs - The information that will be produced is jointly agreed between information users and producers - Data are publicly available and accessible - The information provided is concise and addresses the policy objectives - The information is targeted towards specific audiences - The information combines results from various organisations and is integrated over disciplines - Information is communicated to the information users and a broader stakeholder or public audience and evaluated before being incorporated into policy support
	Additional needs for AM
- The outcomes of the monitoring program (data) are the focus. - The purpose of the monitoring program is to evaluate environmental status set against target values. - Monitoring follows management and policy implementation.	- The monitoring program design and the responses on this design are as important as the results: the focus is on learning. - Monitoring becomes the primary tool for learning, i.e. understanding the system, assessing the effectiveness of management activities evaluating the system changes, and measuring progress towards participatory defined goals. - Monitoring, management and governance are interdependent.

Table 1. Comparison among current, IWRM and AM monitoring

3.1 Learning process using AMIS

Learning aspects in the AMIS are not about the monitoring as a simple process or its data, but about an increase of the system understanding, communication between stakeholders to influence decision making (McIntosh et al., 2006). While giving floor to and later using knowledge, concerns, demands, and expertise from different points of view, which result from a stakeholder involvement, one will indeed achieve better decision making with more alternatives of choice on the one hand, and a broader and more balanced acceptance of the decision making in management.

To initiate and later-on ensure learning processes using a monitoring system, all relevant stakeholder groups need access to it. Being involved when objectives are defined, data and processes transparently observed, stakeholders get enabled to learn about variables and interactions of "their own" systems and "their own" decisions which could lead to a revision or adaptation of management decisions (Pahl-Wostl, 2007. Further, this creates the feeling that stakeholders "buy in" into the product, that the monitoring system is "their" and therefore deserves more credibility (McIntosh et al., 2006). According to recent approach, the involvement of stakeholders can be extended to monitoring activities and not only to the design phase. The use of local knowledge enhances the understanding of environmental system, particularly in data poor areas. Moreover, adopting a community-based approach to monitoring can promote the public awareness of environmental issues.

Thus the intensive dialogue between science and many different stakeholders offers the opportunity for a mutual development, assessment, enhancement and implementation of new or already existing concepts, methods and tools, and helps improve the quality and acceptance of the decisions that are made. Last not least when using success-stories in management, based on the AMIS design, for the further development and enhancement of the monitoring system, the learning cycle is closed.

The following criteria, implemented into an AMIS, are indispensable to serve as a learning tool (cf. McIntosh et al., 2006):

1. **Understandability**: for each group of participants one should use "professional" indicators and perception-oriented "public" indicators to support learning processes for both of them
2. **Representativity** in involvement. Regardless of the method used to solicit user groups of the AMIS, every attempt should be made to involve a diverse group of stakeholders or broad audience that represent a variety of interests regarding the issue addressed. While key stakeholders should be invited to the process of indicator formulation, there should be also an open invitation to all interested parties to join the evaluation of the system. This adds to the public acceptance and respect of the results of the AMIS. If a process is perceived to be exclusive, both key members of the decision-making community and the wider public may reject monitoring.
3. **Scientific credibility**. Although participatory monitoring as it is understood in the AMIS design incorporates values and beliefs, the scientific components of the monitoring system must adhere to standard scientific practice and objectivity. This criterion is essential in order to maintain credibility among all groups, expert-decision-makers, scientists, stakeholders, and the public.
4. **Objectivity**. The stakeholder community must trust the facilitators of a participatory monitoring as being objective and impartial. In this regard, facilitation by university researchers or outside consultants often reduces the incorporation of stakeholder biases into the scientific components of the monitoring system.
5. **Understanding uncertainty**. Understanding scientific uncertainty is critically linked to the expectations of real world results associated with decisions made as a result of the modelling process. This issue is best communicated through direct participation in the modelling process itself.
6. AMIS' own **adaptability** to incorporate new users groups, changed frameworks and newly gained (quantitative and qualitative) data. The monitoring system developed should be relatively easy to use and up-date by the administrators. This requires excellent documentation and a good user interface. If non-scientist users cannot understand the monitoring system as a source to work with, local decision-makers will not apply it to support real management problems.

3.2 Technical adaptability of an AMIS

In this section some technical aspects related to the adaptive degree of AMIS are described. Firstly, AMIS should be flexible and able to incorporate new information and data, of different type and with different formats. Using a relational database (RDBMS) is a sound basis to be open for new information requirements, because it is very flexible and extendable. The information can be well structured and redundancy can be avoided. The user can create new tables and link them to the existing database.

To satisfy the information needs of various user groups according to their knowledge of environmental system behaviour, different types of information for different purposes must be produced. One important aim of the AMIS is to provide the user with various methods and predefined algorithms to produce information. AMIS should provide the user with user-friendly predefined methods and algorithms to produce information, such as data visualisation tools as well as automatically generated information from incoming data.

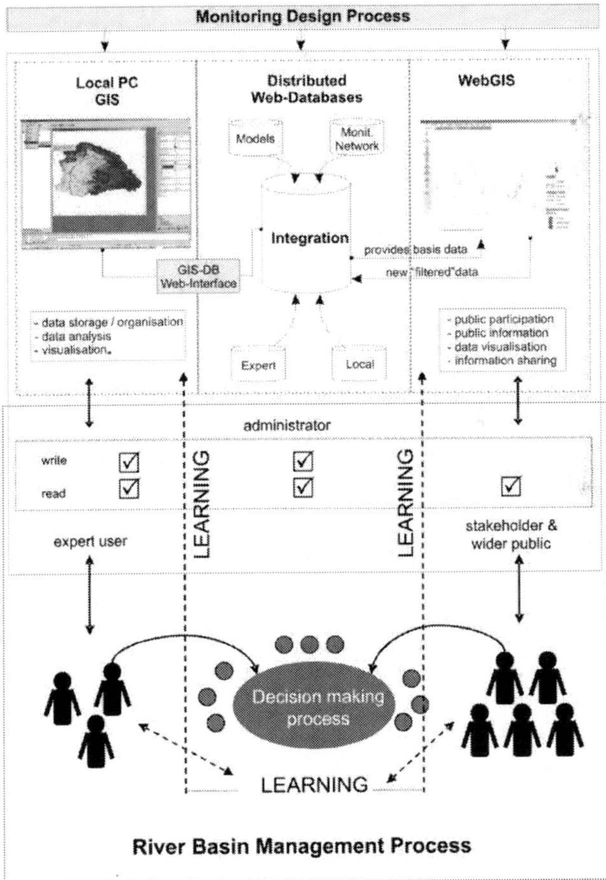

Fig. 2. Technical components of AMIS.

Another aspect of being flexible and extendable is to provide the possibility to add new modules easily, for instance hydrological or economical models, methods to analyse map layers etc. This kind of flexibility is of interest for developers or advanced users with programming skills. A modular or object oriented software structure is necessary to permit this task.

Taking the above mentioned arguments into consideration the information system is quiet flexible and open to include new information. But it is impossible to foresee what kind of requirements will be demanded from the information system in a few years. Thus, it should be possible to improve, maintain, and extend the software for everybody with programming knowledge. To be "technically sustainable" open source software should be used and local IT experts involved in the development process, particularly, if the software prototype will be produced within a project over a certain period and not by a company. One should emphasise the problem here that after a project has finished, often the developers are not available or not in charge for the product anymore. To facilitate future improvements the AMIS must be equipped with a sound documentation of the source code.

4. The adaptability of the groundwater monitoring system in Apulia Region: Main drawbacks and potential improvements

The aim of this work is to criticize the current approaches to monitoring design, highlighting the main drawbacks which hamper the adaptability of monitoring system. Moreover, potential improvements are discussed. To this aim a framework to assess the adaptability degree of monitoring design approach has been developed. The framework is structured as shown in the following table.

Criteria	Meaning
- Information producer/information users interaction	- Is the monitoring system based on the elicitation of the decision-makers' information needs?
- Degree of participation	- How many actors have been involved in the process of monitoring system design? At which level? In which phase?
- Multi-scale monitoring	- Is the monitoring system able to collect information at different spatial and temporal scale?
- Integration of information sources	- Is the monitoring system based on the integration of different sources of data and information?
- Long time sustainability	- Is the monitoring system capable to provide long time series of data?
- Monitoring/modelling interaction	- Is the monitoring system integrated with modelling to support data analysis and interpretation?
- Policy evaluation	- Is the monitoring system capable to support the evaluation of the policy impacts and suggest improvements?
- Monitoring evaluation	- Does the monitoring program provide for an evaluation and adaptation of the monitoring strategy?

Table 2. Comparison among current, IWRM and AM monitoring

This criteria have been used to evaluate the adaptability of the groundwater monitoring system of the Apulia region (Southern Italy).

The groundwater monitoring network of the Apulia Region was established in 2006 to meet the wide range of standards set by the water related national legislation adopted in 1999 (Italian Legislative Decree n. 152/1999). Consequently, the monitoring network was designed, realized and finally used in order to produce water quality and quantity information useful to characterize the environmental status of the main regional groundwater bodies.

The monitoring network has been promoted and financed by the regional offices in charge of the collection, storage and processing of data collected in accordance with relevant regulations. The network design and implementation and the enforcement of the monitoring practices fall within the scope of the project called TIZIANO whose completion is scheduled for the end of 2011.

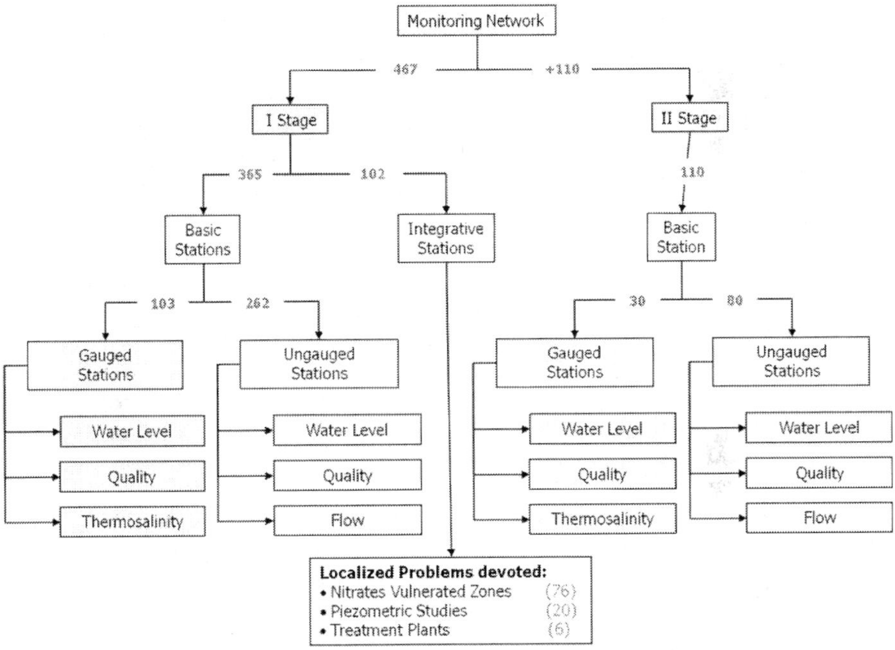

Fig. 3. TIZIANO monitoring design and number of monitoring stations. The process was composed by two main phases to identify the monitoring stations.

The TIZIANO monitoring network is made of more than 600 wells mostly spread within the boundaries of the four main aquifers of the region even if some tens of them have been located within some minor groundwater bodies. About 130 wells have been equipped with automatic probes for continuous measuring of groundwater level. During the last five years hundreds of quality and quantity measures have been made on site and thousands of samples, collected in the wells of the network, have been analyzed in laboratory in order to determine the concentration of the main chemicals, metals, organic compounds, pesticides and level of harmful microorganisms. The huge amount of information, collected during the last five years, was stored in a Geographic Information System (GIS) specifically designed for the project. It allowed regional decision-makers to assess the environmental state of the aquifers and plan and carry out specific actions to improve it, when not good, or reverse worsening trends, when they were to lead to adverse conditions of groundwater quality and quantity.

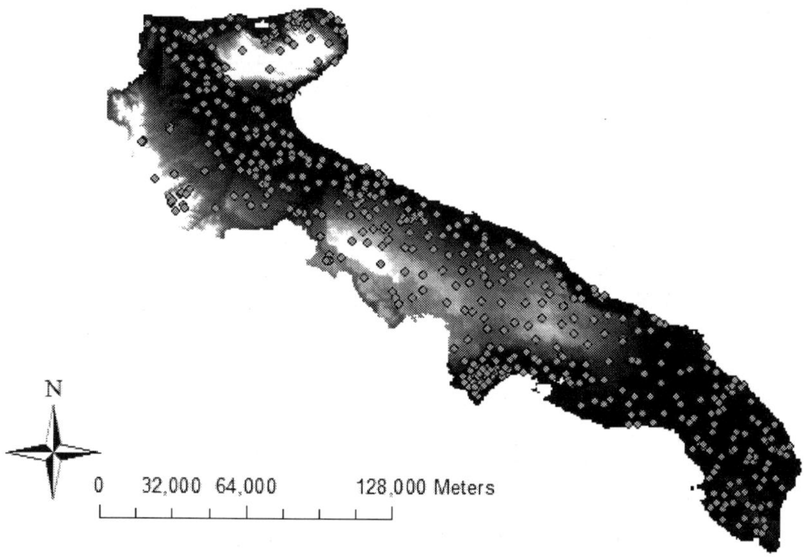

Fig. 4. Distribution of the monitoring station.

As reported above, the TIZIANO monitoring network started late in 2006, but the administrative process which led to its design and funding started several years early, at the turn of the century. In the meantime the European Union issued the Water Framework Directive (2000/60/CE), which was implemented in Italy exactly in 2006 (Italian L.D. n. 152/2006), and the, so called daughter Groundwater Directive in 2006 (2006/118/CE), recently implemented in Italy with the L.D. n. 30/2009. Although the Italian L.D. 152/1999 would herald a number of rules, then enshrined in European directives, it is evident that the future implementation of the decrees of 2006 and 2009 have clarified and modified, sometimes substantially, type, detail and timing of information to be acquired by monitoring and all management activities resulting from its processing.

4.1 Information producer/information users interaction

The Region already had a modest, monitoring network made of about 100 piezometers equipped with water level gauges, where some sporadic sampling was collected during the early 90s. Nevertheless, because of various causes, this network was abandoned after some years of functioning. At this point, within the regional offices in charge of water resources

management and protection, arose the need of recovering and, possibly, potentiate the network.
In the meantime several important water related, European directives (e.g.: the Nitrate Directive, 1991/676/EEC) and national decrees had been promulgated, which forced regional water offices to move toward a detailed knowledge of the qualitative and quantitative state of water resources in order to protect such resources and restore their original natural status.
The evaluation of the institutional, legislative, technical and scientific needs and expectations led to the design of the regional groundwater monitoring network by a small team of super-experts which were careful to meet the requirements coming from various and different parts. Measures of water level and physical-chemical parameters were carried out following rules and times required by national environmental legislation implementing EU rules and a number of scientific measures and controls were preformed in order to give responses to the scientific community.
The information provided by the new monitoring system was essential, among other, in order to assess the environmental state of the Apulian groundwater bodies or delimit Nitrate Vulnerable Areas, and design and plan specific actions of different complexity and socio-economical cost, able to recover and protect groundwater.
Summarizing, measures of water level and physical-chemical parameters were carried out following rules and times required by national environmental legislation implementing EU rules and a number of scientific measures and controls were preformed in order to give responses to the scientific community.

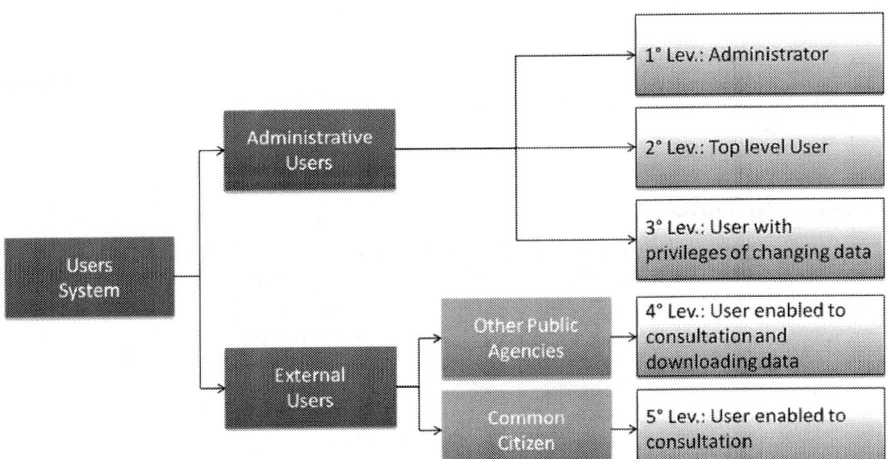

Fig. 5. Information accessibility according to TIZIANO monitoring program.

4.2 Degree of participation

From what said above derives that the position of the decision-makers in the design of the monitoring system was rather weak, i.e. the Apulian Region's role was limited to promote and fund the design. The role of decision-makers in the functioning of the TIZIANO monitoring network is strong and constant. Regional offices are in charge of producing, controlling and processing monitoring information in order to assess environmental indices and plan and execute actions for recovering deteriorated resources.

4.3 Multi-scale monitoring

Given the multi-objective frame of the monitoring each class of data has been collected with different spatial and temporal resolution. Let's have a short description of classes of data and related time-space scale starting from groundwater level.

In order to capture the cyclic behaviour of groundwater levels in the wells, measures are taken on site almost every three months. About 130 wells have been equipped with automatic water level gauges capable of acquiring and transmitting a measurement every 15 minutes. These equipped wells have been located at strategic sites, in order to use them as controlling stations. So, the project database stores groundwater levels measured at different temporal scales at different locations all over the aquifers extension. Nevertheless, there is no analysis of the inter-linkages among the process at different scales.

4.4 Integration of information sources

Given the complexity of the monitoring network, the data collection system is extremely various and includes manual and automatic measures, on site and laboratory analysis, coastal and inland exploration, airborne remote sensing. The whole amount of collected data is stored in GIS after a validation phase. Nevertheless, data coming from different platforms are sporadically integrated. The different measures follow a separated path, which passes through a separated validation step. In conclusion, the monitoring system is not based on a strong integration between sources of data.

4.5 Long time sustainability

The whole monitoring system, as currently conceived, is particularly expensive. Let's report some of the main weakness of the project concerning its own costs.

The monitoring area is objectively wide and the number of monitoring points huge, while the location of the monitoring teams is centralized and, consequently, they need to travel hundreds of kilometres during the monitoring surveys or for maintenance. Instrumentation need to be constantly maintained and often replaced due to theft.

Moreover the costs of the system are rather high due to the frequent outsourcing of monitoring activities. Costs could be reduced dramatically, if most of the monitoring practices were carried out by Regional Agencies and Offices and only very specialized activities were outsourced. In conclusion, only if an intelligent redistribution of activities within public institutions will be put in place, with a consequent cost reduction, the network is likely to become a long-term system.

4.6 Monitoring/modelling interaction

Various statistical, geostatistical, hydrogeological and hydrogeochemical, deterministic and stochastic, simple and complex models have been applied to process data collected and stored into the GIS. Nevertheless, it was not specifically designed to be compliant to any particular model. In fact, given the wide range of expected uses of the different dataset stored the design choice was to keep the organization of data extremely simple, and then easily adaptable to different kinds of models just through a simple pre-processor. In the TIZIANO monitoring project, the monitoring/modeling interaction is one-directional. That is, monitoring provides data to models, but the models are used to support the evaluation and, eventually, the re-design of the monitoring network.

4.7 Policy evaluation

Theoretically speaking, the TIZIANO groundwater monitoring system should be capable of supporting regional decision-makers at each step of the decisional path. In few words, the network should support: 1) Assessing the initial state of the natural system and reporting negative trends; 2) Controlling the effects of environmental actions and politics; 3) Alerting for undesired evolutions.

The spatial extension of the monitoring network and the number of monitoring wells should be revised at each step. Step one should be performed extensively over the monitoring area and step two should focus around risk area. Step three should be suitably designed in order to be capable of capturing any warning signal, at this step the position of the monitoring points, the parameters to be measured and the frequency of measurement need to be carefully evaluated. The TIZIANO monitoring network performed very well the first step (Assess). Unlikely, we have less evaluation elements concerning the monitoring network suitability during the second phase (Control). Finally, concerning the third step (Alert), the monitoring activity have been moved and increased around area considered mostly at risk and reduced in the rest of the region.

4.8 Monitoring evaluation

The monitoring program does not contain an evaluation phase. This means that the second learning process described in figure 1 cannot be supported.

The critical analysis of the TIZIANO groundwater monitoring system can be summarized as shown in table 3.

Criteria	Evaluation
- Information producer/information users interaction	- Mainly based on scientific requirements and legislation
- Degree of participation	- Weak in the design phase, strong during the implementation
- Multi-scale monitoring	- There is no analysis of inter-linkages between different scales
- Integration of information sources	- There is no integration
- Long time sustainability	- The monitoring costs are too high
- Monitoring/modelling interaction	- One-directional flow of information
- Policy evaluation	- The impacts on groundwater are monitored
- Monitoring evaluation	- The learning process is not supported

Table 3. Results of the evaluation

5. Conclusion

Starting from the results of the critical analysis, some drawbacks and potential improvements for the TIZIANO monitoring program have been identified and discussed in the following sections.

5.1 Main drawbacks

According to the results of the critical analysis, we can infer that the TIZIANO groundwater monitoring network cannot be considered as adaptive and it is not suitable to support the adaptive management. Firstly, the excessive cost for collecting and analyzing data have a strongly negative impact on the long term sustainability of the program. This, in turn, would reduce the capability of the monitoring system to detect the long term unintended consequences of the groundwater management policies.

Secondly, the monitoring system is not integrated in a wider program aiming to analyze the different potential impacts of the policies - e.g. socio-economic impacts. The TIZIANO monitoring program is based on the sectorial approach to environmental resources management which is still common is socio-institutional contexts characterized by a centralized and command-and-control regime. A more holistic and systemic approach is required.

Thirdly, there is no integration between different sources of information. This has a negative impact on the flexibility of the monitoring program. In fact, if the data collection is based only on traditional "static" devices - i.e. monitoring stations - then the adaptation of the monitoring program to modified information needs would be difficult: changing sensor is not always easy and/or cheap, the position of the station cannot be modified easily, even the time schedule for data collection cannot be changed easily. Although remote sensing data are mentioned in the program, the integration of this source of data with the traditional information sources is still far from being achieved.

Finally, an adaptive monitoring system requires an evaluation phase. That is, a critical analysis of the suitability of the designed monitoring system is crucial. This phase has not been considered in the current monitoring program. This means that the revision of the program depends exclusively on the political willing of the local authorities and on the availability of further funds.

5.2 Potential improvements

Some improvements to make the TIZIANO monitoring program more suitable to support the adaptive water management were defined:

- Monitoring costs: the current monitoring costs could be reduced only if an intelligent redistribution of activities within public institutions will be put in place. This means that the outsourcing activities have to be strongly reduced. Moreover, since the costs are mainly related to laboratories analysis, the integration of different sources of information would have a positive impact on monitoring costs.
- Systemic analysis of the policy impacts: the increasing awareness of the complexity of the real world forces us to adopt a system dynamic approach to monitor and analyze the different and interrelated policy impacts. Although the aim of the TIZIANO network is to collect data about the physical and chemical state of the groundwater, it has to be integrated in a more systemic monitoring program, able to detect even the socio-economical impacts.
- Integration between different sources of information: The integration of different sources of knowledge seems particularly useful to design a multi - variate and multi - scale monitoring system for adaptive management. The Use of alternative sources of information increases the flexibility of monitoring program and reduce the costs. Among the alternative sources of information, local knowledge is increasingly considered as crucial (see as example the Hyogo Framework for Action). The analysis of

the literature review on this issue allowed us gain some hints. The key to guarantee the long term involvement of local community members in monitoring is to keep the monitoring activities as simple and similar to the traditional methods for environmental assessment as possible. Moreover, the involvement in monitoring is easier if the monitoring activities are incorporated in the community members' daily activities. The key to guarantee the actual usability of local knowledge in monitoring activities is: 1) fully integrating local knowledge into existing traditional institutions; and 2) structuring local knowledge so that it is transformed into meaningful and relevant information for decision-making. The integration between local and scientific knowledge allowed to enhance the reliability of local knowledge.

- Learning process in monitoring activities: as widely discussed in the scientific literature, the design of a monitoring system cannot be considered as a linear process. It is rather a cycle of design - implementation - evaluation - adaptation. The information needs can change due to several reasons. Adaptive monitoring system should be able to follow these changes. To this aim an evaluation phase should be formally included in the monitoring program. The evaluation should be based on the interaction between policy and decision makers (information users) and monitoring system managers (information producers).

6. References

Bossel, H. (2001). Assessing viability and sustainability: a systems-based approach for deriving comprehensive indicator sets. *Conservation Ecology* 5(2): 12. [online] URL: http://www.consecol.org/vol5/iss2/art12/

Brock, W. A., and S. R. Carpenter (2006). Variance as a leading indicator of regime shift in ecosystem services. *Ecology and Society* 11(2): 9. [online] URL: http://www.ecologyandsociety.org/vol11/iss2/art9/

Campbell, B., J. A. Sayer, P. Frost, S. Vermeulen, M. Ruiz Pérez, A. Cunningham, and R. Prabhu (2001). Assessing the performance of natural resource systems. *Conservation Ecology* 5(2): 22. [online] URL: http://www.consecol.org/vol5/iss2/art22/

Checkland, P. (2001). Soft System Methodology. In *Rational Analysis for a Problematic World.* Rosenhead, J., Mingers J. (eds), pp. 61-89. John Wiley and Sons, Chichester, UK.

Cofino, W.P. (1995). Quality management of monitoring programs. In *Proceeding of the international workshop on monitoring and assessment in water management; Monitoring Tailor-Made.* Adriaanse M., J van der Kraats; P.G. Stocks, and R.C. Wards (eds), 20-23 September 1994, Beekbergen, The Netherlands.

Fazey, I., J.A. Fazey, and D.M.A. Fazey (2005). Learning More Effectively from Experience. *Ecology and Society*, 10(2), 4. [online] URL: http://www.ecologyandsociety.org/vol10/iss2/art4/

Holling, C.S. (ed.) (1978). *Adaptive Environmental Assessment and Management.* John Wiley and Sons, New York.

Kolkman, M.J., M. Kok, A. van der Veen (2005). Mental model mapping as a new tool to analyse the use of information in decision-making in integrated water management. *Physics and Chemistry of the Earth,* 30: 317-332.

McIntosh, B S, Giupponi, C, Smith, C, Voinov, A, Matthews, K B, Monticino, M, Kolkman, M J, Crossman, N, van Ittersum, M, Haase, D, Haase, A, Mysiak, J, Groot, J C J, Sieber, S, Verweij, P, Quinn, N, Waeger, P, Gaber, N, Hepting, D, Scholten, H, Sulis,

A, van Delden, H, Gaddis, E, Assaf, H. (2006). Bridging the gaps between design and use: developing tools to support management and policy. In print.

Pahl-Wostl, C. (2007). The implications of complexity for integrated resources management. *Environmental Modelling and Software*, 22: 561-569.

Smit, A.M. (2003). Adaptive monitoring: an overview. *DOC Science Internal Series*, vol. 138. Department of Conservation, Wellington. 16 p.

Timmerman, J.G. and S. Langaas (2004). Conclusions. In *Environmental information in European transboundary water management*. Timmerman, J.G. and S. Langaas (eds.), pp. 240-246. IWA Publishing, London, UK. ISBN: 1 84339 038 8.

Timmerman, J.G., J.J. Ottens, and R.C. Ward (2000). The information cycle as a framework for defining information goals for water-quality monitoring. *Environmental Management* 25(3): 229-239.

Walker, B. H., L. H. Gunderson, A. P. Kinzig, C. Folke, S. R. Carpenter, and L. Schultz (2006). A handful of heuristics and some propositions for understanding resilience in social-ecological systems. *Ecology and Society* 11(1): 13. [online] URL:http://www.ecologyandsociety.org/vol11/iss1/art13/

Walker, B. and J. A. Meyers (2004). Thresholds in ecological and social–ecological systems: a developing database. *Ecology and Society* 9(2): 3. [online] URL: http://www.ecologyandsociety.org/vol9/iss2/art3

Ward, R.C. (1995). Monitoring Tailor-made: what do you want to know? In *Proceeding of the international workshop on monitoring and assessment in water management; Monitoring Tailor-Made.* Adriaanse M., J van der Kraats; P.G. Stocks, and R.C. Wards (eds), 20-23 September 1994, Beekbergen, The Netherlands.

25

Autonomous Decentralized Control Scheme for Long-Term Operation of Large Scale and Dense Wireless Sensor Networks with Multiple Sinks

Akihide Utani
Tokyo City University,
Japan

1. Introduction

Various communication services have been provided. They include environmental monitoring and/or control, ad-hoc communication between mobile nodes, and inter-vehicle communication in intelligent transport systems. As a means of facilitating the above advanced communication services, autonomous decentralized networks, such as wireless sensor networks (Akyildiz et al., 2002; Rajagopalan & Varshney, 2006), mobile ad-hoc networks (Perkins & Royer, 1999; Johnson et al., 2003; Clausen & Jaquet, 2003; Ogier et al., 2003), and wireless mesh networks (Yamamoto et al., 2009), have been intensively researched with great interests. Especially, a wireless sensor network, which is a key network to construct ubiquitous information environments, has great potential as a means of realizing a wide range of applications, such as natural environmental monitoring, environmental control in residential spaces or plants, object tracking, and precision agriculture (Akyildiz et al., 2002). Recently, there is growing expectation for a new network service by a wireless sensor network consisting of a lot of static sensor nodes arranged in a service area and a few mobile robots as a result of the strong desire for the development of advanced systems that can flexibly function in dynamically changing environments (Matsumoto et al., 2009).

In this chapter, a large scale and dense wireless sensor network made up of many static sensor nodes with global positioning system, which is a representative network to actualize the above-mentioned sensor applications, is assumed. In a large scale and dense wireless sensor network, generally, hundreds or thousands of static sensor nodes limited resources, which are compact and inexpensive, are placed in a service area, and sensing data of each node is gathered to a sink node by inter-node wireless multi-hop communication. Each sensor node consists of a sensing function to measure the status (temperature, humidity, motion, etc.) of an observation point or object, a limited function of information processing, and a simplified wireless communication function, and it generally operates on a resource with a limited power-supply capacity such as a battery. Therefore, a data gathering scheme and/or a routing protocol capable of meeting the following requirements is mainly needed to prolong the lifetime of a large scale and dense wireless sensor network composed of hundreds or thousands of static sensor nodes limited resources.

1. Efficiency of data gathering
2. Balance of communication load among sensor nodes
3. Adaptability to network topology changes

As data gathering schemes for the long-term operation of a wireless sensor network, cluster-ing-based data gathering (Heinzelman et al., 2000; Dasgupta et al., 2003; Jin et al., 2008) and synchronization-based data gathering (Wakamiya & Murata, 2005; Nakano et al., 2009; Nak-ano et al., 2011) are under study, but not all the above requirements are satisfied. Recently, bio-inspired routing algorithms, such as ant-based routing algorithms, have attracted a sign-ificant amount of interest from many researchers as examples that satisfy the three require-ments above. In ant-based routing algorithms (Subramanian et al., 1998; Ohtaki et al., 2006), the routing table of each sensor node is generated and updated by applying the process in which ants build routes between their nest and food using chemical substances (pheromon-es). Advanced ant-based routing algorithm (Utani et al., 2008) is an efficient route learning algorithm which shares route information between control messages. In contrast to conven-tional ant-based routing algorithms, this can suppress the communication load of each sen-sor node and adapt itself to network topology changes. However, this does not positively ease the communication load concentration on specific sensor nodes, which is the source of problems in the long-term operation of a wireless sensor network. Gradient-based routing protocol (Xia et al., 2004) actualizes load-balancing data gathering. However, this cannot su-ppress the communication load concentration to sensor nodes around the set sink node. Int-ensive data transmission to specific sensor nodes results in concentrated energy consumpti-on by them, and causes them to break away from the network early. This makes long-term observation by a wireless sensor network difficult.

In a large scale and dense wireless sensor network, the communication load is generally co-ncentrated on sensor nodes around the set sink node during the operation process. In cases where sensor nodes are not placed evenly in a large scale observation area, the communica-tion load is concentrated on sensor nodes placed in an area of low node density. To solve this communication load concentration problem, a data gathering scheme for a wireless sen-sor network with multiple sinks has been proposed (Dubois-Ferriere et al., 2004; Oyman & Ersoy, 2004). In this scheme, each sensor node sends sensing data to the nearest sink node. In comparison with the case of one-sink wireless sensor networks, the communication load of sensor nodes around a sink node is reduced. In each sensor node, however, the destinati-on sink node cannot be selected autonomously and adaptively. In cases where original data transmission rate from each sensor node is not even, therefore, the load of load-concentrated nodes is not sufficiently balanced. An autonomous load-balancing data transmission scheme is required.

This chapter represents a new data gathering scheme with transmission power control that adaptively reduces the load of load-concentrated nodes and facilitates the long-term operati-on of a large scale and dense wireless sensor network with multiple sinks (Matsumoto et al., 2010). This scheme has autonomous load-balancing data transmission devised by consider-ing the application environment of a wireless sensor network as a typical example of compl-ex systems where the adaptive adjustment of the entire system is realized from the local int-eractions of components of the system. In this scheme, the load of each sensor node is auton-omously balanced. This chapter consists of four sections. In Section 2, the above data gather-ing scheme (Matsumoto et al., 2010) is detailed and its novelty and superiority are described. In Section 3, the results of simulation experiments are reported and the effectiveness of our scheme (Matsumoto et al., 2010) is demonstrated by comparing its performances with those of existing schemes. In Section 4, the overall conclusions of this work are given and future problems are discussed.

2. Autonomous decentralized control scheme

To facilitate the long-term operation of an actual sensor network service, a recent approach has been to introduce multiple sinks in a wireless sensor network (Dubois-Ferriere et al., 2004; Oyman & Ersoy, 2004). In a wireless sensor network with multiple sinks, sensing data of each node is generally allowed to gather at any of the available sinks. Our scheme (Matsumoto et al., 2010) is a new data gathering scheme based on this assumption, which can be expected to produce a remarkable effect in a large scale and dense wireless sensor network with multiple sinks. In our scheme, each sensor node can select either of high power and low power for packet transmission, where high power corresponds to normal transmission power and low power is newly introduced to moreover balance the load of each sensor node.

2.1 Routing algorithm

Each sink node has a connective value named a "value to self", which is not updated by transmitting a control packet and receiving data packets. In the initial state of a large scale and dense wireless sensor network with multiple sinks, each sink node broadcasts a control packet containing its own location information, ID, hop counts(=0), and "value to self" by high power. This control packet is rebroadcast throughout the network with hop counts updated by high power. By receiving the control packet from each sink node, each sensor node can grasp the "value to self" of each sink node, their location information, IDs, and the hop counts from each sink node of its own neighborhood nodes.

Initial connective value of each sensor node, which is the connective value before starting data transmission, is generated by using the "value to self" of each sink node and the hop counts from each sink node. The procedure for computing initial connective value of a node (i) is as follows:

1. The value [$v_{ij}(0)$] on each sink node (j=1, … ,S) of node (i) is first computed according to the following equation

$$v_{ij}(0) = vo_j \times dr^{hops_{ij}} \quad (j = 1,\ldots,S) \tag{1}$$

   where vo_j(j=1, … ,S) is the "value to self" of sink node (j), $hops_{ij}$(j=1, … ,S) is the hop counts from sink node (j) of node (i). dr represents the value attenuation factor accompanying the hop determined within the interval [0,1].

2. Then, initial connective value [$v_i(0)$] of node (i) is generated by the following equation

$$v_i(0) = \max v_{ij}(0) \qquad (j = 1, \ldots, S) \tag{2}$$

   where this connective value [$v_i(0)$] can be also conducted from the following equation

$$v_i(0) = vm_i(0) \times dr \tag{3}$$

In the above Equation (3), $vm_i(0)$ represents the greatest connective value before starting data transmission in neighborhood nodes of node (i).

Before data transmission is started, each sensor node computes initial connective value of each neighborhood node based on the above Equations (1) and (2), and stores the computed connective value, which is used as the only index to evaluate the relay destination value of each neighborhood node, in each neighborhood node field of its own routing table.

2.2 Data transmission and connective value update

For a while from starting data transmission, each sensor node selects the neighboring node with the greatest connective value from its own routing table as a relay node, and transmits the data packet to this selected node by high power. In cases where more than one node shares the greatest connective value, however, the relay node is determined between them at random. The data packet in each sensor node is not sent to a specified sink node. By repetitive data transmission to the neighboring node with the greatest connective value, data gathering at any of the available sinks is completed. In our scheme, the connective value of each sensor node is updated by considering residual node energy. Therefore, by repetitive data transmission to the neighboring node with the greatest connective value, the data transmission routes are not fixed.

To realize autonomous load-balancing data transmission, in our scheme (Matsumoto et al., 2010), the data packet from each sensor node includes its own updated connective value. We assume that a node (l) receives a data packet at time (t). Before node (l) relays the data packet, it replaces the value in the connective value field of the data packet by its own renewal connective value computed according to the following connective value update equation

$$v_l(t) = vm_l(t) \times dr \times e_l(t) / E_l \tag{4}$$

where $vm_l(t)$ is the greatest connective value at time (t) in the routing table of node (l). $e_l(t)$ and E_l represent the residual energy at time (t) of node (l) and the battery capacity of node (l), respectively.

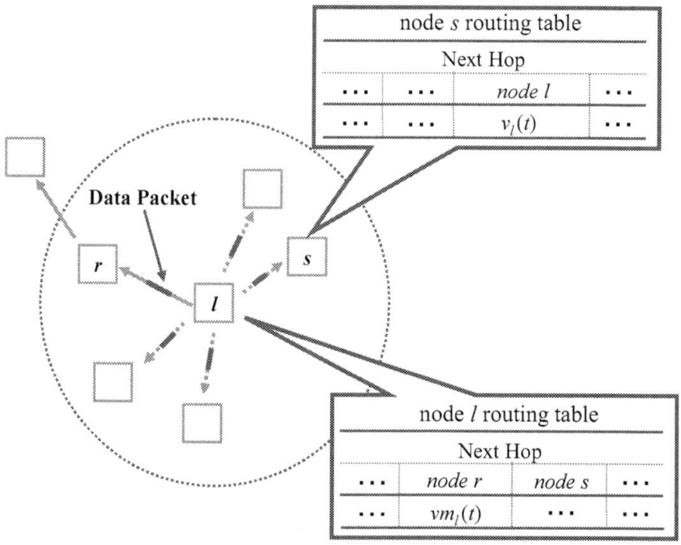

Fig. 1. Data packet transmission and connective value update

In our scheme, the data packet addressed to the neighboring node with the greatest connective value is intercepted by all neighboring nodes. This data packet includes the updated connective value of the source node based on the above Equation (4). Each neighborhood node that intercepts this packet stores the updated connective value in the source node field of its own routing table. Fig.1 shows an example of data packet transmission and its accompanying connective value update. In this example, node (l) refers to its own routing table and addresses the data packet to node (r), which has the greatest connective value [$vm_l(t)$]. When this data packet is intercepted, each neighboring node around node (l) stores the updated connective value [$v_l(t)$] in the data packet in the node (l) field of its own routing table.

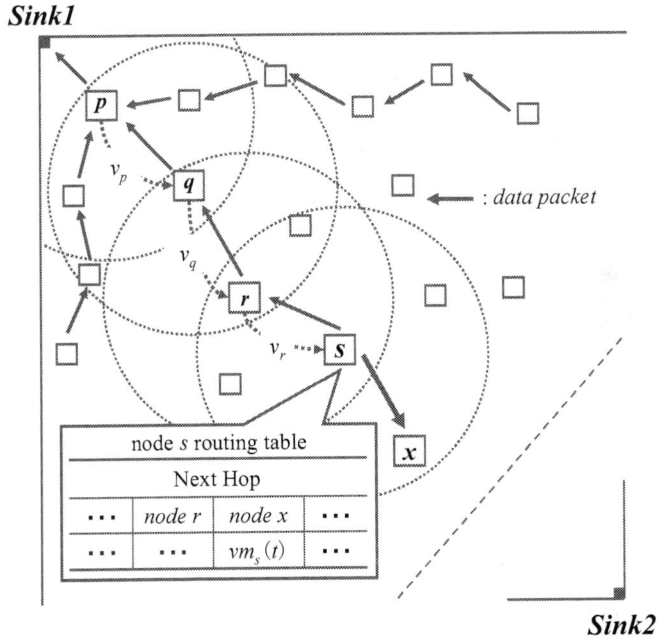

Fig. 2. An example of autonomous load-balancing data transmission to multiple sinks

Our scheme (Matsumoto et al., 2010) requires the construction of a data gathering environment in the initial state of a large scale and dense wireless sensor network with multiple sinks, but needs no special communication for network control. The above-mentioned simple mechanism alone achieves autonomously adaptive load-balancing data transmission to multiple sinks, as in Fig.2. The lifetime of a wireless sensor network can be extended by reducing the communication load for network control and solving the node load concentration problem.

2.3 Transmission power control

For data packet transmission, the transmission power of each sensor node is switched to low power if its own residual energy is less than the set threshold [T_e]. In this case, each sensor node selects the neighboring node with the greatest connective value within range of radio wave of low power as a relay node, and transmits the data packet to this selected node by low power.

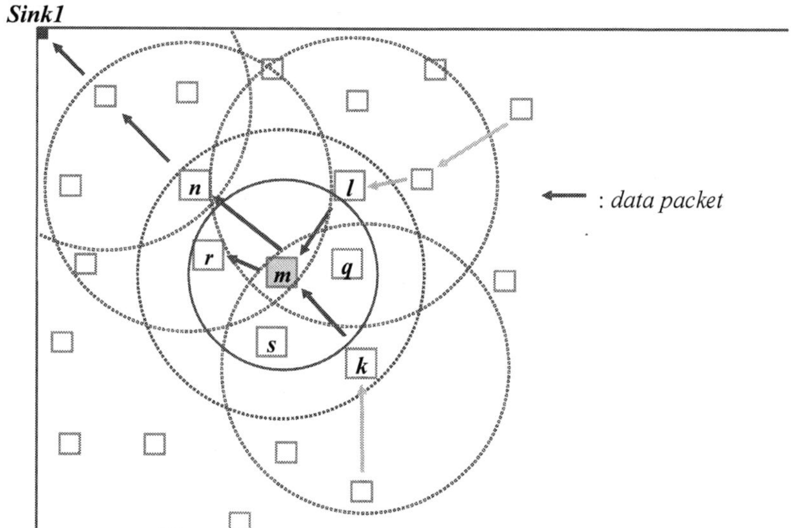

node *m* routing table

Next Hop					
node k	*node l*	*node n*	*node q*	*node r*	*node s*
10.0	20.0	50.0	12.0	25.0	12.0

Fig. 3. An example of transmission power control

Fig.3 shows an example of the above transmission power control, which means that the tra-nsmission power of each sensor node is switched to low power according to the above con-dition. In this example, node (*m*) is a load concentration node. Node (*m*) has autonomously transmitted the data packet to node (*r*) with the greatest connective value within low power range by low power because its own residual energy has become less than the set threshold [T_e]. By switching to low power, the energy consumption of node (*m*) is saved, but node (*k*) and node (*l*) may continue to transmit the data packet to node (*m*) because they cannot grasp the updated connective value of node (*m*). In our scheme, therefore, every tenth data packet from the node switched to low power is transmitted by high power.

3. Simulation experiment

Through simulation experiments on a wireless sensor network with multiple sinks, the perf-ormances of our scheme have been investigated in detail to verify its effectiveness.

3.1 Conditions of simulation

In a large scale and dense wireless sensor network with multiple sinks consisting of many static sensor nodes placed in a large scale observation area, only sensor nodes that detected abnormal data set were assumed to transmit the measurement data. The conditions of the si-mulation which were used in the experiments performed are shown in Table1. In the initial state of the simulation experiments, static sensor nodes are randomly arranged in the set ex-perimental area, and multiple sinks are placed on the boundaries containing the corners of this area. The network configuration is shown in Fig.4. In the experiments performed, the value attenuation factor accompanying hop (*dr*) and the "value to self" of each sink node in-troduced in our scheme were set to 0.5 and 100.0, respectively.

Simulation size	*2400m × 2400m*
Number of sensor nodes	*750, 1000, 1250*
Range of radio wave	*150m or 200m*
Number of sinks	*2 or 3*
Size of each data packet	*18 [bytes]*
Size of each control packet	*6 [bytes]*
Battery capacity of each sensor node	*0.2 [J] or 0.5 [J]*

Table 1. Conditions of simulation

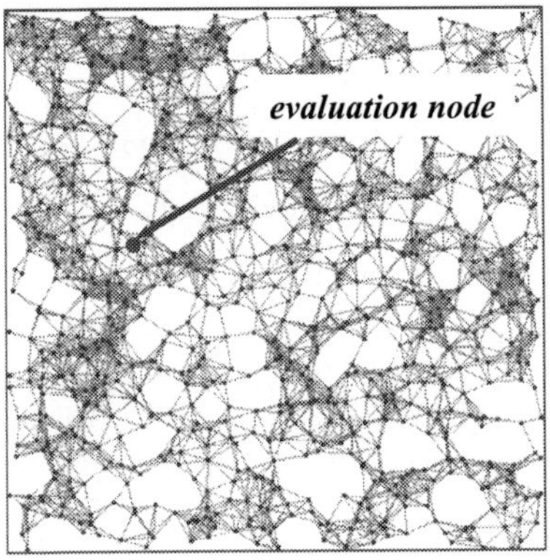

Fig. 4. Large scale and dense wireless sensor network consisting of many static sensor nodes

In the experimental results reported, our scheme (Matsumoto et al., 2010) is evaluated thro-ugh a comparison with existing ones (Dubois-Ferriere et al., 2004; Oyman & Ersoy, 2004; Ohtaki et al., 2006; Utani et al., 2008) where the parameter settings that produced good results in a preliminary investigation were adopted in preference to existing ones.

3.2 Experimental results on simulation model with two sinks

In this subsection, experimental results on the simulation model with two sinks of our sche-me without transmission power control are shown, where the number of sensor nodes was 1000, the range of radio wave and the battery capacity of each sensor node were set to 150m and 0.5J, respectively.

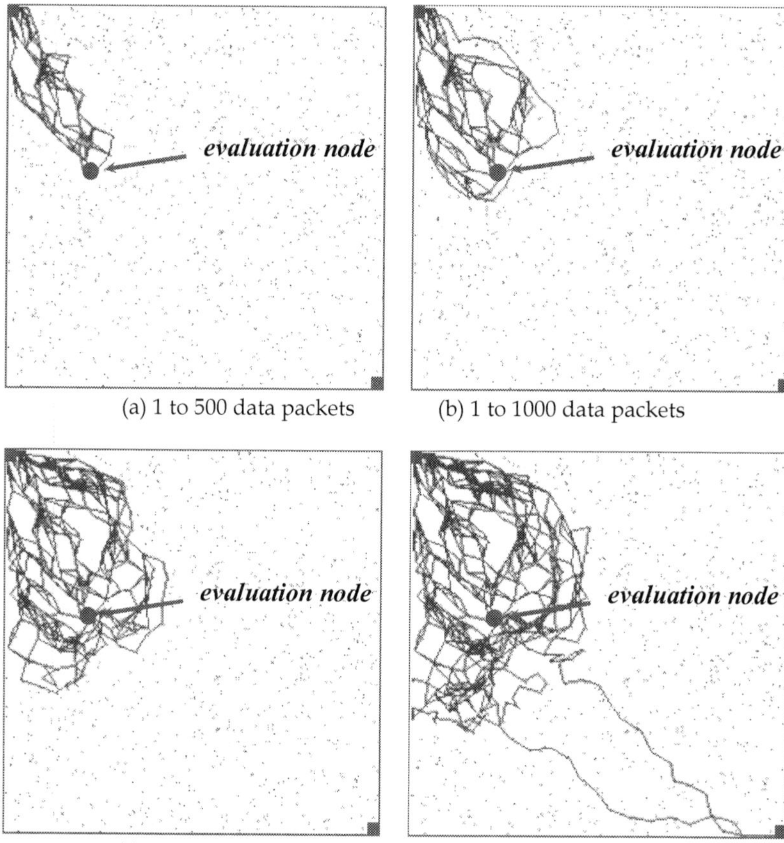

(a) 1 to 500 data packets (b) 1 to 1000 data packets

(c) 1 to 2000 data packets (d) 1 to 3000 data packets

Fig. 5. Routes used by applying our scheme to the simulation model with two sinks

As the first experiment on the simulation model with two sinks, it was assumed that the evaluation node marked in Fig.4 detected an abnormal value and transmitted the data packet with this abnormal value periodically. The routes used by applying our scheme are shown in Fig.5. Of the 3000 data packets transmitted from the evaluation node, the routes used by the first 500 data packets are illustrated in Fig.5(a), those used by the 1000 data packets are in Fig.5(b), those used by the 2000 data packets are in Fig.5(c), and those used by a total of 3000 data packets are in Fig.5(d). From Fig.5, it can be confirmed that our scheme enables the autonomous load-balancing transmission of data packets to two sinks using multiple routes.

Next, it was assumed that data packets were periodically transmitted from a total of 20 sens-or nodes placed in the set simulation area. In Fig.6, the transition of the delivery ratio of the total number of data packets transmitted from a total of 20 randomly selected sensor nodes is shown, and the lifetime of the simulation model with two sinks, as in Fig.5, is compared. In Fig.6, the existing schemes in Ohtaki et al., 2006 and Utani et al., 2008, which belong to the category of ant-based routing algorithms, are denoted as *MUAA* and *AAR*, respectively. The existing scheme in Dubois-Ferriere et al., 2004 and Oyman and Ersoy, 2004, which describe representative data gathering for a wireless sensor network with multiple sinks, is denoted as *NS*. From Fig.6, it can be confirmed that our scheme denoted as *Proposal* in Fig.6 achieves a longer-term operation of a wireless sensor network with multiple sinks than the existing ones because it improves and balances the load of each sensor node by the communication load reduction for network control and the autonomous load-balancing data transmission. Through simulation experiments, it was verified that our scheme (Matsumoto et al., 2010) is substantially advantageous for the long-term operation of a large scale and dense wireless sensor network with multiple sinks.

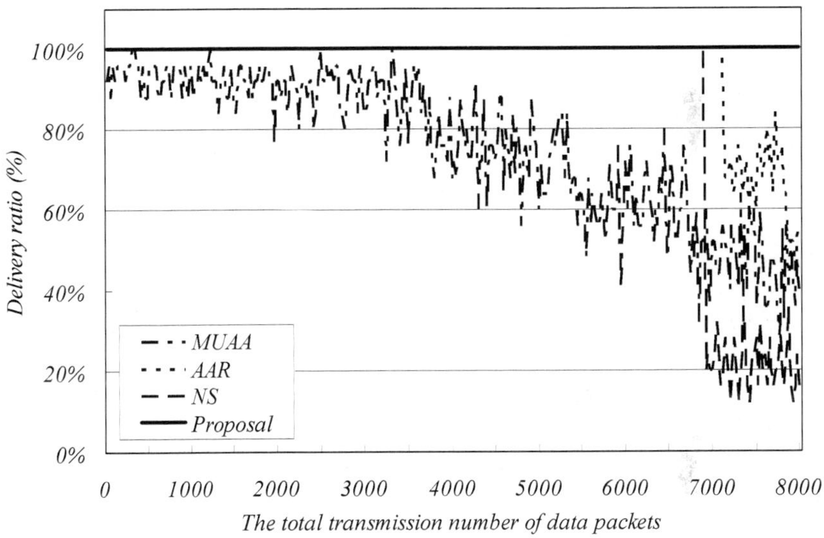

Fig. 6. Transition of delivery ratio

3.3 Experimental results on simulation model with three sinks

In this subsection, through experimental results on the simulation model with three sinks, the effectiveness of the transmission power control introduced in our scheme is evaluated. In the following experimental results, the battery capacity of each sensor node was set to 0.2J, and the range of radio wave of high power transmission in each sensor node was set to 200 m and it of low power transmission in each sensor node was set to 150m.

As the first experiment on the simulation model with three sinks, it was assumed that the evaluation node marked in Fig.4 detected an abnormal value and transmitted the data packet with this abnormal value periodically, as in the above subsection 3.2. The routes used by applying our scheme are shown in Figs.7, 8 and 9, where the number of sensor nodes is 1000. In Figs.7, 8 and 9, T_e was set to 0.0J, E×0.5J, and E×0.9J, where E indicates the battery capaci-ty of each sensor node. Of the 3000 data packets transmitted from the evaluation node, the r-outes used by the first 500 data packets are illustrated in Figs.7, 8 and 9(a), those used by the 1000 data packets are in Figs.7, 8 and 9(b), those used by the 2000 data packets are in Figs.7, 8 and 9(c), and those used by a total of 3000 data packets are in Figs.7, 8 and 9(d). From Figs. 7, 8 and 9, it can be confirmed that the effect of our scheme is extended by early switching to low power.

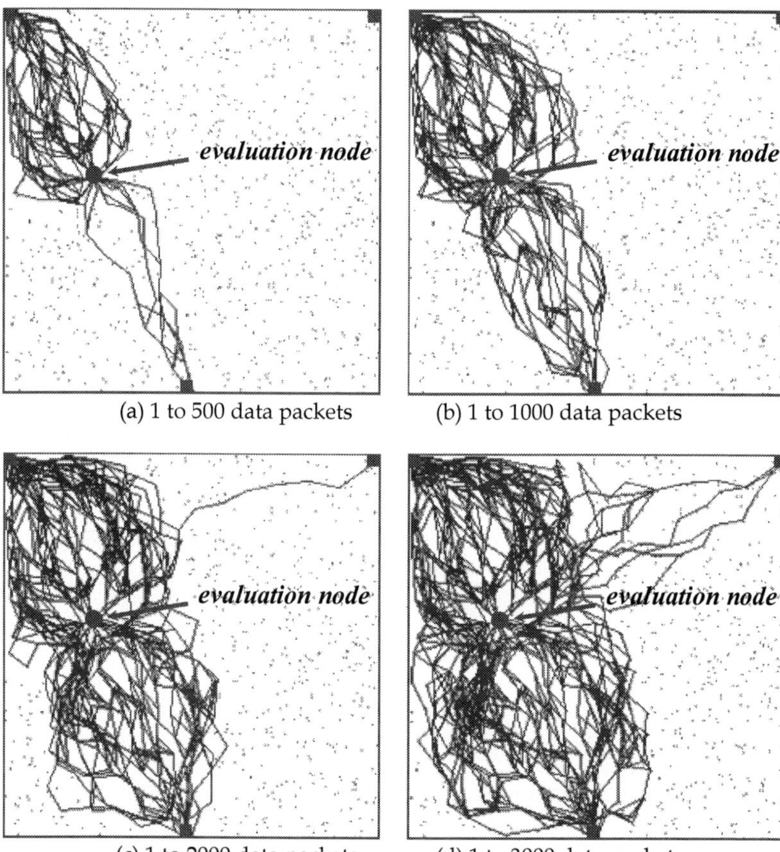

Fig. 7. Routes used by applying our scheme (T_e = 0.0J)

Next, it was assumed that data packets were periodically transmitted from a total of 20 sens-or nodes placed in the set simulation area. In Figs.10, 11 and 12, the transition of the delivery ratio of the total number of data packets transmitted from a total of 20 randomly selected sensor nodes is shown, and the lifetime of the simulation model with three sinks, as in Figs.7, 8 and 9, is compared. From Figs.10, 11 and 12, it can be confirmed that the effect of our scheme is extended by early switching to low power in high node density.

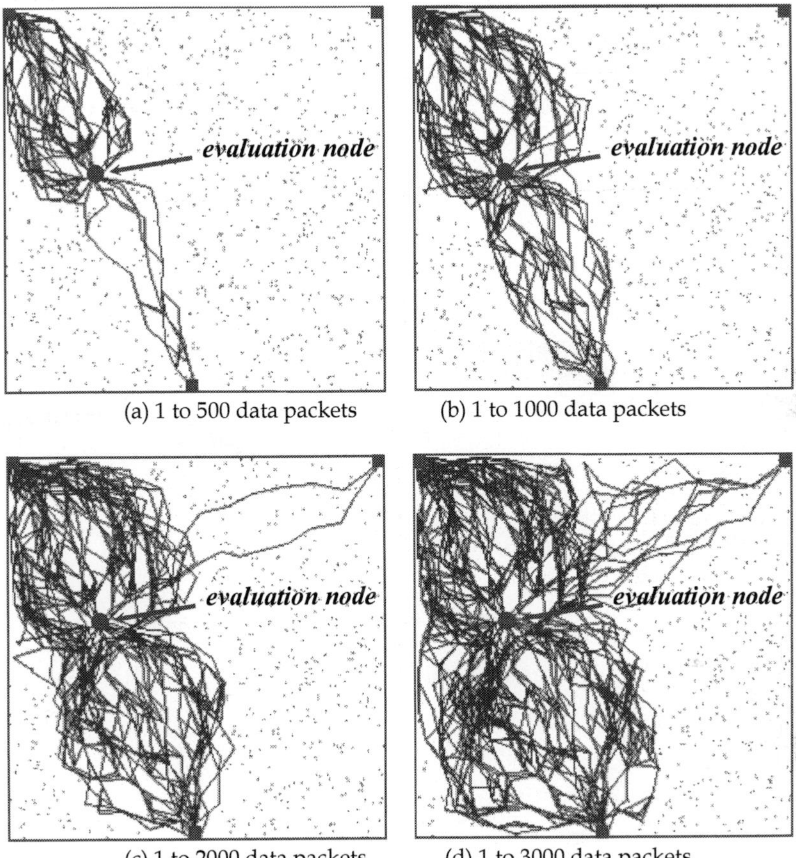

Fig. 8. Routes used by applying our scheme ($T_e = E \times 0.5$J)

3.4 Discussion

To facilitate ubiquitous information environments by wireless sensor networks, their control mechanisms should be adapted to the variety of types of communication, depending on ap-plication requirements and the context. Currently, adaptive communication protocols for the long-term operation of the above ubiquitous sensor networks (Intanagonwiwat et al., 20-03; Silva et al., 2004; Heidemann et al., 2003; Krishnamachari & Heidemann, 2003; Wakabay-ashi et al., 2007) are under study. In addition, the advanced design schemes of wireless sens-or networks, such as sink node allocation schemes based on the particle swarm optimization algorithms aiming to minimize total hop counts in a network and to reduce the energy cons-umption of each sensor node (Kumamoto et al., 2008; Yoshimura et al., 2009; Taguchi et al., 2010), and forwarding node set selection schemes (Nagashima et al., 2009; Sasaki et al., 2010) and forwarding power adjustment scheme (Nagashima et al., 2011) for adaptive and efficie-nt query dissemination throughout a wireless sensor network, are positively researched. By coupling our scheme (Matsumoto et al., 2010) with the above advanced design schemes, it can be expected that the lifetime of a wireless sensor network is moreover prolonged.

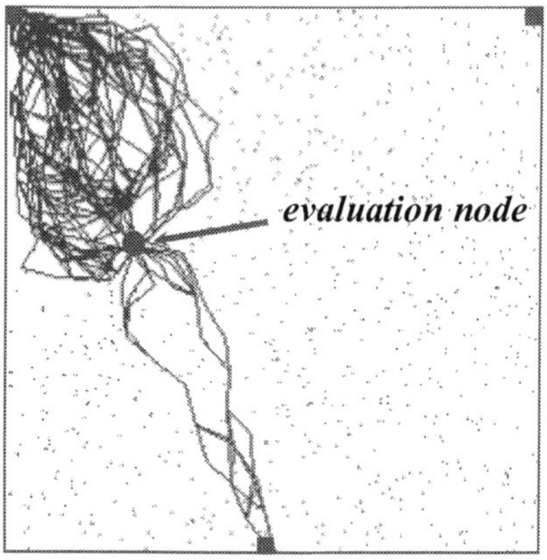

(a) 1 to 500 data packets

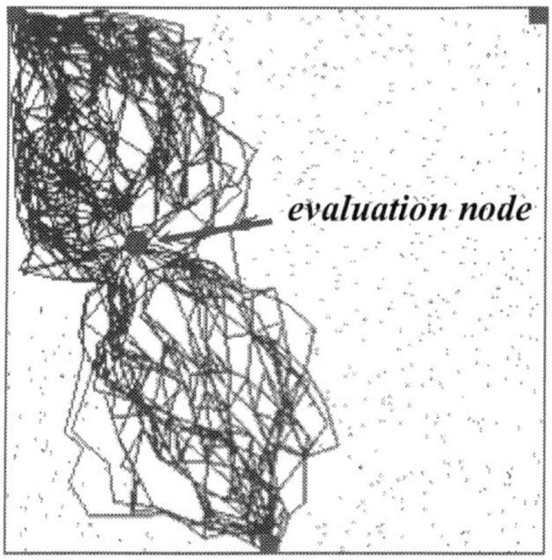

(b) 1 to 1000 data packets

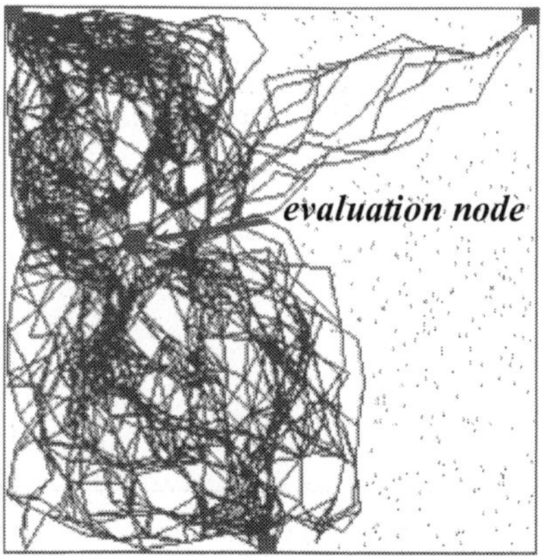

(c) 1 to 2000 data packets

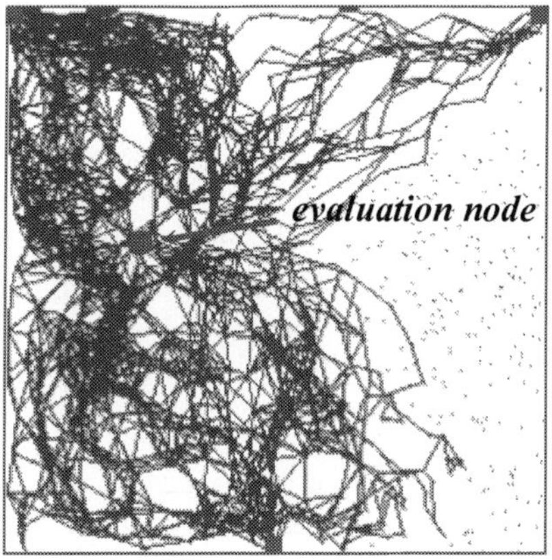

(d) 1 to 3000 data packets

Fig. 9. Routes used by applying our scheme ($T_e = E \times 0.9$J)

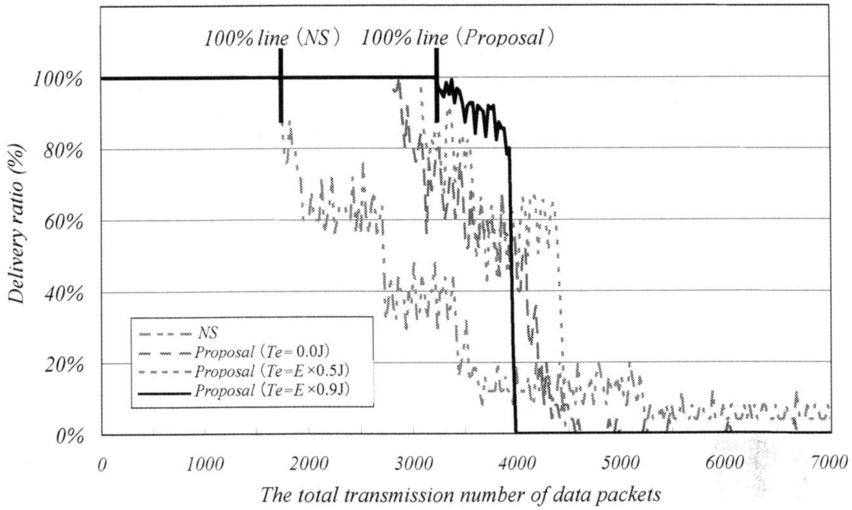

Fig. 10. Transition of delivery ratio (The number of sensor nodes is 750)

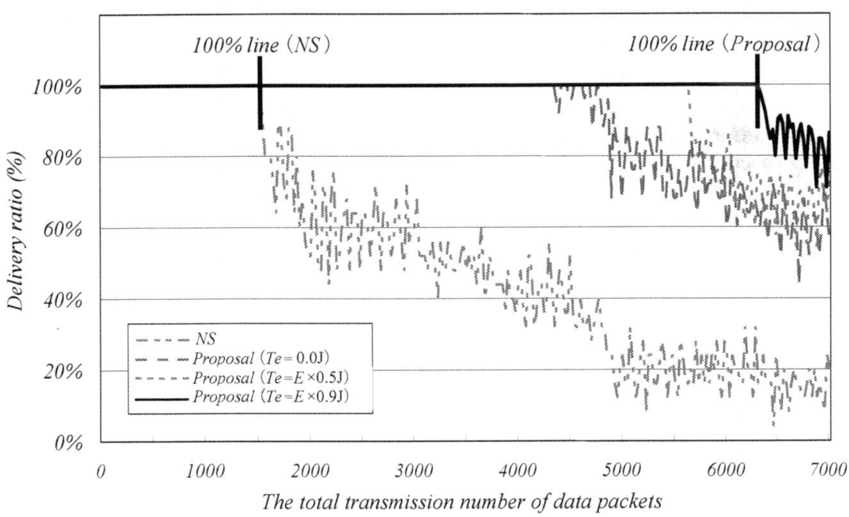

Fig. 11. Transition of delivery ratio (The number of sensor nodes is 1000)

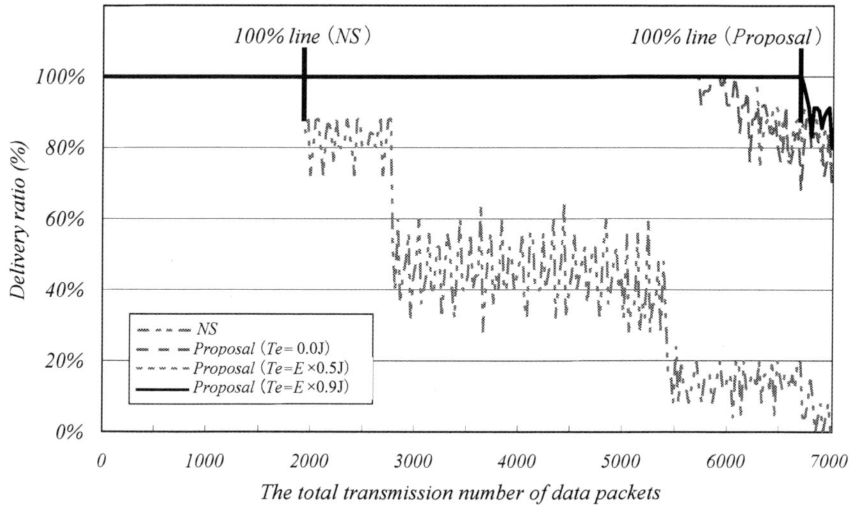

Fig. 12. Transition of delivery ratio (The number of sensor nodes is 1250)

4. Conclusions

In this chapter, a new data gathering scheme with transmission power control that adaptively reduces the load of load-concentrated nodes and facilitates the long-term operation of a large scale and dense wireless sensor network with multiple sinks, which is an autonomous load-balancing data transmission one devised by considering the application environment of a wireless sensor network to be a typical example of complex systems, has been represented. In simulation experiments, the performances of this scheme were compared with those of the existing ones. The experimental results indicate that this scheme is superior to the existing ones and has the development potential as a promising one from the viewpoint of the long-term operation of wireless sensor networks. Future work includes a detailed evaluation of the parameters introduced in this scheme in various network environments.

5. Acknowledgment

The development of a new autonomous decentralized control scheme for the long-term operation of wireless sensor networks with multiple sinks represented in this chapter is supported by the Grant-in-Aid for Scientific Research (Grant No.21500082) from the Japan Society for the Promotion of Science.

6. References

Akyildiz, I.; Su, W.; Sankarasubramaniam, Y. & Cayirci, E. (2002). Wireless sensor networks: A survey, *Computer Networks Journal*, Vol.38, No.4, 393-422

Clausen, T. & Jaquet, P. (2003). Optimized link state routing protocol, *Request for Comments (RFC) 3626*

Dasgupta, K.; Kalpakis, K. & Namjoshi, P. (2003). An efficient clustering-based heuristic for data gathering and aggregation in sensor networks, *Proceedings of IEEE Wireless Communications and Networking Conference*, 16-20

Dubois-Ferriere, H.; Estrin, D. & Stathopoulos, T. (2004). Efficient and practical query scoping in sensor networks, *Proceedings of IEEE International Conference on Mobile Ad-Hoc and Sensor Systems*, 564-566

Heidemann, J.; Silva, F. & Estrin, D. (2003). Matching data dissemination algorithms to application requirements, *Proceedings of 1st ACM Conference on Embedded Networked Sensor Systems*, 218-229

Heinzelman, W.R.; Chandrakasan, A. & Balakrishnan, H. (2000). Energy-efficient communication protocol for wireless microsensor networks, *Proceedings of Hawaii International Conference on System Sciences*, 3005-3014

Intanagonwiwat, C.; Govindan, R.; Estrin, D.; Heidemann, J. & Silva, F. (2003). Directed diffusion for wireless sensor networking, *ACM/IEEE Transactions on Networking*, Vol.11, 2-16

Jin, Y.; Jo, J. & Kim, Y. (2008). Energy-efficient multi-hop communication scheme in clustered sensor networks, *International Journal of Innovative Computing, Information and Control*, Vol.4, No.7, 1741-1749

Johnson, D.B.; Maltz, D.A.; Hu, Y.C. & Jetcheva, J.G.(2003). The dynamic source routing protocol for mobile ad hoc networks, *IETF Internet Draft*, draft-ietf-manet-dsr-09.txt

Krishnamachari, B. & Heidemann, J. (2003). Application-specific modeling of information routing in wireless sensor networks, *Technical Report*, ISI-TR-2003-576, USC-ISI

Kumamoto, A.; Utani, A. & Yamamoto, H. (2008). Improved particle swarm optimization for locating relay-dedicated nodes in wireless sensor networks, *Proceedings of 2008 Joint 4th International Conference on Soft Computing and Intelligent Systems and 9th International Symposium on Advanced Intelligent Systems*, 1971-1976

Matsumoto, K.; Utani, A. & Yamamoto, H. (2009). Adaptive and efficient routing algorithm for mobile ad-hoc sensor networks, *ICIC Express Letters*, Vol.3, No.3(B), 825-832

Matsumoto, K.; Utani, A. & Yamamoto, H. (2010). Bio-inspired data transmission scheme to multiple sinks for the long-term operation of wireless sensor networks, *International Journal of Artificial Life and Robotics*, Vol.15, No.2, 189-194

Nagashima, J.; Utani, A. & Yamamoto, H. (2009). Efficient flooding method using discrete particle swarm optimization for long-term operation of sensor networks, *ICIC Express Letters*, Vol.3, No.3(B), 833-840

Nagashima, J.; Utani, A. & Yamamoto, H. (2011). A study on efficient query dissemination in distributed sensor networks -Forwarding power adjustment of each sensor node using particle swarm optimization-, *Proceedings of 16th International Symposium on Artificial Life and Robotics*, 703-706

Nakano, H.; Utani, A.; Miyauchi, A. & Yamamoto, H. (2009). Data gathering scheme using chaotic pulse-coupled neural networks for wireless sensor networks, *IEICE Transactions on Fundamentals*, Vol.E92-A, No.2, 459-466

Nakano, H.; Utani, A.; Miyauchi, A. & Yamamoto, H. (2011). Chaos synchronization-based data transmission scheme in multiple sink wireless sensor networks, *International Journal of Innovative Computing, Information and Control*, Vol.7, No.4, 1983-1994

Ogier, R.; Lewis, M. & Templin, F.(2003). Topology dissemination based on reverse-path forwarding (TBRPF), *IETF Internet Draft*, draft-ietf-manet-tbrpf-10.txt

Ohtaki, Y.; Wakamiya, N.; Murata, M. & Imase, M. (2006). Scalable and efficient ant-based routing algorithm for ad-hoc networks, *IEICE Transactions on Communications*, Vol.E 89-B, No.4, 1231-1238

Oyman, E.I. & Ersoy, C. (2004). Multiple sink network design problem in large scale wireless sensor networks, *Proceedings of 2004 International Conference on Communications*, Vol. 6, 3663-3667

Perkins, C.E. & Royer, E.M. (1999). Ad hoc on-demand distance vector routing, *Proceedings of 2nd IEEE Workshop on Mobile Computing Systems and Applications*, 90-100

Rajagopalan, R. & Varshney, P.K. (2006). Data aggregation techniques in sensor networks: A survey, *IEEE Communications Surveys and Tutorials*, Vol.8, 48-63

Sasaki, T.; Nakano, H.; Utani, A.; Miyauchi, A. & Yamamoto, H. (2010). An adaptive selection scheme of forwarding nodes in wireless sensor networks using a chaotic neural network, *ICIC Express Letters*, Vol.4, No.5(A), 1649-1655

Silva, F.; Heidemann, J.; Govindan, R. & Estrin, D. (2004). Directed diffusion, *Technical Report*, ISI-TR-2004-586, USC-ISI

Subramanian, D.; Druschel, P. & Chen, J. (1998). Ants and reinforcement learning: A case study in routing in dynamic networks, *Technical Report TR96-259*, Rice University

Taguchi, Y.; Nakano, H.; Utani, A.; Miyauchi, A. & Yamamoto, H. (2010). A competitive particle swarm optimization for finding plural acceptable solutions, *ICIC Express Letters*, Vol.4, No.5(B), 1899-1904

Utani, A.; Orito, E.; Kumamoto, A. & Yamamoto, H. (2008). An advanced ant-based routing algorithm for large-scale mobile ad-hoc sensor networks, *Transactions on SICE*, Vol. 44, No.4, 351-360

Wakabayashi, M.; Tada, H.; Wakamiya, N.; Murata, M. & Imase, M. (2007). Proposal and evaluation of a bio-inspired adaptive communication protocol for sensor networks, *IEICE Technical Report*, Vol.107, No.294, 89-94

Wakamiya, N. & Murata, M.(2005). Synchronization-based data gathering scheme for sensor networks, *IEICE Transactions on Communications*, Vol.E88-B, No.3, 873-881

Xia, L.; Chen, X. & Guan, X.(2004). A new gradient-based routing protocol in wireless sensor networks, *Lecture Notes in Computer Science*, Vol.3605, 318-325

Yamamoto, I.; Ogasawara, K.; Ohta, T. & Kakuda, Y. (2009). A hierarchical multicast routing using inter-cluster group mesh structure for mobile ad hoc networks, *IEICE Transactions on Communications*, Vol.E92-B, No.1, 114-125

Yoshimura, M.; Nakano, H.; Utani, A.; Miyauchi, A. & Yamamoto, H. (2009). An effective allocation scheme for sink nodes in wireless sensor networks using suppression PSO, *ICIC Express Letters*, Vol.3, No.3(A), 519-524

26

Collaborative Environmental Monitoring with Hierarchical Wireless Sensor Networks

Qing Ling[1], Gang Wu[1] and Zhi Tian[2]
[1]*Department of Automation, University of Science and Technology of China*
[2]*Department of Electrical and Computer Engineering, Michigan Technological University*
[1]*China*
[2]*USA*

1. Introduction

In the last decade, advances in wireless communication and micro-fabrication have motivated the development of large-scale wireless sensor networks (Akyildiz et al., 2002; Yick et al., 2008). A large number of low-cost sensor nodes, equipped with sensing, computing, and communication units, organize themselves into a multi-hop network. The wireless sensor network takes measurements from the environment, processes the sensory data, and transmits the sensory data to end-users. Beginning from the seminar work in (Estrin et al., 1999; 2002), the wireless sensor network technology has been well recognized as a revolutionary one that transforms everyday life. Typical applications of wireless sensor networks include military target tracking and surveillance (Simon et al., 2004; He et al., 2006), precise agriculture (Langendoen et al., 2006; Wark et al., 2007), industrial automation (Gungor and Hancke, 2009), structural health monitoring (Li and Liu, 2007; Ling et al., 2009), environmental and habitat monitoring (Zhang et al., 2004; Corke et al., 2010), to name a few.

1.1 Network infrastructure

To organize the large amount of sensor nodes and enable efficient data collection, a wireless sensor network generally adopts one of the following three infrastructures: centralized, decentralized, and hierarchical. In the centralized infrastructure, sensor nodes transmit the sensory data to the fusion center via multi-hop communication. In the decentralized infrastructure, each sensor node firstly refines the sensory data through collaborative and decentralized in-network processing with the neighboring sensor nodes, and secondly transmits the refined data to the fusion center. While in the hierarchical infrastructure, sensor nodes are divided into multiple clusters, and sensor nodes within one cluster send their sensory data to the cluster head. These cluster heads either transmit the collected sensory data to the fusion center, or collaboratively process them and transmit the refined one to the fusion center. These two different implementations of the hierarchical infrastructure, centralized processing and decentralized collaboration, are depicted in Figure 1.

In deploying a wireless sensor network, the choice of its infrastructure is decided by several key factors: energy, bandwidth, robustness, etc. Sensor nodes are often equipped with batteries and recharging is difficult. Since wireless data transmission is the main source of energy consumption of a sensor node (Sadler, 2005), the network infrastructure

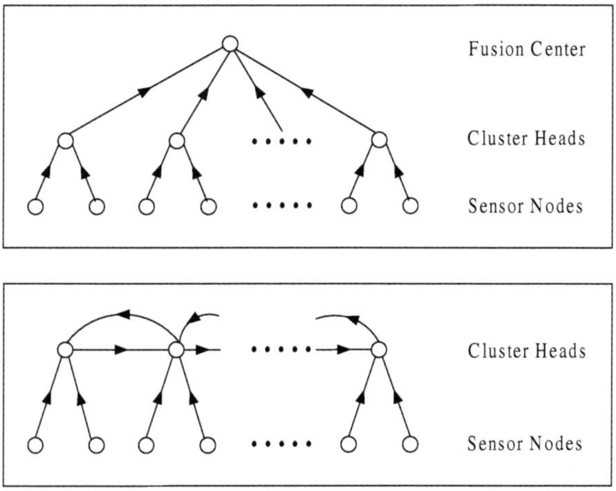

Fig. 1. Two different schemes of implementing the hierarchical infrastructure: (TOP) centralized processing in a fusion center and (BOTTOM) decentralized collaboration among cluster heads.

should guarantee that each sensor node has low data transmission rate while successfully accomplishing the data collection task. Bandwidth is also a kind of precious resource in wireless environment; over-competition of wireless channels leads to frequent retransmission and hence consumes more energy. Further, sensor nodes are often fragile due to being out of batteries or other physical damages. The network infrastructure should be carefully designed such that the failure of few sensor nodes shall not result in the malfunction of the whole network.

When the network size is small, the centralized infrastructure is an acceptable choice. Take a volcano monitoring network containing 3 sensor nodes (Werner-Allen et al., 2005) as an example, these sensor nodes directly connect to a fusion center which collects sensory data and transmits them to the end-user. Later on the network is extended to the scale of 16 sensor nodes (Werner-Allen et al., 2006), and the sensor nodes communicate with the fusion center via multi-hop relays. However, for GreenOrbs (Liu et al., 2011), a large-scale forest monitoring network composed of up to 330 sensor nodes, experiments demonstrate that sensor nodes within some "hot areas" may face higher competition for bandwidth, consume more energy, and be more sensitive to the failure of sensor nodes. The decentralized infrastructure, on the other hand, has great potential to reduce the total amount of transmitted data and hence improve the energy efficiency via in-network collaboration; further, it also enhances robustness of the network since all sensor nodes play equal roles (Ling and Tian, 2010). Nevertheless, collaboration of the sensor nodes brings more difficulty to network coordination, and is subject to the limited processing and communication capabilities of sensor nodes. For this reason, the decentralized infrastructure is still far from practical applications. To the best of our knowledge, most large-scale wireless sensor networks are deployed with the hierarchical infrastructure. Following we give some examples: ExScal, an intrusion detection network with more than 1000 sensor nodes and more than 200 backbone nodes (Arora et al., 2005); VigilNet, a military surveillance network with 200 sensor nodes (He et al., 2006); Trio, a target tracking network with 557 solar-powered sensor nodes

(Dutta et al., 2006); SenseScope, an environmental monitoring network consisting of from 3 to 97 sensor nodes (Barenetxea et al., 2008). In view of this fact, we will focus on the design of a hierarchical wireless sensor network.

1.2 Our contributions

In some hierarchical wireless sensor networks such as ExScal (Arora et al., 2005), the cluster heads are specifically designed, having better data processing and wireless communication abilities than general sensor nodes, and equipped with stronger or even uninterruptible power sources. These cluster heads can directly transmit the collected data to a remote fusion center, without introducing any collaborative processing among cluster heads. However, in most wireless sensor networks, cluster heads are elected from sensor nodes to simplify design, deployment, and maintenance. For example, in the LEACH protocol (Heinzelman et al., 2002), sensor nodes autonomously elect cluster heads, aiming at evenly distributing energy consumption among all sensor nodes so that there are no overly-utilized sensor nodes that will run out of energy before the others. In this case, how to process the collected sensory data in the cluster heads is a critical problem to accomplishing the data collection task while maximizing the network lifetime.

This chapter addresses this problem; specifically, we study a generalized environmental monitoring model with large-scale hierarchical wireless sensor networks, and focus on two questions: for cluster heads in a hierarchical network, *should they collaborate or not collaborate* and *how can they collaborate*. Our contributions are two-fold.

First, through theoretical analysis and simulation validation, we make the following recommendations on whether to collaborate or not: when each cluster head has a large amount of data to process (namely, each cluster contains a large number of sensor nodes) and multi-hop relay is necessary to communicate with a fusion center (namely, cluster heads have limited communication range), decentralized data processing among cluster heads is more efficient; otherwise centralized decision-making with the aid of a fusion center can be advantageous.

Previous work, such as (Rabbat and Nowak, 2004; Aldosari and Moura, 2004), has suggested similar network design principles in the context of decentralized infrastructures: when each sensor node collects a large amount of data or the size of the network is large, collaborative processing is more efficient than centralized decision-making. This paper extends the conclusions to hierarchical networks, and compares decentralized versus centralized processing among cluster heads rather than among all sensor nodes.

Second, we develop a decentralized collaborative algorithm for decision making among the sub-network of cluster heads, after they have collected sensory data from local sensor nodes within their individual clusters. Particularly, we study a typical environment monitoring application, in which a large-scale hierarchical wireless sensor network is deployed to monitor sparsely occurring phenomena over a large sensing field. The monitoring problem is formulated as a non-negative quadratic program, which optimizes a sparse decision vector depicting the spatial map of the phenomena of interest. An optimal iterative algorithm, in which cluster heads iteratively exchange information and make decisions, is proposed based on the alternating direction method of multipliers (ADMM) (Bertsekas and Tsitsiklis, 1997). Our development is permeated with the benefits of compressive sensing (Donoho et al., 2006). Exploiting the sparse nature of the unknown phenomena, we allow the number of sensor nodes to be much smaller than what would have been required in a traditional scheme for sensing at high spatial resolution over a large field. In this sense, our proposed algorithm is also applicable to other compressive sensing problems in distributed systems.

1.3 Chapter organization

The rest of this chapter is organized as follows. We first give a brief survey on the applications of wireless sensor networks in environmental monitoring. Second, we study a generalized environmental monitoring model with large-scale hierarchical wireless sensor networks and develop a decentralized collaborative algorithm for decision making among the cluster heads. Finally we discuss the design consideration, namely, to collaborate or not to collaborate, based on theoretical analysis and simulation results.

2. A brief survey

In this section, we give a brief survey on the applications of wireless sensor networks in environmental and habitat monitoring. Though this overview is far from complete, it reflects the promising future of the wireless sensor network technology in helping us understand and protect natural environment.

For environmental and habitat monitoring applications, one of the first known practical wireless sensor networks was deployed by a group at Berkeley in 2002, on Great Duck Island on the coast of Maine, USA. Two networks with a total of 147 sensor nodes collect data to study the ecology of the Leach's Storm Petrel (Szewsczyk et al., 2004). Later on, the Macroscope system which contains 33 sensor nodes, also developed at Berkeley, was used for microclimate monitoring of a redwood tree (Tolle et al., 2005). Another notable application is ZebraNet, which used GPS technology to record position data in order to track long term animal migrations. In the prototype system, researchers deployed 7 sensor nodes on zebras in Kenya (Zhang et al., 2004). Energy harvesting technologies have also attracted much research interest to address the challenge of energy supply in remote environmental monitoring applications. One successful example is LUSTER, which was developed at University of Virginia, featuring a specifically designed hybrid multichannel energy harvesting device (Selavo et al., 2007). Accompanied with the unprecedented data collection opportunities, data processing also emerges as a new challenge in the wireless sensor network technology. The data processing task is indeed application-oriented. For example, an ellipsoids-based anomaly detection algorithm was designed to monitor unusual and anomalous behaviors in a particular marine ecosystem (Bedzek et al., 2011). The network was deployed in 2009 at the Heron Island, Australia, as part of the Great Barrier Reef Ocean Observation System.

One significant advantage of wireless sensor networks over traditional data collection techniques is that they can be applied in harsh environments. For example, in the GlacsWeb system, researchers at University of Southampton deployed 9 sensor nodes inside a glacier (Martinez et al., 2004). The sensor nodes monitored pressure, temperature, and tilt, in order to monitor subglacial bed deformation. Even on active volcanos, which are often forbidden areas for data collection, wireless sensor networks can still work well. In the work of (Werner-Allen et al., 2005; 2006), one small sensor network with 3 sensor nodes was deployed on Vlcan Tungurahua in Ecuador as a proof of concept in 2004; then in 2005, the network size was extended to 16 sensor nodes. Wireless sensor networks are also fit for aquatic environmental monitoring applications. In (Alippi et al., 2011), a robust, adaptive, and solar-powered network was developed in 2007 for such an application. The network was deployed in Queensland, Australia, for monitoring the underwater luminosity and temperature, information necessary to derive the health status of the coralline barrier. At the same time, sensory data can be used to provide quantitative indications related to cyclone formations in tropical areas.

However, applying wireless sensor networks in environmental monitoring is still a challenging task when the network size is large. When the number of sensor nodes increases, difficulties emerge for system integration (creating an end-to-end system that delivers data to the end-user), performance (reliability, accuracy, and calibration), productivity (how well the sensory data assists the end-user and how to reduce the total cost in implementing the wireless sensor network), etc (Corke et al., 2010). One negative example is reported in (Langendoen et al., 2006), in which researchers at Delft University of Technology deployed a large-scale network in a potato field to improve the protection of potatoes against disease. The application was not successful due to unanticipated issues; nevertheless, the lessons are precious, such as software, hardware, and even team coordination. A systematic discussion, named as "the hitchhiker's guide", is provided in (Barenetxea et al., 2008). Based on the deployment of a wireless sensor network on a rock glacier located at a mountain in the Swiss Alps, this guide covers almost all stages of a project, from hardware and software development, testing and preparation, to deployment. One of the recent efforts to investigate the practical implementations of large-scale wireless sensor networks is the GreenOrbs system (Liu et al., 2011). The network with 330 sensor nodes was deployed in Tianmu Mountian, China, aiming at all-year-around ecological surveillance in the forest. It is shown that many traditional design guidelines for small-scale wireless sensor networks can be questionable for large-scale applications.

3. Problem formulation

In this chapter, we focus on a generalized event detection model for environmental monitoring applications. Let us consider a large-scale wireless sensor network randomly deployed in a two-dimensional area for monitoring sparsely occurring events. The network has a set of L sensor nodes, denoted as $\mathcal{L} = \{v_l\}_{l=1}^{L}$. Sensor nodes are divided into I clusters, each having one cluster head in the set $\mathcal{I} = \{c_i\}_{i=1}^{I}$. Sensor nodes within a cluster are able to directly transmit measurements to the cluster head, and the cluster head is aware of the positions of all sensor nodes within its cluster. Further, the cluster heads have a common communication range r_C such that the sub-network of cluster heads is bi-directionally connected.

3.1 Basic assumptions

Suppose that at each sampling time, multiple phenomena may occur in the sensing field. Our objective is to detect and identify the source locations and estimate their amplitudes from sensory measurements. We make the following basic assumptions for the sensing problem of interest, similar to those in (Bazerque and Giannakis, 2010):

(A1): The sensing field is viewed through a spatial grid with K grid points denoted by $\mathcal{K} = \{g_k\}_{k=1}^{K}$, whose locations are known to the corresponding cluster heads. Each event can occur only at a grid point, indicating the spatial resolution offered by this sensor network. The amplitude of an event occurring at grid point g_k is x_k.

(A2): The influence of a unit-amplitude event at grid point g_k on a sensor point v_l is f_{kl}. Generally speaking, f_{kl} is decided by the distance d_{kl} between g_k and v_l.

(A3): The measurement of one sensor node is the linear superposition of the influences of all phenomena plus random noise. Mathematically, the measurement b_l of sensor node v_l is hence $b_l = \sum_{k=1}^{K} f_{kl} x_k + e_l$ in which x_k is the amplitude of event at $g_k \in \mathcal{K}$ and e_l is measurement noise.

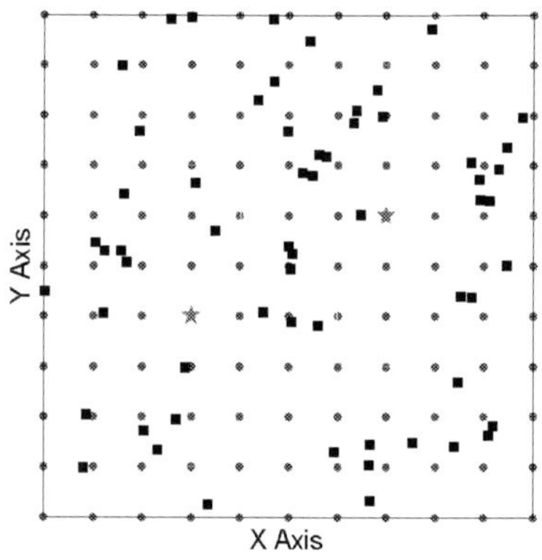

Fig. 2. The sensor nodes denoted as solid squares are uniformly randomly deployed in the monitoring area. The candidate positions for phenomena are grid points denoted as solid dots. Phenomena denoted as pentagrams occur at the current snapshot, and the shadow regions illustrate the influence of phenomena.

These assumptions are depicted in Figure 2. The sensor nodes denoted as solid squares are uniformly randomly deployed in the monitoring area. The candidate positions for phenomena are grid points denoted as solid dots. Phenomena denoted as pentagrams occur at the current snapshot, and the shadow regions illustrate the influence of phenomena.

The assumption **(A1)** simplifies the recovery problem by confining the sources of phenomena to grid points. Without this assumption, an alternative way is to use positions and amplitudes of the sources as decision variables and formulate a least squares problem. However, this formulation is highly nonlinear and intractable, since the number of decision variables is even unknown. Based on **(A1)**, we can formulate the otherwise nonlinear problem as recovering the vector $\mathbf{x} = [x_1, ..., x_K]^T$ from linear measurements $b_l, \forall l$. Entries in $\mathbf{x}$ with nonzero values reveal the locations and amplitudes of the multiple phenomena of interest. This assumption approximately holds when the grid points are dense; namely, the density of the grid points decides the spatial resolution of the recovery algorithm.

The assumption **(A2)** describes the influence of one event on the entire sensing field. For example, in target tracking or nuclear radioactive detection, the influence of a source decreases polynomially as the distance increases. Without loss of generality, we define the influence function as $f_{kl} = \exp(-d_{kl}^2/\sigma^2)$ for grid point g_k and sensor point v_l, where σ is a common constant. This Gaussian-shaped function well approximates the influence of many practical events.

Based on the assumption **(A3)**, we readily have the following least squares formulation for recovering $\mathbf{x}$:

$$\min \quad \sum_{v_l \in \mathcal{L}} (b_l - \sum_{k=1}^{K} f_{kl} x_k)^2, \tag{1}$$

or equivalently in a matrix form:

$$\min \quad ||\mathbf{F}\mathbf{x} - \mathbf{b}||_2^2. \tag{2}$$

Here $\mathbf{b} = [b_1, ..., b_L]^T$ is the measurement vector and $\mathbf{F}$ is the $L \times K$ influence matrix with its l-th row given by $[f_{1l}, ..., f_{Kl}]$.

Nevertheless, the least squares formulation (2) ignores the sparsity of the vector $\mathbf{x}$. Notice that when the grid is dense, the number of events is generally much smaller than the number of grid points; hence the vector $\mathbf{x}$ is a sparse vector with a large amount of zero entries. Without considering this prior knowledge, the least squares formulation (2) leads to a non-sparse solution, which means a non-neglectable number of false alarms. The sparsity of a signal vector can be measured by its ℓ_1 norm (Donoho et al., 2006). Exploiting the sparse nature of $\mathbf{x}$ to alleviate false alarms, we formulate the following ℓ_1 regularized least squares problem (Kim et al., 2007):

$$\min \quad \frac{\lambda}{2}||\mathbf{F}\mathbf{x}-\mathbf{b}||_2^2 + ||\mathbf{x}||_1. \tag{3}$$

Here λ is a non-negative weight.

3.2 Decentralized optimization

In a centralized setting, the ℓ_1 regularized least squares problem (3) has been extensively studied in both signal processing and numerical optimization communities (Donoho et al., 2006; Figueiredo et al., 2007). However, in a large-scale wireless sensor network, centralized processing is not efficient in terms of energy consumption and communication overhead. In contrast, collaborative signal processing among cluster heads is preferred, leading to a robust and scalable network.

We address this issue by developing a collaborative sparse signal recovery algorithm in the chapter. Sensor nodes or cluster heads do not necessarily exchange information with a fusion center; rather, sensor nodes only need to transmit measurements to their cluster heads, and cluster heads iteratively optimize the decision vector $\mathbf{x}$ via exchanging information with their neighboring cluster heads.

For each cluster head c_i, let us collect the local measurements $\{v_l\}$ within this cluster and their corresponding l-th rows in the measurement matrix $\mathbf{F}$ into a sub-vector $\mathbf{b}_i$ and a sub-matrix $\mathbf{F}_i$, for all sensor nodes v_l whose cluster head is c_i. Per assumptions **(A1)** and **(A2)**, each cluster head knows all sensor node locations and grid point locations within its cluster, which means that $\mathbf{F}_i$ is known to c_i. Hence the problem boils down to the following one: *suppose that the local measurement vector $\mathbf{b}_i$ and corresponding measurement matrix $\mathbf{F}_i$ are available to each cluster head c_i, $\forall i$, how can we design a decentralized algorithm to recover the signal $\mathbf{x}$ via collaboration among the cluster heads?*

Let $\mathbf{x}_i$ denote the local copy of the decision vector $\mathbf{x}$ at c_i, $\forall c_i \in \mathcal{I}$. Meanwhile, given the communication range r_C, the set of neighboring cluster heads of c_i is denoted by $\mathcal{N}_i$, with cardinality $|\mathcal{N}_i|$. The formulation in (3) can be transformed to the following consensus optimization problem:

$$\begin{aligned} \min \quad & \textstyle\sum_{i=1}^{I}\left(\frac{\lambda}{2}||\mathbf{F}_i\mathbf{x}_i-\mathbf{b}_i||_2^2 + \frac{1}{I}\mathbf{1}^T\mathbf{x}_i\right), \\ s.t. \quad & \mathbf{x}_i = \mathbf{x}_j, \quad \forall c_i \in \mathcal{I}, c_j \in \mathcal{N}_i. \end{aligned} \tag{4}$$

The $K \times 1$ all-one vector $[1, 1, ..., 1]^T$ is denoted as $\mathbf{1}$. Here, cluster heads optimize their own local copies of $\mathbf{x}$ separately, and these decision vectors are forced to be equal via the consensus constraints. An alternative formulation is to force $\mathbf{x}_i$ to consent with the average of its neighboring decisions, as follows:

$$\begin{aligned} \min \quad & \textstyle\sum_{i=1}^{I}\left(\frac{\lambda}{2}||\mathbf{F}_i\mathbf{x}_i-\mathbf{b}_i||_2^2 + \frac{1}{I}\mathbf{1}^T\mathbf{x}_i\right), \\ s.t. \quad & \mathbf{x}_i = \frac{1}{|\mathcal{N}_i|}\textstyle\sum_{c_j\in\mathcal{N}_i}\mathbf{x}_j, \quad \forall c_i \in \mathcal{I}. \end{aligned} \tag{5}$$

It has been proved that if the sub-network of the cluster heads is bi-directionally connected, then (4) and (5) are equivalent to (3) (Zhu et al., 2007). Both (4) and (5) can be solved similarly, as below.

4. Collaborative environmental monitoring algorithm

We now apply an optimal algorithm, the alternating direction method of multipliers (ADMM) (Bertsekas and Tsitsiklis, 1997), to solve (4).

4.1 Algorithm development

To solve (4) with the ADMM, we first introduce a new block of auxiliary variables. Then (4) can be rewritten as:

$$\begin{array}{ll} \min\limits_{\{\mathbf{x}_i\},\{\mathbf{z}_{ij}\}} & \sum_{i=1}^{I}(\frac{\lambda}{2}||\mathbf{F}_i\mathbf{x}_i - \mathbf{b}_i||_2^2 + \frac{1}{I}\mathbf{1}^T\mathbf{x}_i), \\ s.t. & \mathbf{x}_i = \mathbf{z}_{ij}, \mathbf{x}_j = \mathbf{z}_{ij}, \quad \forall c_i \in \mathcal{I}, c_j \in \mathcal{N}_i. \end{array} \tag{6}$$

Here $\mathbf{z}_{ij}$ is an auxiliary vector attached to $\mathbf{x}_i$ and $\mathbf{x}_j$. The augmented Lagrangian function of (6) is:

$$\begin{array}{l} L_a\left(\{\mathbf{x}_i\},\{\mathbf{z}_{ij}\},\{\boldsymbol{\beta}_{ij}\},\{\boldsymbol{\gamma}_{ij}\}\right) = \sum_{i=1}^{I}(\frac{\lambda}{2}||\mathbf{F}_i\mathbf{x}_i - \mathbf{b}_i||_2^2 + \frac{1}{I}\mathbf{1}^T\mathbf{x}_i) \\ \quad + \sum_{i=1}^{I}\sum_{c_j\in\mathcal{N}_i}\boldsymbol{\beta}_{ij}^T(\mathbf{x}_i - \mathbf{z}_{ij}) + \frac{d}{2}\sum_{i=1}^{I}\sum_{c_j\in\mathcal{N}_i}||\mathbf{x}_i - \mathbf{z}_{ij}||_2^2 \\ \quad + \sum_{i=1}^{I}\sum_{c_j\in\mathcal{N}_i}\boldsymbol{\gamma}_{ij}^T(\mathbf{x}_j - \mathbf{z}_{ij}) + \frac{d}{2}\sum_{i=1}^{I}\sum_{c_j\in\mathcal{N}_i}||\mathbf{x}_j - \mathbf{z}_{ij}||_2^2, \end{array} \tag{7}$$

in which $\{\boldsymbol{\beta}_{ij}\}$ and $\{\boldsymbol{\gamma}_{ij}\}$ are Lagrange multipliers and d is a positive constant. At time t, the ADMM optimizes the augmented Lagrangian function as follows:
Step 1: Optimizing the local copies $\{\mathbf{x}_i\}$:

$$\{\mathbf{x}_i(t+1)\} = \arg\min_{\{\mathbf{x}_i\}} \quad L_a\left(\{\mathbf{x}_i\},\{\mathbf{z}_{ij}(t)\},\{\boldsymbol{\beta}_{ij}(t)\},\{\boldsymbol{\gamma}_{ij}(t)\}\right). \tag{8}$$

Notice that the objective function is separable, $\mathbf{x}_i(t+1)$ can be updated as:

$$\begin{array}{l} \mathbf{x}_i(t+1) = \arg\min_{\mathbf{x}_i} \quad (\frac{\lambda}{2}||\mathbf{F}_i\mathbf{x}_i - \mathbf{b}_i||_2^2 + \frac{1}{I}\mathbf{1}^T\mathbf{x}_i) \\ \quad + \sum_{c_j\in\mathcal{N}_i}\boldsymbol{\beta}_{ij}^T(t)\mathbf{x}_i + \sum_{c_j\in\mathcal{N}_i}\boldsymbol{\gamma}_{ji}^T(t)\mathbf{x}_i \\ \quad + \frac{d}{2}\sum_{c_j\in\mathcal{N}_i}||\mathbf{x}_i - \mathbf{z}_{ij}(t)||_2^2 + \frac{d}{2}\sum_{c_j\in\mathcal{N}_i}||\mathbf{x}_i - \mathbf{z}_{ji}(t)||_2^2. \end{array} \tag{9}$$

Step 2: Optimizing the Auxiliary Variable $\{\mathbf{z}_{ij}\}$:

$$\{\mathbf{z}_{ij}(t+1)\} = \arg\min_{\{\mathbf{z}_{ij}\}} \quad L_a\left(\{\mathbf{x}_i(t+1)\},\{\mathbf{z}_{ij}\},\{\boldsymbol{\beta}_{ij}(t)\},\{\boldsymbol{\gamma}_{ij}(t)\}\right). \tag{10}$$

Here the objective functions is also separable. Therefore:

$$\begin{array}{l} \mathbf{z}_{ij}(t+1) = \arg\min\limits_{\mathbf{z}_{ij}} \quad -\boldsymbol{\beta}_{ij}^T(t)\mathbf{z}_{ij} - \boldsymbol{\gamma}_{ji}^T(t)\mathbf{z}_{ij} \\ \quad + \frac{d}{2}||\mathbf{x}_i(t+1) - \mathbf{z}_{ij}||_2^2 + \frac{d}{2}||\mathbf{x}_j(t+1) - \mathbf{z}_{ij}||_2^2. \end{array} \tag{11}$$

It has an explicit solution:

$$\mathbf{z}_{ij}(t+1) = \frac{1}{2}\left(\mathbf{x}_i(t+1) + \mathbf{x}_j(t+1)\right) + \frac{1}{2d}\left(\boldsymbol{\beta}_{ij}(t) + \boldsymbol{\gamma}_{ij}(t)\right). \tag{12}$$

Step 3: Updating the Lagrange Multipliers $\{\boldsymbol{\beta}_{ij}\}$ *and* $\{\boldsymbol{\gamma}_{ij}\}$:

$$\begin{array}{l} \boldsymbol{\beta}_{ij}(t+1) = \boldsymbol{\beta}_{ij}(t) + d\left(\mathbf{x}_i(t+1) - \mathbf{z}_{ij}(t+1)\right), \\ \boldsymbol{\gamma}_{ij}(t+1) = \boldsymbol{\gamma}_{ij}(t) + d\left(\mathbf{x}_j(t+1) - \mathbf{z}_{ij}(t+1)\right). \end{array} \tag{13}$$

The updating rules of (9), (11), and (13) can be further simplified. Substituting (12) to (13) yields:

$$\begin{aligned}\boldsymbol{\beta}_{ij}(t+1) &= \boldsymbol{\beta}_{ij}(t) + \tfrac{d}{2}\left(\mathbf{x}_i(t+1) - \mathbf{x}_j(t+1)\right) - \tfrac{1}{2}(\boldsymbol{\beta}_{ij}(t) + \boldsymbol{\gamma}_{ij}(t)),\\ \boldsymbol{\gamma}_{ij}(t+1) &= \boldsymbol{\gamma}_{ij}(t) + \tfrac{d}{2}\left(\mathbf{x}_j(t+1) - \mathbf{x}_i(t+1)\right) - \tfrac{1}{2}(\boldsymbol{\beta}_{ij}(t) + \boldsymbol{\gamma}_{ij}(t)).\end{aligned} \tag{14}$$

Since we often set $\boldsymbol{\beta}_{ij}(0) = \boldsymbol{\gamma}_{ij}(0) = \mathbf{0}$ where $\mathbf{0}$ denotes a $K \times 1$ all-zero vector $[0, 0, ..., 0]^T$, (14) implies that $\boldsymbol{\beta}_{ij}(t) = -\boldsymbol{\gamma}_{ij}(t) = \boldsymbol{\gamma}_{ji}(t)$. Then (12) becomes:

$$\mathbf{z}_{ij}(t+1) = \frac{1}{2}\left(\mathbf{x}_i(t+1) + \mathbf{x}_j(t+1)\right). \tag{15}$$

and (13) becomes

$$\begin{aligned}\boldsymbol{\beta}_{ij}(t+1) &= \boldsymbol{\beta}_{ij}(t) + \tfrac{d}{2}\left(\mathbf{x}_i(t+1) - \mathbf{x}_j(t+1)\right) = \boldsymbol{\gamma}_{ji}(t+1),\\ \boldsymbol{\gamma}_{ij}(t+1) &= \boldsymbol{\gamma}_{ij}(t) + \tfrac{d}{2}\left(\mathbf{x}_j(t+1) - \mathbf{x}_i(t+1)\right) = \boldsymbol{\beta}_{ji}(t+1).\end{aligned} \tag{16}$$

Summarizing the three sides of (16) and define a new Lagrangian multiplier $\boldsymbol{\alpha}_i = \frac{2}{|\mathcal{N}_i|}\sum_{c_j \in \mathcal{N}_i} \boldsymbol{\beta}_{ij} = \frac{2}{|\mathcal{N}_i|}\sum_{c_j \in \mathcal{N}_i} \gamma_{ji}$, the updating rule for α_i is:

$$\boldsymbol{\alpha}_i(t+1) = \boldsymbol{\alpha}_i(t) + d\mathbf{x}_i(t+1) - \frac{d}{|\mathcal{N}_i|}\sum_{c_j \in \mathcal{N}_i} \mathbf{x}_j(t+1). \tag{17}$$

Substituting (15) to (9), we have the updating rule for $\mathbf{x}_i$:

$$\begin{aligned}\mathbf{x}_i(t+1) = \arg\min_{\mathbf{x}_i} \quad & \textstyle\sum_{i=1}^{I}\left(\frac{\lambda}{2}||\mathbf{F}_i\mathbf{x}_i - \mathbf{b}_i||_2^2 + \frac{1}{I}\mathbf{1}^T\mathbf{x}_i\right)\\ & + |\mathcal{N}_i|\boldsymbol{\alpha}_i^T(t)\mathbf{x}_i + d\textstyle\sum_{c_j \in \mathcal{N}_i} ||\mathbf{x}_i - \frac{1}{2}\left(\mathbf{x}_i(t) + \mathbf{x}_j(t)\right)||_2^2.\end{aligned} \tag{18}$$

Iteratively solving (18) and updating (17) leads to the optimal solution of (4).
It should be noted that the problem we are discussing is indeed a special case of compressive sensing (Donoho et al., 2006). When the number of sensor nodes is smaller than the number of grid points, down-sampling is achieved via exploiting the sparse nature of the signal $\mathbf{x}$. From this viewpoint, the proposed decentralized sparse signal recovery algorithm is also applicable to other compressive sensing problems, in which distributed sensor nodes hold parts of measurement matrices as well as measurement vectors, and collaboratively make decisions.

4.2 Algorithm outline

The collaborative sparse signal recovery algorithm is summarized as follows. The algorithm is fully decentralized, requiring only collaboration between neighboring cluster heads.

Algorithm: Collaborative environmental monitoring

Step 1: Initialization. At each sampling point, each cluster head collects position information and measurements from sensors within its cluster. Hence cluster head c_i knows the partial measurement matrix $\mathbf{F}_i$ and the partial measurement vector $\mathbf{b}_i$. The number of cluster heads I, the non-negative constant λ, and the positive constant d are also known.
Step 2: Communication. At iteration $t+1$, each cluster head c_i broadcasts to its neighboring

cluster heads to acquire intermediate decision vectors $\mathbf{x}_j(t)$ of iteration t, $c_j \in \mathcal{N}_i$.
Step 3: Optimization. At iteration $t+1$, each cluster head c_i updates its Lagrange multiplier $\boldsymbol{\alpha}_i(t+1)$ and decision vector $\mathbf{x}_i(t+1)$ according to (17) and (18).
Step 4: Iteration. Repeat Step 2 and Step 3 until convergence.

---

5. Performance analysis

In this section we will briefly discuss the impact of parameter settings on the performance of the algorithm, as well as the design choice of a hierarchical wireless sensor network. By performance we are mainly concerning: 1) quality of recovery, which includes the number of false alarms and the gap between the true and estimated amplitudes; and 2) convergence rate, which directly decides the communication burden of the cluster heads.

5.1 Parameter settings

The role of the non-negative weight λ in (3) has been extensively discussed in compressive sensing literature, such as (Donoho et al., 2006; Figueiredo et al., 2007). There is a constant $\lambda_{min} = 1/||\mathbf{F}^T\mathbf{b}||_\infty$, such that if $\lambda \leq \lambda_{min}$ the optimal solution is $\mathbf{0}$. When λ goes to infinity, the optimal solution has the minimum ℓ_1 norm among all points that satisfy $\mathbf{F}^T(\mathbf{Fx} - \mathbf{b} = 0)$, if these points exist. Hence if $\mathbf{b}$ is noise-free and $\mathbf{F}$ is a full rank square matrix, then the optimal solution goes to the true signal. However large λ generally leads to a non-sparse solution under the existence of measurement noise.

In Step 1 of the collaborative environmental monitoring algorithm, we need to know I, the number of cluster heads. This procedure requires multi-hop communications if the cluster heads are not directly connected with each other. However, accurate knowledge of I is not necessary since it is the product λI that decides the optimal solution.

The proposed algorithm converges for any given positive constant d; however, the value of d influences the convergence rate, and thus the communication burden. It is possible to dynamically increase the value of d to infinity to improve the convergence rate during the iterative optimization process (Bertsekas and Tsitsiklis, 1997). Due to the extra burden of updating d, we simply choose d as a constant.

One of the most important advantages of the hierarchical infrastructure is its flexibility to different application scenarios. By setting $I = 1$, the infrastructure turns to be centralized; while with $I = L$ and sensors being cluster heads, the network is a fully decentralized one. This flexibility enables the network to adapt to different application scenarios.

5.2 To Collaborate or not to collaborate

Now we revaluate the order of the required communication load in a hierarchical network without any fusion center. First, at the data acquisition stage, cluster heads need to collect measurements from sensor nodes and construct the local measurement matrix, at communication cost on the order $\sim O(L)$. Second, at the optimization stage, one cluster head needs to transmit its decision vector to each neighboring cluster head at each iteration. The lengths of decision vectors are all K; the average number of neighboring cluster heads varies from $\sim O(1)$ (multi-hop communications) to $\sim O(I)$ (one-hop communications) depending on the communication range r_C of cluster heads. Denote the number of iterations as T, which is in general influenced by the choices of d and λ, as well as the topology of the network. Therefore the overall communication load ranges from $O(KIT)$ to $O(KI^2T)$.

For comparison, we also consider the communication load of a hierarchical network in the presence of fusion center. The communication load at the data acquisition stage is the same as before. Then each cluster head needs to transmit its own part of the measurement vector and the measurement matrix to the fusion center. Transmission of the measurement matrix is necessary since the wireless sensor network is often subject to node failure, node displacement, etc; hence the measurement matrix is generally a dynamic one. Suppose that sensor nodes are evenly divided into clusters; as a result, each cluster head has $1/I$ of the entire measurement matrix with KL entries and the entire measurement vector with L data. Depending on the communication range r_C of cluster heads, average communication load of sending one packet to the fusion center ranges from $O(1)$ (one-hop communications) to $O(I)$ (multi-hop communications). Thus, the overall communication load is from $O(KL)$ to $O(KIL)$.

The analysis provides us guideline on whether to use a fusion center or to implement decentralized collaborative information processing among cluster heads. When the number of sensor nodes within each cluster is large (large amount of data) and when the cluster heads are subject to multi-hop communications (limited communication range), the collaborative algorithm is superior in terms of energy efficiency. For a large-scale wireless sensor network, each cluster generally contains a large number of sensor nodes and the range of the sensing area is much larger than the communication range of cluster heads. Therefore, decentralized collaboration among cluster heads is preferred. In addition, collaborative information processing offers the benefits of robustness to node failure, obviation of multi-hop routing, and alleviated level of congestion.

6. Simulation results

Let us consider a 100×100 sensing area with length from 0 to 100 and width from 0 to 100. The area is divided into 100 squares with 121 grid points. Sensor nodes are uniformly randomly deployed in the sensing area. Two sources of events, one with amplitude 1 and another with amplitude 0.5, occur at grid points $(20, 80)$ and $(50, 50)$ respectively. We assume the influence function to be $f_{kl} = \exp(-d_{kl}^2/\sigma^2)$, with $\sigma = 20$. Two parameters of the decentralized algorithm are set to $\lambda = 100$ and $d = 1$.

First, we consider 100 sensor nodes, which are divided into 5 clusters with 20 sensor nodes in each cluster. The cluster heads are bi-directionally connected to each other by properly setting the communication range r_C. As an ideal case, the measurements are assumed to be noise-free. The convergence property of one cluster head is depicted in Figure 3. The decision variables corresponding to the two events converge to the optimal values, while the amplitudes of other grid points converge to 0. Due to the consensus constraints, decision vectors of different cluster heads converge to the same solution.

Next, we study the influence of the number of cluster heads I on the convergence time. By convergence time we mean the minimum iteration number with which the differences between all decision variables and their optimal values are within 0.01. According to Figure 4, it is not surprising that the fully centralized infrastructure ($I = 1$) achieves the best convergence rate while the fully decentralized one ($I = 100$) converges slowly. Figure. 4 indicates that the convergence time is $\sim O(I^2)$ for the decentralized algorithm.

Further, to compare the communication load between the two schemes, we assume all sensor nodes and cluster heads have a common communication range $r_C = 10$. In the centralized scheme, one sensor node is chosen as the fusion center. In the decentralized scheme, each cluster head is supposed to have at least one neighboring cluster head, since

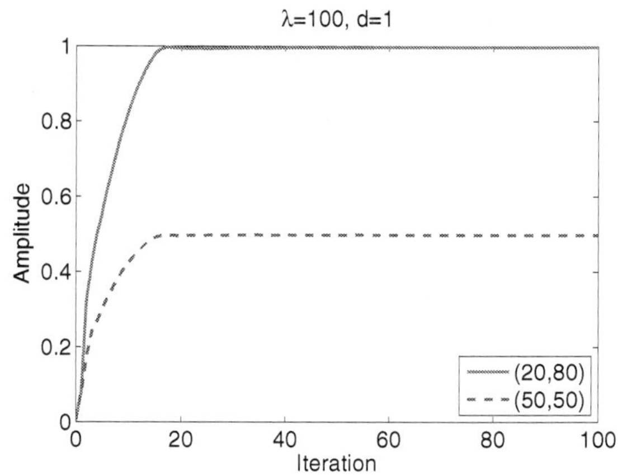

Fig. 3. Convergence of the proposed algorithm.

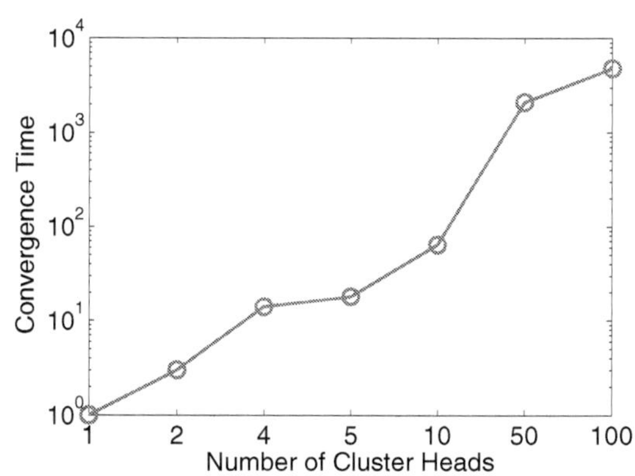

Fig. 4. Number of cluster heads I versus convergence time.

the connectivity of the sub-network of cluster heads generally cannot be satisfied when the number of cluster heads is small. The communication loads are depicted in Figure. 5. It is shown that when the cluster heads collect a large amount of data, collaborative optimization is energy-efficient. When the number of cluster heads increases, the communication load for consensus optimization dominates, leading to poor efficiency. This fact suggests us to properly select the cluster size and number of clusters in order to be energy efficient. It also points out an important but challenging research topic for future work, namely, improving the energy-efficiency of hierarchical networks via accelerating convergence rate of the decentralized collaborative algorithm.

Finally we simply discuss the compressive ratio of the proposed algorithm, namely, the ratio of the number of sensor nodes versus the number of grid points. We maintain the parameter settings in the previous simulation; the number of cluster heads is not necessary since any

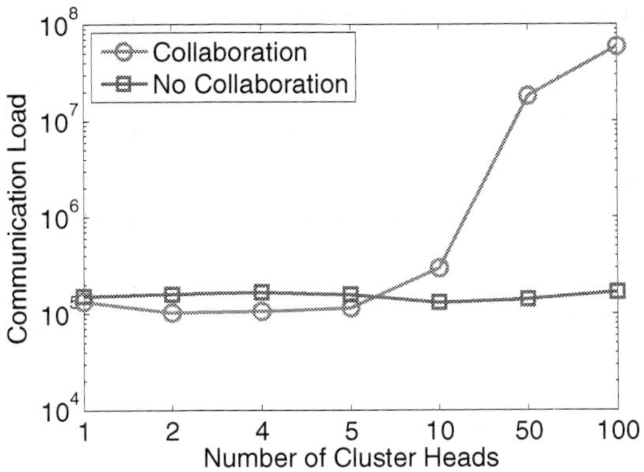

Fig. 5. Number of cluster heads I versus communication load.

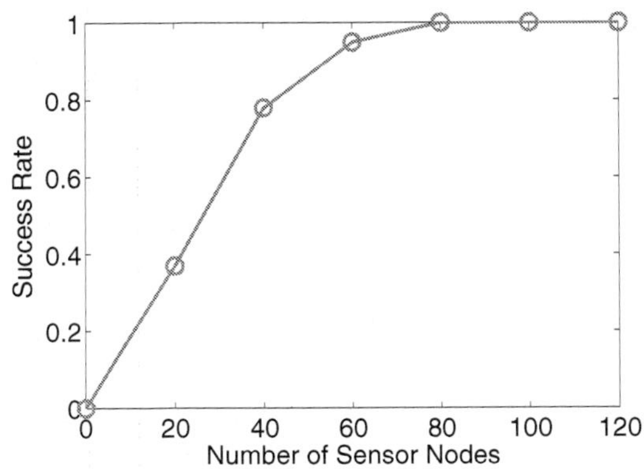

Fig. 6. Number of sensor nodes L versus the success rate of recovery.

settings will lead to global convergence. The relationship between the compressive ratio and the probability of successful recovery is shown in Figure 6. The simulation is repeated for 100 times with randomly deployed sensor nodes for each time. When the number of sensor nodes is larger than nearly half of the number of grid points, the recovery is successful with high probability.

7. Conclusion

This chapter discusses the design of hierarchical wireless sensor networks for environmental monitoring applications. Specifically, we focus on a generalized event detection model which is able to discover sparse events based on sensory data. Both positions and amplitudes of the events can be recovered from a convex program. Then we elaborate on an optimal decentralized algorithm which requires no fusion center but only collaboration of neighboring cluster heads. Through theoretical analysis and simulation experiments, we suggest when the cluster heads need to collaborate and when not; this provides a design guideline for hierarchical wireless sensor networks.

8. Acknowledgement

Qing Ling is supported in part by National Nature Science Foundation under grant 61004137 and the Fundamental Research Funds for the Central Universities under grant WK2100100007.

9. References

I. Akyildiz, W. Su, Y. Sankarasubramaniam, and E. Cayirci, "Wireless sensor networks: a survey," Computer Networks, vol. 38, pp. 393–422, 2002

J. Yick, B. Mukherjee, and D. Ghosal, "Wireless sensor network survey," Computer Networks, vol. 52, pp. 2292–2330, 2008

D. Estrin, R. Govindan, J. Heidemann, and S. Kumar, "Next century challenges: scalable coordination in sensor networks," In: Proceedings of MOBICOM, 1999

D. Estrin, D. Culler, K. Pister, and G. Sukhatme, "Connecting the physical world with pervasive networks," IEEE Pervasive Computing, vol. 1, pp. 59–69, 2002

G. Simon, M. Maroti, A. Ledeczi, G. Balogh, B. Kusy, A. Nadas, G. Pap, J. Sallai, and K. Frampton, "Sensor network-based countersniper system," SENSYS 2004

T. He, S. Krishnamurthy, L. Luo, T. Yan, L. Gu, R. Stoleru, G. Zhou, Q. Cao, P. Vicaire, J. Stankovic, T. Abdelzaher, J. Hui, and B. Krogh, "VigilNet: an integrated sensor network system for energy-efficient surveillance," ACM Transactions on Sensor Networks, vol. 2, pp. 1–38, 2006

K. Langendoen, A. Baggio, and O. Visser, "Murphy loves potatoes: experiences from a pilot sensor network deployment in precision agriculture," In: Proceedings of IPDPS, 2006

T. Wark, P. Corke, P. Sikka, L. Klingbeil, Y. Guo, C. Crossman, P. Valencia, D. Swain, and G. Bishop-Hurley, "Transforming agriculture through pervasive wireless sensor networks," IEEE Pervasive Computing, vol. 6, pp.. 50–57, 2007

V. Gungor and G. Hancke, "Industrial wireless sensor networks: challenges, design principles, and technical approaches," IEEE Transactions on Industrial Electronics, vol. 56, pp. 4258–4265, 2009

M. Li and Y. Liu, "Underground structure monitoring with wireless sensor networks," In: Proceedings of IPSN, 2007

Q. Ling, Z. Tian, Y. Yin, and Y. Li, "Localized structural health monitoring using energy-efficient wireless sensor networks," IEEE Sensors Journal, vol. 9, pp. 1596–1604, 2009

P. Zhang, C. Sadler, S. Lyon, and M. Martonosi, "Hardware design experiences in ZebraNet," In: Proceedings of SENSYS, 2004

P. Corke, T. Wark, R. Jurdak, W. Hu, P. Valencia, and D. Moore, "Environmental wireless sensor networks," Proceedings of the IEEE, vol. 98, pp. 1903–1917, 2010

B. Sadler, "Foundamentals of energy-constrained sensor network systems," IEEE Aerospace and Electronic Systems Magazine, vol. 20, pp. 17–35, 2005

G. Werner-Allen, J. Johnson, M. Ruiz, J. Lees, and M. Welsh, "Monitoring volcanic eruptions with a wireless sensor network," In: Proceedings of EWSN, 2005

G. Werner-Allen, K. Lorincz, M. Welsh, O. Marcillo, J. Johnson, M. Ruiz, and J. Lees, "Deploying a wireless sensor network on an active volcano," IEEE Internet Computing, vol. 10, pp. 18–25, 2006

Y. Liu, Y. He, M. L, J. Wang, K. Liu, L. Mo, W. Dong, Z. Yang, M. Xi, J. Zhao, and X. Li, "Does wireless sensor network scale? a measurement study on GreenOrbs," In: Proceedings of INFOCOM, 2011

Q. Ling and Z. Tian, "Decentralized sparse signal recovery for compressive sleeping wireless sensor networks,ąś IEEE Transactions on Signal Processing, vol. 58, pp. 3816–3827, 2010

A. Arora, R. Ramnath, E. Ertin, P. Sinha, S. Bapat, V. Naik, V. Kulathumani, H. Zhang, H. Cao, M. Sridharan, S. Kumar, N. Seddon, C. Anderson, T. Herman, N. Trivedi, C. Zhang, M. Nesterenko, R. Shah, S. Kulkarni, M. Aramugam, L. Wang, M. Gouda, Y. Choi, D. Culler, P. Dutta, C. Sharp, G. Tolle, M. Grimmer, B. Ferriera, and K. Parker, "ExScal: elements of an extreme scale wireless sensor network," In: Proceedings of RTCSA, 2005

P. Dutta, J. Hui, J. Jeong, S. Kim, C. Sharp, J. Taneja, G. Tolle, K. Whitehouse, and D. Culler, "Trio: enabling sustainable and scalable outdoor wireless sensor network deployment," In: Proceedings of IPSN, 2006

G. Barrenetxea, F. Ingelrest, G. Schaefer, and M. Vtterli, "The hitchhiker's guide to successful wireless sensor network deployment," In: Proceedings of SENSYS, 2008

W. Heinzelman, A. Chandrakasan, and H. Balakrishnan, "An application-specific protocol architecture for wireless microsensor networks," IEEE Transactions on Wireless Communications, vol. 1, pp. 660–670, 2002

M. Rabbat and R. Nowak, "Distributed optimization in sensor networks," In: Proceedings of IPSN, 2004

S. Aldosari, J. Moura, "Fusion in sensor networks with communication constraints," In: Proceedings of IPSN, 2004

D. Bertsekas and J. Tsitsiklis, *Parallel and Distributed Computation: Numerical Methods*, Second Edition, Athena Scientific, 1997

D. Donoho, M. Elad, V. Temlyakov, "Stable recovery of sparse overcomplete representations in the presense of noise," IEEE Transactions on Information Theory, vol. 52, pp. 6–18, 2006

R. Szewcszyk, A. Mainwaring, J. Polastre, J. Anderson, and D. Culler, "Lessons from a sensor network expedition," In: Proceedings of EWSN, 2004

G. Tolle, J. Polastre, R. Szewczyk, D. Culler, N. Turner, K. Tu, S. Burgess, T. Dawson, P. Buonadonna, D. Gay, and W. Hong, "A macroscope in the redwoods," In: Proceedings of SENSYS, 2005

L. Selavo, A. Wood, Q. Cao, T. Sookoor, H. Liu, A. Srinivasan, Y. Wu, W. Kang, J. Stankovic, D. Young, and J. Porter, "LUSTER: wireless sensor network for environmental research," In: Proceedings of SENSYS, 2007

K. Martinez, J. Hart, and R. Ong, "Environmental sensor networks," Computer, vol. 37, pp. 50–56, 2004

J. Bezdek, S. Rajasegarar, M. Moshtaghi, C. Leckie, M. Palaniswami, and T. Havens, "Anomaly detection in environmental monitoring networks," IEEE Computational Intelligence Magazine, vol. 6 pp. 52–58, 2011

C. Alippi, R. Camplani, G. Galperti, and M. Roveri, "A robust, adaptive, solar-powered WSN framework for aquatic environmental monitoring," IEEE Sensors Journal, vol. 11, pp. 45–55, 2011

J. Bazerque and G. Giannakis, "Distributed spectrum sensing for cognitive radio networks by exploiting sparsity," IEEE Transactions on Signal Processing, vol. 58, pp. 1847–1862, 2010

S. Kim, K. Koh, M. Lustig, S. Boyd, and D. Gorinevsky, "An interior-point method for large-scale ℓ-1 regularized least squares," IEEE Journal of Selected Topics in Signal Processing, vol. 1, pp. 606–617, 2007

M. Figueiredo, R. Nowak, and S. Wright, "Gradient projection for sparse reconstruction: application to compressed sensing and other inverse problems," IEEE Journal of Selected Topics in Signal Processing, vol. 1, pp. 586–597, 2007

H. Zhu, A. Cano, and G. Giannakis, "Consensus-based distributed MIMO decoding using semidefinite relaxation," In: Proceedings of CAMSAP, 2007

27

Environmental Monitoring WSN

Ittipong Khemapech
University of the Thai Chamber of Commerce, Thailand

1. Introduction

Energy conservation is currently growing in importance. This chapter focuses on the issue of energy conservation within the domain of Wireless Sensor Network (WSN). There are also specific reasons why energy conservation is more important for WSN than for other types of networks. A WSN consists of multiple sensors which are able to sense some aspect of their environment and communicate their readings to a base station or sink without being physically connected to it. Sensors are often also resource constrained, being small in size and relying on small batteries for power. Consequently, the efficient utilisation of energy should be an important priority for designing WSN network protocols. This differs from the traditional approach to designing network protocols where issues like survivability, maximising throughput or reliability have been prioritised. Making energy conservation an important design priority is a new approach.

Wireless sensor network (WSN) is an important research area with a major technological impact. With significant breakthroughs in "Micro Electromechanical Systems", or MEMS, (Warneke & Pister, 2002), sensors are becoming smaller. It is feasible to fit them into a smaller volume with more power and less production costs. WSN may be deployed in a wide range of different environments. These include remote and hostile environments as well as local and friendly ones. A major driving force behind research in WSN has been military and surveillance applications. Recently, however diversification has occurred with the development of civil applications. One example which is used as a reference point throughout this work is Great Duck Island (GDI). Sensors were scattered over a remote island to monitor the seabird's migration (Mainwaring et al., 2002). In another example WSN was deployed around volcanoes (Allen et al., 2006). Such applications illustrate the usefulness of WSN which make data collection feasible from remote and hostile environments with minimal human intervention.

One of the main objectives of WSN power conservation is to minimise energy usage whilst other functional requirements such as reliability or time synchronisation are still achieved. Some authors argue that multi hop communication allows for deployment in scenarios where direct communication with a base station is not practical (Arora et al., 2004; Allen et al., 2006; Chintalapudi et al., 2006). However, the spread of the Internet means that wireless devices may often communicate directly with a device that is connected to the Internet and has a reliable power supply. This work focuses on the design of wireless sensor networks protocols where direct communication with a powered base station is feasible and data is sent from the sensors to the base station at regular intervals. There are several important scenarios where such two assumptions hold.

This research work specifically looks at WSN where direct communication is possible and beneficial. A protocol for WSN, Power & Reliability Aware Protocol (PoRAP), is developed and provides energy efficient data delivery, without increasing packet loss. In designing PoRAP several experiments were conducted to establish the relationship between transmission power, reception signal strength and packet reception success. These showed a strong correlation between Received Signal Strength Indicator (RSSI) and Packet Reception Rate (PRR). In PoRAP, the RSSI is monitored at the base station. If the RSSI is too high the base station signals the sensor to reduce its transmission level, thereby saving power. If the RSSI is too low the base station signals the sensor to increase its transmission level so that packet loss is avoided.

PoRAP adopts a schedule based scheme for the sources' transmissions. It is assumed that nodes will be reporting measurement data regularly back to the base station. Each reporting interval consists of three time periods. In the first the base station sends a configuration packet. This informs nodes whether they are to increase, decrease or leave unaltered their transmission levels. There are then slots, each of which contains a data slot within which a sensor may transmit its data to the base station. There may then be a period of quiet before the start of a new cycle. Delays and clock drifts are measured so that nodes know when to wake up to listen and transmit. Delays depend upon payload size.

The design aims to optimise energy conservation rather than system throughput, in many sensing scenarios high throughput is not required. Sensors collect and report some physical data such as temperature and humidity. In such cases, the data reporting rate may be in minutes or hours. Two packet structures are used in PoRAP. The control packet is used in control and setup phase. It contains essential information for transmission power adaptation and scheduling. The data packet is used to deliver the collected physical data back to the base station.

The remaining parts of this chapter is organised as follows: Section 2 addressed application specific WSN. At present, WSN has been used in both military and civil applications. Each application category has particular characteristics and its own set of requirements. Hence, there are significant challenges in a generic protocol design for a variety of applications. Resource constraint issues are provided in Section 3. Apart from limited power resources, sensors also have constrained communication ranges for indoor and outdoor environments. The distance between the source and destination is crucial to employing an appropriate underlying communication paradigm. Section 4 describes the experimental details and their results which motivate the design of PoRAP. There are several factors which affect the link quality metrics such as distance between source and base station and time of day. The design of PoRAP is outlined in Section 5. PoRAP consists of several TinyOS components at the source and base station. The results shown in Section 4 motivates the design. The results of PoRAP evaluation in terms of energy conservation are presented in Section 6. Lower transmission power can be used to save the power whilst the reliability is in the desired range. Finally, Section 7 concludes the chapter.

2. Application specific WSN

Apart from being used in military or surveillance, WSN has been deployed in several civil applications which have different requirements. Periodic sensing is required in some habitat and environmental monitoring systems whilst event sensing is the norm in surveillance systems. Network lifetime and data reporting rates are therefore major concerns for the former and latter cases, respectively. To be application specific results in a more complicated design process, especially in the case of designing a generic power-aware protocol.

In total, seven groups of applications have been categorised by us based upon their functionalities including habitat monitoring (HM) (Juang et al., 2002; Mainwaring et al., 2002; Szewczyk et al., 2004), environmental monitoring (EM) (Allen et al., 2006; Martinez et al., 2005), health monitoring (HEM) (Jovanov et al., 2003, Otto et al., 2006), structural health monitoring (SHM) (Chintalapudi et al., 2006; Kottapalli et al., 2003; Paek et al., 2005, Schmid et al., 2005), event detection and tracking (EDT) (Arora et al., 2004; Dreicer et al., 2002; Simon et al., 2004), transport monitoring (TM) (Coleri et al., 2004) and location-aware system (LAS) (Brignone et al., 2005). Specific capabilities and underlying communication paradigms have been outlined. For example, data encryption may be required in some health monitoring systems for transmitting a patient's diagnosis data to the main server located at the hospital. Furthermore, data correctness is also required in this case. In some applications such as event tracking and detection systems, several intermediate nodes are required for forwarding the sensed data to the base station. However, a direct communication from source to base station is found in some health monitoring systems. This section addresses application specific characteristics of WSN applications by detailing the differences in their requirements.

2.1 Event/periodic based

The "Event/Periodic Based" aspect demonstrates how often data reporting is conducted. There are three main types including event-based, periodic-based and hybrid. Each sensor is triggered to operate by the occurrence of an event in the case of an event-based application. An example of this application type is the Event Detection and Tracking. Congestion is one of the major concerns designing a protocol to support event-based networking as it is caused by a lot of traffic generated by all sources in an event area. The key idea of congestion avoidance is to control data reporting rate of such sensors (Sankarasubramaniam et al., 2003). However, the main assumption is that all data packets have the same priority. Packet loss is therefore tolerantly acceptable. There are several works on congestion control specifically developed for WSN (Ee & Bajcsy, 2004; Hull et al., 2004, Lu et al., 2002, Wan et al., 2003). The congestion control approach focuses on channel monitoring to dynamically adjust the data forwarding rate. CODA (Wan et al., 2003) has been designed to cover two types of problems corresponding to the deployed sensors and their data rate. However, it does not provide any queue occupancy monitoring. Sending an ACK (Acknowledgement) in the case of persistent congestion, even if it is small in size, may increase the number of traffic. This mechanism also requires feedback signalling which results in higher cost. Only packet prioritisation could be found in (Lu et al., 2002). However, it proposes the VMS (Velocity Monotonic Scheduling) policy which supports both static and dynamic computation of the requested velocity and it also solves the fairness problem. Both channel and queue occupancy monitoring are provided in (Hull et al., 2004) and (Ee & Bajcsy, 2004). A child node can transmit packets only when its parent does not experience congestion problems and some help from the MAC (Medium Access Control) layer to shift the transmitting time to avoid interference are proposed in (Hull et al., 2004). A similar concept also exists in (Ee & Bajcsy, 2004) by comparing the normalised rate of a node and its parents.

Each sensor periodically performs its operation. Some examples of data collected by the sensors are temperature and humidity. The significant change in readings may be used to identify the presence of seabirds (Mainwaring et al., 2002) and intruders (Arora et al., 2004). Instead of heavily generated traffics, both sensor and network lifetimes are the core requirement of this application type. Finally, both event and periodic sensing operations may be desired in some applications such as SHM (Structural Health Monitoring) and EDT systems. For example, the displacement of construction elements is periodically reported for maintenance purposes whilst an event-based operation is applied for warning and evacuating notifications during an earthquake.

This work focuses on developing a power-aware protocol which supports an efficient data delivery in periodic based applications such as health, habitat and environmental monitoring where the data reporting rate is in minutes or hours. Sensors may be scattered over a remote and hostile area to collect and report physical data and they should have to operate for months. Hence, battery lifetime is important and one of the main goals is to conserve communication energy.

2.2 Mobility of sources

The mobility of sources or sensors can be found in some particular applications such as HM (Habitat Monitoring, HEM (HEalth Monitoring and LAS (Location-Aware System). In some cases, sensors are attached to the targeted objects or location (Jovanov et al., 2003; Juang et al., 2002, Martinez et al., 2005) in order to monitor the data of interest or current location. Mobile sensor networks have a different set of supporting infrastructures compared to the traditional WSN. It is essential for each mobile sensor to know its own location. The GPS (Global Positioning System) is used for locating sensors which are attached to the goods. Alternatively, several nodes with known locations may be used as references for the others to calculate their own locations [Brignone et al., 2005]. The issues of sensor mobility are beyond the scope of this work.

2.3 Mobility of sources

Wireless sensor network (WSN) consists of sensors which are wirelessly connected. The main objective of WSN development is to collect physical data from an area of interest. Therefore, communication between sensors is a key aspect. Normally there are two node types in WSN including the source and base station. Sources are ordinary sensors having limited resources whereas base stations are assumed to have more power and other resources. The main duty of sensors is collecting and transmitting data to the destination or base station. The sensors are probably required to cover a large area and direct communication between sources and base station is unlikely due to limited communication range. Several intermediate sensors responsible for forwarding data packets to the base station are therefore required. This is known as multi-hop communication. Each sensor also acts as a routing node in order to find the shortest or cheapest path by means of power consumption. Several applications deploy multi-hop communication (Allen et al., 2006; Chintalapudi et al., 2006; Schmid et al., 2005; Dreicer et al., 2002; Simon et al., 2004). The multi-hop approach has several advantages. For example, a new path is discovered when some sensors die. Deploying a large number of cheap sensors over a large area is feasible as the sensors can act as routing nodes and the collected data is forwarded to the destination. However, one of its drawbacks is each node has to listen to the channel most of the time in order to detect if a message is arriving. The sensors have to conduct some computations in order to discover the cheapest path. Moreover, communication with its neighbours is another requirement to set up a selected path. Such processes require a significant amount of power, taken from the battery power available. Introducing several intelligent features to each sensor is also limited due to the power constraint.

Each source can transmit the data directly to the base station if the sources are located within the base station's communication range. Some examples of existing applications deploying single-hop communication (Mainwaring et al., 2002; Martinez et al., 2005; Jovanov et al., 2003; Otto et al., 2006). For single-hop, the sources are located within the base station's range. Direct communication is therefore feasible and several benefits are realised. One of the advantages is the ability to introduce a variety of intelligent features to the base station as it is assumed to have more power and computational capabilities compared to an ordinary sensor. Each source does not require the power necessary for routing. Idle listening can be minimised as the sources can be switched to sleep mode if they do not transmit data or receive the control packet. The base station controls the communication schedule of its sources to avoid data collisions. Power for carrier sensing is not desired. In multi-hop, each source is responsible for sensing, data reporting and routing. The number of transmissions and receptions increases with the number of intermediary nodes required for data forwarding.

This work looks at protocol development for single-hop. A scenario where the single-hop is viable is Environmental Monitoring (EM). Sources and base stations are distributed and several clusters or patches are formed. A power-aware, single-hop protocol can thus be used in each of the clusters (Mainwaring et al., 2002). A low duty cycle is the norm in EM so the communication cycle of each source can be scheduled by the base station. A time slot is allocated to each source to perform data transmissions. Carrier sensing is thus not required in this scheme. The sources synchronise to the base station by checking the information included in the control packet.

2.4 Reliability

Wireless sensor network (WSN) has been currently deployed in several civil applications. The physical data is collected and transmitted for further analysis. The issue of reliability in data delivery is important for providing complete reliability consumes a significant proportion of power. Applying the Transmission Control Protocol (TCP) protocol to WSN is expensive because of its three-way handshake mechanism and packet header size. The User Datagram Protocol (UDP) is considered to be more suitable for sensors although it was designed to provide unreliable data transport. In some applications, data loss may be not a serious problem because of the large amount of deployed sensors. Reliable data transport is important for some types of data such as control messages delivered by the base station (Wan et al., 2002). The following paragraphs provide some details of reliable transport protocol for WSN researches including PSFQ (Pump Slowly, Fetch Quickly) (Wan et al., 2002), ESRT (Event-to-Sink Reliable Transport) (Sankarasubramaniam et al., 2003), and RMST (Reliable Multi-Segment Transport) (Stann & Heidemann, 2003).

One of the main goals to achieve reliable data transport is to orchestrate data receiving and forwarding processes to lessen the packet loss due to buffer overflow. PSFQ proposes three different operations including pump, fetch and report. Data generated from a source node is injected slowly into the network in order to allow such nodes experiencing data loss to fetch the missing packets very aggressively. Timing is a core process in order to avoid operational synchronisation. A hop-by-hop recovery is applied to avoid exponential error accumulation as occurs in the end-to-end scheme. Data delivery status information can be sent back to users or applications in a piggyback fashion.

Focusing only on the forward or sensor-to-sink direction, ESRT was designed to provide a reliable data transport by inspecting current network state in terms of reliability and congestion. The state result is categorised and the reporting frequency is then repetitively adjusted to reach an optimal point. ESRT provides both reliable data transport and congestion control. Local buffer level monitoring is used to detect congestion.

Directed Diffusion (Intanagonwiwat et al., 2003) is a routing protocol which provides a multipoint-to-multipoint communication. A sink firstly indicates an interest and propagates it to the nodes. Interest and node information is kept as gradients. The optimised reinforced path is then established to send the attribute-value pairs data. RMST is implemented as a filter to provide some information about the data fragment such as ID and total number of fragments to detect loss. A NACK (Negative ACKnowledgement) is sent via a back-channel to upstream neighbouring nodes in case of data loss.

According to the above fundamental protocol descriptions, several conclusions can be made. In a densely deployed environment, data loss may be accepted. However, this condition may apply only in the case of sensor-to-sink traffic. The sink or base station plays a major role in the network by broadcasting several control packets to the sensors. Such packets should not be lost. Moreover, there are various types of sensing data, such as structural displacement due to wind or earthquake (Xu et al., 2004), which need some combination from different nodes to create usable data before forwarding that data to the sink. PSFQ designing concepts are more complicated but can be applied to a broader area of application. The data retransmission mechanisms are not mentioned in ESRT as it focuses on statistical reliability. However, PSFQ does not provide congestion control schemes as ESRT does. RMST is designed to run over the Directed Diffusion routing protocol. Although it may take the least effort compared to the other two, it is not generic enough.

3. Resource constraint issues

This section introduces several issues of resource constraint in WSN. A sensor can be considered as a small computing device which is capable of sensing, data processing, storage and communication. Sensors are deployed in an area of interest and they may have to operate without maintenance throughout their lifetimes. Power is thus one of the limited resources. Unless an external source of energy is provided, power for all operations comes from batteries. Two AA batteries are required in the widely used platforms such as Tmote, Telos and Mica. The capacity of the AA battery is approximately 2,000 to 3,000 milli-ampere-hour (mAh). In order to calculate the battery life, the capacity is divided by the actual load current and the obtained lifetime is in hours. An equation for calculating sensor's lifetime is given in (Polastre et al., 2004) where the lifetime is equal to the product between capacity (mAh) and voltage (3V) divided by total energy consumption in micro-joules. The default capacity defined in (Polastre et al., 2004) is set at 2,500mAh.

Communication accounts for a significant proportion of energy consumption. There are four main modes of communication including sending, receiving, sleeping and listening. The transceiver is one of the major sensor components and it makes them capable of communicating with other nodes. Recent transceivers or radio chips such as CC1000 and CC2420 provide programmable transmission power. Sensors consume less power when they send at a lower power level. Hence, transmission power control is one of power-aware schemes in WSN. The sensors do not always send at the maximum power. Tmote platform is chosen in this study and it employs CC2420 transceiver. For the CC2420 mote the minimum and maximum transmission power is 8.5 and 17.4 milli-amperes (mA). Over 50% of the power can be saved if the minimum power is always used.

Sensors equipped with CC2420 radio chips consume a greater amount of power when they receive data. According to the data sheet, 19.7mA is required for reception. Listening and sleeping consume 365 and 20 micro-amperes (μA), respectively. Hence, in the case of the CC2420 mote, data reception consumes more energy than transmission. The base station is the destination and it may be connected to a desktop or laptop computer. In such cases, the base station has extra power from the connected machine. However, the sensors which act as intermediary nodes between source and destination have to receive and forward packets resulting in sensor's lifetimes being decreased. The listening power is approximately 17 times greater than sleeping. In some applications such as environmental monitoring, the data sampling interval may be in minutes or hours. The transceivers should be switched to sleep mode instead of listening. Scheduling issues occur when two nodes communicate with each other. The data is not received if the receiver is in sleep mode. The nodes have to agree upon the same scheduling. Another scheme based upon contention-based can be used; the receiver can periodically listen to the signal propagated over the medium to inspect whether the incoming message is destined for it.

WSN is also a shared medium system. Each of the sources and base station has to engage the medium to perform data communication. Data collisions occur if the sources transmit at the same time and energy will be wasted by unsuccessful data delivery. A Medium Access Control (MAC) protocol is required to resolve the contention. The features of the MAC protocol together with the application behaviour determine when a node is idle, when it is listening and when it is sending. As each of these states have different power requirements the MAC protocol impacts upon the efficiency of operation and the power consumption. There are two main MAC schemes; the contention and the schedule based. The medium is sensed prior to transmission and the sensors have to backoff if the medium is declared busy. This work focuses on the single-hop where the sources send data directly to the base station. Another scheme, schedule based, is adopted. A data slot is allocated to each node. No carrier sensing and corresponding energy is required. The main issue is that the slot must be long enough for completing data delivery, otherwise, data collisions are likely. Experimentations required to determine the duration required for both sending and receiving together with the effective factors such as data payload size. Each node is switched to sleep mode to spend the least amount of power when its slot does not arrive.

The buffering capacity of CC2420 is limited to 128 bytes. Taking the header's and footer's sizes into account, the allowable data payload size is thus less than 128 bytes. Apart from sensed data, some control information is required in the packet such as identification and timestamp. Additional packet structures may be required if all the information cannot be stored in one packet. Control overhead is considered as one of the costs and should be minimised in order to decrease transmission and reception energy.

Wireless sensor network (WSN) has been currently deployed in several surveillance and civil applications. Sensors may be scattered over an area of interest which can be very large. The communication range is thus important and depends upon the selected transceiver. For example, the CC2420 mote has 50m and 125m indoor and outdoor ranges. Under some circumstances, the maximum transmission power may not produce the maximum ranges. Furthermore, sending data to the node located at farther distances requires higher transmission power. Multi-hop is therefore usually used in WSN. Several intermediary sensors are required for data forwarding from the source to destination. Single-hop communication is feasible if the destination is located within the source's range. Multiple transmissions and receptions are not required if direct communication applies. However, the same transmission power cannot always be used as the link quality changes over time. The next section describes several sources of variability in radio frequency

4. Motivation of PoRAP development

This work aims at building a communication protocol for WSN. The targeted scenario is the periodic-based where a low duty cycle is required. The network consists of a fixed set of sources and a base station. Furthermore, direct data communications between the base station and its sources are feasible. The communication protocol to be developed will effectively support the single-hop WSN. Such a structure forms a network cluster which can be used in some environmental or habitat monitoring system such as (Mainwaring et al., 2002) and (Tolle et al., 2005). As the number of sources is fixed throughout the communications, the data reporting rate is fairly constant. The communication of the sources can be therefore scheduled and controlled by the base station. A time slot is allocated to each source and will be used for data communication. Only one source can use the shared medium whilst the others switch to sleep mode by turning their radios off and consuming the least amount of energy. Data collision can be avoided and idle listening can be minimised.

4.1 Sensor node power consumption

This section establishes the significance of network communication as a consumer of energy within a wireless sensor network. In doing so a careful reading of sensor data sheets is used to inform calculations based upon the sensor's parameters and simulations. What proportion of the power is used for communication is investigated and how power may be conserved is identified.

In order to investigate how power is consumed by a sensor, a simulation study has been established. The results are validated by the CC1000 transceiver data sheet. As the sensor operating system used in this work is TinyOS, the selected simulator is TOSSIM which is a TinyOS library. TinyOS is an operating system specifically designed for embedded devices such as sensors. It has been widely used in both research and commercial communities. The selected release of the simulator is TOSSIM 1 and it does not provide power usage measurement capability. PowerTOSSIM, an extension module developed for analysing power consumption of hardware components (Shnayder et al., 2004) is used to address the investigation on power consumption and it is included in Tiny 1.1.11. The only sensor platform supported in PowerTOSSIM is Mica2 which employed the CC1000 radio chip. The PowerTOSSIM supports an operating frequency of 400 Megahertz (MHz) and a voltage of 3 Volt. The energy model file of PowerTOSSIM adopts the required transmission current for each power level. According to the CC1000 datasheet, 31 output power levels ranging from -20 to +10dBm can be programmed. The dBm is the measurement of power loss in decibels (dB) using 1 milli-watt (mW) as a reference value.

4.1.1 Simulation parameters

A sensor node was created in the simulation and performs as a transmitting node. An experiment was conducted to obtain the current consumption required by each transmission power level. In total five transmission powers including -20, -10, 0, +6 and +10dBm were used. The corresponding current consumption was measured by (Shnayder et al., 2004) and their results are shown in Table 1. A simulation duration of 60 seconds and a total of 30 runs were conducted at each power level. A higher current will be consumed if the sensor transmits at a higher power.

Transmission Power (dBm)	Required Current (mA)
-20	5.21
-10	6.10
0	8.47
+6	13.77
+10	21.48

Table 1. Current consumption measured by Shnayder et al., 2004

The results shown in Table 1 were used to compute the energy consumption required by each transmission power level. Fig. 1 shows error-bar plots of radio and total energy consumption at -20, -10, 0, +6 and +10 dBm. An analysis of power usage and conservation with respect to the maximum power level is described in Table 2.
According to Fig. 1, several observations can be made. Firstly, an increase in transmission power results in a higher energy consumption. Transmitting data at lower power uses less energy. For example, over 75% of energy can be conserved if the minimum power is used for transmission instead of the maximum. Secondly, the radio unit consumes a significant amount of energy. Up to 56% and 84% of energy are used by the radio if the sensor transmits at minimum and maximum power levels, respectively. The results are validated by the CC1000 data sheet which is the employed radio in Mica2. According to the CC1000 datasheet, the required current consumption for -20 and +10 dBm are 6.9 and 26.7 milli-amp (mA), respectively. Therefore, over 74% can be conserved and this is close to the 75% which is obtained from PowerTOSSIM.

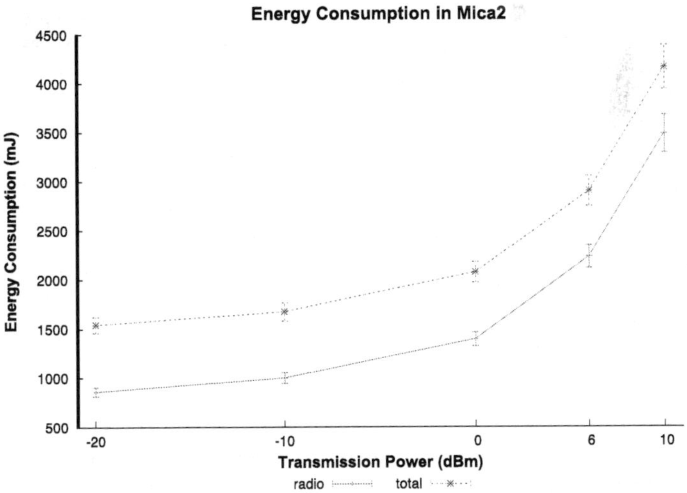

Fig. 1. Radio and total energy consumption at various transmission power levels

Transmission Power (dBm)	Average of Radio Power Consumption (mJ)	Percentage of Used Power	Percentage of Saved Power
-20	861.52	24.67	75.33
-10	1000.33	28.64	71.36
0	1396.44	39.98	60.02
+6	2236.90	64.05	35.95
+10	3492.48	100	0

Table 2. Average radio power consumption (mJ) and percentages of used and saved power

Two key motivations are established with respect to the simulation results. Firstly, transmission power considerably affects radio power consumption. The power-aware approach based upon power adaptation is Transmission Power Control (TPC). PoRAP adopts the TPC concepts in order to achieve the power conservation goal. The selected sensor platform in this work is Tmote and it employs the CC2420 radio instead of the CC1000. Like the CC1000, the CC2420 also supports transmission power adaptation but it provides a different range of power levels. Table 3 shows some of the possible power levels and the corresponding current consumption. An analysis of power conservation with respect to the maximum level is also shown.

Transmission Power (dBm)	Current Consumption (mA)	Percentage of Used Current	Percentage of Saved Current
-25	8.5	48.85	51.15
-15	9.9	56.90	43.10
-10	11.2	64.37	35.63
-7	12.5	71.84	28.16
-5	13.9	79.89	20.11
-3	15.2	87.36	12.64
-1	16.5	94.83	5.17
0	17.4	100	0

Table 3. Transmission power levels provided by CC2420 and analysis of power conservation

According to Table 3, over 50% of power can be saved if the minimum power is used for data transmission. The transmission power is one of the main factors which produces different reception strengths. The power adaptation is based upon the current link quality in order to maintain a good link. However, power adaptation is based upon several factors affecting link quality such as distance and time-of-day.

Secondly, according to Fig. 1, the radio unit accounts for a significant amount of power compared to the total consumed by all hardware components. Keeping the radio in sleep mode after the sensor has transmitted the data may establish an enhancement in power conservation. This is feasible if the single-hop network sensors do not listen to transmissions from other nodes in order to discover optimal data paths. The schedule-based MAC (Medium Access Control) approach suits the direct communication scenario as each of the sources wake up for control reception and data transmission. Otherwise, they are in sleep mode and consume the least amount of communication energy.

4.2 Environmental investigation of transmission power and reliability

This section provides details of experimental studies aimed at establishing effects of transmission power, distances and time-of-day on link quality metrics. In total three metrics including RSSI (Received Signal Strength Indicator), LQI (Link Quality Indication) and PRR (Packet Reception Rate) are used to describe the effects. The relationships between the metrics are also investigated and will be used for establishing power adaptation policies.

4.2.1 Link quality metrics

There is a variety of sources which cause variability in link quality in wireless communication. Unlike wired communication, environmental factors such as climatic conditions and time-of-day also affect the degree of signal attenuation. A significant degree of signal attenuation or interference may lead to unsuccessful data transmission. Link quality measurement is therefore one of the major issues in wireless network communication.

A transmitter sends data packets at a specific transmission power wirelessly over a medium to a receiver. The transmission power level is programmable and this capability is provided by a transceiver or radio unit which is a component responsible for data transmission and reception. A sensor communicates with the other node by sending and receiving messages via wireless channel which is normally air. Several signals are generated from various sources such as electronic appliances and they are dissipated to the air. A wireless channel may then have background noise which is capable of interfering with data delivery between a pair of nodes. Moreover, time-of-day and climatic conditions such as fog and rain affects the wireless link quality. In order to determine link quality characteristics, all causes of signal strength reduction are considered as sources of signal attenuation. The reduced magnitude in signal strength is therefore defined as signal attenuation. If the transmission power is less than signal attenuation, the message cannot be successfully received. When the receiver is not able to receive the sent packet and the number of received packets is not increased, the reliability requirement defined by an application may not be met. Transmission power should be adjusted in response to the changing link quality.

A radio unit provides several mechanisms to measure received signal power. The measured values are categorised as received signal strength (RSS). In total two attributes including RSSI (Received Signal Strength Indicator) and LQI (Link Quality Indication) are in the RSS category. The RSS can be used to indicate link quality. The reliability requirement specified by an application indicates a required number of packets received at the base station. The percentage of data receptions can be used to describe the link quality. The packet reception rate (PRR) is therefore introduced. Relationships amongst transmission power (TX), received signal strength (RSS) based attributes and PRR is useful for mapping application requirements to link quality measurements. Thus, the transmission power is adapted in order to provide reliability of packet reception.

Received Signal Strength Indicator (RSSI) is defined as a measurement of the signal strength of an incoming message. The transmitted signal strength or transmission power reduces as the signal propagates through the medium. The RSSI is measured at the receiver and it demonstrates the received signal strength. Therefore, signal attenuation is approximately the difference between the transmission power and the RSSI. Link Quality Indication (LQI) is another metric in the RSS-based category. According to the definition outlined in IEEE 802.15.4 Standard for Local and Metropolitan Area Networks, the LQI measurement is a characterisation of the strength and/or quality of received packet. Each received packet has its own LQI measurement and the integer value ranges from 0 to 255. Therefore, the minimum and maximum values of LQI for each packet are 0 and 255, respectively. The IEEE standard recommends at least eight unique values of LQI should be used in order to yield a uniform distribution between the two limits. The following details of LQI are based upon the CC2420 radio unit as it is used in both Tmote Sky and Tmote Invent which are the chosen platforms in this research. Apart from RSSI and LQI, PoRAP determines an additional link quality index. The main reason is that both RSSI and LQI are not transparent to the user or application. Mapping mechanisms are required in order to convert an application requirement to the ranges of RSSI and LQI values the base station should have. This subsection aims to describe the Packet Reception Rate (PRR) which is more closely related to the application requirement. In this research, the PRR is defined as a percentage of the number of correctly received to that of transmitted packets. The PRR value is in the range of 0% to 100%. The 100% PRR indicates complete reliability. Each received packet has its own measured RSSI and LQI which can be used to predict the PRR. Models representing relationships amongst metrics are therefore required and demonstrated later in this chapter.

4.2.2 Experimental setup

In our implementation-based experiments, Tmote Invent and Tmote Sky are used as the sensor and base station, respectively. Both of them employ the CC2420 radio which has working frequency band from 2,400 to 2,483 Megahertz (MHz). The radio transmission data rate is 250 kilobits per second (kbps). The random access memory (RAM) and program flash sizes are 10 and 48 kilobytes (Kbytes). The main difference between both platforms is that the Tmote Invent provides built-in sensor and battery boards. The minimum and maximum transmission power levels are -25 and 0dBm, respectively. Tmote sensors consume 8.5 and 17.4 milli-amps (mA) for transmitting a data packet at minimum and maximum power levels, respectively. A current of 19.7mA is required for radio receiving. This indicates that receiving accounts for a large radio power usage. Listening removal in PoRAP may enhance power conservation in WSN. Each Tmote sensor includes an internal Inverted-F antenna which is a wire monopole. The top section of the antenna is folded down to be parallel with the ground plane. The communication ranges for indoor and outdoor are 50m and 125m, respectively.

The experiments were conducted in the 16m x 20m indoor environment. The base station was plugged into a desktop computer and received data from sensors. Three sensors were used and they were placed at the same locations. In total 10 locations including 1, 2, 3, 4, 5, 7, 10, 13, 16 and 20m were used. The sensors and base station had the same antenna orientation and height above floor level. The payload size was 12 bytes. In total 8 transmission power levels including 3, 7, 11, 15, 19, 23, 27 and 31 associated to -25, -15, -10, -7, -5, -3, -1 and 0 dBm were used. The sensors transmitted one packet every second. At each power, the sensors transmitted 50 packets for statistical analysis. Upon data reception, the base station measured RSSI and LQI. The number of received packets was counted in order to compute PRR.

4.2.3 Experiments on location as a determination of necessary transmission power

The significance of the locations of the sending and receiving motes to determine the relationship between transmission power (TX) and reception quality is established. In this experiment, the base station location was the same whilst three sensors were placed at 10 different locations in the same direction with clear line-of-sight (LOS) including 1, 2, 3, 4, 5, 7, 10, 13, 16 and 20m. Each power adaptation cycle was ended after the maximum power had been reached. The other experimental parameters such as power levels, data sending rate and number of runs are stated in Section 4.2.2.

Fig. 2 shows the average RSSI readings of the three sensors at various locations and transmission power levels. The missing data indicate that the power provides RSSI reading less than -95dBm which is the minimum value reported by TinyOS. Fig. 3 shows average LQI readings of a sensor at various locations and transmission power levels. The missing data indicate unsuccessful data delivery.

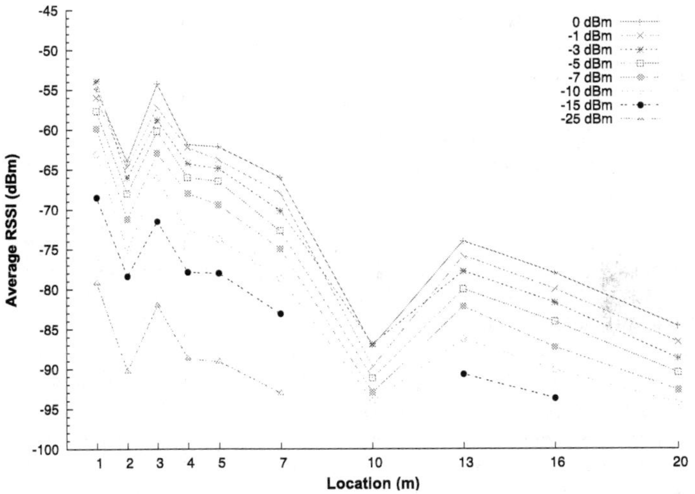

Fig. 2. Effects of sensor locations on RSSI

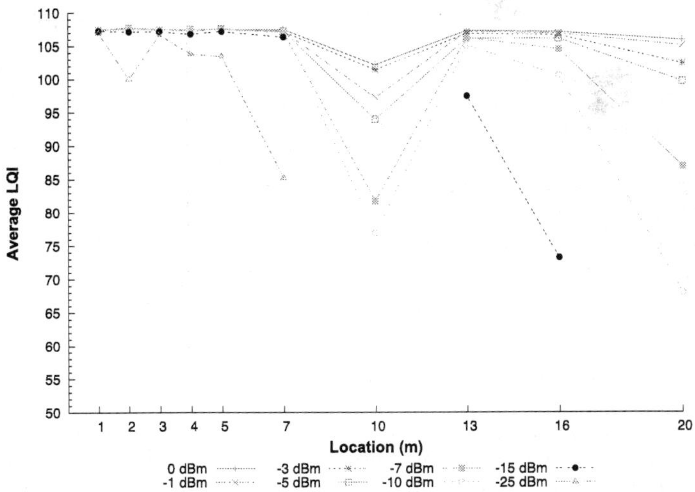

Fig. 3. Effects of sensor locations on LQI

According to Fig. 2 and Fig. 3, most of the RSSI measurements proportionally increased with the transmission power levels. Unlike the RSSI, the LQI measurements were stable at closer locations especially when higher power was used for transmission. Most of the LQI values decreased at greater distances. The minimum power level of -25dBm could be used to successfully deliver data to the base station only when the locations were within 7m. The decrease in received signal strength with increasing distances assumed in the prediction models do not apply in the results. For example, in the case of 2m, the sensor provides a weaker strength compared to a distance of 3m. The experimental results given in (Lin et al., 2006) and (Stoyanova et al., 2007) demonstrate similar observations on location effects. The RSSI and LQI are measured only when the base station receives data. The observed minimum RSSI values higher than -95 dBm indicate data reception.

4.2.4 Fluctuation in link quality metrics over time of day

This section investigates on how RSSI, LQI and PRR fluctuate over the time of day. The same base station and Sensor 1 were used. The sensor was located at 20m in the same environment. It transmitted one packet every second at 0 dBm for 1,440 minutes or 24 hours. The experiment was started in the morning before the office hour.

Fig. 4 demonstrates fluctuation of the RSSI, LQI and PRR over time of day. The RSSI fluctuated during the first half of the experiment. It was stable during the night time and the fluctuation was back later in the experiment. Unlike the RSSI, the LQI fluctuated throughout the experiment. At the beginning the PRR siginificantly decreased. This observation was resulted from the presence of people around the lab. The PRR increased during the night time as there were no staff and student in the lab.

In summary, apart from transmission power, location and heterogeneity in the manufacture, the link quality metrics are affected by the time-of-day. The presence of people around the lab is the main factor in this experiment and is considered as temporary physical barrier. Radio communication in WSN requires a line-of-sight. Some packets may be lost if there are some people in the sending path.

4.2.5 Relationship between metrics

This section aims to describe the relationships between RSSI, LQI and PRR. During packet reception, the base station measures RSSI and LQI. Apart from RSSI and LQI, the standard message type of TinyOS includes the CRC field which is a Boolean data type. The base station also looks at the CRC field to see if the data packet is received correctly. The numbers of data transmissions and receptions are counted to compute the PRR. This scheme can be used in a long-term operation.

However, the PRR may be estimated from the RSSI or LQI measurements. This concept suits a short term operation. The base station does not count the numbers of sent and received packets. Hence, the relationship between metrics needs to be established. Fig. 5 shows relationships between the link quality metrics at 5m, 12m and 19m. The average RSSI and LQI are computed at each transmission power level. The number of received packets is counted in order to calculate the PRR.

According to Fig. 5, several observations can be made as follows:

1. The PRR steeply increases with RSSI up to a certain point followed by more stable reliability measurements. Significant variations in reception rates are found when the RSSI readings are between -95 and -90 dBm. At least 95% PRR may be achieved at all distances if the sensor transmits data at the power producing RSSI greater than -90 dBm.

2. The higher LQI results in a more stable PRR. The relationship between LQI and PRR shown in Fig. 5 (b) is less clear than Fig. 5 (a). Similar results are also addressed in (Lin et al., 2006). According to these observations, RSSI should be used to relate to the PRR.
3. The LQI significantly increases with the RSSI. Convergence to particular LQI values is then observed. A lower bit error rate is observed when the base station receives packets with higher RSSI measurements.

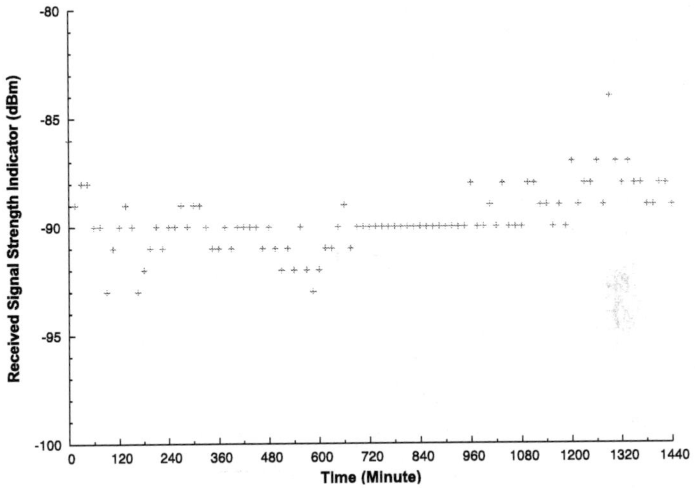

Fig. 4. (a) Fluctuation in link quality metrics over 24 hours RSSI

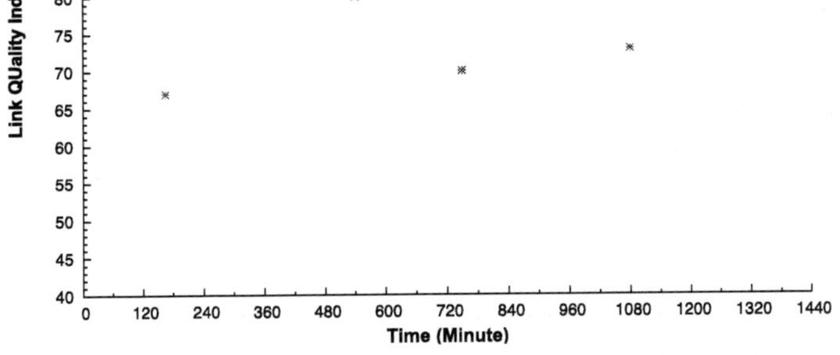

Fig. 4. (b) Fluctuation in link quality metrics over 24 hours LQI

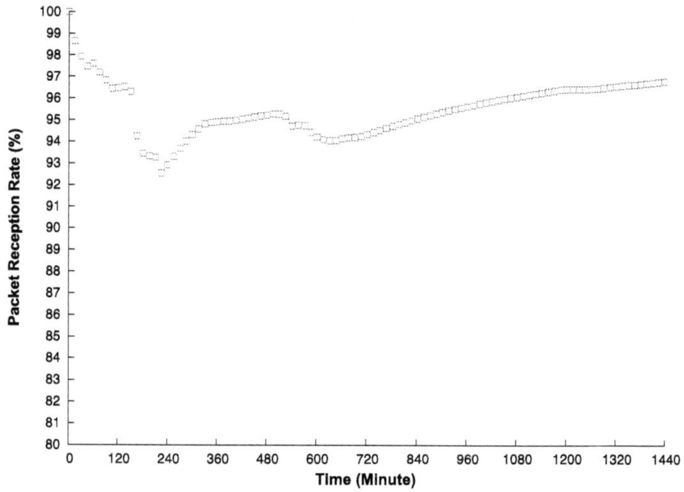

Fig. 4. (c) Fluctuation in link quality metrics over 24 hours PRR

The relationship between link quality metrics can be used to estimate an observed reliability from the measured receiving strength. This observation is addressed in (Lin et al., 2006) and (Srinivasan et al., 2006). After measuring the metrics, the base station determines whether the current transmission power requires an adaptation. The PRR steeply increases with the RSSI followed by significantly more stable measurements. The PRR should not be estimated from the RSSI between -95 to -90dBm as transmission power adaptation based upon this region will not be accurate. The measurements demonstrate that the network should operate at levels taken from an appropriate region.

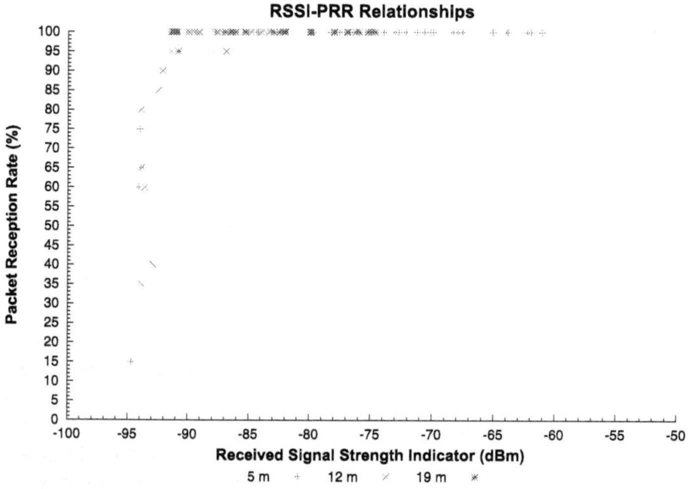

Fig. 5. (a) Relationships between metrics RSSI-PRR

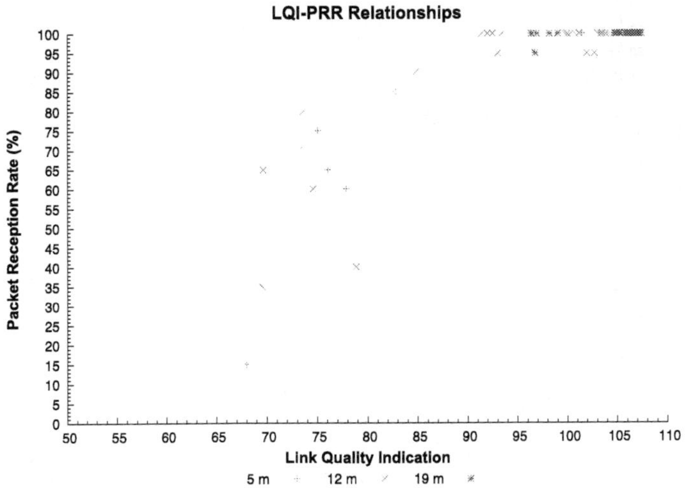

Fig. 5. (b) Relationships between metrics LQI-PRR

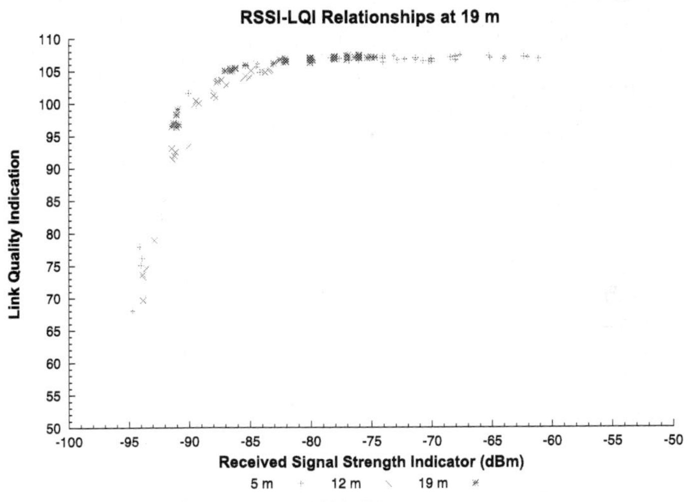

Fig. 5. (c) Relationships between metrics RSSI-LQI

4.3 Delays in wireless sensor network

This section provides some experimental results on delays in wireless sensor network (WSN) which affects PoRAP architecture development. Communication is represented by a frame structure which consists of several slots. A slot is assigned to each source and it transmits data when the allocated slot arrives. The slot length should be long enough to avoid data collisions at the base station where two packets from two different sources arrive approximately at the same time. Several experiments have been conducted in order to investigate some factors which affect the delays, including heterogeneity in sensor manufacturing and payload sizes.

4.3.1 Timestamp measurements and delay calculations

Details of timestamping scenario and delay calculations are given. As the base station does not know when the source is booted, at the beginning it broadcasts the control packet periodically. The periodic broadcast was set to 1 second. After the source is booted, it starts its transmission after the packet has been received. Similarly, the base station starts the next transmission after it has received the packet back from the source. Packet timestamping mechanisms and delay calculations are respectively illustrated in Fig. 6 and Table 4.

According to Fig. 6, the base station is booted at x_0. When the base station is ready to send, the timer is set to be fired at x_1 and send command is called at x_2. A timer is used in order to trigger packet transmission. Prior to transmission, the base station sets some fields in the message structure such as its id and transmission power. The SFD (Start of Frame Delimiter) transmission occurs at x_3. The timestamp is created and the packet payload content is modified to include the time of the transmission. Therefore, the fire-to-send and send command delays of the base station are equal to $x_2 - x_1$ and $x_3 - x_2$. The packet is completely transmitted by the radio at x_4 and the transmission delay is $x_4 - x_3$.

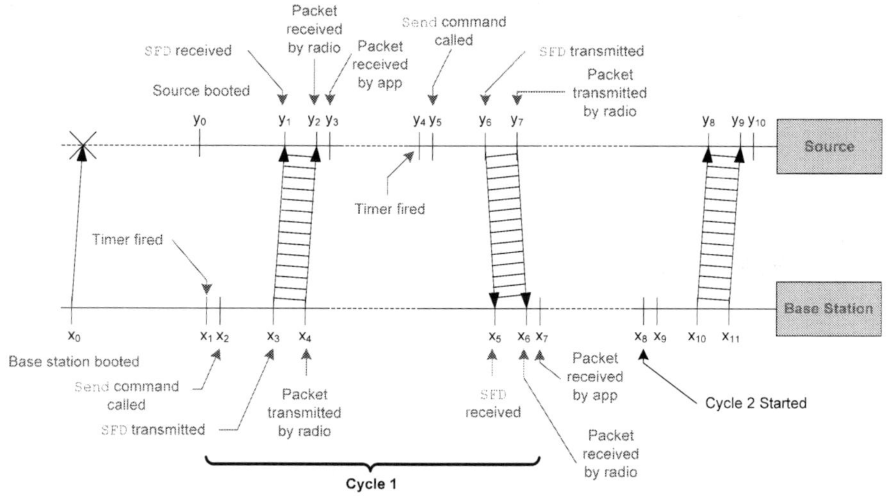

Fig. 6. Timestamp at various events

After being booted at y_0, the source receives the SFD at y_1. The receive event of the radio and application are signalled at y_2 and y_3 when the source receives the packet. The reception and receive delays of the base station are therefore $y_2 - y_1$ and $y_3 - y_2$. Once the packet has been received, the source requires some duration to process the information obtained from the packet. It then sets up its own transmission and the bits of packet are loaded into the radio buffer. The timer is fired at y_4 and the send command is called at y_5. The SFD is transmitted at y_6. Hence, the send command delay of the source is equal to $y_6 - y_5$. The transmission delay is $y_7 - y_6$. Table 4 summarises the delay calculations.

Delays	Calculations
Base Station	
• Fire-to-Send	$x_2 - x_1$
• Send Command Delay	$x_3 - x_2$
• Transmission	$x_4 - x_3$
• Reception	$x_6 - x_5$
• Receive	$x_7 - x_6$
Source	
• Reception	$y_2 - y_1$
• Receive	$y_3 - y_2$
• Fire-to-Send	$y_5 - y_4$
• Send Command Delay	$y_6 - y_5$
• Transmission	$y_7 - y_6$
Two-Way Propagation	$(x_5 - x_3) - (y_6 - y_1)$

Table 4. Summary of delay calculations

According to Table 4, the transmission and reception delays are calculated based upon when the events take place. The transmission delay is defined as the duration required for the radio to transmit the packet. In TinyOS 2.x, the CC2420Transmit interface provides a sendDone() event which notifies packet transmission completion. The reception delay is the duration required for packet reception by the radio, and the receive event is used for the timestamp. The fire-to-send delay indicates the desired interval for starting packet transmission after the timer is fired.

One Tmote Sky base station and one Tmote Invent source were used. The source was located at 0.5 m away from the base station. The base station was plugged into a desktop computer. In total 1,000 cycles of message exchange were run for each source. After the packet had been received, the node waited for 128ms and initiates its data transmission.

4.3.2 Experimental results

In order to consider the effects of payload size, an additional experiment was conducted. The scenario shown in Fig. 6 was used. All settings are the same except the payload sizes. In total five payload sizes were used including 39, 55, 75, 95 and 115 bytes. Note that the maximum payload for the CC2420 radio is limited to 117 bytes whilst the header size is 11 bytes. Send command and transmission delays of the source were determined. Two-way propagation delays were also computed. In the case of 39 bytes, reception and receive delays of source and base station were observed whilst all delays were observed for the larger payload sizes.

Statistical analysis of fire-to-send, send and transmission delays in milliseconds were conducted. The relationships between the 50th percentiles or medians of all sending delays and payload sizes are shown in Fig. 7. Note that "Send Command" delay is represented as "Send" in the figure. The results show that all delays increase with increasing payload sizes. The source requires more time to deliver larger packets to the radio. Similarly, larger packets require a longer duration for transmission. Increases in send command and transmission delays are greater than those of fire-to-send delay.

Statistical analyses of reception and receive delays in milliseconds were also made. The relationships between the 50th percentiles or medians of both receiving delays and payload sizes are shown in Fig. 8. Linear relationship between reception delay and payload size is also observed in Fig. 8. The receive delays are constant for all payload sizes.

The 32-KHz clock has been used in this experimental study and provides 32,768 ticks per second. There are 32 ticks in one millisecond. Therefore, the finest precision is approximately 0.03125 millisecond or 31.25 microseconds. The two-way propagation delays for all payload sizes are calculated and frequencies of the delay occurrences in ticks are shown in Table 5.

According to Table 5, frequencies of the 0-tick decrease with increasing payload sizes. Larger packets require more time to travel from source to destination. However, the two-way propagation delays are significantly less than the other delays.

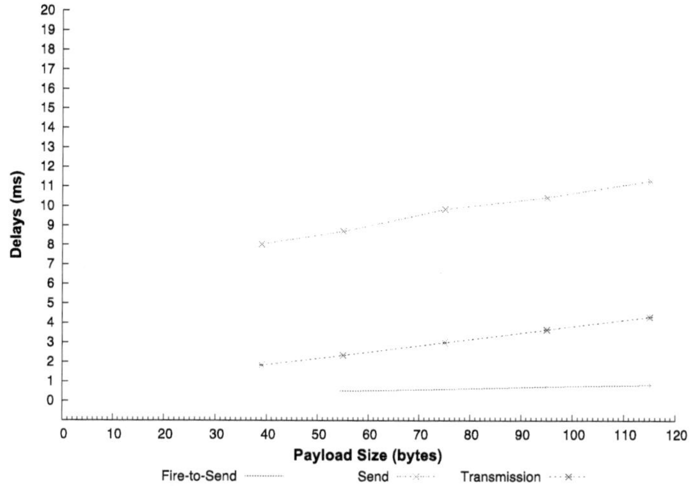

Fig. 7. Relationships between source sending delays and payload sizes

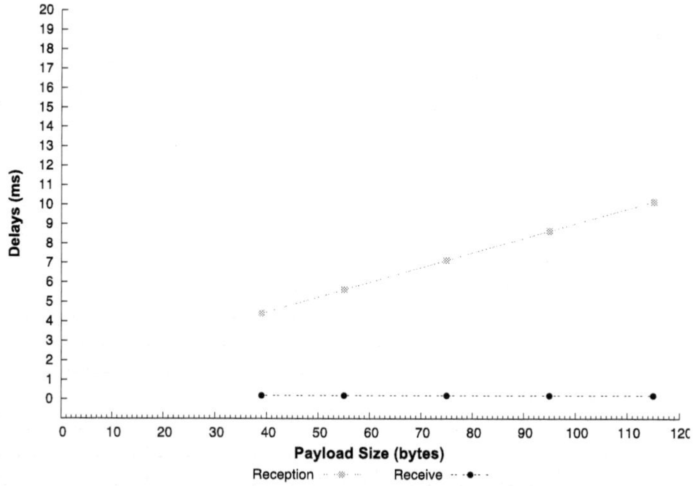

Fig. 8. Relationships between source receiving delays and payload sizes

Attribute		Payload Size (bytes)				
		39	**55**	**75**	**95**	**115**
Frequencies	0	858	807	785	755	740
	1	141	193	212	245	259
	2	0	0	3	0	0
Cycles		999	1,000	1,000	1,000	999

Table 5. Frequencies of two-way propagation delays

5. Design of PoRAP

This section describes the design of PoRAP (Power & Reliability Aware Protocol) which aims at minimising communication energy in wireless sensor network (WSN). The experimental results stated in previous section inform the design.

5.1 PoRAP main capabilities

In PoRAP, power can be conserved via transmission power adaptation and efficient medium access management. The selected link quality index is Received Signal Strength Indicator (RSSI) and it is measured by the base station during data reception. Along with the awareness of data loss, the adjusted power will often maintain the network operating at the region where data loss is minimised.
Additional communication can be saved by adopting the schedule-based MAC approach. Sending and receiving delays can be estimated as they are dependent upon packet size whilst two-way propagation delay is significantly small. Data transmissions are scheduled and the sources are mostly in sleep mode to conserve energy. Only one source engages the shared medium at a time for data transmission. Thus, data collision can be avoided and idle listening can be minimised. More explanations on PoRAP key capabilities are given as follows:

5.1.1 Schedule-based protocol

In the single-hop networks, sources are capable of communicating with their base station directly. This scenario is feasible when the sources and base station are located within communication range of each other. The base station may be connected to several sensors which require an access to the shared medium. Uncontrolled medium access possibly leads to data collisions at the base station. Collision is one of the main sources of power wastage in the WSN shared medium system. The medium access control (MAC) approach attempts collision avoidance. There are currently two main approaches proposed for WSN. Firstly, the medium is sensed to detect any ongoing activities in the medium before conducting data transmission and reception. This scheme is named contention-based.
PoRAP employs another approach in which each node is assigned a specific duration to use the shared medium. This scheme is called schedule-based. The other sensors cannot access and use the medium whilst a sensor is communicating within its time slot. Sources listen to the base station only once in a frame. Idle listening is therefore minimised. Moreover, data collisions at the base station can be avoided as there is only one source sending at a time. The slot length should be long enough to let the source and base station complete data transmission and reception. This scheme may not be suitable in the case of multi-hop WSN where each resource-constrained sensor has to maintain slot information

of its neighbours. Furthermore, time synchronisation is required as both sender and receiver have to orchestrate the data communications to avoid collision caused by the other receivers.
Centralised scheduling control by the base station is feasible in PoRAP. Slot arrangement information can be sent to all sensors located in the range. The base station broadcasts a packet to all sources located in its range. Slot information such as number of slots, slot length and start time of first slot are included in the payload. Once the first frame is finished, the base station broadcasts again with the transmission power adaptation notification.

5.1.2 Communication power conservation

Power constraint should be taken into account when designing a protocol for WSN. Sensors may be left unattended after being deployed in the remote or hostile environment where battery recharge or replacement may be costly or infeasible. Communication accounts for power consumption in WSN. Several sensor platforms provide adaptation to the transmitting power and the concept of Transmission Power Control (TPC) has been adapted to WSN. The CC2420 radio employed by Tmote platform, which is used in this research, supports transmission power (TX) setting. The TX levels are stated by a 5-bit number. There are therefore 32 possible TX settings provided by the CC2420. In TinyOS, the setPower() command provided by CC2420Packet interface accepts a value between 0 to 31 for TX setting. However, the CC2420 datasheet specifies programmable TX in 8 steps from approximately -25 to 0dBm which are respectively equivalent to the power levels of 3 and 31. The Tmote datasheet follows guidelines given by the CC2420.
Transmission power adaptation policies in WSN should take application specifics into account. Different applications may require the sources to transmit data at different rates. For example, an environmental monitoring system may require the current temperature hourly whilst a surveillance system may require the data every second when an intrusion is detected. The sensors should be switched to sleep mode after transmission in order to minimise the idle listening. In a multi-hop network, each node is responsible for routing. It has to communicate with its neighbours to discover the best path by means of the least power utilisation. An amount of power is therefore required for listening in the multi-hop. However, a sensor in the single-hop scenario is capable of transmitting data directly to the base station. It may be switched to sleep mode after transmission. However, the source has to listen during the control slot transmission from the base station.
The power adaptation mechanisms in PoRAP do not require historic entries of RSSI and associated transmission power. The main reason is the limitation of buffering capacity of the radio chip. The base station should support a significant number of sources. In the CC2420 radio, the maximum buffer size is 128 bytes. Some bytes are required for the header and other controlling details. Only two bits are used to notify the power adaptation. The RSSI-PRR relationship obtained from the experimental studies is considered for adaptation as it suggests the operating region for WSN. In the case of power adaptation, the base station sets particular bits to notify the source. The sources get the bits and set their transmission power accordingly.

5.1.3 Link quality monitoring

Radio communication uses air as the transmission medium. There are several attributes ranging from differences in hardware components to environmental factors such as physical barriers which affect signal attenuation. Received signal strength estimation is unlikely as sensors can be placed in various areas of interest. An estimation model should not only determine distance between sender and receiver as an input, location should also be taken into account. A shorter distance may not always provide a higher received strength if a physical barrier appears in the communication line-of-sight (LOS). Moreover, the link quality metrics fluctuate over the time of day. The observed strength in an indoor environment may be lower during the nighttime. Applying the simple received signal strength estimation models, focusing mainly on distance and hardware properties, may not be sufficient. Therefore, PoRAP employs the measurement-based approach in order to more accurately adapt the transmission power.

Two link quality metrics are used in PoRAP. The RSSI is obtained by the radio chip whilst the PRR is specified by the applications. The relationship between RSSI and PRR can relate the application requirement to the observed link quality. As shown in Section 4.2.5, a clear relationship between the two metrics is established. The PRR steeply increases with the RSSI up to a certain point. The PRR is then stable after a certain value of RSSI and a lower RSSI or TX can be used to obtain the required PRR.

The range of required RSSI is obtained from the reliability requirement and the RSSI-PRR relationship. This range is recognised by the base station. Upon data reception, the base station measures the RSSI and compares it to the RSSI thresholds. The adaptation bits are set with respect to the comparison result. There are three available patterns of bit settings; the transmission power will be increased if the measured RSSI is lower than require and it will be decreased if the RSSI is higher. The sources will be notified to retain the current power if the RSSI is within the range.

5.2 PoRAP architecture

This section aims to describe PoRAP architecture. PoRAP aims at an efficient data delivery in WSN by means of energy conservation. Input of PoRAP comes from two external components, the user/application and the monitored phenomenon. PoRAP recognises the duty cycle and the awareness of data loss. The sensed data is another input and it will be sent from the source to the base station. In order to achieve the goals, the base station controls the sources whereas the sources send data to the base station. Required functionalities of the base station and the sources are then stated. The interactions between them are described and they are used to address the required components within the source and the base station. Moreover, the interactions between such components are also given in this section.

5.2.1 Overview of PoRAP

The main objective of PoRAP development is to provide an efficient data communication in WSN where the user/application has his/its own requirements such as reliability and duty cycle. The development of a generic network protocol for WSN is challenging as WSN are application specific. Fig. 9 shows an overview of PoRAP architecture in terms of the interactions between its main components.

According to Fig. 9, four main components are addressed including the user/application, sensed phenomenon, base station and sources. As WSN is application specific, the user/application has its own set of requirements. The base station directly interacts with the user/application whilst the sources collect physical directly from the phenomenon. The functionalities required at the base station and source can be listed as follows:

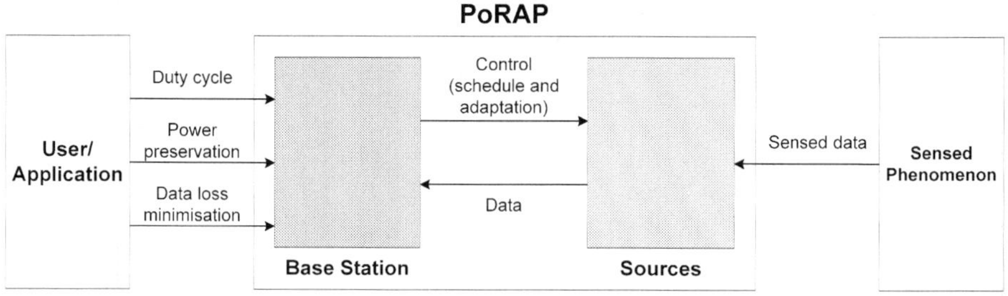

Fig. 9. Overview of PoRAP

Base station:

- **Recognise the requirements of user/application:** PoRAP aims at the low duty cycle application where the key objective is power conservation instead of throughput. Examples of this application category are habitat and environmental monitoring systems.
- **Control the source's operation:** This work focuses on the single-hop network where direct communication between sources and base station is feasible. No routing is required at each source and its operation is controlled by the base station in two aspects. Firstly, the base station determines whether transmission power used by the source needs to be adjusted by looking at the RSSI. Secondly, the communication cycle of each source is scheduled in order to avoid data collision and minimise idle listening.

Source:

- **Collect physical data:** WSN has been deployed to collect physical data from the targeted environment such as temperature and humidity. This work looks at how an efficient data delivery can be achieved by using lower transmission power whilst data loss is minimised. The processes of data collection are outside the scope of this study.
- **Data transmission:** After receiving the control information, the source sets two parameters. Firstly, it synchronises the communication schedule. Thus it will know when to start the radio for control reception and data transmission. Secondly, the source adapts its transmission power level according to the included notification. Lower power can be used and a significant amount of transmission power can be conserved.

Several interactions between the source and base station are required to achieve the functional requirements and they are addressed in Fig. 10.

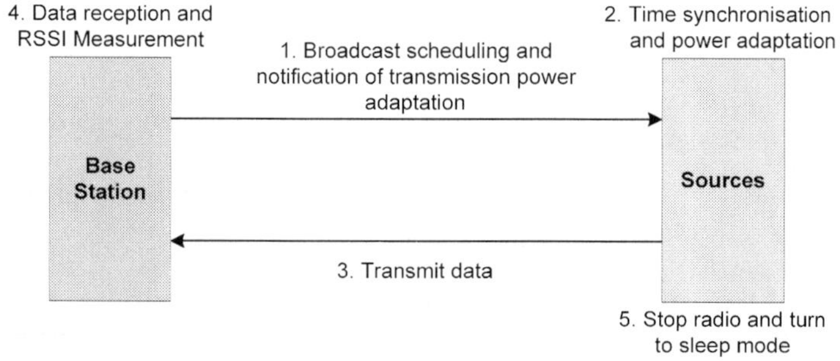

Fig. 10. Interaction between sources and base station

1. PoRAP focuses on the set of fixed sources which are located within communication range of the base station. The control packet includes scheduling and power adaptation notification and is broadcast to the sources using the maximum power level. This is feasible as the base station obtains extra power from the connecting computer.
2. Once the control packet is received by the source. Information on scheduling and notification is read. The source synchronises its schedule with the other nodes together with adjusting its transmission power accordingly.
3. After conducting time synchronisation and transmission power adaptation, the source waits for its slot to conduct data transmission using the adjusted transmission power. The radio must be started for communication.
4. The base station measures the RSSI during data reception. The observed RSSI is compared to the desired range which includes minimum and maximum values. The setting of the RSSI thresholds is obtained from the RSSI-PRR relationship. The selected RSSI should be obtained from the region where significant stability in the PRR is observed. The base station then decides whether transmission power adaptation is required. The notification is set accordingly.
5. The source stops its radio after transmission to save power. The amount of power consumption is the least when the source is in sleep mode. Timing is required for the source to start the radio again for the next communication cycle.

5.2.2 Components

The previous section points out several essential functions which are required to achieve the objectives of PoRAP development. This section aims to describe the essential components which give rise to this functionality. The selected operating system for WSN in this work is TinyOS which already provides several useful components and PoRAP takes those in TinyOS and adds some further modifications. The main components are determined from the interactions including the user/application, the observed phenomenon, the base station and source. Several components required at the base station and source are then considered. Moreover, the interactions between each component are demonstrated.

A) Components at base station and sources

The base station recognises the requirements of the user/application and controls the sources based upon the requirements. As PoRAP aims at the direct communication, the control information is broadcast to the sources which are located within the communication range. After physical data collection, the sources set their communication parameters prior to data transmissions. Fig. 11 depicts several components required at the base station and sources.

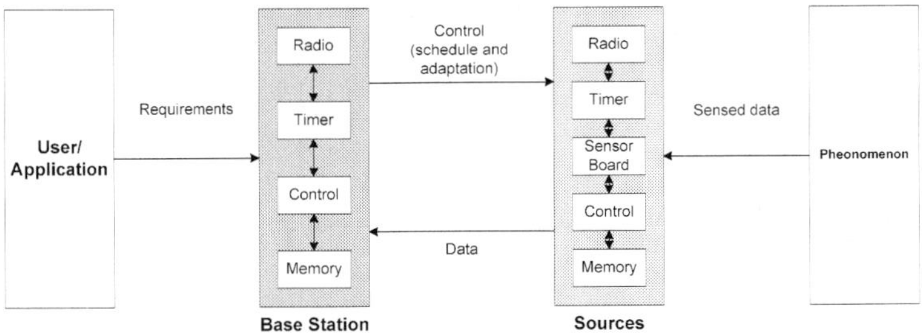

Fig. 11. Components at base station and sources

Each of the required components is described as follows:

- **Radio:** Each sensor employs the radio communication for wirelessly communicating with its neighbours or destinations. The radio has four major functions as follows:
  - o *Data communications:* Control information is sent by the base station's radio chip and is received by the source's radio chip. Data is sent by the source's radio chip and is received by the base station's radio chip.
  - o *Data buffering:* Prior to forwarding the received data to the higher layers or transmitting the data through the medium, the data is buffered. The buffering capacity is limited and dependent upon the radio chip. The capacity is important to the design of packet structures. For example, the control packet must not be longer than the allowable capacity but it has to carry all the required information.
  - o *Received signal strength measurement:* The received signal strength is important as it can reflect the current link quality. The latest radio chip provides the measurement of received signal strength such as Received Signal Strength Indicator (RSSI) and Link Quality Indication (LQI). RSSI is used in this work as it can be obtained from several radio models and its relationship with the Packet Reception Rate (PRR) is clear.
  - o *Transmission power adaptation:* The RSSI changes with transmission power and several factors such as location, time-of-day and environment. One of the main features in PoRAP is transmission power adaptation. The key concept is adjusting the current transmission power to achieve the power conservation and data loss minimisation. The latest radio model supports programmable transmission power.
- **Timer:** WSN is considered a share-medium system as all nodes have to access the medium prior to transmission. PoRAP aims at single-hop WSN where direct communication between source and base station is feasible. The sources are not responsible for routing. Instead of applying the contention-based scenario, the transmissions are scheduled. A slot is allocated for each source so that it can send only when its slot arrives. Otherwise, the radio is stopped and the source is switched to sleep mode for minimum energy consumption. A timer is therefore required for scheduling the radio start and stop.
- **Control:** It is used to control the other components especially when there is no control mechanism provided for some components. For example, an additional control interface is required for the radio and the interface is used to start and stop the radio.
- **Memory:** This component is the basic one which is also included in the sensor. Several variables along with their values and measurements are stored in the memory. For example, the required RSSI range which is obtained from the RSSI-PRR relationship. This range is stored in the memory and will be compared to the observed RSSI to determine whether any transmission power adaptation is required.
- **Sensor board:** This component is crucial for the sensors as it is responsible for collecting the physical data from the environment. The sensor board consists of several sensors such as temperature and humidity.

B) Interactions between components

This section aims at addressing the interactions between the components, and they are described in Fig. 12. The interactions within the base station and source can be separately described as follows:

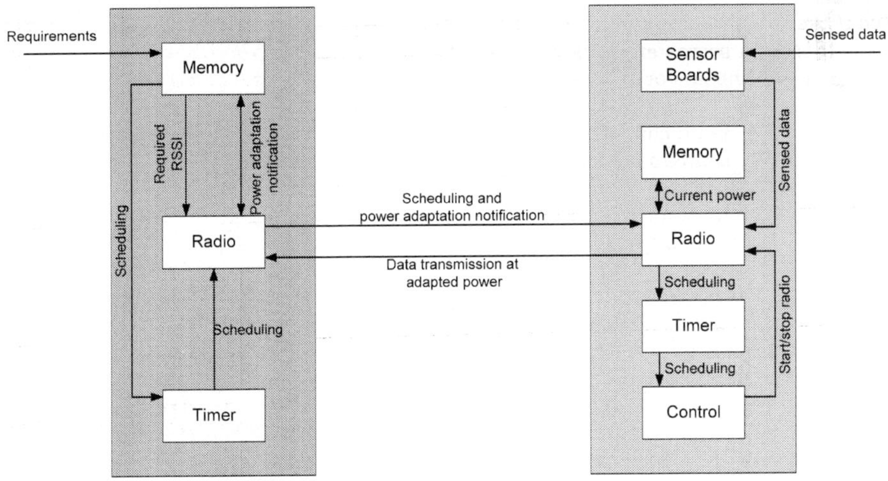

Fig. 12. Interactions between components

Base station

The base station acts as a destination for the data. The requirements are stored in the memory and they are used to set required RSSI range and the data sending rate. In PoRAP, the schedule-based scheme is adopted where each source has its own slot for data transmission. The slot must be large enough to accommodate several communication delays. According to the results in Section 4.3.2, sending and receiving delays are mainly dependent upon the packet size whereas the two-way propagation delay is significantly small. Models are required for estimating the slot size and they will be described later in this chapter. The next transmission begins after the other sources have already transmitted. Hence, PoRAP suits the applications which require a low duty cycle. The timer is used for scheduling the communications so it also uses this requirement from the application.

The required RSSI range can be obtained from the RSSI-PRR relationship which is dependent upon different conditions such as time-of-day, environment and location of deployment. The PRR is also used as an additional link quality metric as it is close to the reliability requirement. The main objective of PoRAP is to conserve communication energy whilst data loss is minimised. In the short term, the base station measures the RSSI when it receives the data packet. It uses the observed RSSI to determine whether power adaptation is required. The notification bits which are reserved for each source are then set. In the medium or longer term, the base station measures the PRR and uses that to determine what the upper and lower RSSI bounds should be. If more packets are lost, the RSSI bounds are increased. However, the bounds are slowly lowered to reduce power expenditure if the loss is low or non-existence. The number of notification bits is crucial as the base station has to communicate with all the sources in its range. Using too many bits may lead to a control packet which is larger than the buffering capacity of the radio chip.

The base station radio is not started or stopped as it has to continually receive the data packets from its sources. Data packet receptions occur after broadcasting the control packet at the maximum transmission power level. This concept is feasible as the base station has an extra source of power from its connecting computer. In PoRAP, the power conservation goal is mainly located at the sources.

Source

In WSN, the source is responsible for physical data collection. The data is then transmitted to the base station. The key objective of PoRAP is to conserve communication power of the source. Prior to transmission, the source determines whether it has to adapt its current power. The notification is included in the control packet and it is received by the radio of the source. As the buffering capacity of the radio is limited, the base station notifies what the source should do to its current power instead of specifying the appropriate power level. Thus, the source has to store the current power in the memory. For example, the current power is increased if a lower RSSI is measured by the base station. Moreover, the source should recognise the limitations of the transmission power adaptation. The base station may need its source to increase the power even if the maximum has already been reached. The minimum and maximum power levels are dependent upon the selected radio chip.
Apart from the power adaptation signaling, the scheduling is also included in the control packet. Time synchronisation is crucial in the schedule-based approach. The local clock of each node may run at different speeds. In PoRAP, the sources synchronise with their base station. The synchronisation refers to several timestamps which are conducted at the MAC layer where hardware and operating system dependent delays can be disregarded. The scheduling is also recognised by timer and controls components. Several timers are required as they are responsible for timing the sending and receiving communications. The timers operate closely with the control in order to start and stop the radio. For example, the radio is stopped after the data packet is sent. The source knows when it has to wake up to receive the next control packet. The timer is then started, counting the generated ticks. A control interface is used to start the radio for control reception when the scheduled time has come.

5.2.3 Transmission power adaptation policies

A sensor consists of hardware components working together to facilitate sensing, processing and communicating tasks. Amongst these components, the transceiver or radio unit is responsible for data communication. Normally, the radio unit supports programmable transmission power and the possible adaptable range is given in the datasheet. For example, the Tmote sensor platform which is chosen for this work employs the CC2420 radio. The minimum and maximum powers are 0 and -25dBm, respectively. There are two main factors which should be taken into account when transmission power adaptation is required. Several hardware limitations of the radio unit include the allowable minimum, maximum transmission power and base noise. The environmental factors leading to signal strength attenuation should be determined. The selected transmission power should be high enough to produce the associated receiving strength which is not discarded by the receiving node. The maximum power allowed by the radio unit is used as the upper limit. In PoRAP, sources use maximum power for their first transmissions. This policy ensures that the packets will likely be transmitted to the base station. However, both base noise and attenuation are respectively hardware and environment dependent. It is difficult to specify an accurate power adaptation level which can be generally used. Moreover, additional resources will be required if the sources periodically measure and send their base noise to the base station. Attenuation is hard to predict as link quality changes over time. Hence,

PoRAP repetitively increases or decreases the transmission power within an allowable range instead of discovering the right power.

5.2.4 Frame structure and slot decomposition

In PoRAP, a frame is used to represent a communication cycle which consists of one control slot at the beginning followed by several data slots. Its structure is shown in Fig. 13.
G indicates the guard of the frame and is used to protect frame overlapping. A control slot is used by the base station for broadcasting control data which includes scheduling information and transmission power (TX) adaptation notification to its sources. The slot information is required by the sources in order to synchronise themselves to the base station. The time of starting the first data slot is required so that the sources know when data is sent. In PoRAP, each slot has the same length which should accommodate a specific data payload size to be completely transmitted and received.

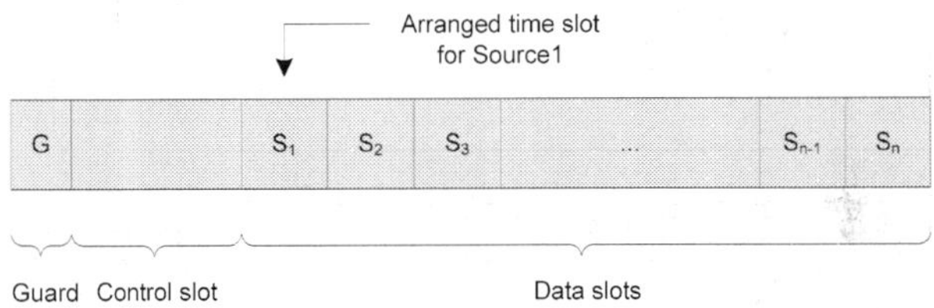

Fig. 13. Frame structure

According to Fig. 13, the sources firstly turn their radios on during the control slot to receive the control information. If they are not assigned to the first data slot, they stop the radios after knowing when their slots start. When their slots arrive, the radios are re-started to send the data. Unlike sources, the base station listens to the medium for data packet reception all the time. The decomposition of a slot is depicted in Fig. 14.

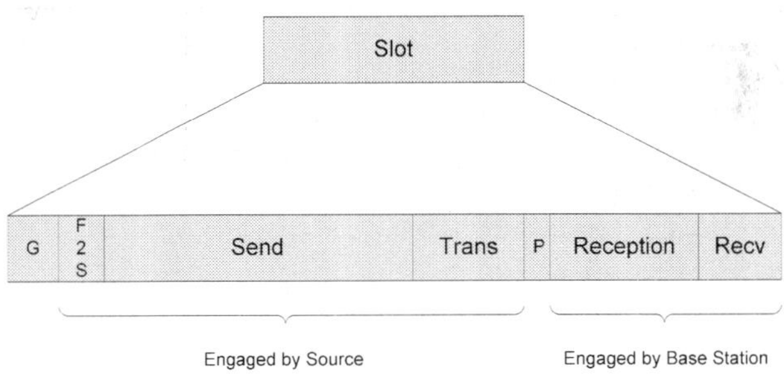

Fig. 14. Data slot decomposition

There are four main delay components in Fig. 14. The *G* and *P* are respectively the guard time and propagation delay. The first component is the guard length which prevents the slots from overlapping. Feasible overlapping scenarios together with guard time consideration are provided later in this section. The second component consists of fire-to-send (F2S), send and transmission delays and this is the sending delay component. This component is caused by the source. The third one is propagation delay which is considerably smaller than the other delays. Finally, the receiving delay component includes the reception and receive delays. This component is considered during packet arrival at the base station.

5.2.5 Estimation of communication delays

A schedule-based approach is adopted in PoRAP. The base station allocates and manages several time slots. In this work, a set of fixed nodes is determined. The number of data slots is therefore equal to the number of booted sources which are able to receive the control packet broadcast by the base station. The source initiates transmission when its assigned slot arrives. Apart from data slots, a frame also contains a control slot which is used by the base station. The slot must be large enough to accommodate sending and receiving delays to avoid feasible data collisions. As shown in Section 4.3.2, the delays are dependent upon packet sizes. This section analyses these relationships for delay estimations.

The experimental results on delays described in Section 4.3.2 demonstrate linear relationships between delays and data packet sizes. The key objective in this part is to discover the two coefficients obtained from linear regression analyses. The coefficients will be used to establish the models providing estimated delays where payload sizes are input. In total 5 payload sizes including 39, 55, 75, 95 and 115 bytes were varied to investigate changes in delays. Regression analyses have been applied to the results of the sending and receiving delays of the source and the base station. As linear relationships between delays and payload sizes are observed, two coefficients of the linear equation (c_0 and c_1) are the required output where c_0 is the y-intercept and c_1 is the slope. The Table 6 summarises the coefficients of each delay.

According to Table 6, the coefficients for the base station do not significantly differ from those for the source. The fire-to-send delays of the base station are constant whilst the source provided a linear relationship.

Delays	Measured at	Coefficients	
		c_0	c_1
1. Fire-to-send (F2S)	Base station	Constant delays of 0.50 ms	
	Source	0.204	0.025
2. Send	Base station	11.367	0.043
	Source	11.263	0.043
3. Transmission	Base station	0.490	0.033
	Source	0.552	0.033
4. Reception	Base station	1.521	0.076
	Source	1.521	0.076
5. Receive	Base station	Constant delays of 0.22 ms	
	Source	Constant delays of 0.22 ms	

Table 6. Coefficients obtained from experimental results at 99th percentile

In the case where the payload size is zero, a specific duration is still required for header transmission and reception. For CC2420, the header is approximately 11 bytes and requires 0.352ms for the delivery. An additional duration is required for transmitting processes which can be considered as an overhead. The send delay is the largest of the experimental results. It is an interval from calling the send() command until capturing the SFD. Several mechanisms undertaken by the application software and operating system to facilitate the sending also require time and are included in the send delay. For example, when the send() command is called by the application, an interrupt is signaled to TinyOS. The packet is buffered and the CC2420 is switched to transmitting mode. This sending overhead due to software manipulation and hardware setup is regardless of payload size. Increases in payload size require additional delays. For example, for every byte increase in the payload size, the send and reception delays of a source respectively increase by 0.043 and 0.076ms. However, the payload size does not affect receive delay. The coefficients can be used to estimate the communication delays.

5.2.6 PoRAP scenario

PoRAP is developed to effectively support data communication in single-hop wireless sensor network (WSN). The base station communicates with its sources for controlling and data collection purposes. As the base station does not know when each source is booted, a setup process is required at the beginning of frame structure. Acting as a data receiver, the base station always listens to the medium for incoming messages after broadcasting the control packet. Hence, the base station desires extra power which can be obtained from external sources such as a desktop or laptop computer.

A) Control and setup phase

Prior to data transmission, the sources have to setup their parameters based upon the control information received from their base station. The information such as number of slots, slot length and slot start time is used to control the sources in order to send data within an allocated slot at an adapted transmission power. As the base station has no information on when the sources join the network, it has to discover which sources are booted and ready for communication. In the control and setup phase, the base station periodically broadcasts control packets to all sources located in its communication range. The broadcasted packet is received by the booted sources and they use the received information to setup the communication parameters.

There are three main parts to the control information included in the control packet. The first attribute indicates the identification of the base station. This field supports a future enhancement of PoRAP which supports the multiple base station system. It can be also used to differentiate between the control and data packets. The second attribute is schedule related. Some information is required by the sources in order to synchronise with their base station. These parameters include the number of slots, slot length and the start time of the first slot. The base station specifies the slot start time with respect to the Start of Frame Delimiter (SFD) transmission in order to minimise the effects of application and hardware processing delays. The source assigned to the first data slot sets its timer to fire and sends data when the time arrives. Other sources start at different times and they compute the starting times from the slot information. The transmission parameters are required to be completely set before the phase begins. Slot length determination for data slot can therefore be applied to the control slot. The base station periodically broadcasts its control packet. There are two main objectives of periodic broadcasting are maintaining synchronisation between nodes and supporting changes in network topology. Additional sources may be booted during the frame and some sources may be running out of energy. The number of sources is therefore modified by the base station.

B) Data delivery phase

Slots are allocated by the base station in order to facilitate data transmissions of the sources. The data delivery phase starts after the control packet is received by the sources. The number of slots is fixed as it assumes that the base station communicates with the fixed number of sources and the number is constant throughout the operation. Data collected by the sources is stored in the data packet and is delivered to the base station.

The Received Signal Strength Indicator (RSSI) is measured when the base station receives the data. The RSSI linearly relates to the transmission power and the RSSI-PRR relationship is established in Fig. 5 (a). The PRR steeply increases with the RSSI up to a certain point. The increase in PRR then becomes insignificant or it becomes constant after this point. The RSSI is monitored and compared to the desired range. Power adaptation notification is conducted by the base station. The sources are notified by control packet reception in the next frame.

Apart from data, the identification (id) of a source is also included in the data packet. Specifying source id is an important issue and it may be done in several ways. For example, the SFD of the control packet reception time may be modified to obtain the id. However, sensors are considered resource constrained. Simple calculations should be included in the sources. Within the 128-byte buffering limitation in CC2420, one to two bytes should be enough to represent the id. Furthermore, the id can be assigned at installation time. Prior to deployment, a particular id is allocated to the source. For example, an id of 1 may be used for installing PoRAP in the first source in the network. Additional power conservation is introduced during the data delivery phase. The strategy benefits from adopting the time-slot based concept. As sources know when to receive control and to transmit data packets, it is possible to periodically turn the radio on for such periods. Fig. 15 describes the mode switching concept during the data delivery phase. The C&S, R, S and G represent control and setup, receive, send and guard, respectively.

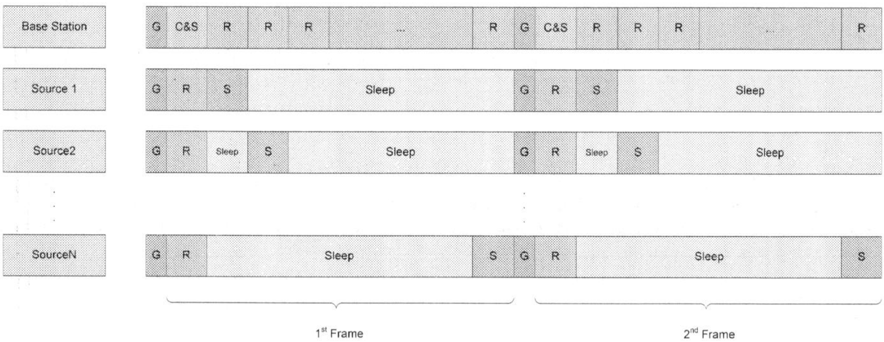

Fig. 15. Mode switching during the data delivery phase

According to Fig. 15, each source is in wakeup mode when its radio is turned on for two reasons; control packet reception and data packet transmission. Otherwise, its radio is turned off and the source is switched to sleep mode. However, the base station radio is always turned on. This strategy minimises idle listening power at the sources.

6. PoRAP energy conservation evaluation

An experiment was conducted in a 16m x 20m indoor environment to evaluate the energy conservation of PoRAP. A network consisting of 20 sources and a base station was set up. Tmote Sky motes were used as both sources and base station. The sources were placed at 20 different locations with 14 different distances and the base station was connected to a desktop machine. All motes had the same height above ground level and had the same antenna orientation. The minimum and maximum distances are 1 and 22.5m, respectively. Initially, the base station broadcast its 18-byte control packet to the sources. The sources then transmitted the 48-byte data packets back to the base station. A communication cycle was completed after the base station had received the data from all sources. Apart from the maximum power settings, four additional RSSI settings are included. The minimum RSSI thresholds were set to -90, -80, -70 and -60dBm whereas the corresponding maximum thresholds were -80, -70, -60 and -50dBm, respectively. The power is not adapted if the measured RSSI is between the thresholds and the aim is to obtain nearly 100% PRR. Each mote transmitted every 5 minutes and the experiment lasted for 24 hours. The results are shown in Table 7.

Dist. (m)	-90 < RSSI < -80		-80 < RSSI < -70		-70 < RSSI < -60		-60 < RSSI < -50		Max TX	
	Saved Trans Current	Packet Loss (%)	Saved Trans Current	Packet Loss (%)	Saved Trans Current	Packet Loss (%)	Saved Trans Current	Packet Loss (%)	Saved Trans Current	Packet Loss (%)
1	51.2	0	51.2	0	51.2	0	35.6	0	0	0
2	51.2	0.3	35.6	0.7	0	0	0	0	0	0
4	43.1	2.3	43.1	0.7	28.2	0	0	0	0	0
6	43.1	4.7	28.2	0	0	0.3	0	0	0	0
8	51.2	5	0	0.7	0	0.3	0	0	0	0
10	51.2	5.3	35.6	0	0	0	0	0	0	0
12	51.2	5.7	20.1	0	0	0	0	0	0	0
14	0	14	28.2	0	0	0	0	0.4	0	3.7
16	28.2	5.7	20.1	0	0	0	0	0	0	1.2
20	43.1	3.7	0	0.7	0	0	0	1.2	0	2.1

Table 7. Conserved transmitting current and data packet loss

According to Table 7, lower RSSI settings result in higher percentage of packet loss and conserved transmitting power. Lower power is used to produce the required RSSI range. A significant amount of power up to 50% can be yielded. However, the highest packet loss is obtained when the RSSI is between -90 and -80 dBm.

7. Conclusion

This chapter describes several aspects which should be considered during developing a network protocol for wireless sensor network (WSN). WSN has been used in both surveillance and civil applications. It is considered application specific as each application has its own set of requirements. Two main categories are proposed including event-based and periodic-based application. Throughput is the key requirement in the event-based whilst lifetime is the key in the periodic-based. Moreover, one of major drawbacks of WSN is resource constraint. The power for all operations comes from tiny batteries. Under some circumstances, it is uneconomical or impractical to change or recharge the batteries. In WSN, the data is delivered via wireless link which is susceptible to the surrounding environments. The radio unit is responsible for data delivery has a limited buffering capacity. Control information should be minimised to be included in a packet.

The Power & Reliability Aware Protocol (PoRAP) is developed and its main objective is to provide an efficient data communication by means of energy conservation whilst reliability is maintained. Its three key elements include direct communication, adaptive transmission power and intelligent scheduling. With adaptive transmission power and intelligent scheduling, the power consumption is minimised as a result of a lower transmitting power, collision avoidance and minimised idle listening without unnecessary data losses. The key capabilities of PoRAP make it suitable for use in the periodic-based WSN applications with regular reporting patterns where maximising bandwidth is not the prime concern. PoRAP thus applies to some of the WSN applications such as environmental and habitat monitoring where the sources often remain at their positions throughout the operation. Slots are allocated to the sources for data transmissions. In PoRAP, it is assumed that the number of allocated slots is equal to that of sources. A low duty cycle application is more efficient using PoRAP when the percentage of slot usage is high. The evaluation results indicate up to 50% of power can be yielded whilst the reliability is within the desired range. However, PoRAP is not applicable if a source has to wait longer until the next cycle is started. Therefore, a limitation of PoRAP arises when there is a high slot overhead because there are many sources in the network.

8. References

Warneke, B. & Pister, K.S.J. (2002). MEMS for Distributed Wireless Sensor Networks, *Proceeding of the 9th International Conference on Electronics, Circuits and Systems*. Dubrovnik, Croatia

Mainwaring, A.; Polasrte, J. ; Szewczyk, R.; Culler, D. & Anderson, J. (2002). Wireless Sensor Networks for Habitat Monitoring, *WSNA'02*, Atlanta, Georgia, USA.

Allen, G.W.; Lorincz, K.; Ruiz, A.; Marcillo, O.; Johnson, J.; Lees, J. & Welsh, M. (2006). Deploying a Wireless Sensor Network on an Active Volcano. *IEEE Internet Computing*, Vol.10, No.2, pp.18-25

Essa, I.A. (2000). Ubiquitous Sensing for Smart and Aware Environments. *IEEE Personal Communications*

Srivastava, M.; Muntz, R. & Potkonjak, M. (2001). Smart Kindergarten: Sensor-Based Wireless Networks for Smart Developmental Problem-Solving Environments. *ACM SIGMOBILE*, Rome, Italy

Jovanov, E.; O'Donnell Lords, A.; Raskovic, D.; Cox, P.G.; Adhami, R. & Andrasik, F. (2003). Stress Monitoring Using a Distributed Wireless Intelligent Sensor System. *IEEE Engineering in Medicine and Biology Medicine*

Otto, C.; Milenković, A.; Sanders, C. & Jovanov, E. (2006). System Architecture of a Wireless Body Area Sensor Network for Ubiquitous Health Monitoring, *Journal of Mobile Multimedia*, Vol.1, No.4, pp.307-326

Arora, A.; Dutta, P.; Bupat, S.; Kulathumani, V.; Zhang, H.; Naik, V.; Mittal, V.; Cao, H.; Demirbas, M.; Gouda, M.; Choi, Y.; Herman, T.; Kulkarni, S.; Arumugam, U.; Nesterenko, M.; Vora, A. & Miyashita, M. (2004). A Line in the Sand: A Wireless

28

Standardised Geo-Sensor Webs for Integrated Urban Air Quality Monitoring

Bernd Resch, Rex Britter, Christine Outram, Xiaoji Chen and Carlo Ratti
Massachusetts Institute of Technology, USA

1. Introduction

'In the next century, planet earth will don an electronic skin. It will use the Internet as a scaffold to support and transmit its sensations. This skin is already being stitched together. It consists of millions of embedded electronic measuring devices: thermostats, pressure gauges, pollution detectors, cameras, microphones, glucose sensors, EKGs, electroencephalographs. These will probe and monitor cities and endangered species, the atmosphere, our ships, highways and fleets of trucks, our conversations, our bodies – even our dreams.' (Gross, 1999)

Following this comprehensive vision by Neil Gross (1999), it can be assumed that sensor network deployments will increase dramatically within the coming years, as pervasive sensing has recently become feasible and affordable. This enriches knowledge about our environment with previously uncharted real-time information layers.

However, leveraging sensor data in an ad-hoc fashion is not trivial as ubiquitous geo-sensor web applications comprise numerous technologies, such as sensors, communications, massive data manipulation and analysis, data fusion with mathematical modelling, the production of outputs on a variety of scales, the provision of information as both hard data and user-sensitive visualisation, together with appropriate delivery structures. Apart from this, requirements for geo-sensor webs are highly heterogeneous depending on the functional context.

This chapter addresses the nature of this supply chain; one overarching aspect is that all elements are currently undergoing both great performance enhancement combined with drastic price reduction (Paulsen & Riegger, 2006). This has led to the deployment of a number of geo-sensor networks. On the positive side the growing establishment of such networks will further decrease prices and improve component performance. This will particularly be so if the environmental regulatory structure moves from a mathematical modelling base to a more pervasive monitoring structure.

Of specific interest in this chapter is our concern that most sensor networks are being built up in monolithic and specific application-centred measurement systems. In consequence, there is a clear gap between sensor network research and mostly very heterogeneous end user requirements. Sensor network research is often dedicated to a long-term vision, which tells a compelling story about potential applications. On the contrary, the actual implementation is mostly not more than a very limited demonstration without taking into account well-known issues such as interoperability, sustainable development, portability or the coupling with established data analysis systems.

Therefore, the availability of geo-sensor networks is growing but still limited. Deborah Estrin pointed out in 2004 that no real sensor network applications exist, apart from short-lived prototypical and very domain-specific demos (Xu, 2004). Recently, some examples of urban geo-sensor networks arose, but the clear social, health and economic benefits have not been demonstrated or described in a manner that would compel this sort of investment in particular for urban environments.

The goal of the Common Scents project is that its highly flexible architecture will bring sensor network applications one step further towards the realisation of the vision of a 'digital skin for planet earth' and have particularly far-reaching impacts on urban monitoring systems through the deployment of ubiquitous and very fine-grained sensor networks. In other words, the broad goal of the project is to develop an overarching infrastructure for various kinds of sensor network applications.

After this brief introduction the chapter provides a discussion of related work. Thereafter, we present the Common Scents approach and implementation for urban monitoring applications. Then, we illustrate some possible application areas for the system. We treat challenges for the deployment of sensor networks that are specific to the city and analyse how such systems can change the city as the functional context by adding new unseen information layers. We conclude with a short summary, discussion and future outlook.

2. Related work

The first domain of related work is **sensor network** development for environmental monitoring. The Oklahoma City Micronet (University of Oklahoma, 2009) is a network of 40 automated environmental monitoring stations across the Oklahoma City metropolitan area. The network consists of 4 Oklahoma Mesonet stations and 36 sites mounted on traffic signals. At each traffic signal site, atmospheric conditions are measured and transmitted every minute to a central facility. The Oklahoma Climatological Survey receives the observations, verifies the quality of the data and provides the data to Oklahoma City Micronet partners and customers. One major shortcoming of the system is that it is a very specialised implementation not using open standards or aiming at portability. The same applies to CORIE (Center for Coastal and Land-Margin Research, 2009), which is a pilot environmental observation and forecasting system (EOFS) for the Columbia River. It integrates a sensor network, a data management system and advanced numerical models.

Paulsen (2008) presents a sensing infrastructure called PermaGIS that attempts to combine sensor systems and GIS-based visualisation technologies. The sensing devices, which measure rock temperature at ten minute intervals, focuses on optimising resource usage, including data aggregation, power consumption, and communication within the sensor network. In its current implementation, the infrastructure does not account for geospatial standards in sensor observations. The visualisation component uses a number of open standards (OGC WMS, WFS) and open-source services (UMN Map Server, Mapbender).

There are a number of approaches to **leveraging sensor information in GIS** applications. Kansal et al. (2007) present the SenseWeb project, which aims to establish a Wikipedia-like sensor platform. The project seeks to allow users to include their own sensors in the system and thus leverage the 'community effect', building a dense network of sensors by aggregating existing and newly deployed sensors within the SenseWeb application. Although the authors discuss data transformation issues, data fusion, and simple GIS analysis, the system architecture is not based on open (geospatial) standards, only standard web services. The web portal implementation, called SensorMap (Nath et al., 2006), uses the Sensor Description Markup Language (SDML), an application-specific dialect of the OGC SensorML standard.

A GIS mash-up for environmental data visualisation is presented in the nowCOAST application (National Oceanic and Atmospheric Administration, 2011). Data from providers such as National Oceanic and Atmospheric Administration (NOAA), U.S. Geological Survey (USGS), Department of Defence (DOD) or local airports, are integrated in a web-based graphical user interface. nowCOAST visualises several types of raw environmental parameters and also offers a 24-hour sea surface temperature interpolation plot with 0.5 degree spatial resolution, produced using a two-dimensional variation interpolation scheme. The system does not make use of open standards for sensor measurements and data provision.

The second related research area is real-time data integration and **sensor fusion** for geospatial analysis systems. Harrie (2004) and Lehto & Sarjakoski (2005) present web services based on the classic request/response model. Although both methods widely use open GIS standards, they are not suitable for the integration of real-time data for large volumes of data. Sarjakoski et al. (2004) establish a real-time spatial data infrastructure (SDI), which performs several application-specific steps, such as coordinate transformation, spatial data generalisation, query processing or map rendering and adaptation. However, the implemented system accounts neither for event-based push mechanisms nor for the integration of sensor data.

Other approaches for real-time data integration try to tackle the issue from a database perspective. Oracle's system, presented by Rittman (2008), is essentially a middleware between (web) services and a continuously updated database layer. Like Sybase Inc. (2008), the Oracle approach is able to detect database events in order to analyse heterogeneous data sources and trigger actions accordingly. Rahm et al. (2007) present a more dynamic method of data integration and fusion using on-the-fly object matching and metadata repositories to create a flexible data integration environment.

The third related research area is the development of **an open data integration system architecture** in a non-application specific infrastructure. Srivastava et al. (2006) and Balazinska et al. (2007) present general concepts in a systems architecture and data integration approach but there are no concrete conclusions as to how the final goal of establishing such an infrastructure could be achieved. A more technical approach for ad-hoc sensor networks is described by Riva & Borcea (2007), where the authors discuss challenges to making heterogeneous sensor measurements combinable through the creation of highly flexible middleware components. The method is application-motivated and thus very detailed as far as specific implementation details are concerned.

3. Common Scents measurement infrastructure

The *Common Scents* project aims at developing an interoperable open standards based infrastructure for providing fine-grained air quality data to allow users to assess environmental conditions instantaneously and intuitively. The goal is to provide citizens with unseen up-to-date information layers in order to support their short-term decisions. To achieve this vision, we utilise the CitySense sensor testbed and technologically build our system upon the *Live Geography* approach presented by Resch et al. (2009a). It should be stated that the term 'real-time' in our case is not defined by a pre-set numerical time constant, but more by qualitative expressions such as 'immediately' or 'ad-hoc', i.e. information layers have to be created in a timely manner to serve application-specific purposes. (Resch et al., 2009b)

3.1 Design principles

In the conception of our technical infrastructure we accounted for different design principles (Service Oriented Architectures - SOA, modular software infrastructures etc.) to ensure flexibility, reusability and portability of the components and the overall infrastructure. Figure 1 shows the modular architecture and the standardised service interfaces that are used to connect the different components in the workflow.

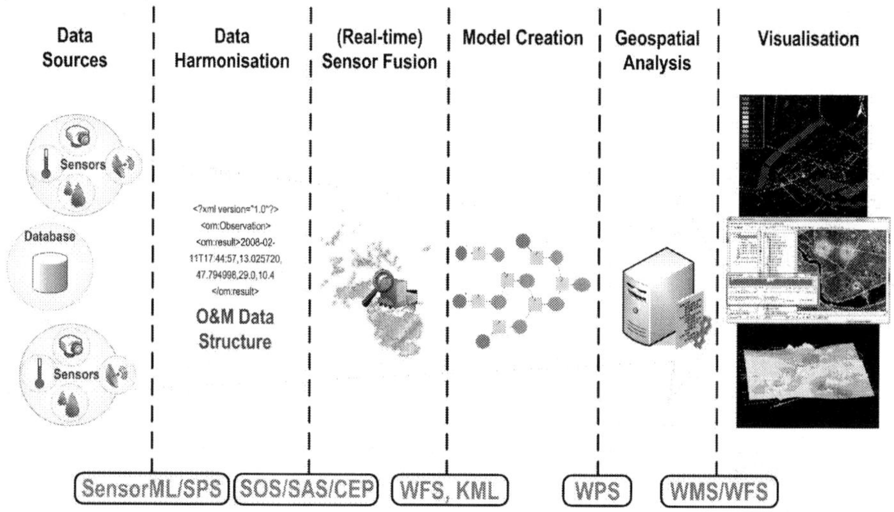

Fig. 1. Common Scents - Modular Standardised Infrastructure

According to principles of SOA and sustainable infrastructure development, we conceived data collection, processing and information provision architecture, which covers the whole process chain from sensor network development via measurement integration, data analysis and information visualisation, as shown in Figure 1. Thus, our approach builds the architectural bridge between domain-independent sensor network development and use case specific requirements for end user tailored information output for environmental monitoring purposes.

3.2 Standardised measurement infrastructure

The modules of the workflow shown in Figure 1 are separated by several interfaces, which are defined using open standards. The first central group of standards is subsumed under the term Sensor Web Enablement (SWE), an initiative by the OGC that aims to make sensors discoverable, query-able, and controllable over the Internet (Botts et al., 2007). Currently, the SWE family consists of seven standards comprising data models and service interfaces:

- *Sensor Model Language (SensorML)* – This standard provides an XML schema for defining the geometric, dynamic and observational characteristics of a sensor. Thus, SensorML assists in the discovery of different types of sensors, and supports the processing and analysis of the retrieved data, as well as the geo-location and tasking of sensors.
- *Observations & Measurements (O&M)* – O&M provides a description of sensor observations in the form of general models and XML encodings. This framework labels several terms for the measurements themselves as well as for the relationship between them. Measurement results are expressed as quantities, categories, temporal or geometrical values as well as arrays or composites of these.
- *Transducer Model Language (TML)* – Generally speaking, TML can be understood as O&M's pendant or streaming data by providing a method and message format describing how to interpret raw transducer data.
- *Sensor Observation Service (SOS)* – SOS provides a standardised web service interface allowing access to sensor observations and platform descriptions.
- *Sensor Planning Service (SPS)* – SPS offers an interface for planning an observation query. In effect, the service performs a feasibility check during the set up of a request for data from several sensors.
- *Sensor Alert Service (SAS)* – SAS can be seen as an event-processing engine whose purpose is to identify pre-defined events such as the particularities of sensor measurements, and then generate and send alerts in a standardised protocol format.
- *Web Notification Service (WNS)* – The Web Notification Service is responsible for delivering generated alerts to end-users by E-mail, over HTTP, or via SMS. Moreover, the standard provides an open interface for services, through which a client may exchange asynchronous messages with one or more other services.
- *Sensor Web Registry* – The registry serves to maintain metadata about sensors and their observations. In short, it contains information including sensor location, which phenomena they measure, and whether they are static or mobile. Currently, the OGC is pursuing a harmonisation approach to integrate the existing CS-W (Web Catalogue Service) into SWE by building profiles in ebRIM/ebXML (e-business Registry Information Model).

The functional connections between the SWE standards are illustrated in Figure 2.

It should be mentioned that the OGC is currently establishing the so-called *SWE Common* namespace specification, which aims at grouping elements that are used in more than one standard of the SWE family. In effect, this will minimise redundancy, and optimise re-usability and efficiency of the standards. SWE Common will mainly comprise very general elements such as counts, quantities, time elements or simple generic data representations.

More information on the Sensor Web Enablement initiative, the incorporated standards and the efforts to embed it into the OGC standard service development can be found on the OGC web site[1].

Aside from the SWE standard family, which is used throughout the sensor network process chain, other OGC standards are employed to build up the overall infrastructure. At first, the Web Processing Service (WPS) is used for integrating measurement data with thematic process models in order to generate contextual information layers. WPS (Schut, 2007) basically allows for the implementation and execution of pre-defined analysis processes

[1] http://www.opengeospatial.org

with dedicated input and output parameters. It supports synchronous and asynchronous data processing to enable sophisticated processing of large amounts of vector and raster data. The WPS standard has been discussed by Resch et al. (2010a) including issues such as input/output data definition, WPS profiling, asynchronous processing, and others.

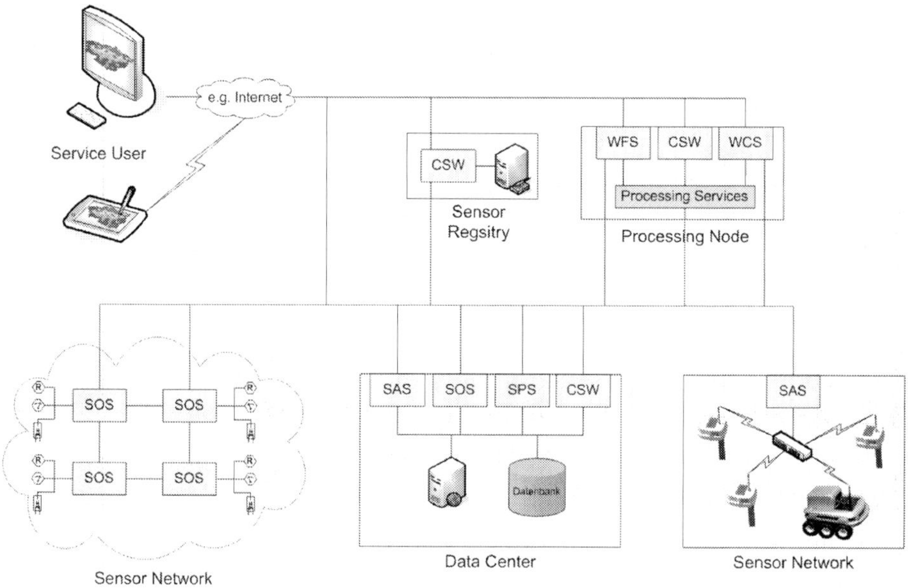

Fig. 2. Functional Connections Between the SWE Standards. (adapted from Botts et al., 2006)

The OGC developed a set of service interfaces for standardised data provision and visualisation dealing with various kinds of GIS data types. The Web Feature Service (WFS), the Web Map Service (WMS) and the Web Coverage Service (WCS) standards allow for access to geo-data such as vectors (point, line, polygon), raster images, and coverages (surface-like structures). More information about these standards and service implementations can be found on the OGC web site.

The essential benefit of using the OGC processing and data provision services mentioned above is the wide variety of standardised (GML, KML etc.) and custom output formats (GeoRSS, PDF etc.). This allows for the integration of the OGC service outputs into other processing, visualisation or decision support services including legacy COTS and open-source GIS analysis tools.

3.3 Data source: Geo-sensor web

In the Common Scents project, two pilot studies have been conducted. The first one used the *CitySense* sensing network (Murty et al., 2008) as the underlying sensing and data collection infrastructure. The main goal of the ongoing CitySense project is to build an urban sensor network to measure environmental parameters and is thus the data source for further data analysis. The project focuses on the development of a city-wide sensing system using an optimised network infrastructure. Currently, the network consists of 16 nodes deployed around the city of Cambridge measuring different environmental parameters such as CO_2 concentrations, air temperature, wind speed and direction, or precipitation. The final CitySense deployment will comprise up to 100 sensing nodes to build the basis for pervasive urban monitoring. (Resch et al., 2009b)

Another pilot experiment, which aimed at the deployment of a mobile sensor network, was conducted in the city of Copenhagen, Denmark. Ten bicycle mounted sensors[2] were used to collect environmental data (CO, NO_X, noise, air temperature and relative humidity) together with time and the geographic location using GPS - from which velocity and acceleration can be calculated. In this experiment of ubiquitous mobile sensing, we used the Sensaris City Senspod[3], a relatively low-cost sensor pod. The deployment in Copenhagen was a combined effort of the MIT SENSEable City Laboratory, and Københavns Kommune, Denmark.

To comply with the standardised infrastructure described in sub-section 3.1, we implemented several standardised services on top of these sensor networks, in accordance with the Live Geography approach (Resch et al., 2009a). For data access, we developed a Sensor Observation Service (SOS), which supplies measurement data in the standardised O&M format. It builds the O&M XML structure dynamically according to measured parameters and filter operations. To generate alerts e.g. in case of exceedance of a threshold, we implemented an XMPP (Extensible Messaging and Presence Protocol) based Sensor Alert Service (SAS). It is able to detect patterns and anomalies in the measurement data and generate alerts and trigger appropriate operations such as sending out an email or a text message, or to start a pre-defined GIS analysis operation.

3.4 Event-based alerting

Within the workflow described in sub-section 3.1, event recognition and processing happens in two different stages: 1.) at sensor level, Complex Event Processing (CEP) is used to detect errors in measurement values by applying different statistical operations such as standard deviations, spatial and temporal averaging, or outlier detection. Thus, it can be considered a mechanism for quality control and error prevention. To be able to detect condition changes in measurement values, we enhanced CEP and Event Stream Processing (ESP) techniques by the location parameter. Thus, these methods can also serve for the federal organisation of pre-defined geographical domain violations like geo-fences, and for tracing and analysing spatial patterns. 2.) after the data harmonisation process, CEP serves for spatio-temporal pattern recognition, anomaly detection, and alert generation in case of threshold exceedance.

Figure 3 shows the components of the CEP-based event processing component, which is built up in a modular structure.

Generally speaking, the event processing component serves as a connection between the data layer (sensor measurements) and the data analysis and data visualisation components, i.e. it prepares raw data in order to be process-able in the analysis and the visualisation layers. The first module is the data transportation, which connects different real-time and non-real-time data sources, i.e. it serves as an entry point into the event processing layer. Next, the retrieved data is passed on to the data transformation module, which prepares the data to be further processed. This 'processing' shall not be seen as data analysis, but more as data preparation. Basically, the transportation module converts the byte input stream to objects.

---

[2] http://senseable.mit.edu/copenhagenwheel

[3] http://www.sensaris.com

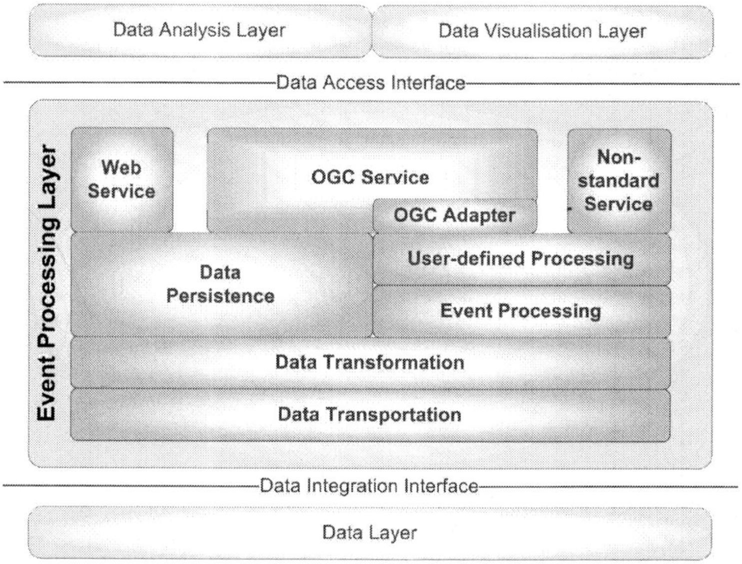

Fig. 3. CEP-based Event Processing Component Architecture

These objects can then be used by two higher-level components; firstly, by the data persistence component, which establishes a static data structure from the source data; secondly, by an event processing engine, which processes a real-time stream, identifies/selects events and pushes them to the user-defined processing module. The latter performs a kind of 'event filtering' and sends the resulting message to the service components.

One particularity, which shall be mentioned at this point, is the connection between the processing and the data persistence modules. The idea behind this functional link is that data, which have been prepared by the processing components, can either be pushed to the service components or they can be temporarily stored to be accessed by OGC services.

The two service-related components in Figure 3 (Web Service and Non-standard Service) serve as the direct interfaces to the data integration and data analysis layers. They offer the pre-processed raw data via a defined data structure, e.g. in a standardised output format such as GML, KML, geoTIFF etc. or in a custom output format.

For the OGC service component, all data are served over well-established open and standardised interfaces (OGC WFS, WMS and WCS). These XML web interfaces enable standardised data access and guarantee combinability of the various kinds of used data for further automated processing, as described in sub-section 3.2. In this way, pre-defined aggregation services can be implemented in the data analysis layer offering the results to a range of different users, i.e. platforms.

In addition to the standardised interfaces, also a non-standard service has to be created as existing OGC services don't support push mechanisms per se. A longer term option will be to replace these non-standard interfaces by push-capable standard services. More detailed information about the event processing component can be found in Resch et al. (2010b).

3.5 Real-time sensor fusion

We are currently facing a drastic increase in the availability of geospatial real-time data sources, and this applies especially to rapid developments and price reduction in sensing technologies. To make use of this immense amount of data within environmental monitoring systems, real-time data integration mechanisms have to be developed, which harmonise and fuse the different kinds of data. Furthermore, these data have to be provided in standardised formats in order to allow interoperability and collaboration between different institutions and data providers.

Most current data integration systems make use of a temporary database to combine different kinds of raw data, as stated in section 2. This approach has two distinctive disadvantages. Firstly, it manifests data into a physical structure and thus severely limits real-time capabilities. Secondly, the laborious operation of creating and filling a database table adds another step in the overall workflow, which decreases performance and expands implementation complexity and costs.

To overcome these shortcomings, we implemented the real-time data integration component in a custom data store for the open-source server GeoServer. This solution offers two main advantages: at first, data are fused on-the-fly in a highly dynamic, fast and parallelised process. At second, GeoServer provides standardised WFS, WMS and WCS outputs, as described above, which allows for simple integration into analysis and visualisation software. More about implementation details can be found in Resch et al. (2009a).

4. Results of spatio-temporal data analysis

Using the sensor web deployments described in section 3, we implemented two spatio-temporal data analysis modules.

The first pilot deployment has been carried out in the course of the *Copenhagen Wheel* project. This project was unveiled in Copenhagen on 15 December 2009 as part of 15th Conference of the Parties during the 2009 United Nations Climate Change Conference meeting. The Copenhagen Wheel is capturing information about carbon monoxide (CO), NO_x (NO + NO_2), noise, ambient temperature, relative humidity in addition to position, velocity and acceleration.

The environmental sensors were originally intended to be placed within the hub of the bicycle wheel however due to logistical pressure they were placed on bicycles ridden by couriers in Copenhagen going about their normal daily routine. Thus the testing was essentially a proof-of-concept. Ten cycles were instrumented and tested over 2 December 2009. It is believed that this was the first time multiple mobile sensors had been used in the field with such a large variety of environmental sensors.

The analysis component, which processes the collected data, performs a spatial Inverse Distance Weighting (IDW) interpolation (for a comparison with Kriging interpolation, s. Zimmermann et al., 1999) on temperature measurements, which will be used in further research efforts for correlation operations with emission distribution or traffic emergence, and for the detection of urban heat islands.

Moreover, the processing module analyses the CO distribution throughout the city of Copenhagen. The basic CO map containing the GPS traces and the output of the interpolation process – a navigable 3D map – are shown in Figure 4.

The second geo-processing component uses ArcGIS's Tracking Analyst tool to perform spatio-temporal analysis on measurement data over a period of time. In order to achieve a coarse overview of pollutant variability, we used CO_2 data captured by the CitySense network in Cambridge. This allows for correlating temporal measurement data fluctuation to traffic density, weather conditions or day-time related differences in a very flexible way. The lower left part of Figure 5 shows the temporal gradient of the measurement values. Running the time series then changes symbologies in the map on the right side accordingly in a dynamic manner in real-time.

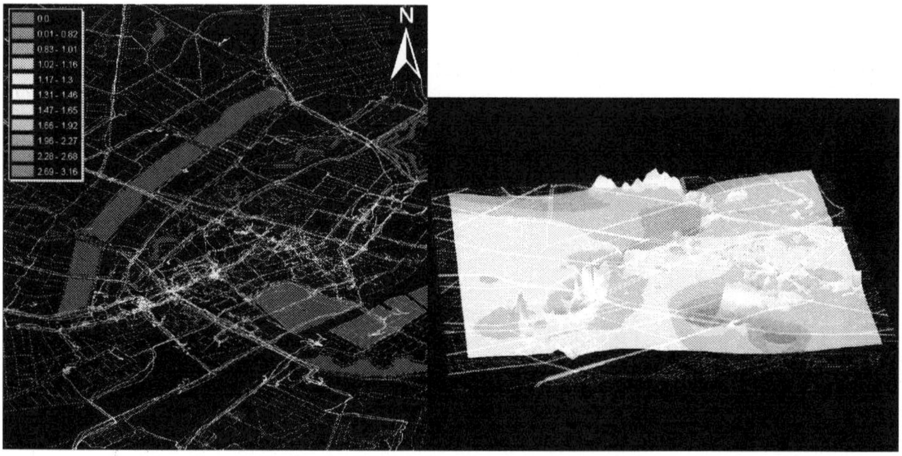

Fig. 4. Spatial Distribution of CO Values in the City of Copenhagen.

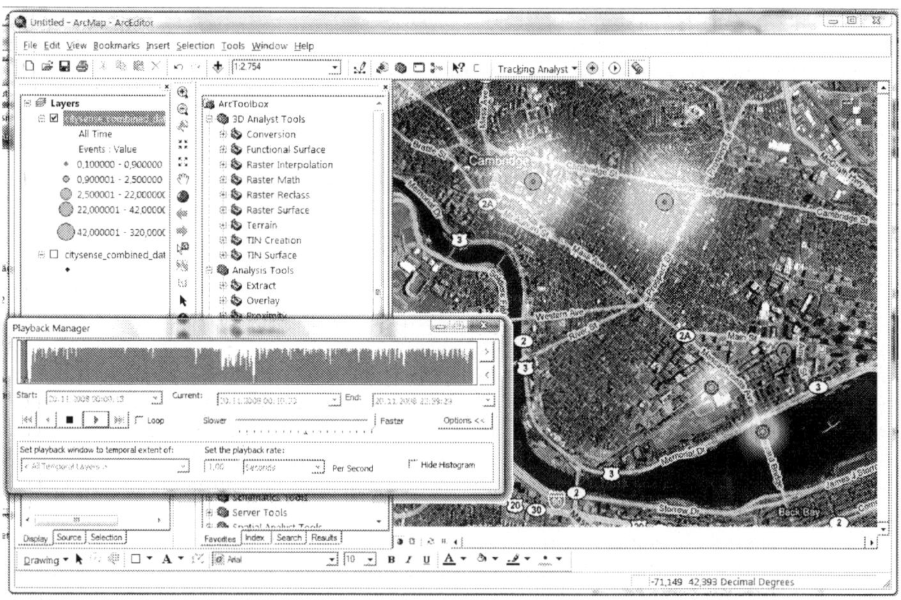

Fig. 5. Spatio-temporal CO_2 Data Analysis Using Tracking Analyst.

Figure 6 illustrates a time series representation of the measured parameters ambient temperature, CO, NO_x and noise. These measurements were taken in Copenhagen over a period of five hours on 2 December 2009. A first assessment shows that there are strong correlations between ambient temperature, CO and NO_x values.

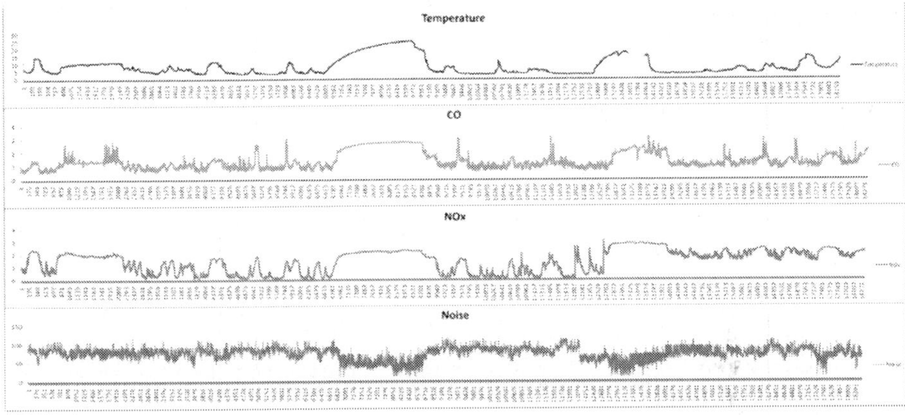

Fig. 6. Time Series Representation of Environmental Measurements.

Preliminary findings show that both CO and CO_2 are characterised by very high temporal and spatial fluctuations, which are induced by a variety of factors including temperature variability, time during the day, traffic emergence or 'plant respiration' – the fact that plants release major amounts of CO_2 over night. Also, CO is a measure of the efficiency of combustion in vehicles and may be used to reflect changing driving patterns and the sensitivity of air quality to larger scale environmental features such as the wind speeds over the city. However, the detailed interplay of these parameters still has to be investigated in a next step. Especially CO values measured in the Copenhagen pilot have to be normalised over humidity and temperature to perform further quantitative (absolute amounts) and qualitative (impact on public health) analysis.

5. Potential application areas

More than ten years ago, Al Gore articulated a vision of 'Digital Earth' as a multi-resolution, three-dimensional representation of the planet that would make it possible to find, visualise, and make sense of vast amounts of geo-referenced information on the physical and social environment (Gore 1998, for a comprehensive discussion see Craglia et al. 2008). Google Earth, NASA World Wind and other geo-browsers brought high resolution imagery to hundreds of millions of internet users and a major industry developed ways to explore data geographically, and visualise overlaid information provided by both the public and private sectors, as well as citizens who volunteer new data (Goodchild, 2007).
Generally speaking, fine-grained urban sensing greatly enhances our knowledge of the environment by adding objective and non-visible data layers in real-time (Resch et al., 2008). In other words, these systems help us increase our capacity to observe and understand the city, and the impacts on and by society. This seems to be a very desirable state because more

accurate data about local air temperature, atmospheric humidity, gaseous and particulate air pollution, and traffic emissions can positively influence areas such as public health, traffic management or emergency response. Apart from this information enrichment, accurate sensor measurements also have a much broader influence: considering for example that 'air quality' is only a surrogate for the effects of pollutants on humans makes a fine-grained air quality map a very sensitive information layer, as discussed in section 4.

Within the Common Scents project, we focus on the use case of air quality monitoring for use in the public health sector. However, we designed the monitoring infrastructure in such a modular way that it is not bound to one single application area. Below, several practically motivated fields of real-world applications are described, which could use the same infrastructure presented in section 3.

Public Health has been asked to participate in policymaking on 'quality of life' issues increasingly over the past decade. The superimposing of the medical model to describe the impact of conditions that have traditionally been regarded as nuisances has created a great challenge, particularly in the field of environmental health.

One pollutant often used to serve as a proxy is NO_x, which technically represents various gaseous species comprised of oxygen and nitrogen molecules. Another indicator of near-roadway effects that has gained recent attention is ultrafine particulates (UFPs), particles that are less than 0.1 microns (100 nm) in diameter. Thus, air quality measurements of hazardous air pollutants can be widely associated with traffic (non-point sources). A pervasive sensor network could help capture measurements in high spatial and temporal resolution to take short-term measures (dynamically adapt traffic management or send out alerts to citizens in case of threshold exceedance). Also, it could support traditional long-term studies on the impact of certain pollutants on public health.

The use case of noise mapping has received a lot of attention recently. Many disputes within the research field emerge from noise impacts associated with construction, excavation or some other commercial or industrial enterprise. These disputes also arise from use of domestic landscaping equipment, like leaf blowers and snow blowers. The limits imposed by the city on noise generation are intended to assess the background noise levels. A source cannot be held responsible for noise levels that exceed the city's allowable limits if the ambient noise in that area already exceeds those limits. The development of noise 'maps' may not immediately result in satisfaction from aggrieved residents, but it can be used to consider the noise impact of future development and zoning policies. It may also contribute to efforts to reduce the number of cars travelling across the city by adding the noise impact dimension to the discussion. This is much more likely to be given full consideration if it can be demonstrated with highly resolved data maps, which can be generated in near real-time using the Common Scents infrastructure.

The urban heat island effect describes the contribution of the built environment to the ambient temperature within urban areas. While this is not likely to become a primary public health concern, it has great bearing on efforts to limit the loss of heating energy across the city. Different agencies have been established to work on a long-term strategy to reduce overall energy use (e.g. Cambridge Energy Alliance: http://www.cambridgeenergyalliance.org) and to encourage individual homeowners and building owners to evaluate their energy loss. It is quite possible that small changes in heat loss, as described through a detailed heat map of the city over time, could show progress towards energy efficiency in a materials way. This could be used both as an evaluation tool in tracking the city's progress, and as a means to engage the public in the energy goals of the community.

6. Conclusion

Ubiquitous and continuous environmental monitoring is an enormous challenge, and this is particularly true in the urban context, which poses very specific challenges as well technologically as socially and politically. In this chapter we discussed several of these issues, and outlined how our approach can meet future requirements for urban sensing.

The focus is to contribute to a 'complete' picture of a living city for decisions makers, planners and operators beyond locational analysis. This may be seen distinct from a number of citizen-centred sensor approaches and context-aware systems. While a number of people-centric pervasive sensing systems are notable successes (Campbell et al. 2006), most of these examples focus on localising people and objects in a defined environment to enable context aware applications. In such projects, the notion of sensing is confined to supporting location-based context-awareness. In our *Live Geography* approach a more general integrated sensing architecture to support the diversity of applications and hardware platforms has been developed.

Based on the Live Geography approach, we outlined the Common Scents concept, which tries to establish an interoperable, modular and flexible sensing and data analysis infrastructure, as opposed to hitherto monolithic sensor networks. To prove our system's portability, we did implementations in two different pilot deployments (Cambridge, MA US and Copenhagen, Denmark) using the same data integration and analysis infrastructure.

Further exploitation of this approach is planned for other cities. We see more future challenges in the socio-political domain rather in the technological development necessary. It becomes more and more obvious that a cross-disciplinary group of researchers and technologists needs to persistently interact with end users. Only then we may achieve a wide appreciation of sensing which is needed to support future civic, cultural, and community life in cities. In many parts of the world, notably Germany and some Western European countries, attempts to 'completely' map cities are very sensitive. Google faces great problems with its StreetView approach. An integrated Common Scents must provide a clearly recognisable benefit to the citizens in order to be appreciated by all societal groups. Public Health applications may have a good chance to get accepted although some of the capabilities, for instance the ability to measure remotely the conditions of people in real time, raise social concerns centred on privacy issues. Methods for sensor data fusion and designs for human-computer interfaces are both crucial for the full realisation of the potential of integrated and pervasive sensing.

We also believe that the impact of pervasive sensing in the city has to be carefully assessed. We found that e.g. providing very fine-grained information layers might on the one hand be a powerful decision support instrument, but on the other hand too detailed environmental information might also have negative effects. As the term 'air quality' is just a surrogate for more personal impacts such as life expectation or respiration diseases, this information could yield a very broad impact in various kinds of areas such as housing market, the insurance sector or urban planning in general.

As the Common Scents concept has been developed and implemented together with the Public Health Department of the City of Cambridge, MA US as concrete end users, we believe that our approach can respond to dedicated needs of the city management. Therefore, the longer-term goal is to enhance people's perception of their environment by adding unseen information layers and thus changing their short-term behaviour by providing real-time decision support.

7. Acknowledgement

The Common Scents project is a concerted effort between the Research Studio iSPACE, the MIT SENSEable City Lab, the City of Cambridge's Public Health Department and the Harvard University School of Engineering and Applied Sciences. We would like to thank all internal and external collaborators for making this project happen.

Several technical parts of this research have been funded through the European Commission in the course of the FP7 GENESIS project. Furthermore, the Austrian Ministry of Science and Research (BMWF) funded pieces of the efforts in the Research Studio iSPACE.

The research described in this project by one of the authors (REB) was funded in part by the Singapore National Research Foundation (NRF) through the Singapore-MIT Alliance for Research and Technology (SMART) Center for Environmental Sensing and Monitoring (CENSAM).

The Copenhagen Wheel team from the SENSEable City Laboratory, MIT is composed of Christine Outram (Project Leader), Rex Britter, Andrea Cassi, Xiaoji Chen, Jennifer Dunnam, Paula Echeverri, Myshkin Ingawale, Ari Kardasis, E Roon Kang, Sey Min, Assaf Biderman and Carlo Ratti The courier cyclists were organized by Signe Gaarde in Copenhagen.

8. References

Balazinska, M., Deshpande, A., Franklin, M.J., Gibbons, P.B., Gray, J., Hansen, M., Liebhold, M., Nath, S., Szalay, A., Tao, V. (2007) Data Management in the Worldwide Sensor Web. IEEE Pervasive Computing, vol. 6, no. 2, pp. 30-40, Apr-Jun, 2007.

Botts, M., Robin, A., Davidson, J. and Simonis, I. (Eds.) (2006) OpenGIS Sensor Web Enablement Architecture Document. http://www.opengeospatial.org, OpenGIS Interoperability Project Report OGC 06-021r1, Version 1.0, 4 March 2006. (12 June 2011)

Botts, M., Percivall, G., Reed, C., and Davidson, J. (Eds.) (2007) OGC Sensor Web Enablement: Overview and High Level Architecture. http://www.opengeospatial.org, OpenGIS White Paper OGC 07-165, Version 3, 28 December 2007. (17 June 2011)

Campell, A., Eisenman, S.B., Lane, N.D., Miluzzo and E., Peterson, R.A. (2006) People-Centric Urban Sensing. Proceedings of the 2nd annual international Workshop on Wireless Internet, Boston, 2006.

Center for Coastal and Land-Margin Research (2009) CORIE. http://www.ccalmr.ogi.edu/CORIE. (14 June 2011)

Goodchild, M. (2007). Citizens as sensors: The world of volunteered geography. Geo-journal, 69: 211-221.

Gore, A. (1998) The Digital Earth: Understanding Our Planet in the 21st Century. Speech given by Vice President Al Gore at the California Science Center, Los Angeles, California, on January 31, 1998, http://www.isde5.org/al_gore_speech.htm. (27 June 2011)

Gross, N. (1999) 14: The Earth Will Don an Electronic Skin. http://www.businessweek.com, BusinessWeek Online, 30 August 1999. (20 May 2011)

Harrie, L. (2004) Using Simultaneous Graphic Generalisation in a System for Real-Time Maps. Papers of the ICA Workshop on Generalisation and Multiple Representation, August 20-21, 2004, Leicester (electronic version available at http://ica.ign.fr/Leicester/paper/Harrie-v2-ICAWorkshop.pdf).

Kansal, A., Nath, S., Liu, J. and Zhao, F. (2007) SenseWeb: An Infrastructure for Shared Sensing. IEEE Multimedia. vol. 14, no. 4, pp. 8-13, October-December 2007.

Lehto, L. and Sarjakoski, L.T. (2005) Real-time Generalisation of XML-encoded Spatial Data for the Web and Mobile Devices. International Journal of Geographical Information Science, vol. 19, no. 8-9, pp. 957-973.

Murty, R., Mainland, G., Rose, I., Chowdhury, A., Gosain, A., Bers, J. and Welsh, M. (2008) CitySense: A Vision for an Urban-Scale Wireless Networking Testbed. In Proceedings of the 2008 IEEE International Conference on Technologies for Homeland Security, Waltham, MA, May 2008.

Nagel, D. (2003) Pervasive Sensing. Proceedings of the SPIE, vol. 4126, no. 71, doi:10.1117/12.407543, 2003.

Nath, S., Liu, J. and Zhao, F. (2006) Challenges in Building a Portal for Sensors World-Wide. First Workshop on World-Sensor-Web: Mobile Device Centric Sensory Networks and Applications (WSW'2006), Boulder CO, 31 October, 2006.

National Oceanic and Atmospheric Administration (2011) nowCOAST: GIS Mapping Portal to Real-Time Environmental Observations and NOAA Forecasts. http://nowcoast.noaa.gov. (15 June 2011)

Paulsen, H. and Riegger, U. (2006). SensorGIS - Geodaten in Echtzeit. In: GIS-Business, vol. 8/2006, pp. 17-19, Cologne.

Paulsen, H. (2008) PermaSensorGIS - Real-time Permafrost Data. In: Geoconnexion International Magazine, vol. 02/2008.

Rahm, E., Thor, A. and Aumueller, D. (2007) Dynamic Fusion of Web Data. XSym 2007, Vienna, Austria, pp.14-16.

Resch, B., Calabrese, F., Ratti, C. and Biderman, A. (2008) An Approach Towards a Real-time Data Exchange Platform System Architecture. Sixth Annual IEEE-IARIA International Conference on Pervasive Computing and Communications, Hong Kong, 17-21 March 2008.

Resch, B., Mittlboeck, M., Girardin, F., Britter, R. and Ratti, C. (2009a) Live Geography – Embedded Sensing for Standardised Urban Environmental Monitoring. International Journal on Advances in Systems and Measurements, 2(2&3), ISSN 1942-261x, pp. 156-167.

Resch, B., Mittlboeck, M., Lipson, S., Welsh, M., Bers, J., Britter, R. and Ratti, C. (2009b) Urban Sensing Revisited – Common Scents: Towards Standardised Geo-sensor Networks for Public Health Monitoring in the City. In: Proceedings of the 11th International Conference on Computers in Urban Planning and Urban Management - CUPUM2009, Hong Kong, 16-18 June 2009.

Resch, B., Sagl, G., Blaschke, T. and Mittlboeck, M. (2010a) Distributed Web-processing for Ubiquitous Information Services - OGC WPS Critically Revisited. In: Proceedings of the 6th International Conference on Geographic Information Science (GIScience2010), Zurich, Switzerland, 14-17 September 2010.

Resch, B., Lippautz, M. and Mittlboeck, M. (2010b) Pervasive Monitoring - A Standardised Sensor Web Approach for Intelligent Sensing Infrastructures. Sensors - Special Issue 'Intelligent Sensors 2010', 10(12), 2010, pp. 11440-11467.

Rittman, M. (2008) An Introduction to Real-Time Data Integration. http://www.oracle.com/technology/pub/articles/rittman-odi.html, 2008. (22 May 2011)

Riva, O. and Borcea, C. (2007) The Urbanet Revolution: Sensor Power to the People!. IEEE Pervasive Computing, vol. 6, no. 2, pp. 41-49, Apr-Jun, 2007.

Sarjakoski, T. et al. (2004) Geospatial Info-mobility Service by Real-time Data-integration and Generalisation. http://gimodig.fgi.fi. (22 June 2011)

Schut, P. (ed.) (2007) Web Processing Service. OpenGIS Standard, version 1.0.0, OGC 05-007r7, 8 June 2007.

Srivastava, M., Hansen, M., Burke, J., Parker, A., Reddy, S., Saurabh, G., Allman, M., Paxson, V. and Estrin D. (2006). Wireless Urban Sensing Systems. Technical Report 65, Center for Embedded Network Sensing, UCLA, April 2006.

Sybase Inc. (2008) Real-Time Events Data Integration Software. http://www.sybase.com/products/dataintegration/realtimeevents. (07 July 2011)

University of Oklahoma (2009) OKCnet. http://okc.mesonet.org. (12 March 2011)

Xu, N. (2004) A Survey of Sensor Network Applications. http://courses.cs.tamu.edu, Computer Science Department, University of Southern California, 2004. (10 July 2011)

Zimmerman, D., Pavlik, C., Ruggles, C. and Armstrong, M.P. (1999) An Experimental Comparison of Ordinary and Universal Kriging and Inverse Distance Weighting. Mathematical Geology, 31(4), pp. 375-390.